The German Language and the Real World

Sociolinguistic, Cultural, and Pragmatic Perspectives on Contemporary German

Edited by
PATRICK STEVENSON

PF
3087
.G47
1995

CLARENDON PRESS · OXFORD
1995

Oxford University Press, Walton Street, Oxford OX2 6DP
Oxford   New York
Athens   Auckland   Bangkok   Bombay
Calcutta   Cape Town   Dar es Salaam   Delhi
Florence   Hong Kong   Istanbul   Karachi
Kuala Lumpur   Madras   Madrid   Melbourne
Mexico City   Nairobi   Paris   Singapore
Taipei   Tokyo   Toronto
and associated companies in
Berlin   Ibadan

Oxford is a trade mark of Oxford University Press

Published in the United States
by Oxford University Press Inc., New York

© The several contributors, and in this collection
Oxford University Press 1995

All rights reserved. No part of this publication may be reproduced,
stored in a retrieval system, or transmitted, in any form or by any means,
without the prior permission in writing of Oxford University Press.
Within the UK, exceptions are allowed in respect of any fair dealing for the
purpose of research or private study, or criticism or review, as permitted
under the Copyright, Designs and Patents Act, 1988, or in the case of
reprographic reproduction in accordance with the terms of the licences
issued by the Copyright Licensing Agency. Enquiries concerning
reproduction outside these terms and in other countries should be
sent to the Rights Department, Oxford University Press,
at the address above

British Library Cataloguing in Publication Data
Data available

Library of Congress Cataloging in Publication Data
The German language and the real world : sociolinguistic, cultural,
and pragmatic perspectives on contemporary German
edited by Patrick Stevenson.
Includes bibliographical references and index.
1. German language — 20th century.   2. German language — Political
aspects.   3. German language — Social aspects.   4. Sociolinguistics.
I. Stevenson, Patrick.
PF3087.G47   1995   306.4'4'094309049 — dc20   94-22023
ISBN 0-19-824054-6

10   9   8   7   6   5   4   3   2   1

Typeset by Rowland Phototypesetting Ltd.
Printed in Great Britain on acid-free paper by
Biddles Ltd., Guildford and King's Lynn

INDIANA-
PURDUE
WITHDRAWN
SEP 6 1996
FORT WAYNE

FTW
AHK 3216

For my parents, Molly and Gerry Stevenson,
and for Jo, Rosie, and Jack

Preface

THE early 1990s have been turbulent times in central Europe, a period of *Umbruch* or radical change. As the twentieth century draws towards its close, some of its creations (such as Yugoslavia and Czechoslovakia) have dissolved themselves into smaller entities, but two others (the Federal Republic of Germany and the German Democratic Republic) have bucked this trend and formed, or (as some would have it) re-formed, a single German state. As language has always been a major player in the debates on cultural identity in the German-speaking countries, it is not surprising that this momentous development should have occasioned a flurry of activity amongst professional observers of 'the German language', much of which has consisted of fevered attempts to capture the details of linguistic confrontation and change before the specificity of this historical moment is swept away, a form of linguistic 'rescue archaeology' of the present.

This book was also driven by the recent upsurge of interest in the German language, but although it addresses the 'what now?' question from various perspectives, it seeks to develop a more wide-ranging and less ephemeral agenda. Its central topic is the contemporary language, but its authors have tried to place their particular concerns in a broader context. The two main objectives will be apparent from the titles and the names on the contents page: first, it offers descriptive, theoretical, and analytical contributions to the study of the forms, functions, and uses of contemporary German, and secondly it offers some insights into the interests and approaches of German-speaking linguists.

The overall aim was therefore to appeal to several different but partly overlapping constituencies: to readers interested in the German language itself, to those interested in approaches to the study of the German language, and to those interested more generally in the study of the uses of language and language in use. Readers should find answers to questions such as 'what is happening to the German language?', but also to questions like 'what do German-speakers make of their language?', 'what kind of work do German-speaking (socio) linguists do?', 'what have they adopted from other traditions and what can they contribute?' In order to achieve this aim, the text has been written entirely in English and (where practicable) the examples and illustrations have been given in both German and English.

The contributors themselves, while writing in an individual capacity, nevertheless represent a wide spectrum of approaches to the study of what I am calling 'real language' (see Chapter 1). Some would call themselves either sociolinguists or applied linguists, others simply linguists. For the purposes of this book, I have deliberately avoided all such occupational labels, and I hope that its very diversity will be part of its appeal.

Ulrich Ammon and Florian Coulmas begin by looking at two contrasting functions of the language as a whole: Ammon uses the concept of an 'international language' to assess changes in the relative importance or 'value' of the language in global terms, while Coulmas traces the various attempts to construct a sense of nationhood on the basis of a 'common language' from the Enlightenment to the present day.

Wolfgang Sauer and Helmut Glück also adopt a historical perspective in their account of the socio-cultural project of 'fixing' the form of the language. Focusing on the emergence of a standardized spelling system and continued attempts to reform it, they show how this apparently esoteric academic debate is also a significant and hotly disputed public issue. In their second chapter, they extend their argument for a flexible and tolerant approach to norms and variations by outlining a number of current linguistic developments that conventional 'authorities' (grammars and dictionaries) ignore or discount.

The role of linguistic change in the enactment of social and cultural change is demonstrated graphically by Peter Schlobinski and Helmut Schönfeld. During the 1980s, these two linguists worked separately on the urban vernacular of Berlin, based at institutes that were a few miles apart but to the west and east of the Berlin Wall respectively. They are now able to collaborate, and their joint work on sociolinguistic change in the new capital city exemplifies the kind of 'joint venture' advocated by Norbert Dittmar in his chapter on theories of sociolinguistic variation in the German context. Dittmar isolates the central theoretical issues in the development of sociolinguistics in the Federal Republic (and Austria) on the one hand and the GDR on the other, and shows how the strengths of both traditions could be incorporated in new approaches in the future. He also shows why theoretical work on sociolinguistic variation has developed in rather different ways in Germany and in the USA and the UK, despite the fact that the two traditions have many areas of common interest.

Long before the dismantling of the Berlin Wall in 1989/90 and of internal borders within the European Community in 1993, Europe in general and Germany in particular had been developing into multicultural and multilingual societies. One of the great challenges for this 'new' Europe will be to overcome communication barriers in increasingly heterogeneous and highly mobile populations: Martina Rost-Roth discusses the problems of intercultural communication in the German context, both between 'native' and 'non-native' German-speakers and between native-speakers from West and East. Ruth Wodak also considers problems of communication, but focuses specifically on institutional settings and shows how 'critical linguistics' can both reveal and help to counteract the unequal division of power in social relationships.

Siegfried Jäger and Sylvia Moosmüller deal with political language and the language of politicians respectively. Like Wodak, Jäger has developed an approach to critical discourse analysis that has much in common with similar work in the USA, the UK, and France. He applies it here to the analysis of political discourse, concentrating on the language of political journals and other publications of the right in Germany, with particular emphasis on the discourse of racism. He also discusses changes in *Neues Deutschland*, the former official newspaper of the ruling East German Socialist Unity Party, and the problems posed to the left by the Gulf War. Moosmüller investigates attitudes to language variation in public domains, looking in particular at perceptions of the language behaviour of Austrian politicians against the background of the general evaluation of regional dialects and the standard variety.

The last three chapters also deal with aspects of language which enjoy a high public profile, albeit in very different ways. Marlis Hellinger discusses the debates on the androcentricity of German grammar and gender-related language behaviour, drawing comparisons with (and contrasts to) English and language use in anglophone contexts, and considers feminist programmes for planned language change. In the public debates on the supposed 'decline' in the 'quality' of German (or rather its use), two culprits in particular are often identified: youth and television. Peter Schlobinski seeks to undermine this view, as well as the conventional 'taxonomic' approach to the study of so-called *Jugendsprache*, arguing that a pragmatic approach focusing on speech styles rather than individual words and phrases shows that youth language is a highly skilled and creative activity. Werner Holly also sees no support for the 'language decline' thesis in relation to the consumption of television, and concentrates instead on television as a special form of communication in which the viewer is actively engaged, illustrating his analysis with discussions of four TV genres (news bulletins, soap operas, quiz and game shows, and commercials).

In the first chapter I would like to introduce not so much the remaining contributions specifically as the whole subject of the study of language use in the German-speaking context. My intention is to provide a kind of prism or filter through which the reader can view the rest of the book in a coherent way. Although I shall take a partially historical approach, my real concern is to pull together what seem to me to be the crucial strands of this complex area of study. While discussing (the relevance and scope of) particular academic disciplines, I shall avoid the potentially arid pastime of determining demarcation lines and propose instead that the 'active reader' draw his/her own conclusions from the contents of the book as to what might or should constitute appropriate and fruitful areas of research on 'language in use'. Some of what I shall say will be picked up again and dealt with in more detail in later chapters: my intention is to show how these and other aspects fit into the larger picture of real language study in German-speaking countries.

                                                                    P.S.

Acknowledgements

As with any such undertaking, there are many people to whom I am indebted for their help in the planning and preparation of this volume.

First and foremost, I would like to thank the contributors for their co-operation in meeting often quite tight deadlines and for putting up with my constant correspondence.

Many people in Southampton, colleagues, family, and friends, helped, encouraged, and guided me in numerous ways and made the whole process bearable. Amongst my immediate colleagues, I am particularly grateful to Alison Piper, Rodney Ball, Christopher Brumfit, and Rosamond Mitchell for their interest, comments, and support; Ian McComb, Jean Watts, and Hazel Paul for their invaluable help with technical nightmares and typing; and Monika Mott for help with correspondence. Katharina Hall and Mark Moss translated Chapters 4 and 5 respectively. Above all, though, I want to thank my best friends, Jo, Rosie, and Jack, for constantly reminding me that there are other things in life than books.

I would like to acknowledge the constructive suggestions made by Anthony Fox, and I am deeply grateful to Christopher Wells for his meticulous reading of the text and detailed comments; naturally, I must take responsibility for any weaknesses that remain.

Finally, I would like to record my gratitude to the Institut für deutsche Sprache, Mannheim, especially to the Directors Gerhard Stickel and Rainer Wimmer, and to the Librarian Eva Teubert; and to the British Academy, the British Council, and the Committee for Advanced Studies at the University of Southampton for making it financially possible for me to carry out my own work in Mannheim and elsewhere in Germany.

P.S.

Contents

Notes on Contributors   xv

1   The Study of Real Language: Observing the Observers   1
Patrick Stevenson

2   To What Extent is German an International Language?   25
Ulrich Ammon

3   Germanness: Language and Nation   55
Florian Coulmas

4   Norms and Reforms: Fixing the Form of the Language   69
Wolfgang Werner Sauer and Helmut Glück

5   Directions of Change in Contemporary German   95
Helmut Glück and Wolfgang Werner Sauer

6   After the Wall: Social Change and Linguistic Variation in Berlin   117
Helmut Schönfeld and Peter Schlobinski

7   Theories of Sociolinguistic Variation in the German Context   135
Norbert Dittmar

8   Language in Intercultural Communication   169
Martina Rost-Roth

9   Critical Linguistics and the Study of Institutional Communication   205
Ruth Wodak

10   Political Discourse: The Language of Right and Left in Germany   231
Siegfried Jäger

11   Evaluation of Language Use in Public Discourse: Language Attitudes in Austria   257
Sylvia Moosmüller

xiv   *Contents*

12   Language and Gender   279
     Marlis Hellinger

13   *Jugendsprachen*: Speech Styles of Youth Subcultures   315
     Peter Schlobinski

14   Language and Television   339
     Werner Holly

Index   375

Notes on Contributors

ULRICH AMMON is professor in the *Fachbereich Sprach- und Literatur-wissenschaft* (Department of Linguistic and Literary Studies) at the Gerhard Mercator Universität Duisburg. His main research interests are in sociolinguistics, international communication, dialectology, and the teaching of German, both as a native and as a foreign language. His recent publications include: (with N. Dittmar and K. J. Mattheier) *Sociolinguistics: An International Handbook of the Science of Language and Society* (1987–8, de Gruyter); *Studienmotive und Deutschenbild australischer Deutschstudenten und -studentinnen* (1991, Steiner); *Die internationale Stellung der deutschen Sprache* (1991, de Gruyter).

FLORIAN COULMAS is professor of sociolinguistics and general linguistics at Chuo University, Tokyo. His main research interests are the socio-economics of language, phraseology, writing systems and written language, and Japanese sociolinguistics. His recent publications include: *Sprache und Staat* (1985, de Gruyter); *Language Adaptation* (1989, Cambridge University Press); *The Writing Systems of the World* (1989, Blackwell); *Language and Economy* (1992, Blackwell).

NORBERT DITTMAR is professor of linguistics at the Freie Universität Berlin. His main research interests are in sociolinguistics, second language acquisition, semantics, and discourse analysis. His recent publications include: (with P. Schlobinski and I. Wachs) *Berlinisch. Studien zum Lexikon, zur Spracheinstellung und zum Stilrepertoire* (1986, Berlin Verlag); (with U. Ammon and K. J. Mattheier) *Sociolinguistics: An International Handbook of the Science of Language and Society* (1987–8, de Gruyter); (with P. Schlobinski) *The Sociolinguistics of Urban Vernaculars* (1988, de Gruyter).

HELMUT GLÜCK is professor of linguistics at the Universität Bamberg. His main research interests are writing and its use, sociolinguistics and language policy, and current developments in German grammar and vocabulary. His recent publications include: *Schrift und Schriftlichkeit. Eine sprach- und kulturwissenschaftliche Studie* (1987,

Metzler); (with W. W. Sauer) *Gegenwartsdeutsch* (1990, Metzler); *Metzler Lexikon Sprache* (1993, Metzler).

MARLIS HELLINGER is professor in the *Englisches Seminar* (Department of English Studies) at the Universität Hannover. Her main research interests are contrastive linguistics, sociolinguistics, pidgin and creole linguistics, and feminist linguistics. Her recent publications include: *Sprachwandel und feministische Sprachpolitik. Internationale Perspektiven* (1985, Westdeutscher Verlag); *Kontrastive feministische Linguistik* (1990, Hueber); (with Ulrich Ammon) *Status Change of Languages* (1992, de Gruyter); (with Christine Bierbach) *Eine Sprache für beide Geschlechter. Richtlinien für einen nicht-sexistischen Sprachgebrauch* (1993, Deutsche UNESCO-Kommission).

WERNER HOLLY is professor of Germanic Linguistics at the Technische Universität Chemnitz-Zwickau. His main research interests are pragmatics, conversation analysis, political language, and language and the media. His recent publications include: (with Peter Kühn and Ulrich Püschel) *Politische Fernsehdiskussionen* (1986, Niemeyer); (also with P. Kühn and U. Püschel) (eds.) *Redeshows* (1989, Niemeyer); *Politikersprache* (1990, de Gruyter).

SIEGFRIED JÄGER is professor of linguistics at the Gerhard Mercator Universität Duisburg and Director of the Duisburger Institut für Sprach- und Sozialforschung (DISS). His main research interests are discourse theory and discourse analysis, right-wing extremism, and racism. His recent publications include: *Rechtsdruck. Die Presse der Neuen Rechten* (1988, Dietz); *BrandSätze. Rassismus im Alltag* (1992, 1993, DISS); (with J. Link) *Die vierte Gewalt. Rassismus und die Medien* (1993, DISS).

SYLVIA MOOSMÜLLER holds a research position at the Forschungsstelle für Schallforschung der Österreichischen Akademie der Wissenschaften in Vienna. Her main research interests are phonetics, phonology, sociophonology and sociophonetics, and feminist linguistics. Her recent publications include: *Soziophonologische Variation im gegenwärtigen Wiener Deutsch* (1987, Steiner); *Hochsprache und Dialekt in Österreich* (1991, Böhlau); 'Assessment and Evaluation of Dialect and Standard in Austria', in I. Werlen (ed.), *Verbale Kommunikation in der Stadt* (1994, Narr).

MARTINA ROST-ROTH is lecturer in Linguistics at the Freie Universität Berlin. Her main research interests are intercultural communication, second language acquisition, German as a Foreign Language, and conversation analysis. Her recent publications include: *Sprechstrategien in 'freien Konversationen': Eine linguistische Untersuchung zu*

Interaktionen im zweitsprachlichen Unterricht (1989, Narr); 'Reparaturen und Foreigner Talk—Verständnisschwierigkeiten in Interaktionen zwischen Muttersprachlern und Nichtmuttersprachlern', *Linguistische Berichte*, 125 (1990); 'Verständigungsprobleme in der interkulturellen Kommunikation. Ein Forschungsüberblick zu Analysen und Diagnosen in empirischen Untersuchungen', *Zeitschrift für Literaturwissenschaft und Linguistik* (forthcoming).

WOLFGANG WERNER SAUER is professor in the *Seminar für deutsche Literatur und Sprache* (Department of German Literature and Language) at the Universität Hannover. His main research interests are current developments in German grammar and vocabulary, German sociolinguistics, and lexicography. His recent publications include: *Der 'Duden', Geschichte und Aktualität eines Volkswörterbuchs* (1988, Metzler); (with H. Glück) *Gegenwartsdeutsch* (1990, Metzler).

PETER SCHLOBINSKI is professor in the *Institut für Deutsche Philologie* (Department of German Linguistics) at the Universität München. His main research interests are sociolinguistics, functional grammar, spoken German, and Chinese. His recent publications include: *Funktionale Grammatik und Sprachbeschreibung* (1992, Westdeutscher Verlag); (with G. Kohl and I. Ludewigt) *Jugendsprache—Fiktion und Wirklichkeit* (1993, Westdeutscher Verlag); (with Michael Dürr) *Einführung in die deskriptive Linguistik* (2nd edn. 1994, Westdeutscher Verlag); and *Empirische Sprachwissenschaft* (forthcoming, Westdeutscher Verlag).

HELMUT SCHÖNFELD, formerly of the Zentralinstitut für Sprachwissenschaft der Akademie der Wissenschaften der DDR, is now working on a research project on Berlin speech at the Humboldt-Universität Berlin. His main research interests are dialects, regional language varieties, and urban speech. His recent publications include: *Sprache und Sprachvariation in der Stadt. Zu sprachlichen Entwicklungen und zur Sprachvariation in Berlin und anderen Städten im Nordteil der DDR* (1989, Zentralinstitut für Sprachwissenschaft); 'Die berlinische Umgangssprache im 19. und 20. Jahrhundert', in J. Schildt and J. Schmidt (eds.), *Berlinisch. Geschichtliche Einführung in die Sprache einer Stadt* (1992², Akademie-Verlag).

PATRICK STEVENSON is lecturer in German in the School of Modern Languages at the University of Southampton. His main research interests are sociolinguistic variation in German, public language, language and identity, and contact linguistics. His recent publications include: (with S. Barbour) *Variation in German* (1990, Cambridge University Press); 'Political Culture and Intergroup Relations in Plurilingual

Switzerland', *Journal of Multilingual and Multicultural Development*, 3 (1990); 'The German language and the construction of national identities', in J. L. Flood *et al.*, *Das unsichtbare Band der Sprache* (1993, Stuttgarter Arbeiten zur Germanistik).

RUTH WODAK is professor of Applied Linguistics at the Universität Wien. Her main research interests are sociolinguistics, communication in institutions, gender studies, discourse and politics, discourse and racism, and media studies. Her recent publications include: *The Language of Love and Guilt* (1986, Benjamins); *Language, Power and Ideology* (ed.) (1989, Benjamins); *'Wir sind alle unschuldige Täter'* (1990, Suhrkamp); *The Disorders of Discourse* (forthcoming, Longman).

1 The Study of Real Language: Observing the Observers

PATRICK STEVENSON

I   THE REDISCOVERY OF REAL LANGUAGE AND THE RADICALIZATION OF LINGUISTICS IN THE FEDERAL REPUBLIC

The traffic in borrowed words and phrases between English and German is by no means one-way, even if the balance of trade heavily favours English exports into German. Perhaps because of their relative rarity, German words used in English often appear more conspicuous and more exotic: consider, for example, *Schadenfreude* or *Berufsverbot*. One of the most recent additions to this inventory is the concept of the *Wende* (meaning 'radical change of direction' or 'turning point'). In the immediate past, the term *Wende* was used to refer to the dramatic historical developments in the former GDR in its last days, and it is probably true to say that in virtually everything that has been written about the twelve months between the first pro-democracy marches in Leipzig in October 1989 and the official day of unification in October 1990 the word *Wende* has appeared, either in its own right or in one of many newly coined compounds (*Wendesprache*, 'language of the *Wende*'; *Wenderede*, 'speech about the *Wende*', *Wendesprüche*, '*Wende* sayings or slogans'; *Wendehals*, 'turncoat'; etc.).

As one of the great media events of the time, this *Wende* was recorded and analysed in unprecedented detail and at great length, and along with the political commentators linguists of various descriptions leapt with alacrity on to the bandwagon. Even before unification was a legal reality, the *Wende* had spawned myriad linguistic investigations, from the anecdotal (Wotjak 1991) and ironic or even whimsical (Röhl 1990*a*, 1990*b*) to the earnest and systematic (Hellmann 1990); the most ambitious project to date is the *Gesamtdeutsche Korpusinitiative* currently being conducted by the Institut für deutsche Sprache in Mannheim, consisting of a vast corpus of linguistic data drawn from 10,000 newspaper texts from East and West (Hellmann 1991, Herberg and Stickel 1992, Herberg 1993).

Just as it seemed as if the long-running academic soap opera based on the study of linguistic contrasts between East and West was on its last legs, the *Wende* appeared like a *deus ex machina* to open up a whole new chapter. Indeed, after the critical transition period a long phase of social and psychological adjustment will follow, providing countless opportunities for linguistic research on many different levels: intercultural communication, sociolinguistic variation, political discourse, media language, and so on. For what is sometimes still seen as the parent discipline of formal linguistics developments such as these may be of little consequence, but they give a decisive impetus to the study of real language, the language that is in daily use for all conceivable purposes: in personal interaction, for self-expression, for reflection, and so forth. The *Wende* in the geopolitical situation of what we may now call simply 'Germany' (Press Departments of German Embassies announced shortly after unification that this would be acceptable practice although the country's official name remains the Federal Republic of Germany) inevitably brings about a *Wende* in real language study, albeit of a less dramatic nature.

However, the term *Wende* itself has an interesting and important history. Before the demise of the GDR looked an even remote possibility, the word was used in West German political discourse throughout the 1980s to refer to the shift (back) to the right after the years of social democratic–liberal alliance in the 1970s. Thus, for example, a book published in 1986 by Hans Uske entitled *Die Sprache der Wende* considers linguistic manifestations of this switch to conservative dominance, particularly the language of Christian Democrat politicians. The way had been paved for this crucial turning-point by a *Tendenzwende* in the 1970s, a more general and diffuse trend away from the more liberal, less authoritarian ethos of the years following the traumatic upheaval in the educational world in the late 1960s, itself perhaps the single most profound development in West Germany's social history. As well as 1989, therefore, both 1982 (the year when Chancellor Helmut Kohl's first government came to power) and 1968 (the year that is synonymous with student protest in Germany) were major turning-points not just in German history but in the study of the German language.

The explosion of academic interest in language use and social reality in the late 1960s that is now often referred to as the 'pragmatische Wende' in German linguistics (see e.g. Steger 1980: 352, and Hartung 1991*a*: 24–36) was an abrupt departure from previously entirely formal/theoretical concerns. Admittedly, this new preoccupation was not without precedent in the history of language study in the German context. In the long tradition of dialectology, an empirical discipline

par excellence, language was always seen as embedded in social relationships: the term *soziallinguistisch* was first used by Ferdinand Wrede at
the beginning of this century (see Wrede 1903) and Adolf Bach (1934/
1969) talks of a 'sozial-linguistisches Prinzip'. However, 'social
aspects' are not really integral to the studies of dialectologists, but
tend to be tacked on as an extra dimension, and although they are
the natural precursors of contemporary sociolinguists, their work is
not a kind of sociolinguistics *avant la lettre* (Hartig 1985: 18). Furthermore, as Löffler (1985: 13–14) argues, the necessary socio-cultural
and socio-historical conditions for the emergence of an integrative
discipline of 'social linguistics' simply did not exist until very recently.

The turning-point in the study of language around the beginning
of the 1970s saw the emergence of what initially at least were seen as
two discrete (sub-)disciplines: sociolinguistics and linguistic pragmatics (also referred to, among other things, as pragmalinguistics).
The latter was and to a large extent continues to be more a portman-
teau term for a broad range of approaches, which Schlieben-Lange
(1979: 11–22) subsumes under three main headings: pragmatics as a
theory of the use of signs, as the linguistics of dialogue, and as speech
act theory. The roots of some of these approaches lie more in philosophical, of others more in linguistic traditions, but together they
constitute a science of 'speaking as an activity', whose ultimate aim
is to establish 'die universellen Bedingungen der Möglichkeit von
Kommunikation und dann die jeweils einzelsprachlichen und einzelge-
sellschaftlichen Typen sprachlicher Tätigkeiten' (the universal conditions that make communication possible and then particular types
of linguistic activity in specific languages and societies); (Schlieben-
Lange 1979: 22).

Sociolinguistics, on the other hand, was a much more sharply cir-
cumscribed discipline, and had a much narrower focus than the sociolinguistics developing in North America and in Britain at that time.
Of all the various strands of sociolinguistic research being conducted
in these environments (for an overview, see Trudgill 1983), virtually
the only one which impinged on the German context was the British
sociologist Basil Bernstein's theory of linguistic codes (for a detailed
discussion, see Barbour and Stevenson 1990, chap. 6, and Dittmar,
this volume). Although other aspects have since been adopted, it is
important to note that for many people in Germany today who studied
virtually any 'philological' discipline in the 1970s, sociolinguistics continues to be synonymous with code theory and the subsequent *Sprachbarrieren* controversy that engulfed all levels of the educational world.

Trends in academic research are rarely entirely independent of
influences from the socio-political environment in which the research

takes place, and the emergence of sociolinguistics and linguistic pragmatics in the Federal Republic provides a classic example of how
developments within a discipline may be contingent on events outside
the ivory tower. On the one hand, the deeply conservative institution
of *Germanistik* was in a state of crisis after decades of isolation from
international trends, and on the other hand the profoundly disillu-
sioned post-war generation in the Federal Republic perceived both
academic and political establishments as irredeemably compromised
by their refusal to abandon the 'old order'. At the same time, debates
were raging about the whole structure of the education system, which
was also seen by many as maintaining élitist traditions and failing to
deliver equality of opportunity. The educationalist Georg Picht (1964)
pricked the bubble of optimism and self-confident expansion with his
prediction of a *Bildungskatastrophe* (educational disaster).

A floundering academic discipline in search of 'ein neues Selbstverständnis' (a new self-image) (Schlieben-Lange 1991: 56), a student
population frustrated and alienated from its studies, a society deeply
divided over educational principles: the ground was well prepared for
the reception of a controversial theory that appeared to show how
class-based differences in socialization led to different degrees of
access to the dominant linguistic codes in the classroom and thus to
different life chances. Whether or not Bernstein was actually saying
what his enthusiastic readers in Germany supposed (see again Barbour
and Stevenson 1990, chap. 6) is historically less important than the
fact that the code theory provided a concrete focus for the various
aspects of the conflict: academic, social, and educational. In particular,
it appeared to offer students a radical alternative to the conventional
canon of 'philological studies', a sense of purpose and 'social relevance', a real feeling of self-justification. Linguists were divided
between those who continued to adhere to traditional areas of phil-
ology, and those who turned to (American) structuralism and either
persevered with it or rejected its abstraction and its silence on questions of language behaviour in favour of a linguistics of everyday
life. More significantly, perhaps, the nature of these new ideas about
language made them both accessible and appealing to academics from
other disciplines: sociology, philosophy and above all the new discipline of educational studies (*Pädagogik*). (For an important early discussion of these issues and more besides, see Wunderlich 1971.)

Ultimately the euphoria and intense enthusiasm surrounding these
debates gave way to renewed disillusion, when they failed to usher in
a brave new world and the years of expansion in higher education
were succeeded in the second half of the 1970s by a chillier climate
of retrenchment and stagnation. Nevertheless, it is important to hold

on to the picture of a radical departure that many believed was being spearheaded by the study of language in the heady post-1968 years. It was not simply a different way of looking at language, nor even merely one that offered a more adequate account of actual language use: both the tenor and the words of early West German socio-linguistic studies were imbued with a sense of almost missionary zeal. Wunderlich (1971: 317–18), for example, argues that the main role of sociolinguistics should be to contribute to a kind of consciousness-raising: not simply working towards equality of opportunity, but developing public awareness, fighting discrimination and manipulation, avoiding stereotypes, questioning assumptions, and so on. In the same vein, in addition to commonly expressed hopes that socio-linguistics could offer both a socially valuable activity for linguists and concrete social benefits especially in the emancipation of the proletariat, Ammon and Simon (1975: 10–15) also seek the development of a critical social awareness among the participants in the educational process: students, teachers, and schoolchildren.

It is also important to appreciate the broader significance of the debates on linguistic codes and *Sprachbarrieren*, and to a lesser extent the new work in linguistic pragmatics, in the development of language study in the Federal Republic. On the one hand, although the discussion of these issues may now be dormant, the issues themselves have not entirely gone away. On the other hand, the sudden and turbulent arrival of sociolinguistics in particular was not a transitional phase but a catastrophic moment in the history of language study. Even if the early hopes later proved unfounded, there was no going back, and subsequent developments could not have happened without this decisive break with the past (see also Dittmar 1983: 22).

What happened to the study of real language after this critical period is the subject of Section 4 below, but we might conclude this section on a sceptical note. Some of the leading figures in American sociolinguistics have always expressed a certain distaste for the term, preferring to see their work as 'the kind of thing all linguists ought to be doing' (cf. Labov 1972: p. xix). For them, sociolinguistics should have an inspirational function rather like pace-setters in middle-distance running, who sacrifice themselves in order to enhance the performance of the others. Dell Hymes, for instance, famously made the morbid declaration that 'the final goal of sociolinguistics must be to preside over its own liquidation' (Hymes 1977: 206). Similar views have been expressed by German linguists. Heinrich Löffler (1985: 19), for example, argues that by its nature sociolinguistics '[kann] nur eine Übergangs-Disziplin sein . . . und keine bleibende Wissenschaft' (can only be a transitional discipline and not a lasting science), it is just one framework for approaching the study of human activity. Peter Auer (cited in Schlieben-Lange 1991: 137) even contends that there is no longer any such thing as sociolinguistics, as it has splintered into a whole range of discrete topics that are not bound together in any coherent way.

This apparently growing fragmentation of sociolinguistic study has always been characteristic of linguistic pragmatics: Dieter Wunderlich (in Funk-Kolleg Sprache 1973: 102), for example, prefers to pose a set of pragmatic questions rather than try to define a discipline of pragmatics. However, as we shall see, although the emergent disciplines of the early 1970s may have since 'presided over their own liquidation', one of the outcomes has been a widespread cross-fertilization of ideas and techniques of analysis, leading to a range of new approaches to the study of real language.

2   LANGUAGE, LINGUISTICS, AND THE SOCIALIST STATE

In the GDR, the study of language, like any other academic enterprise, was always subservient to the 'needs of the socialist nation' (see e.g. Uesseler 1982: 119, Schönfeld 1983: 213): Marxist-Leninist linguistics stressed the dialectical relationship of language and society and therefore established its essential tasks as the identification and resolution of social problems associated with language (Ising 1974 and Große and Neubert 1974*a* are two of many classic formulations of this position). However, although there was no social upheaval in the GDR in the late 1960s comparable to the developments in the Federal Republic, linguistics in the socialist state became much more concretely concerned with sociolinguistic issues at around the same time.

Both American work such as that of Labov and his followers and the Bernstein/*Sprachbarrieren* controversy in the Federal Republic were registered by GDR linguists, but were considered to have little to offer as they derived directly from social problems peculiar to capitalist societies. Some of the very early sociolinguistic work in the GDR was in fact devoted to a critique of Western approaches, particularly the (justifiable) charge that it had no theoretical foundation in any explicit model of society (e.g., Große and Neubert 1974*a*: 9; note also the recent 'assaults from within' in Cameron 1990, Romaine 1984, and Williams 1992). Nevertheless, despite the obvious differences in social context, ideological motivation, and academic objectives, there is in fact a lot of common ground in terms of concepts and views on the nature of linguistic variation: to give just one example, the substance of the discussion of linguistic and communicative competence in

Große and Neubert (1974*a*) would be readily accepted by 'Western' sociolinguists.

The self-image of GDR sociolinguistics was based on a dual conception of its position in the history of language sciences. On the one hand, it was eager to stress its role in re-establishing the continuity of the 'acceptable' German tradition of studying 'social' aspects of language use (especially dialectology) developed in the nineteenth century, and indeed a number of major studies in social dialectology were published during the 1960s and early 1970s (e.g. Rosenkranz and Spangenberg 1963, Schönfeld 1974). On the other hand, as West German dialectologists could also lay claim to this same 'heritage', considerable emphasis was laid on the special role of constructing a progressive discipline that would contribute to the development of a harmonious and integrated socialist society.

Unlike North American and western European sociolinguistics, the sense of a 'grand design' permeates GDR work in this area. This is, of course, partly a function of the respective academic systems: in the West, individual researchers sought to make their mark so that advances in theory or methodology would be associated with their name, while researchers in the East typically worked as members of academic collectives whose work was directed towards the achievement of agreed goals. One inevitable result of this is that the cut and thrust of academic debate conducted openly on the pages of journals and monographs was lacking in the GDR. Another consequence, however, is that it is genuinely possible to characterize GDR sociolinguistics as a coherent academic programme: it had several distinct strands, but they were all clearly related to a central objective.

As linguists in the Federal Republic looked west for ways out of their impasse in the 1960s, GDR sociolinguists not surprisingly drew much of their inspiration and theoretical apparatus from the east, especially the Soviet Union but also Czechoslovakia. First, for example, sociolinguistic variation was described in terms of the Soviet model referred to in German as the *Gefüge der Existenzformen* (literally 'structure of varieties'), which embraced not only the *Literatursprache* (standard variety), *Umgangssprache* (colloquial speech), and *Dialekt*, but also 'social varieties' such as technical registers and other group-specific forms; this *Gefüge* was seen as a dynamic system that was in constant flux responding to changing social and political structures and communicative needs (Schönfeld 1985: 209–10). Secondly, the concept of *Tätigkeit* (activity) was adopted and developed as part of the theoretical basis for studying language use. The first task was to identify the means through which links between complex social processes and language were mediated, and this was only possible 'wenn die sprachliche Kommunikation als eine gesellschaftliche Tätigkeit verstanden wird, die in ein System übergeordneter Tätigkeiten eingeordnet ist'; (if linguistic communication is understood as a social activity, which is embedded in a system of superordinate activities) (Schönfeld 1983: 214). It was on the basis of these concepts that the central notion of GDR sociolinguistics was developed: the *soziolinguistisches Differential*, which is an analytical framework based on the four key factors of code, speaker, interlocutor, and communicative situation, and which also incorporates regional, social, and functional variability (see Große and Neubert 1974*a*: 13–16, Schönfeld 1983: 215, Uesseler 1982: 121; all of these aspects are dealt with in more detail by Dittmar, this volume).

Fundamental to the whole approach is what Hartung (1981) calls the 'Gesellschaftlichkeit der Sprache' (the 'socialness' or social nature of language): language and society are not independent categories which linguists should seek to relate to each other, but rather language is partly constitutive of society. Communication, therefore, is not merely a 'realization of language' but is part of the reality surrounding language; it is the process by which individuals are linked to each other and through which a concrete objective is achieved (Hartung 1991*a*: 25, 37). The emphasis on communication derives directly from the ultimate purpose of GDR sociolinguistics, to identify speakers' 'communicative knowledge' in order to promote efficient social intercourse, and one of the significant concrete outcomes of this was a series of painstaking empirical studies of language use in the workplace (see e.g. some of the papers in the volumes *Aktuelle Probleme* 1974; *Normen in der sprachlichen Kommunikation* 1977, *Kommunikation und Sprachvariation* 1981; also Herrmann-Winter 1979, Schönfeld and Donath 1978). We may also note in passing that this approach implies a pragmatic component, which is why there was no perceived separation between sociolinguistics and linguistic pragmatics in the GDR.

In the course of the 1970s, the overriding social objective of GDR sociolinguistics found a name for itself: *Sprachkultur*. This notion had first been developed by the Prague School of linguists in the 1920s and 1930s, but was only adopted in the GDR at the time when cultural policy in the form of developing 'socialist personalities' (Hartung 1981: 293) was given a high political profile. The idea of 'cultivating' language use is a rather delicate issue in the German context, as it has a number of connotations that many linguists would wish to distance themselves from (I shall return to the important complex of issues associated with terms such as *Sprachpflege*, *Sprachlenkung* and *Sprachkritik* as well as *Sprachkultur* in Section 5 below). In the first half of the GDR's history, efforts that might come under this heading were devoted first to the eradication of 'Fascist elements' and then to the promotion of the *Literatursprache* (standard variety) as a universal means of communication that would ensure equal access to all important social processes. This represented a sharp volte-face in official attitudes towards the standard variety, which in the early years had been seen as a powerful weapon in maintaining the dominant social position of the élite in bourgeois societies. However, it also entailed a rather simplistic and heavy-handed approach to the status and function of non-standard varieties. The development of concepts such as the *soziolinguistisches Differential* made it possible to conceive a more refined approach, that Große and Neubert (1974*a*: 16) for example refer to as a 'gesunde [healthy] Sprachkultur', taking a middle road between prescriptivism and linguistic *laissez-faire*.

In fact, an essential part of this new understanding of *Sprachkultur* was precisely a resistance to older notions of fixed, prescriptive norms based on the 'correctness' of the standard variety. The role of *Sprachkultur* was to promote the knowledge and use of the standard variety while fostering a view of other forms as situational or functional rather than social varieties, in recognition of the complex needs of a modern industrial society. The concept of linguistic norms still played a central role in this; however, norms were no longer 'yardsticks of quality' but parameters which permitted creativity without impeding communication (Nerius 1985). The new watchword was 'communicative adequacy' (see especially Techtmeier 1977), derived from the notion of 'appropriateness' in Soviet linguistics (in many respects this is another point of contact between the sociolinguistics of East and West).

Overall, GDR sociolinguistics was long on programmes of research and theoretical deliberations but relatively short on answers other than rather vague generalizations. Nevertheless, it had the considerable strength of being driven by an explicit social theory and a clear sense of social purpose. Furthermore, one of the last major publications (Hartung 1991*b*) showed not only further refinements of the GDR model but also a greater openness to work conducted in the West.

3 EAST MEETS WEST: LANGUAGE IN TRANSITION

During the years of separation, sociolinguists in East and West followed each other's work and maintained such contact as the obvious restrictions allowed. Despite the differences outlined in the previous two sections, most of those engaged in real language study on both sides of the ideological divide were concerned with similar issues. The question of whether 'the German language' itself was reinforcing the political conflict by developing into two distinct entities was really a side issue for most sociolinguists. Some did choose to specialize in this area and there is no doubt that it threw up many interesting linguistic questions (see e.g. Clyne 1984: chap. 2, Hellmann 1985), but many of the protagonists in this debate were journalists, politicians, and writers, for whom linguistic phenomena were metaphors for broader cultural issues.

There has been a tendency amongst linguists to play down the extent of linguistic differences between 'East' and 'West' German on the basis that the vast majority are lexical or semantic and have no structural consequences. There are two ways of looking at this. On the one hand, we can say that this is true up to a point although it overlooks both the profound cumulative effect of these 'superficial' differences and the importance of pragmatic contrasts in speech behaviour such as the realization or performance of specific speech acts (see Schlosser 1990, Wachtel 1991). On the other hand, it is arguable that the real historical significance of the debate was that it was one of the primary focal points in the ideological struggle over national identities (see Polenz 1988, Dieckmann 1989; in the broader context, Coulmas, this volume, Stevenson, 1993, Townson 1992; Hellmann 1989 gives a very clear account of the 'phases' of the debate and their dependency on changes in the German–German political context).

In any analysis of this question, whether in relation to the past or to the future, it is important to distinguish between official discourse and everyday speech. For obvious reasons, comparative research before 1989 derived almost entirely from the former, although this distinction was often not made explicit, with the result that conclusions were drawn about 'the state of the language' on the basis of highly restricted sources and types of data (if any at all). However, the political *Wende* opened up opportunities to study both categories in great detail. At the centre of the flurry of empirical observations since 1989 has again been the search for changes, this time from the perspective of the assumption that all forms of linguistic expression in the former GDR would be assimilated to the patterns prevailing in the (old) Federal Republic. The general label *Sprache der Wende* covers all manifestations of this period of upheaval: the voice of organized opposition, made public for the first time, and the many voices of protest on the streets; the desperate attempts by the regime to 'regain the confidence of the people' (that is, to cling on to power) by building key concepts of the reformers, such as 'dialogue', into the 'wooden language' of Party discourse (see Good 1991, Teichmann 1991, Schlosser 1990:

184–8); and the changing substance and tone of the media, especially *Neues Deutschland*, formerly the official organ of the SED (Socialist Unity Party) and now published by its successor the PDS (Party of Democratic Socialism) (see Hellmann 1990 and Jäger, this volume).

Much of this is now a matter of retrospective interest, in the sense that it is a closed process which may or may not be of consequence in other contexts, but there is a very serious field of future sociolinguistic work in the study of individual speech behaviour (see the chapters by Rost-Roth, and Schönfeld and Schlobinski, in this volume). The pathetic story recounted by Schlosser (1990: 194) about a GDR business manager declaring during a visit to Frankfurt am Main in March 1990 that 'es wird alles besser werden, wenn wir erst *nach der Marktwirtschaft planen*' (everything will get better when we start planning according to the market economy) may have an apocryphal ring to it, but anyone with experience of both societies will confirm that there are many psychological and communicative barriers still to be overcome by the citizens of the former GDR.

4 NEW DIRECTIONS IN PRAGMATIC SOCIOLINGUISTICS

The predominant approach to sociolinguistics in the GDR, with its emphasis on communicative practices and processes, almost by definition implied a strong pragmatic component. Some West German linguists would now also argue that there is no reason in principle to distinguish between linguistic pragmatics and micro-sociolinguistics (that is, excluding those areas often referred to as the sociology of language, which deal with large-scale phenomena such as language maintenance and shift or language contact and diglossia; see e.g. Hinnenkamp 1989: 3). However, the explosive and chaotic growth of these two disciplines in the Federal Republic in the 1970s resulted in sufficient confusion for a leading practitioner in both fields to call for a pooling of resources (Schlieben-Lange 1979: 112–20). On the one hand, Schlieben-Lange argues the need for a stronger empirical basis for linguistic pragmatics, suggesting greater use of data from face-to-face interactions in natural dialogues and giving promising examples of work on institutional speech from domains such as education, the law, and psychotherapy. On the other hand, she sees another way forward through what she calls 'dialectical sociolinguistics', which would look at how social interaction produces meanings and specific forms of action, which in turn produce further interactions.

Along the same lines and around the same time, Schlieben-Lange and Weydt (1978) set out a concrete agenda for the 'pragmatization of dialectology'. They argue that the traditional industry of German dialectology has produced many answers over the years, but that it keeps asking the same questions. The result is that we know a great deal about regional differences in phonetics and phonology, morphology, syntax, and lexis, but very little about characteristic *Sprechweisen* (speech styles or manners of speaking). Yet speakers do have intuitions about this level of speech and these are precisely the features of regional varieties that create the biggest obstacles to outsiders. The programme of research Schlieben-Lange and Weydt propose would address fundamental questions of utterance interpretation in the context of specific interactions between speakers from different regions. The key general questions would be:

- How are utterances such as X interpreted by (say) Bavarian speakers?
- Which utterances do I choose in (say) Bavarian so that my interlocutor interprets them in the way I intend?

To illustrate their proposal, they discuss eight concrete speech events, including 'reactions to compliments', 'making (apparent) promises', and 'forms of greeting'.

So at the beginning of the 1980s, several different traditions in the study of real language in the Federal Republic had reached at least a potential turning-point. The question, then, is whether (and if so, how) this opportunity for a change of direction was taken up. As far as dialectology is concerned, some research projects that could genuinely come under the heading of 'social dialectology' had already been carried out or at least started during the previous two decades, and Klaus Mattheier's book *Pragmatik und Soziologie der Dialekte* (1980) provided a coherent, partially historical rationale and a strong impetus for further research of this type (see Barbour and Stevenson 1990: chap. 4 for an account of this work). Projects based on major cities (such as Berlin: see Dittmar *et al.* 1986, 1988; and Mannheim: see Kallmeyer and Keim 1988) and small semi-rural communities transformed by the process of urbanization in the post-war years (see especially Besch *et al.* 1981 and Hufschmidt *et al.* 1983 on the 'Erp Project') constitute a clear positive response to the challenge of incorporating pragmatic (and indeed social-psychological) elements into social dialectology. But this response is limited to relatively few projects, and there has never been an extensive series of empirical studies in the German context of the type that mushroomed especially in North America and the UK in the 1960s and 1970s.

The work of Labov and other urban dialectologists was by no means ignored in Germany, but it has always been marginal: frequently referred to but seldom imitated. There are a number of reasons for this. Formal, technical considerations certainly played their part (for a detailed discussion, see Dittmar 1983 and in this volume), but just as important were the context of the way in which sociolinguistics was developing in the Federal Republic and the concerns of those interested in a social linguistics. On the one hand, the reception of Labov's work coincided with the boom period of *Sprachbarrieren* research, so that his contributions to the 'linguistic deficit' debate were taken on board but at the expense of his broader treatment of variation. By the time this debate had subsided, other issues were beginning to take centre stage, above all what Hinnenkamp (1990) calls *Gastarbeiterlinguistik*. On the other hand, Labov's approach appeared unsatisfactory as his concept of 'style' (see Labov 1972) was perceived as a static notion, a predetermined variable that failed to account for the unpredictability and spontaneous creativity of individual interaction. This meant that changes in style were simply explained in a mechanical, deterministic fashion as a result of changes in context. What the post-1968 generation of linguists in Germany demanded, however, was precisely a concept of style that was actively involved in the constitution or construction of contexts. Thus Auer (1989: 29) argues that 'es gibt linguistische Variation als solche, aber Stil immer nur in Beziehung zu einem interpretierenden Teilnehmer der Kultur und in Beziehung zu einem Anderen' (there is linguistic variation as such, but style exists only in relation to an interpreting participant of [a particular] culture and in relation to an 'other'), and styles in this sense are understood as:

dynamische und in der Situation selbst immer wieder erneut hergestellte und gegebenenfalls modifizierte und auf den Rezipienten zugeschnittene . . . Mittel der Signalisierung und Herstellung gemeinsam geteilter, relevanter sozialer und interaktiver Bedeutungen. (Selting and Hinnenkamp 1989: 6)

(dynamic, constantly created afresh (and if necessary modified) within the situation itself and geared directly to the recipient . . . a means of signalling and producing commonly shared, relevant social and interactive meanings.)[1]

What did appeal to many linguists in the 1980s therefore was the

---

[1] It is interesting to note here that, as far as I am aware, the so-called dynamic paradigm of linguistic variation associated with linguists such as Charles-James Bailey and Derek Bickerton (see e.g. Bailey 1973 and Bickerton 1971) has made little if any impact in Germany. This may be because despite its claim to offer an explanation of the actual process of linguistic change, it is predicated on a rather mechanistic and abstract role of the individual. Key notions such as implicational relationships between 'lects' are descriptive methodological constructs that are no better than the 'variable rules' of Labov and others at highlighting the active role of speakers in constructing meanings: they are essentially ways of dealing with data, not ways of explaining the dynamic process of interaction.

work of another American linguist, John Gumperz, with his emphasis on interaction and above all the process of contextualization (again, see Dittmar, in this volume, for a detailed account of this concept and its modification by German linguists). It is probably fair to say that what has emerged as the dominant trend in this area of real language study is what is called either 'interactional' or 'interpretive socio-linguistics' (see e.g. Auer and di Luzio 1984, Hinnenkamp 1989, and Hinnenkamp and Selting 1989). With its focus on the detailed analysis of face-to-face communication in concrete social settings and its objective of showing how social realities and relationships are produced and reproduced in the process of interaction, this emergent discipline is in many respects a direct response to Brigitte Schlieben-Lange's call for a coherent co-operation between pragmatics and socio-linguistics (see above). Furthermore, as several of the chapters in this book show, these central concerns have influenced developments in a number of related fields, such as gender and media studies, research on youth language, institutional discourse, and above all intercultural communication (both between Germans East and West and between Germans and Turks, Germans and Italians, etc.). A good example of this is the way in which research on *Gastarbeiterdeutsch* (GAD) has moved on from the original interest in the special features of GAD and the extralinguistic factors conditioning the acquisition process, to seeing GAD as an 'interactional product': rather than 'what is GAD?' or 'how is GAD acquired?', the question now is 'how do communi-cative processes between native and non-native speakers condition acquisition and the process of mutual comprehension?' (Hinnenkamp 1990: 284).

Moreover, the converse now also applies: the dominance of the interactional approach and the incorporation of pragmatic analytical techniques into sociolinguistic studies has encouraged a greater degree of interdisciplinarity. Consider, for example, gender studies, or more broadly the whole field of feminist linguistics. This has been a major discipline in its own right since the 1970s, and many of the concerns that gave rise to it in the first place still dominate its agenda and motivate further research (see Hellinger, in this volume). However, some recent work has begun to open the field to new influences. For instance, Frank (1992) is critical of what she sees as the sweeping generalizations and the ready acceptance of 'established truths' in the writing of other feminist linguists and implies that publications such as Pusch (1990) and Trömel-Plötz (1991) add little that is new to the debate. Others, such as Gräßel (1991) and Günthner and Kotthoff (1992) seek to expand in different directions (notably interaction between men and women in institutional contexts) and incorporate both analytical methods and actual perspectives on the topic under study from other areas. In a different direction again, Gdaniec (1987) investigates the use of particular political discourses in an attempt to find out how women try to affect (their role in) politics: what happens when women get involved in (male) political discourses? how far do women accept or reject these discourses and how does this affect their understanding of politics? can the dominant political discourses be changed by changing the political agenda?

5   VOICES OF AUTHORITY AND RESISTANCE

To conclude this chapter, I would like to move the discussion in a different direction, to introduce a perspective that cuts across the various strands of the debate on social linguistics dealt with in the previous sections. In addition to the internal wranglings within *Sprachwissenschaft* (the 'science' of linguistics, including therefore both formal linguistics and areas such as sociolinguistics and pragmatics), real language as a public issue has long been the subject of dispute outside the academic discipline of linguistics under the rubrics *Sprachpflege* (literally 'caring for the language', a kind of watchdog function concerned with maintaining the 'quality' of the language) and *Sprachkritik* (on this whole area, see Wells 1985: chap. 10, and Kolde 1986). The names are significant, because this kind of linguistic analysis has generally been dismissed or at best marginalized by academic linguists on the grounds that it is not *wissenschaftlich*. This is understandable in the context of the anxiety of linguists in post-war (West) Germany both to shake off any possible association with the linguistic barbarities of German fascism and to establish the credentials of their discipline as a serious academic enterprise. However, as Heringer (1988*a*: 13) argues, an exclusively descriptive linguistics is ultimately sterile as it lacks a critical dimension, and Kolde (1986: 181–6) shows that there are many areas of potentially common interest.

Both terms are open to various interpretations. *Sprachpflege* is either a euphemism for *Sprachpurismus* (and related terms like *Sprachreinigung* and *Sprachlenkung*, 'linguistic cleansing' and 'linguistic manipulation'), or a synonym for *Sprachkultur* (see Section 2 above). In the first sense it is at best what we would call 'preservation'; in the second sense it is a form of 'conservation'. *Sprachkritik* may then be defined in relation to *Sprachpflege*: either it is a form of resistance to the first sense, or it informs the second sense (or it may even fulfil both functions). What is common to both approaches is that they are concerned with a critical appraisal of actual language use in public domains.

Sprachpflege as an individual or concerted project has a long tradition and can be traced back at least to the *Sprachgesellschaften* (language societies) of the seventeenth century. Its protagonists have always seen themselves as linguistic standard-bearers, in every sense of the term, but its greatest significance lies in its contribution to the establishment of institutions, especially the Bibliographisches Institut (originally in Leipzig, then also in Mannheim), which publishes the series of major reference works known collectively as the *Duden*. Although there has never been any direct equivalent to the Académie Française, the *Duden* has acquired an authoritative status and is widely revered as the ultimate arbiter in all linguistic matters. However it may perceive its own role, the public perception of the *Duden* is thus of a prescriptive institution that determines right from wrong (the two chapters by Helmut Glück and Wolfgang Sauer in this volume deal with various aspects of this issue).

Sprachkritik also has a long tradition, but it is its development since the Second World War that is relevant here. Much of the work in the immediate post-war years, which came under fire from the linguistics establishment, was a continuation of the trenchant critical polemics of writers and publicists such as Kurt Tucholsky and Karl Kraus in the first half of the century. More recently, many linguists (including even some of the earlier sceptics: see e.g. Polenz 1989) have subjected many forms of public language, especially from politics and the media, to critical scrutiny (Heringer 1988*c* contains a selection of characteristic pieces from the 1960s to the 1980s). Although this work is now more linguistically informed, the vast majority of studies in this mould are conducted from a fundamentally resistant, anti-authoritarian perspective. By contrast with *Sprachpfleger*, therefore, *Sprachkritiker* see themselves as either subversive, destabilizing, and awkward (Heringer 1988*a*) or democratizing, liberating, and enabling (Wimmer 1988), but at all events as opponents of prescriptive norms (Heringer 1988*b*) and as promoters of active (self-)reflection on language use as a universal practice (see also Moosmüller, in this volume, on a rather different type of public evaluation of political/politicians' language).

There are many lines of enquiry that a modern *Sprachkritiker/in* could pursue (for a good selection, see the papers in Liedtke *et al.* 1991, Klein 1989; and from a feminist perspective, Günthner and Kotthoff 1992). I would like to suggest though that there are three broad approaches, that we can illustrate with reference to work from both the German and the anglophone contexts. First, there is a large body of literature that deals discursively with (more or less random) issues, features, and tendencies in language use, with minimal linguistic apparatus: obvious examples relating to English are the writings of George Orwell and Raymond Williams (see also in this context Crowley 1989, Milroy and Milroy 1991); for German, consider in addition to Tucholsky and Kraus also Heringer (1988*c*, 1990), Pörksen (1989, 1991), or in a more journalistic vein Zimmer (1986, 1990). Secondly, there is a more linguistically orientated approach, that deals analytically with actual discourse: for example, with respect to English, Fowler *et al.* (1979) and Fairclough (1992*b*) or from a pragmatic perspective Wilson (1990); and in Germany Jäger (1991) and Link (1991), or from a feminist perspective Gdaniec (1987) and Günthner and Kotthoff (1992) (see Section 4 above). Thirdly, there are some attempts at least to advocate a '*Sprachkritik* from below' (Holly 1985: 203), a kind of applied *Sprachkritik* that aims to provide broad sectors of the population with the ability to undertake their own analysis of public language: for example Fairclough (1989, 1992*a*) for English; and for German Heringer (1988*a*) and Wimmer (1988), or again from a feminist perspective Hellinger *et al.* (1985) and Pusch (1984), especially the *Sprachglossen* (short critical analyses of individual words). This categorization may be somewhat arbitrary and the examples certainly are, but the important points are that *Sprachkritik* is a complex and wide-ranging field and that critical perspectives now more than ever are being applied to many aspects of language study.

6 CONCLUSIONS

I have used the term 'real language' here in preference to any of the more precise or more technical terms available for a number of reasons. The first is to permit discussion under one rubric of approaches to language study that are typically assigned to several separate areas or disciplines. The second is to enable a discussion of these approaches and their interrelationships without having to become too closely embroiled in definitions. Finally, and most importantly, I wanted to construct a framework for looking at a range of topics that all have to do with aspects of language use in the real world (whether theoretical, descriptive, or critical) and that form the substance of the following chapters:

- the use of German in relation to other languages: its status and function in the world;
- the appeal to the German language as a constitutive factor in the establishment of national identities: its socio-historical symbolism;
- the constant changes in the shape and form of German in the

hands and mouths of its users: its dynamism and vitality, but also
its potential for controversy;
- and the use of the German language as a means of constructing,
articulating, and analysing social realities: its cultural plasticity.

The choice of the term 'real language' to suit my purposes here was
neither accidental nor original: I deliberately borrowed it from
Eugenio Coseriu, a linguist who has been very influential in German
linguistics but perhaps rather less so in the anglophone world, who
says 'daß in der wirklichen Sprache das Systematische, das Kulturelle,
das Soziale und das Geschichtliche zusammenfallen (that in real language the systematic, the cultural, the social and the historical
coincide) (Coseriu 1974: 53, cited in Schlieben-Lange 1991: 16).

Further Reading

Dittmar and Schlobinski (1988)
Heringer (1988c, 1990)
Hinnenkamp (1989)
Liedtke *et al.* (1991)
Schlieben-Lange (1979, 1991)
Wimmer (1985)

References

Aktuelle Probleme der sprachlichen Kommunikation (1974) (Berlin: Akademie-
Verlag).
ALTHAUS, H.-P., HENNE, H., and WIEGAND, H. E. (eds.) (1980), *Lexikon der
germanistischen Linguistik*, 2nd edn, (Tübingen: Niemeyer).
AMMON, U., and SIMON, G. (1975), *Aspekte der Soziolinguistik* (Weinheim,
Basle: Beltz).
AUER, P. (1989), 'Natürlichkeit und Stil', in Hinnenkamp and Selting (1989),
27–59.
—— and DI LUZIO, A. (eds.) (1984), *Interpretive Sociolinguistics* (Tübingen:
Narr).
BACH, A. (1969), *Deutsche Mundartforschung*, 3rd edn. (1st edn. 1934) (Heidel-
berg: Winter).
BAILEY, C.-J. N. (1973) *Variation and Linguistic Theory* (Washington, DC:
Center for Applied Linguistics).
BARBOUR, S., and STEVENSON, P. (1990), *Variation in German: A Critical
Approach to German Sociolinguistics* (Cambridge: Cambridge University
Press).
BESCH, W., HUFSCHMIDT, J., KALL-HOLLAND, A., KLEIN, E., and MATTHEIER,
K. J. (1981), *Sprachverhalten in ländlichen Gemeinden: Ansätze zur Theorie
und Methode* (Berlin: Schmidt).

BICKERTON, D. (1971), 'Inherent Variability and Variable Rules', *Foundations of Language*, 7: 457–92.

BOLTEN, J. (1989), 'Zum Umgang mit dem Begriff *konservativ* in der politischen Diskussion der Bundesrepublik', in Klein (1989), 277–96.

CAMERON, D. (1990), 'Demythologizing Sociolinguistics: Why Language does not Reflect Society', in Joseph and Taylor (1990), 79–93.

CLYNE, M. G. (1984), *Language and Society in the German-Speaking Countries* (Cambridge: Cambridge University Press).

CROWLEY, T. (1989), *The Politics of Discourse* (Basingstoke: Macmillan).

COSERIU, E. (1974), *Synchronie, Diachronie und Geschichte* (Munich: Fink).

DIECKMANN, W. (1989), 'Die Untersuchung der deutsch-deutschen Sprachentwicklung als linguistisches Problem', *Zeitschrift für germanistische Linguistik*, 17: 162–81.

DITTMAR, N. (1982), 'Soziolinguistik, Teil I', *Studium Linguistik*, 12: 20–52.

—— (1983), 'Soziolinguistik, Teil II', *Studium Linguistik*, 14: 20–57.

—— and SCHLOBINSKI, P. (eds.) (1988), *The Sociolinguistics of Urban Vernaculars* (Berlin: de Gruyter).

—— —— and WACHS, I. (1986), *Berlinisch: Studien zum Lexikon, zur Spracheinstellung und zum Stilrepertoire* (Berlin: Spitz).

—— —— —— (1988), 'Berlin Urban Vernacular Studies: Contributions to Sociolinguistics', in Dittmar and Schlobinski (1988), 1–144.

DITTRICH, E., and RADTKE, F. O. (eds.) (1990), *Ethnizität: Wissenschaft und Minderheiten* (Opladen: Westdeutscher Verlag).

FAIRCLOUGH, N. (1989), *Language and Power* (London: Longman).

—— (1992*a*), *Critical Language Awareness* (London: Longman).

—— (1992*b*), *Discourse and Social Change* (Cambridge: Polity Press).

FLOOD, J., SALMON, P., SAYCE, O., and WELLS, C. J. (eds.) (1993), *Das unsichtbare Band der Sprache. German Language in History and Society: Memorial Studies for Leslie Seiffert* (Stuttgart: Akademischer Verlag).

FOWLER, R., HODGE, R. V., KRESS, G., and TREW, A. (eds.) (1979), *Language and Control* (London: Routledge).

FRANK, K. (1992) *Sprachgewalt: die sprachliche Reproduktion der Geschlechterhierarchie* (Tübingen: Niemeyer).

FUNK-KOLLEG SPRACHE (1973), *Eine Einführung in die moderne Linguistik* (Frankfurt: Fischer).

GDANIEC, C. (1987), 'Politische Diskurse von Frauen', in Vogt (1987), 202–24.

GOOD, C. (1991), 'Language and Totalitarianism: The Case of "East Germany" ', inaugural lecture, University of Surrey.

GRÄßEL, U. (1991), *Sprachverhalten und Geschlecht: eine empirische Studie zu geschlechtsspezifischem Sprachverhalten in Fernsehdiskussionen* (Pfaffenweiler: Centaurus).

GROßE, R., and NEUBERT, A. (1974*a*), 'Thesen zur marxistisch-leninistischen Soziolinguistik', in Große and Neubert (1974*b*), 9–22.

—— (eds.) (1974*b*), *Beiträge zur Soziolinguistik* (Munich: Hueber).

GÜNTHNER, S., and KOTTHOFF, H. (eds.) (1992), *Die Geschlechter im Gespräch. Kommunikation in Institutionen* (Stuttgart: Metzler).

HARTIG, M. (1985), *Soziolinguistik* (Bern: Lang).

HARTUNG, W. (1981), 'Eine hohe Sprachkultur: Aufgabe in der sozialistischen Gesellschaft der DDR', *Deutschunterricht*, 6: 292–303.

—— (1991*a*), 'Linguistische Zugänge zur sprachlichen Kommunikation', in Hartung (1991*b*), 13–90.

—— (ed.) (1991*b*), *Kommunikation und Wissen* (Berlin: Akademie-Verlag).

HELLINGER, M., KREMER, M., and SCHRÄPEL, B. (1985), *Empfehlungen zur Vermeidung von sexistischem Sprachgebrauch in öffentlicher Sprache* (Universität Hannover).

HELLMANN, M. (1985), 'Bemerkungen zur Entwicklung und zur gegenwärtigen Lage des Arbeitsgebietes "Ost–West-Sprachdifferenzierung"', *Mitteilungen 11 des Instituts für deutsche Sprache (Mannheim)*, 76–93.

—— (1989), 'Die doppelte Wende: Zur Verbindung von Sprache, Sprachwissenschaft und zeitgebundener politischer Bewertung am Beispiel deutsch-deutscher Sprachdifferenzierung', in Klein (1989), 297–326.

—— (1990), 'DDR-Sprachgebrauch nach der Wende: Eine erste Bestandsaufnahme', *Muttersprache*, 100/2–3: 266–86

—— (1991), 'Die deutsche Sprache nach der Wende', *Sprachreport*, 1/91: 4.

HERBERG, D. (1993), 'Die Sprache der Wendezeit als Forschungsgegenstand. Untersuchungen zur Sprachentwicklung 1989/90 am IDS', *Muttersprache*, 103/3: 264–6.

—— and STICKEL, G. (1992), 'Gesamtdeutsche Korpusinitiative. Ein Dokumentationsprojekt zur Sprachentwicklung 1989/90', *Deutsche Sprache*, 20/2: 185–92.

HERINGER, H.-J. (1988*a*), 'Sprachkritik – die Fortsetzung der Politik mit besseren Mitteln', in Heringer (1988*c*), 3–34.

—— (1988*b*), 'Normen? Ja – aber meine!', in Heringer (1988*c*), 94–105.

—— (1988*c*), *Holzfeuer im hölzernen Ofen. Aufsätze zur politischen Sprachkritik*, 2nd edn. (Tübingen: Narr).

—— (1990), *'Ich gebe Ihnen mein Ehrenwort'. Politik, Sprache, Moral* (Munich: Beck).

HERMANN-WINTER, R. (1979), *Studien zur gesprochenen Sprache im Norden der DDR* (Berlin: Akademie-Verlag).

HINNENKAMP, V. (1989), *Interaktionale Soziolinguistik und interkulturelle Kommunikation* (Tübingen: Niemeyer).

—— (1990), ' "Gastarbeiterlinguistik" und die Ethnisierung der Gastarbeiter', in Dittrich and Radtke (1990), 277–97.

—— and SELTING, M. (eds.) (1989), *Stil und Stilisierung. Arbeiten zur interpretativen Soziolinguistik* (Tübingen: Niemeyer).

HOLLY, W. (1985), 'Politische Kultur und Sprachkultur. Wie sich der Bürger politische Äußerungen verständlich machen kann', in Wimmer (1985), 196–210.

HUFSCHMIDT, J., KLEIN, E., MATTHEIER, K. J., and MICKARTZ, H. (1983), *Sprachverhalten in ländlichen Gemeinden: Dialekt und Standardsprache im Sprecherurteil* (Berlin: Schmidt).

HYMES, D. (1977), *Foundations in Sociolinguistics: An Ethnographic Approach* (London: Tavistock).

Ising, G. (1974), 'Struktur und Funktion der Sprache in der gesamtgesellschaftlichen Entwicklung', in *Aktuelle Probleme* (1974), 9–36.

Jäger, S. (1991), *Text- und Diskursanalyse. Eine Anleitung zur Analyse politischer Texte*, 3rd. edn. (Duisburg: Duisburger Institut für Sprach- und Sozialforschung).

Januschek, F. (1985*a*), 'Zum Selbstverständnis politischer Sprachwissenschaft', in Januschek (1985*b*), 1–20.

—— (ed.) (1985*b*), *Politische Sprachwissenschaft* (Opladen: Westdeutscher Verlag).

Joseph, J. E., and Taylor, T. (eds.) (1990), *Ideologies of Language* (London: Routledge).

Kallmeyer, W., and Keim, I. (1988), 'The Symbolization of Social Identity: Ethnography and Analysis of Linguistic Variation in a Project about Urban Communication in Mannheim', in Dittmar and Schlobinski (1988), 232–57.

Klein, J. (ed.) (1989), *Politische Semantik* (Opladen: Westdeutscher Verlag).

Klein, W., and Wunderlich, D. (eds.) (1971), *Aspekte der Soziolinguistik* (Frankfurt: Athenäum).

Kolde, G. (1986), 'Sprachkritik, Sprachpflege und Sprachwissenschaft', *Muttersprache*, 96/2: 171–89.

Kommunikation und Sprachvariation (1981) (Berlin: Akademie-Verlag).

Labov, W. (1972), *Sociolinguistic Patterns* (Oxford: Blackwell).

Liedtke, F., Wengler, M., and Böke, K. (eds.) (1991), *Begriffe besetzen* (Opladen: Westdeutscher Verlag).

Link, J. (1991), ' "Der irre Saddam setzt seinen Krummdolch an meine Gurgel!" Fanatiker, Fundamentalisten, Irre und Trafikanten—Das neue Feindbild Süd', in Jäger (1991), 73–92.

Löffler, H. (1985), *Germanistische Soziolinguistik* (Berlin: Schmidt).

Milroy, J., and Milroy, L. (1991), *Authority in Language*, 2nd edn. (London: Routledge).

Nerius, D. (1985), 'Zur Geschichte und Bedeutung des Begriffs Sprachkultur in der Linguistik der DDR', in Wimmer (1985), 55–69.

Normen in der sprachlichen Kommunikation (1977) (Berlin: Akademie-Verlag).

Picht, G. (1964), *Die deutsche Bildungskatastrophe* (Olten: Walter).

Polenz, P. von (1988), ' "Binnendeutsch" oder plurizentrische Sprachkultur?', *Zeitschrift für germanistische Linguistik*, 16: 198–218.

—— (1989) 'Verdünnte Sprachkultur. Das Jenninger-Syndrom aus sprachkritischer Sicht', *Deutsche Sprache*, 4: 289–316.

Pörksen, U. (1989), *Plastikwörter. Die Sprache einer internationalen Diktatur* (Stuttgart: Klett-Cotta).

—— (1991), 'Die totale Entwirklichung. Zur Sprache der Kriegsberichterstattung', *Sprachreport*, 2: 6–9.

Pusch, L. (1984), *Das Deutsche als Männersprache* (Frankfurt: Suhrkamp).

—— (1990), *Alle Menschen werden Schwestern. Feministische Sprachkritik* (Frankfurt: Suhrkamp).

Röhl, E. (1990*a*), 'Deutsche Sprache der DDR—nostalgischer Rückblick', *Sprachpflege und Sprachkultur*, 3/90: 83–5.

RÖHL, E. (1990*b*), 'Deutsch in der Ex-DDR', *Sprachpflege und Sprachkultur*, 4/90: 116.

ROMAINE, S. (1984), 'The Status of Sociological Models and Categories in Explaining Linguistic Variation', *Linguistische Berichte*, 90: 25–38.

ROSENKRANZ, H., and SPANGENBERG, K. (1963), *Sprachsoziologische Studien in Thüringen* (Berlin: Akademie-Verlag).

SCHLIEBEN-LANGE, B. (1979), *Linguistische Pragmatik*, 2nd edn. (Stuttgart: Kohlhammer).

—— (1991), *Soziolinguistik*, 3rd edn. (Stuttgart: Kohlhammer).

—— and WEYDT, H. (1978), 'Für eine Pragmatisierung der Dialektologie', *Zeitschrift für germanistische Linguistik*, 63: 257–82.

SCHLOSSER, H. D. (1990), *Die deutsche Sprache in der DDR zwischen Stalinismus und Demokratie* (Cologne: Verlag Wissenschaft und Politik).

SCHÖNFELD, H. (1983), 'Zur Soziolinguistik in der DDR. Entwicklung, Ergebnisse, Aufgabe', *Zeitschrift für Germanistik*, 2: 213–23.

—— (1985), 'Varianten, Varietäten und Sprachvariation', *Zeitschrift für Phonetik, Sprachwissenschaft und Kommunikation*, 38: 206–24.

—— and DONATH, J. (1978), *Sprache im sozialistischen Industriebetrieb* (Berlin: Akademie-Verlag).

SELTING, M., and HINNENKAMP, V. (1989), 'Einleitung: Stil und Stilisierung in der interpretativen Soziolinguistik', in Hinnenkamp and Selting (1989), 1–23.

Sprachliche Kommunikation und Gesellschaft (1974) (Berlin: Akademie-Verlag).

STEGER, H. (1980), 'Soziolinguistik', in Althaus *et al.* (1980), 349–58.

STEVENSON, P. (1993), 'The German Language and the Construction of National Identities', in Flood *et al.*, (1993).

TECHTMEIER, B. (1977), 'Die kommunikative Adäquatheit sprachlicher Äußerungen', in *Normen* (1977), 102–62.

TEICHMANN, C. (1991), 'Von der "langue de bois" zur "Sprache der Wende" ', *Muttersprache*, 101/3: 252–65.

TOWNSON, M. (1992), *Mother-Tongue and Fatherland: Language and Politics in German* (Manchester, New York: Manchester University Press).

TRÖMEL-PLÖTZ, S. (1991), *Vatersprache, Mutterland. Beobachtungen zu Sprache und Politik* (Munich: Verlag Frauenoffensive).

TRUDGILL, P. (1983), *Sociolinguistics: An Introduction to Language and Society*, 2nd edn. (Harmondsworth: Penguin).

UESSELER, M. (1982), *Soziolinguistik* (Berlin: Deutscher Verlag der Wissenschaften).

USKE, H. (1986) *Die Sprache der Wende* (Berlin: Dietz).

VOGT, R. (ed.) (1987), *Über die Schwierigkeiten der Verständigung beim Reden* (Opladen: Westdeutscher Verlag).

WACHTEL, S. (1991), 'Deutsch sprechen: Zu den Sprechkulturen in Ost- und Westdeutschland', *Muttersprache*, 101/2: 157–65.

WELLS, C. J. (1985), *German: A Linguistic History to 1945* (Oxford: Oxford University Press).

WILLIAMS, G. (1992), *Sociolinguistics: A Sociological Critique* (London: Routledge).

WILSON, J. (1990), *Politically Speaking* (Oxford: Blackwell).

WIMMER, R. (ed.) (1985), *Sprachkultur* (= *Jahrbuch 1984 des Instituts für deutsche Sprache*) (Düsseldorf: Schwann).

—— (1988), 'Überlegungen zu den Aufgaben und Methoden einer linguistisch begründeten Sprachkritik', in Heringer (1988*c*), 290–313.

WOTJAK, B. (1991), 'Rede-"Wendungen" in "Wende"-Reden', *Deutsch als Fremdsprache*, 1/91: 47–51.

WREDE, F. (1903), 'Der Sprachatlas des deutschen Reiches und die elsässische Dialektforschung', *Archiv für das Studium der neueren Sprachen und Literaturen*, 111: 29–48.

WUNDERLICH, D. (1971), 'Zum Status der Soziolinguistik', in Klein and Wunderlich (1971), 297–321.

ZIMMER, D. (1986), *Redens Arten* (Zurich: Haffmans Verlag).

—— (1990), *Die Elektrifizierung der Sprache* (Zurich: Haffmans Verlag).

2 To What Extent is German an International Language?

ULRICH AMMON

I INTRODUCTION

If a language can be used widely in international communication, its speakers have numerous advantages: they can use their native language (mother tongue) for negotiating international business contracts or political treaties, for lecturing and publishing internationally as scientists or scholars, or as tourists, while others have to resort to a foreign language for these activities. The use of a foreign language not only requires considerable additional learning but, as a rule, remains a more strenuous and less effective means of communication than the use of one's native language. In extreme cases, the non-native user of a language may resemble a baby, with respect to his/her verbal skills, as compared to the 'adult' native speaker. Given these practical advantages it is not surprising that most language communities try to spread their language internationally if they see any chance of success, and national pride in their own language further stimulates such endeavours. In this chapter, I shall concentrate on the question of the degree to which the German language actually is international. However, before this question can be assessed systematically, a few remarks on definitions and methods are necessary.

Though the term *international language* occurs quite often in socio-linguistic literature there is no consensus about its meaning. Furthermore, the term is quite uncommon in works of reference for linguists. Where it does occur, it tends to be defined in a way that would not be useful for the present investigation, namely as a language specifically intended for international communication ('created or suggested for adoption for purposes of international communication': Pei 1966: 128, 131). In contrast, I would like to specify the meaning of the term as 'a language actually used in international communication'. Further consideration of *international communication* so defined will help to avoid misunderstanding and illuminate the method by which I intend to assess the degree of internationality of the German language in comparison to other languages.

A language may be considered international if it is used for communication between different nations, or rather their citizens. The term *nation* is, however, commonly used with two different meanings (see Ammon 1990*a*: 136):

1. in the sense of a political unit, held together by a common government, currency, legal system, etc.: roughly synonymous with country or state;
2. in the sense of a cultural and linguistic unit, held together by a common history, culture, and language: roughly synonymous with nationality.

On the one hand therefore one can define communication as international in relation to (1), if it occurs between citizens of different countries or states. On the other hand communication can be considered international in relation to (2), if it occurs between members of different nationalities, that is, different language communities. A third possibility is to consider communication international if both conditions coincide. Such a combination of conditions could be termed *international communication in the narrower sense*. Accordingly, I shall call communication between citizens of different countries *international (only) in the wider sense*, and communication between members of different nationalities or language communities *interlingual* (that is, bridging two different languages). Only if a language is used for international communication in the narrower sense can it, in my opinion, seriously be considered an *international language*.

Thus if on the one hand, for instance, a German and an Austrian (whose native language is in both cases German) communicate in German, they communicate internationally only in a wider sense, which is of limited interest for our topic. If on the other hand a German-speaking and a French-speaking Swiss (their native language being German and French respectively) communicate in one of their languages, their communication is only interlingual, since both are citizens of the same country. If, however, a French-speaking Swiss and a German (their native language being French and German respectively) communicate in one of their languages, it is an instance of international communication in the narrower sense. These conceptual distinctions are illustrated in Figure 2.1.

Figure 2.1 contains the further distinction between *asymmetric* use of a language, in which case the language used for communication is native for one (or some) of the communicators but not the other(s), and *lingua franca* use, in which case the language used is native for

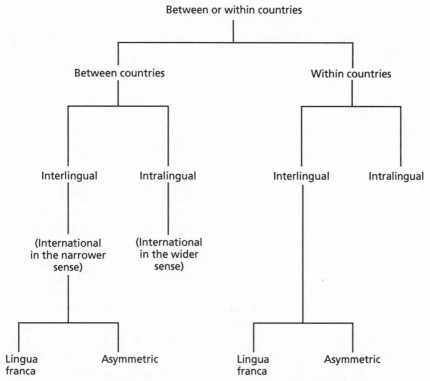

Fig. 2.1. Language choice in communication between speakers from different countries and different nationalities or language communities

none of the communicators. This distinction is important for a precise analysis of international languages; a real international language, one could postulate, has to be used as a lingua franca, not only asymmetrically. For the delimitation of borderline cases, it may be important to specify that 'native language' (or 'mother tongue') should be defined in terms of skills and ontogenetic period of learning (childhood), that is, as really native, in contrast to a language that is merely claimed as a 'mother tongue'. Thus, for an Irishman who grew up with English and speaks it fluently, English is his native language, in spite of the fact that he claims Irish (Gaelic) as his only 'mother tongue'; Irish may, of course, be his second native language if he also learned it in childhood and has a full command of it.

We can now specify that a language is 'more international' the more it is used for international communication in the narrower sense, either asymmetrically or, *a fortiori*, as a lingua franca. Following this line of thought, we can in principle rank languages according to their degree of internationality or even compare them metrically, that is, on the basis of an interval scale, and not only classify them as either international or non-international, as is sometimes done (e.g. Braga 1979).

It should perhaps be pointed out that these remarks do not yet imply precise scales for ranking or for measuring languages according to their degree of internationality. For a precise procedure, one would, among other considerations, have to specify what counts as a single as opposed to two international communicative events. This is obviously a prerequisite for counting such events reliably, in order to rank or compare languages. However, I shall not attempt this here, not only for want of space, but also because of the limited practical use of such an attempt at the present stage of research. Lacking virtually any data on international communicative events, or at least lacking representative data, we have to rely on mere indicators of them, which have so far not been validated and which cannot in fact be validated in the absence of adequate data for what they supposedly indicate (international communicative events). Their value as indicators can at present only be assessed by bits and pieces of evidence or on an intuitive basis by plausibility arguments. An example of such an indicator is the number of scientific publications in a given language. If they are more numerous in language La than in Lb, we may then assume that in the domain of science more international written communication occurs in La than in Lb — that is, if we do not have any reason to believe that publications in La are, on average, less widely read than those in Lb. Though such reasoning may appear questionable at times, we have no alternative for the time being but to base a good deal of our evidence on it. It should, however, be noted that I shall present data not just on a single but a whole number of such hypothetical indicators. One could, therefore, argue that the inadequacy of one indicator might, to some degree, be compensated for by the others.

2   SOME BASIC FACTORS DETERMINING THE INTERNATIONALITY OF A LANGUAGE

2.1   *Numerical Strength*

It seems plausible to assume that all else being equal the language of a large community has a better chance of becoming an international language than does that of a small community. A large community's language is more likely to be studied as a foreign language, because it opens up more opportunities for contacts than the language of a small community. Even a superficial consideration reveals that practically all international languages have, or—as for instance in the case of Latin—once had, large communities of native speakers. This is reinforced by the fact that there are several thousand languages in the world (Grimes (1984: p. xvii), for instance, counts 5,781), most of which have very few native speakers and are not used internationally at all.

Table 2.1 gives an overview of how German compares to other languages in this respect, according to different counts or estimates. The divergencies are in part due to difficulties in defining precisely what a 'native speaker' of a language is, and in part to lack of reliable data (for instance up-to-date censuses).

According to Table 2.1, German ranges between rank 7 and rank 11. The difference between Grimes and the two other estimates is partly due to the time span between them (1984—1987—1990), although this is by no means the only reason. A somewhat earlier estimate (Muller 1964) ranks German sixth (above Japanese, Arabic, Bengali, and Portuguese), and if we step back further in history German rises to still higher ranks in numerical strength among the languages of the world. Thus, around 1920 German ranks on a par with Russian, and around 1800 even exceeds all the other European languages including English (Jespersen 1926: 229). Therefore the factor 'numerical strength' must once have worked more in favour of German as an international language than it does today. It is, incidentally, not very difficult to find some of the reasons why German declined in relative (as opposed to absolute) numerical strength in

Table 2.1. *Number of native speakers of German in comparison to other languages* (millions)

Grimes (1984)		Comrie (1987)		Finkenstaedt and Schröder (1990)	
1. Chinese	700	Chinese	1,000	Chinese	770
2. English	391	English	300	English	415
3. Spanish	211	Spanish	280	Hindi	290
4. Hindi–Urdu	194	Russian	215	Spanish	285
5. Russian	154	Hindi–Urdu	200	Arabic	170
6. Portuguese	120	Indonesian	200	Bengali	165
7. **German**	**119**	Arabic	150	Portuguese	160
8. Arabic	117	Portuguese	150	Indonesian	125
9. Japanese	117	Bengali	145	Japanese	120
10. Indonesian	110	Japanese	115	Russian	115
11. Bengali	102	**German**	**103**	**German**	**92**
12. French	63	French	68	French	55

recent times. German declined vis-à-vis the languages of some developing countries because the population growth of developed countries is generally slower, and it declined vis-à-vis the languages of some European countries because the German-speaking countries did not spread their language beyond Europe by way of conquest and colonialism (with the exception of Namibia).

2.2   Economic Strength

The language of an economically strong community spreads internationally to a greater extent than an economically weak community's language. Economic strength of a language (or rather of a language community) seems to carry even more weight than numerical strength, as may for instance be concluded from the noticeable spread of Japanese in recent times (see Coulmas 1989) as compared to Chinese. Japanese, whose language community is numerically much weaker but economically stronger than the Chinese language community, has recently spread more than Chinese. The language of an economically strong community is attractive to learn because of its business potential. Knowledge of the language potentially opens up the market of that community: it is easier for producers to penetrate a market if they know the language of the potential customer.

Table 2.2 shows how German compares to other languages in economic strength. The figures were calculated on the basis of the two sources indicated (Grimes 1984; Haefs 1989). First, for each country in the world which contains any speakers of the language in question (according to Grimes 1984), the GNP of these speakers was calculated, assuming the same GNP, on average, for each citizen of the country. Then, these figures were added together for all the countries in the world. While it may be assumed that the first ten languages or so are really the ten economically strongest languages in the world, the others were included because they count among the numerically strongest languages in the world. This therefore shows the discrepancy between numerical and economic strength in these cases.

As can be seen from the table, German ranks third among all the languages in the world, behind English and Japanese. Its relative economic strength is therefore considerably higher than its relative numerical strength. Only English is far stronger economically (about four times as strong), while most of the numerically stronger languages are economically weaker, often even considerably weaker. It may, therefore, be assumed that economic strength is among the factors which work in favour of the status of German as an international language.

Table 2.2. *Economic strength of German in comparison to other languages, after Grimes 1984 and Haefs 1989 (US$ bn)*

1.	English	4,271
2.	Japanese	1,277
3.	**German**	**1,090**
4.	Russian	801
5.	Spanish	738
6.	French	669
7.	Chinese	448
8.	Arabic	359
9.	Italian	302
10.	Portuguese	234
11.	Dutch	203
12.	Hindi–Urdu	102
13.	Indonesian	65
14.	Danish	60
15.	Greek	49

2.3 *Number of Countries in which a Language has Official Status on a National or Regional Level ('Political Strength')*

If a language has official status in several countries, either on a national or on a regional level, it will—other circumstances being equal—be more likely to be studied as a foreign language than if it has official status in only very few or no countries. It seems more worth while studying such a language, since its potential for communication with different countries is greater. As a consequence, it also tends to be used more in international communication. One could call the number of countries in which a language has official status the language's 'political strength'.

German has official status in the following seven countries (see map, Figure 2.2):

- Germany, Austria, Liechtenstein (sole official language on the national level);
- Switzerland, Luxembourg (co-official on the national level);
- Italy (South Tyrol), Belgium (German-speaking community) (regional level).

FIG. 2.2.   Countries with German as an official language

Until 1990 German had official status in two more countries, namely in the German Democratic Republic on the national level (before the unification of the two Germanies on 3 October 1990), and in Namibia on a regional level (before the country won its independence on 21 March 1990). These are still included in Table 2.3 which compares German to other languages with respect to number of countries where the languages have official status, since more recent comparative data were not available to me; however, for German the revised new figures are added. While German formerly ranked fifth among all the languages in the world with respect to countries in which it had official status, it now shares rank five with Portuguese. If one weights sole official status on the national level (first figure given in parentheses) more heavily than co-official or regional status (second figure in parentheses), Portuguese even ranks above German, which then takes sixth place. Since for the year 1991 no comparative data with other languages were available, the former rank order is still given in Table

2.3. The rank of German is not affected by the differences between the sources, which are quite considerable for some of the other languages. It would have been useful to distinguish further between national and regional official status for the comparison between the languages; however, this distinction is not made consistently in the sources.

Table 2.3. *The six most widespread national official languages according to numbers of countries*

		Banks (1987)	Haefs (1989)	1991
1.	English	63 (19 + 44)	59 (30 + 29)	
2.	French	34 (11 + 23)	27 (15 + 12)	
3.	Spanish	23 (15 + 8)	21 (17 + 4)	
4.	Arabic	22 (14 + 8)	23 (18 + 5)	
5.	**German**	**8 (4 + 4)**	**9 (4 + 5)**	**7 (3 + 4)**
6.	Portuguese	7 (6 + 1)	7 (7 + 0)	

2.4 Number of Learners of the Language as a Foreign Language ('Study Strength')

The extent to which a language is studied as a foreign language could be called its 'study strength'. German is among the most widely studied foreign languages in the world. It is studied in the schools of about half the countries in the world, though sometimes only in a small proportion of the country's schools. In 1982/3, for example, it was studied in 83 of the then 172 countries of the world (cf. *Bericht* 1985, Ammon 1991: 433). During the twentieth century German has probably always ranked behind English and French as a foreign language in schools, with respect to the total number of students as well as with respect to the number of countries in which it has been a school subject. Today it may even rank behind Spanish in numbers of students, mainly as a consequence of the vast number of students of Spanish in North America; overall comparative figures are, however, not currently available. There is no doubt that German is outnumbered by French, and even more so by English. Drawing on various sources, I found or calculated the following numbers of foreign-language students for these three languages in primary and secondary schools for 1974 (English), 1985 (French), and 1982/3 (German): English: 117.7 million; French: 50.9 million; German: 15.1 million (see Ammon 1991: 437ff.). It may be assumed that the numbers for English were even greater in the 1980s, the years to which the figures for German and French relate.

The proportions on the tertiary level are probably roughly comparable. Instead of comprehensive figures, which were not available to me, I shall present figures for the number of foreign students from countries of other languages in the mother-tongue countries of the languages in question. These figures were taken from the *Statistical Yearbook*, published by Unesco. In this context, a country is considered to be a 'mother-tongue country' of a given language only if a substantial proportion of the country's population are native speakers of that language. For German, for instance, this means the following countries (only those with tertiary institutions were included): the Federal Republic, GDR (as the figures used refer to the situation before the unification of Germany in 1990), Austria, and Switzerland (75 per cent), or for French: France, Canada (29 per cent), Belgium (33 per cent), Switzerland (21 per cent). As the student figures were only available for the entire countries they were scaled down in proportion to percentages of native speakers in each case. In the Unesco *Yearbooks*, the numbers of foreign students in each country are broken down according to countries of origin; only students coming from countries with other mother tongues (other than the mother tongue of the countries where they studied) were included in our calculation.

Relatively few such learners of a foreign language actually study it in one of the mother-tongue countries for the language in question. They are, however, among those who acquire a particularly solid command of the language; otherwise as a rule, they would not be able to study successfully at the tertiary institutions of the mother-tongue countries. To some extent at least, this justifies focusing on them in our attempt to compare languages according to the degree to which they are studied as foreign languages on the tertiary level. Table 2.4 gives the numbers of these students for various languages. As can be seen, German ranks third among all the languages, following English which is way ahead, and French. The proportion for Spanish may be lower in Table 2.4 than it would be for the entirety of students who study it as a foreign language on the tertiary level, since the numbers of foreign students in some Spanish-speaking countries are missing altogether in the Unesco *Yearbooks*.

In fact, figures from private language schools, which are mainly attended by adult learners, show a higher proportion of students of Spanish, as compared to German, than we have in Table 2.4. Table 2.5 gives the percentage of classes per language for various languages studied at the Berlitz schools, which operate in all parts of the world. As can be seen, the overall difference between French, Spanish, and German has become quite small in recent times (1989), in contrast

Table 2.4. *The 'study strength' of German in relation to other major languages*

Language	1967	1977	1986	Total	Growth rate
1. English	156,403	283,859	409,920	850,182	2.62
2. French	44,079	111,181	142,480	297,739	3.23
3. **German**	**39,178**	**68,979**	**96,172**	**204,329**	**2.45**
4. Spanish	25,161	22,492	10,821	58,474	0.43
5. Italian	16,957	31,283	34,720	82,960	2.04
6. Russian	16,100	?	?	?	?
7. Japanese	10,086	14,737	14,960	39,783	1.48

Notes: The figures in the table refer to the number of students from countries of other languages studying the seven given languages in countries where these languages are the mother tongue ('mother-tongue countries').
Growth rate = number for 1986 : number for 1967.

Table 2.5. *Percentage of classes per language at the Berlitz language schools*

	Total		Distribution by region of 1989 totals			
	Early 1970s	1989	Europe	North America	Latin America	Far East
1. English	42	63	37	12	21	30
2. French	25	11	54	34	5	7
3. Spanish	12	9	24	62	12	2
4. **German**	**12**	**8**	**64**	**23**	**6**	**7**
5. Italian	–	3	58	36	3	3
6. Japanese	–	2	9	53	–	38
7. Dutch	–	1	96	4	–	–
8. Portuguese	–	1	28	36	30	6
9. Others	9	2	25	36	25	14

to earlier years, where French had a greater share (1970); the leading position of English has, however, increased significantly (from 42 per cent in 1970 to 63 per cent in 1989).

The columns on the right side of Table 2.5 show how the various languages are distributed over four regions; the percentages for these regions are based on the figures for 1989 (second column on the left, = 100 per cent). It is quite obvious from these figures that German and French are predominantly studied in Europe (64 and 54 per cent respectively), while for Spanish the emphasis is in North America (62 per cent). (Remember that these figures only relate to teaching as a *foreign* language.) English, by contrast, is spread more evenly world-wide, with a low figure only in North America, which is due to the fact that English is the dominant native language there and conse-quently not studied so much as a foreign language. This observation can be taken as an indication that English is a world international language while German, French, and Spanish are only regional inter-national languages.

A closer look at Europe reveals an interesting regional distribution of German and French there: as can be seen from Table 2.6, French is clearly a more important foreign language in western Europe, rep-resented here by the EC countries, which are all western European except Greece. It is studied by over three times as many students as German, while German itself is still studied in western Europe by more than twice as many students as Spanish.

This relationship between German and French is reversed in east-ern Europe. With the exception perhaps of Romania, German is studied by more students than is French everywhere in eastern Europe, often, indeed, by very many more. Thus, for instance, in Czechoslovakia German was chosen by 30–50 per cent of the secondary-level students in 1990, depending on type of school (French 1–3 per cent, English 40 per cent). In Poland, 44 per cent of students at secondary level chose German in the same year (French 16 per cent, English 53 per cent). In Hungary, German was chosen by about as many students as was English, way above the figures for French. In Bulgaria, a choice between English, French, or German was introduced in 1991 for pupils of the 5th grade: about 30–35 per cent chose German, 15–20 per cent French, and 50 per cent English.

Table 2.6. Number of foreign-language students in schools in the EC countries (Eurydice 1989: 2–13)

1.	English	18,133,320	(10 countries: all non-mother-tongue countries)
2.	French	9,088,163	(11 countries: all non-mother-tongue countries)
3.	**German**	**2,888,011**	(11 countries: all non-mother-tongue countries)
4.	Spanish	1,385,801	(9 countries: all non-mother-tongue countries except Greece, Portugal)
5.	Italian	215,840	(8 countries: all non-mother-tongue countries except Greece, Netherlands, Portugal)
6.	Dutch	212,214	(4 countries: Belgium, France, Germany, Luxembourg)
7.	Portuguese	13,708	(3 countries: France, Germany, Spain)
8.	Modern Greek	80	(1 country: France)
9.	Danish	0	

In the USSR German was studied by 34 per cent of secondary-level students in 1989, English by 55 per cent; the figures for French were much lower (see Ammon 1991: 143–6). German is also studied more than French in all the Scandinavian countries and in the Netherlands. Generally speaking, it is studied more than French in all the countries with Slavic and Finno-Ugric languages and all countries with Germanic languages with the exception of the British Isles, while French is studied more in all the other European countries. English is practically always the leading foreign language, with the exception of a few special cases like Luxembourg or Switzerland (German and French more than English), and perhaps the former Czechoslovakia and Hungary (German on about the same level as English).

It may be assumed that the extent to which a language is studied (its study strength) has consequences for the extent to which it is used. Though the formal study of a language is not a strict prerequisite for its use, since a language can also be acquired informally by mere contact with its speakers, it does as a rule enhance its use. As we shall see, German is also used more often as an international language in those regions where it is a preferred subject of study. There are, however, notable exceptions to this rule: German is also studied by a considerable number of students in some East Asian countries, particularly Japan (see Bauer 1989, Ammon 1992b), South Korea, and also Indonesia, but it is hardly used there as an international language. One of the reasons is that the knowledge of German acquired in these countries is mostly quite limited due to the enormous linguistic distance of the L1s (Japanese, etc.) from L2 (German), which makes the language very difficult to learn. German is studied more for traditional reasons in these countries than for the purpose of actual international communication, in the same way as Latin is still studied in a number of countries. One of the reasons for this tradition is that German was once a great language of science, as will be shown in Section 3.2.

3   THE USE OF GERMAN FOR INTERNATIONAL COMMUNICATION

3.1   *Trade*

For the investigation of language choice it makes sense to distinguish between different sectors, or perhaps 'domains', of society. It has been observed for a long time that international languages can be used to different degrees in different domains. Thus there was once a not entirely unjustified view that German was the international language of science, English of trade, and French of diplomacy. It seems useful for investigations of international languages to follow this rough distinction, though others would be possible; it should be pointed out, though, that these domains can on the one hand show considerable overlap, and on the other hand provide quite incomplete coverage of the various segments of society depending on how they are defined. In addition, these domains are so vast that only small sections of them can actually be investigated empirically and these sections simply have to be taken as more or less representative of the whole domain for the time being. Such problems would, however, occur with any other segmentation of society.

A first overview of the international languages of trade, and particularly the role of German, is provided by the German chambers of commerce, which regularly issue recommendations on which languages should be used for trade with each country in the world. These recommendations are based on the experience of the chambers' members, that is, practically all the German firms engaged in international business, as well as the consulates abroad supporting them in their endeavours. These recommendations are made for sellers of goods, who have to be cautious and polite with their language choice, rather than for buyers, to whose language the foreign sellers will tend to adjust if possible. Therefore, it may be assumed that the chambers' recommendations for using German tend to be rather restrictive: German could be used more extensively for buying goods. Table 2.7 lists the languages recommended by the German chambers of commerce together with the number of countries for which they can be used. It should be noted that German is always only a co-language (with one exception, namely in Austria), which means German firms can in no case count on the possibility of using German for selling goods but have to decide from case to case depending on the circumstances.

Let us now have a look at the individual countries for which German can be used. They are listed in Table 2.8. As can be seen, German can be used practically everywhere in Europe except in some western European countries (Britain, Ireland, France, Spain, Portugal), though — as has been pointed out above — German firms cannot rely on the possibility of using German. Outside Europe, however, there is very little scope for using German. Namibia, Chile, and Israel have substantial German-speaking minorities; in Afghanistan and Mongolia German has been taught extensively by the former GDR, in connection with business relations and transfer of technology. It need hardly be pointed out that in Israel the use of German is a highly sensitive issue, as a consequence of the Holocaust. Some readers may therefore

Table 2.7. *Number of countries for which particular languages are recommended for West German trade (Handelskammer Hamburg 1989)*

		Total	Sole language	Co-language
1.	English	122	64	58
2.	French	57	25	32
3.	Spanish	26	17	9
4.	**German**	**26**	**1**	**25**
5.	Arabic	12	–	12
6.	Portuguese	8	–	8
7.	Italian	4	–	4
8.	Dutch	4	–	4
9.	Indonesian	1	–	1
	Czech			
	Danish			
	Finnish			
	Norwegian			
	Polish			
	Russian			
	Slovene			
	Swedish			

be surprised that Israel even appears in the list; not all German-speaking Israelis, however, identify the German language with Nazi Germany. In view of the countries included in the list, it is surprising that some others are missing, like Brazil, Argentina, or Paraguay, which have substantial German-speaking minorities (see Born and Dickgießer 1989). Especially in the case of Brazil, it is widely known in Germany that many business contacts with the southern states (Rio Grande do Sul, Santa Catarina, Paranà, Espirito Santo) are maintained in German. It may reflect an overly cautious stance of the German chambers of commerce not to recommend the use of German even in such cases.

Another way of assessing the international use of German in trade is via job advertisements in non-German-speaking countries. As far as the available data are concerned, this approach is quite unspecific, since they do not distinguish between jobs in trade and others. Nevertheless, they seem to provide rough indicators of what we are looking for. In a research project at the University of Duisburg, newspaper job advertisements in 1991 were analysed for six European countries. The results are given in Table 2.9. Only job advertisements with

Table 2.8.   *Countries for which German is recommended for West German trade (Handelskammer Hamburg 1989)*

Western and northern Europe	Eastern and southern Europe	Other regions
Austria	Albania	Afghanistan
Belgium	Bulgaria	Chile
Denmark	Czechoslovakia	Israel
Finland	Greece	Mongolia
Iceland	Hungary	Namibia
Luxembourg	Italy	
Netherlands	Poland	
Norway	Romania	
Sweden	Soviet Union	
Switzerland	Turkey	
	Yugoslavia	

foreign-language requirements were included in the calculation of percentages. The results confirm the strong position of German in eastern as compared to western Europe, where French ranks before German.

As to the actual use of German or other languages in international trade, only bits and pieces of information are available so far. The results of one of the more representative investigations, which however is limited to a single country, are presented in Table 2.10. The data were collected by means of questionnaires from forty-four Dutch business negotiators. Other data are available on the use of foreign languages at the workplace, although these are obviously not identical with their use in trade. Thus in England, France, Denmark, and Sweden German has been found to rank second as a foreign language used at the workplace, following French (in England) or English (in

Table 2.9.   *Foreign-language requirements in job advertisements in newspapers in six European countries, according to Glück 1992* (% per language)

	German	English	French	Spanish
Hungary	40	37	3	<1
Poland	26	46	7	<1
France	11	71	–	5
Britain	7	–	15	6
Spain	7	60	21	–
Italy	6	69	9	<1

Table 2.10. *Percentages of language use by Dutch managers in business negotiations (Ulijn and Gorter 1989: 495)*

	One of the two most important languages of negotiation	One of the three most important languages of negotiation
English	95	98
German	**74**	**95**
French	25	82
Spanish	–	16

the other three countries); it ranks third in Belgium (following French and English) and probably also in Finland (following English and Swedish), though the data can be interpreted differently there (see Ammon 1991: 182–95).

3.2   *Science*

The importance of German as 'an international language of science' (Ostrower 1965: 148) has often been pointed out – in older literature, one has to say. In more recent times, however, the decline of German in this function has been the topic of publications (e.g. Skudlik 1990) and even of a special conference (see Kalverkämper and Weinrich 1986). If one assumes that the mere number of scientific publications in a given language is a reasonably valid indicator of its role in international scientific communication, at least in written communication, then Tables 2.11 and 2.12 and Figures 2.3 and 2.4 reveal the present standing of German as an international language of science as well as its development in the course of this century.

Natural sciences and social sciences are kept apart in Tables 2.11 and 2.12, because there are reasons to believe that the relative importance of the languages differs between the two groups of sciences; in the social sciences the traditional languages of publication seem to retain some of their importance, while in the natural sciences they have been replaced to a greater extent by English as the modern world language of science. These differences are apparent in the comparison between Tables 2.11 and 2.12, though the evidence is somewhat weakened by the time span between the data; during the five years between 1976 and 1981 English may well have gained a still greater share of publications even in the social sciences. There are various reasons why international communication may not be reduced to a single language to the same degree in the social as in the natural sciences.

Table 2.11. *Languages of publication in five natural sciences in 1981, according to Baldauf and Jernudd 1983: 99* (%)

		Chemistry	Biology	Physics	Medicine	Mathematics
1.	English	66.9	85.7	84.6	73.1	69.3
2.	Russian	12.7	3.9	3.8	5.9	18.1
3.	**German**	**5.5**	**2.5**	**3.9**	**5.5**	**3.6**
4.	Japanese	9.9	1.9	1.5	3.0	0.3
5.	French	1.9	2.1	2.0	4.0	4.8
6.	Chinese	0.9	0.2	0.6	0.7	0.5
	Others	2.2	3.6	3.6	7.8	3.0

For example, some findings in the social sciences may be of only regional interest and therefore do not have to be published in the world language; the technical registers of the social sciences are less formalized than in the natural sciences, so that individual social scientists may find it harder to operate in a different language; and the degree of specialization is not as great in the social sciences, which is why there are still enough experts within the more limited languages, like French or German, with whom communication seems relevant for the individual researcher.

The rank order in Tables 2.11 and 2.12 follows the arithmetical means of all the sciences listed in each table. German still ranks third in the natural as well as in the social sciences, if we assume that the sciences selected here adequately represent the whole group in both cases. In some sciences, however, German clearly ranks lower. Particularly striking examples are chemistry, but also medicine, since German was once a very important language of publication in these fields. As late as in the 1930s textbooks in German were, for instance, used at

Table 2.12. *Languages of publication in four social sciences in 1976, according to Thogmartin 1980* (%)

		Sociology	Economics	Political science	Anthropology
1.	English	46.3	38.5	51.3	46.9
2.	French	14.3	16.6	16.3	26.0
3.	**German**	**5.7**	**9.6**	**12.4**	**10.1**
4.	Russian	11.3	8.0	2.8	7.8
5.	Japanese	7.1	—	—	—
	Others	15.3	27.3	17.2	9.2

American universities, since they contained the most advanced information. The distance of German, but not only of German, from English in number of publications today is striking in all the fields covered by Tables 2.11 and 2.12. In view of this enormous distance, it has been suggested that scientific publications in languages other than English serve international communication only to a very limited degree, even more limited than the figures suggest, and in fact are mainly produced for the 'home market', that is, for 'consumption' (communication) within that particular language community. However, the extent to which this is true has, to my knowledge, not been investigated.

Figures 2.3 and 2.4 show how the relative quantities of scientific publications in particular languages have changed in the course of the last hundred years. The figures are averages of publications in biology, chemistry, physics, medicine, and mathematics which have been calculated on the basis of a study by Tsunoda (1983). Whereas the numbers of scientific publications in German, French and English were about the same around the turn of the century, English more or less steadily increased its share of the 'market' in the years following the First World War, while the shares of German and French declined, in the case of French even earlier than in the case of German.

The two graphs in Figure 2.3 also show an interesting difference which should be a warning against too uncritical an interpretation of data of that sort. According to the American bibliographies and databases (Figure 2.3*b*), English has been the predominant language of science (in terms of number of publications) throughout the last hundred years. However, if one averages the bibliographies and databases from various countries, in this case Germany, France, Russia or the Soviet Union, and the USA (Figure 2.3*a*), one finds that German overtook English in the period shortly before and after the First World War. There are reasons to assume that the bibliographies and databases of each country are at least to some extent biased towards their own language, even if they try to be internationally as representative as possible, for the simple reason that publications in the language of the country are more readily available as well as easier to read. Therefore it seems likely that data which are drawn from bibliographies and databases of different countries are more objective than those from a single country. On the other hand, the US bibliographies have the reputation of being more comprehensive than those of any other country — an assumption which, however, appears to be more justified in recent than in former times. It seems to be impossible, without additional evidence, to decide which of the two versions in Figure 2.3 is closer to reality, that is, to know whether German actually overtook English as the leading language of science (in terms of number of publications) at the beginning of this century or not.

When I specified the turning-point after which English started to outstrip the other two European languages, German and French, as the period after the First World War, I introduced a clue to an explanation, at least a partial explanation, of this development. The countries of both languages, Germany together with Austria, and France together with Belgium, were virtually ruined by the First World War; Britain was not much better off either. The USA, however, emerged from the war practically unscathed to become the leading economic power in the world, which enabled it also to develop into the leading power in science. Germany, which before the First World War was probably the leading country in the world in terms of scientific research, had no resources left to continue research on a similar level

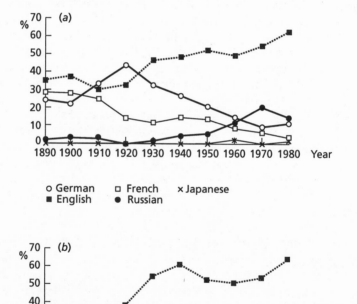

FIG. 2.3. Languages of publication in the natural sciences between 1890 and 1980 (percentages). (*a*) Averages from French, German, Russian, and US bibliographies and databases. (*b*) According to US bibliographies and databases

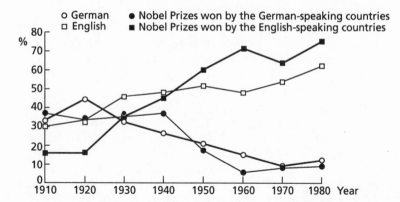

Fig. 2.4.   Percentages of Nobel Prizes in the natural sciences won by
the German-speaking and the English-speaking countries in relation to
percentages of publications in German and in English in the natural sciences

as before, though of course some of the old skills and knowledge were
carried over. In addition, Germany's political choice of Nazism in the
aftermath of the First World War resulted, among other atrocities,
in the expulsion and mass murder of many of her best scientists, and
most of those who were able to escape went to the USA.

These events are mirrored in a very abstract way in Figure 2.4,
which shows, if one interprets it in a straightforward and simplified
way, how the decline of the German language vis-à-vis English accompanied the decline of German science vis-à-vis science in the English-
speaking world, particularly the USA. The Nobel Prizes are taken
here as an indicator of the standing of the sciences in the German-
speaking and the English-speaking countries. It seems hardly necessary to point out that this indicator is very rough indeed. It should,
in particular, be noted that the Nobel Prizes are often given belatedly,
that is, a considerable time after the scientific achievement. This might
be one of the reasons why the decline of German as a language of
science seems to start earlier, according to Figure 2.4, than does the
decline of science in the German-speaking countries. In reality, the
decline of science there, as compared to the English-speaking world,
might also have started earlier than the share of Nobel Prizes for the
German-speaking countries indicates.

Even today, German scientists have not yet fully adjusted to the
change in international status of their language. Particularly scientists
of the older generation often have only a quite inadequate command
of the modern world language of science, namely English. In a questionnaire investigation among scientists at the University of Duisburg

and researchers in industry in that city, 25 per cent confessed to having difficulty in reading English texts, 38 per cent in understanding spoken English, and 57 per cent in writing in English. Furthermore, 19 per cent stated that they sometimes do not participate in conferences, 25 per cent would not engage in contacts with colleagues, and 33 per cent would not accept offers of publication if the use of the English language were required (Ammon 1990*b*). This problem is recognized by many German publishers and German research foundations. However, they offer very little or no help to scientists with inadequate English language skills. The research foundations even seem to refuse help intentionally, either in order not to accelerate the changeover from German to English as the language of publication of German scientists or not to 'waste' resources on what is not considered to be an intrinsic part of scientific research.

3.3   *Diplomacy and International Organizations*

German has never been a very important language in diplomacy, as it has been in science, a fact noted by Ostrower, who deals with a whole number of languages which have played an important role in diplomacy at some time in history. 'The main reason for the failure of German as a language of political importance was the international organization of the Holy Roman Empire, which strove to create the appearance of political continuity with the ancient empire of Rome. The official language of the Empire was Latin, and German linguistic advancement in international relations consequently suffered' (Ostrower 1965: 145–6). Only in the second half of the nineteenth century was German promoted by its 'home countries' as a language of diplomacy, indeed only after German unification in 1871. The peace treaty concluded in that year with France to end the Franco-Prussian War was still written only in French, in spite of the fact that Prussia, in alliance with other German-speaking countries, had been victorious. Thereafter, however, the newly created Germany tried to introduce its language into the world of diplomacy, for instance through persistently corresponding in German with France, in spite of the fact that French was then still the generally acknowledged language of diplomacy. The change in attitudes can also be seen from the regular use of German in international treaties towards the end of the nineteenth and the beginning of the twentieth century, whenever Germany was among the signatory countries (see Ammon 1992*a*).

The further rise of German as a language of diplomacy was, however, stopped short by the First World War, or rather the defeat of Germany and also Austria, the other power which promoted the

German language. In particular, German did not become one of the official languages of the League of Nations, which was founded after the First World War. Ostrower (1965: 360) sees Germany's defeat in the war as one of the reasons for the exclusion of German from the League of Nations' official languages, and also accounts for the exclusion of Russian, another potential candidate for the status of an official language: 'Russia was in a state of revolutionary turmoil and the Germans were defeated in the field of battle, thus the Russian and German languages were out of contest.' It should be added that Germany was only granted membership of the League as late as 1926 and that it withdrew again under Nazi rule as early as 1933. The First World War also had a long-lasting impact on the status of German as a language of diplomacy for another reason, namely Germany's loss of its colonies. As a consequence, German as an official language was practically reduced to its European homelands, in contrast to other European languages, which were spread world-wide through persistent colonialism (see Section 2.3 above). Finally, Nazism and the Second World War contributed decisively to discrediting the German language even further as a language of diplomacy.

In particular, it was out of the question after the Nazi atrocities and the defeat in the Second World War for German to become one of the official languages of the United Nations. Only the rather limited status of a 'documentary language' was granted to German in 1974, one year after the two Germanies, the Federal Republic and the GDR, were admitted as member states. This status implies that the more important documents of the General Assembly, the Security Council, and the Economic and Social Council are translated into German. These translations are, however, funded by the German-speaking member countries of the UN, which are presently Germany, Austria, and Liechtenstein — Switzerland is still not a member even today. It seems unlikely that German will ever become an official language of the UN, since the official languages of the organization surpass German either in number of countries in which they are official languages (French) or in numerical strength (Chinese, Russian) or in both (English, Spanish, Arabic). Only in economic strength and in size of the financial contribution to the UN do the German-speaking members surpass the member states of most of the official languages of the UN, namely of Chinese, French, Spanish, Arabic, and in future perhaps also Russian. Economic considerations have, however, never been a decisive argument of the UN in the choice of an official language; otherwise Japanese would now rank before German as a candidate.

The role which German as compared to other languages plays in international organizations can be inferred to some degree from Table

48   *Ulrich Ammon*

2.13. It shows that German ranks seventh among all the languages
with respect to the number of international organizations in which it
either has official status or is a working language. Table 2.13 contains
figures from two different sources, which show, among other things,
the extent to which counts of international organizations can diverge.
It seems less problematic to follow Banks's (1987) rank order of the
languages, since the *Bericht* (1985) was prepared under the auspices
of the government of the Federal Republic and might, therefore, be
biased towards the German language. Following Banks, it is interesting to note that precisely the six official languages of the UN precede
German in number of international organizations in which they have
a privileged status (official or working language). This is presumably
not only due to the fact that the UN adds to the number of international organizations in which these languages have a privileged
status but also to the fact that official status in the UN enhances a
language's chances of acquiring official status in other international
organizations.

The only institution in which German, according to Banks, has the
status of a working language is the Council of Europe. It should be
pointed out, though, that the terminology of the Council is confusing.
Normally, the status of a working language is higher than that of a
(merely) official language—if this distinction is made in an organization. In the case of the Council of Europe, however, it is the other

Table 2.13. *Working or official languages in international organizations (no. of organizations)*

		Banks 1987		Bericht 1985: 83–90	
		Working	Official	Full status	Partial status
1.	English	16	35	61	3
2.	French	12	37	59	2
3.	Spanish	9	19	23	5
4.	Russian	5	10	18	7
5.	Arabic	2	5	10	7
6.	Chinese	1	4	8	11
7.	**German**	**1**	**3**	**12**	**18**[a]
8.	Italian	1	2	2	3
9.	Portuguese	0	3	–	–
10.	Danish	0	2	2	0
	Dutch	0	2	2	4

[a] Five of these funded by German-speaking countries.

way round: English and French are the 'official languages', in which in this case all the proceedings can be conducted without limitation, while German, Italian, and Spanish are (only!) the Council's 'working languages', whose use is limited to certain functions within the organization. Thus, for instance, interpreting of speeches is only granted *from* the 'working languages', not *into* them, while for the 'official languages' interpreting is guaranteed in both directions.

The most important international organization in which German has an official status is probably the European Union (EU), as the former 'European Community' (EC) has been referred to since the implementation of the Treaty of Maastricht in November 1993. German is one of the nine official languages of this organization of twelve member states (in 1993). According to its statutes, all its official languages are equal. It is, however, generally known that this is not the case in reality. The possibility of differences among the official languages is, in fact, in a subtle way permitted by regulations which recommend 'pragmatic' solutions to problems of language choice in certain situations if necessary, for instance in the choice of interpreters (see Hoof-Haferkamp 1991). It has often been observed informally, and has now also been corroborated in more systematic investigations, that in reality French and English are the dominant working languages of the political bodies of the organization. German ranks third, followed by Spanish and Italian, but the difference in extent of use between the first two and the other languages is enormous. Table 2.14 shows the proportions. Haselhuber (1991) investigated how many of the young practitioners (not yet firmly employed officials) in the EC Commission, perhaps the most important political body of the organization, use the various languages regularly; Gehnen (1991) found out which proportions of the communications in the EC (in its General Directorates) were carried out in the various languages.

The German government has stressed repeatedly that it is not content with the role which the German language currently plays in the political bodies of the EU. Its arguments were stimulated and supported by German business organizations, which contended that German companies are linguistically disadvantaged in the competition for EU business opportunities, since EU calls for tenders usually appear later in German than they do in French or English, and in bad translations at that; in addition it is often even expected that offers be made in French or English. Other language communities could, of course, raise complaints of the same sort. However, the German government has pointed out that Germany's financial contribution to the EU is by far the greatest of all the member states and that the German language community is the numerically strongest within the EU.

Table 2.14.   *Frequency of use of languages in the General Directorates of the EC Commission* (%)

		In writing		In speaking	
		Haselhuber 1991	Gehnen 1991	Haselhuber 1991	Gehnen 1991
1.	French	92.5	64.0	90.1	62.0
2.	English	73.3	35.0	60.8	31.0
3.	**German**	**18.3**	**1.0**	**15.0**	**6.0**
4.	Spanish	6.7	–	9.2	–
5.	Italian	8.3	–	6.7	–

It seems impossible to foresee at the present moment whether German will eventually play a more prominent role in the procedures of the EU. The non-German-speaking member countries do not seem to be very interested in the upgrading of German; they would have to handle yet another language, and for some the advance of the German language would even imply a decline in status for their own language. An example of British feelings can be found in a 1992 press campaign in the United Kingdom, in which attempts at strengthening the position of German were answered by some newspapers by calling German a 'horrid, guttural language', alluding to Chancellor Kohl as the new German 'Führer' and maintaining that the Germans were 'cracking the whip' on the rest of Europe (cf. the report in the *New York Times* of 24 February 1992: 'Thus Spake Helmut Kohl: Auf Deutsch'). Incidentally, the entire press campaign was started on the basis of a hoax. The British newspapers claimed that Kohl had written a letter to EC Commission President Delors demanding that the German language be 'elevated' within the EC. There had, however, been no such letter from Kohl around that time; the only letter Kohl ever wrote to Delors about the German language dates back to 1988.

In spite of such adverse feelings on the part of some other member countries, it seems likely that German will play a more important role in the political bodies of the EU in the future. The economic importance of Germany and the size of its population will probably have an effect in that direction, particularly since the German government has become aware of the language question and actually tries to promote German where possible. The role of German will most likely also be strengthened vis-à-vis French, though not necessarily vis-à-vis English, when eastern European countries including Austria enter the EU.

4   CONCLUSIONS

The data presented here show clearly that German ranks among the world's more important international languages, though different parameters reveal various ranks of internationality for it in different domains (ranging from rank three to rank ten). The data also reveal that German has lost ground in the course of the twentieth century vis-à-vis other international languages, especially English. While German was formerly an international language particularly in the domain of science, its present international standing seems to be based mainly on the economic strength of the German-speaking countries. It may be assumed that on this basis it will play a more prominent role than today in future international communication in Europe, which in turn will help to stabilize its international standing elsewhere.

The bulk of the data presented is, as I pointed out at the beginning of this chapter, only indicative of international communication and is not based on the direct observation of actual communicative behaviour, which remains one of the desiderata of future investigations. Such empirical data would permit a better assessment of the internationality of the various languages than merely using indicators.

Moreover, my presentation is largely restricted to description, and offers only occasional hints towards explanations as to why German, or the other languages involved, are international to their respective degree. It will be an enormously complex research task for the future to isolate and to weigh factors which determine the degree of internationality of a language, that is, to explain the actual degree of internationality of various languages (for an attempt in that direction focusing on the English language see Fishman, Cooper, and Conrad 1977).

Further Reading

Ammon (1991)
Ammon and Haarmann (1991)
Born and Dickgießer(1989)
Coulmas (1991)
Skudlik (1990)
Sturm (1987)

References

AMMON, U. (1990*a*), 'German as an International Language', *International Journal of the Sociology of Language*, 83: 135–70.

—— (1990*b*), 'German or English? The Problems of Language Choice Experienced by German-Speaking Scientists', in Nelde (1990), 33–51.

—— (1991), *Die internationale Stellung der deutschen Sprache* (Berlin: de Gruyter).

—— (1992*a*), 'On the Status and Changes in the Status of German as a Language of Diplomacy', in Ammon and Hellinger (1992), 421–38.

—— (1992*b*), 'Zur Stellung der deutschen Sprache in Japan', *Muttersprache*, 102/3: 204–17.

—— and HAARMANN, H. (eds.) (1991), 'Status und Funktion der Sprachen in den Institutionen der Europäischen Gemeinschaft/Focus: Status and Function of the Languages in the Political Bodies of the European Community/Thème principal: Status et fonction des langues dans les organes de la Communauté Européenne', *Sociolinguistica*, 5.

—— and HELLINGER, M. (eds.) (1992), *Status Change of Languages* (Berlin: de Gruyter).

BALDAUF, R., and JERNUDD, B. (1983), 'Language of Publications as a Variable in Scientific Communication', *Australian Review of Applied Linguistics*, 6/1: 97–108.

BANKS, A. S., *et al.* (eds.) (1987), *Political Handbook of the World 1987: Governmental and Intergovernmental Organizations as of March 15, 1987* (Binghamton, NY: CSA Publications).

BAUER, H. L. (ed.) (1989), *Deutsch als zweite Fremdsprache in der gegenwärtigen japanischen Gesellschaft* (Munich: Iudicium).

Bericht der Bundesregierung über die deutsche Sprache in der Welt (1985), Bundestagsdrucksache V/2344, Bonn.

BORN, J., and DICKGIEßER, S. (1989), *Deutschsprachige Minderheiten. Ein Überblick über den Stand der Forschung für 27 Länder* (Mannheim: Institut für deutsche Sprache).

BRAGA, G. (1979), 'International Languages: Concept and Problems', *International Journal of the Sociology of Language*, 22: 27–49.

COLEMAN, H. (ed.) (1989), *Working with Language* (Berlin: Mouton de Gruyter).

COMRIE, B. (ed.) (1987), *The World's Major Languages* (London: Croom Helm).

COULMAS, F. (1989), 'The Surge of Japanese', *International Journal of the Sociology of Language*, 80: 115–31.

—— (ed.) (1991), *A Language Policy for the European Community: Prospects and Quandaries* (Berlin: Mouton de Gruyter).

EURYDICE (1989), *Teaching of Languages in the European Community: Statistics*, Working Document, March 1989 (Brussels: Eurydice).

FINKENSTAEDT, T., and SCHRÖDER, K. (1990), *Sprachschranken statt Zollschranken? Grundlegung einer Fremdsprachenpolitik für das Europa von morgen* (Essen: Stifterverband für die deutsche Wissenschaft).

FISHMAN, J. A., COOPER, R. L., and CONRAD, A. W. (1977), *The Spread of English: The Sociology of English as an Additional Language* (Rowley, Mass: Newbury House).

GEHNEN, M. (1991), 'Die Arbeitssprachen in der Kommission der Europäischen Gemeinschaft unter besonderer Berücksichtigung des Französischen. Eine Fragebogenerhebung in den Generaldirektionen, konzipiert von Hartmut Kleineidam', *Sociolinguistica*, 5: 51–63.

GLÜCK, H. (1992), 'Die internationale Stellung des Deutschen auf dem europäischen Arbeitsmarkt. Eine Vorstudie zu Stellenangeboten in Tageszeitungen aus sechs Ländern', in Weiß (1992), 47–75.

GRIMES, B. F. (ed.) (1984), *Languages of the World: Ethnologue*, 10th edn. (Dallas: Wycliffe Bible Translation).

HAEFS, H. (ed.) (1989), *Der Fischer Weltalmanach '88/'90*. (Frankfurt: Fischer).

HANDELSKAMMER HAMBURG (ed.) (June 1989), *Konsulats- und Mustervorschriften*, 28th edn. (Hamburg: C. H. Dieckmann).

HASELHUBER, J. (1991), 'Erste Ergebnisse einer empirischen Untersuchung zur Sprachensituation in der EG-Kommission (Februar 1990)', *Sociolinguistica*, 5: 37–50.

HOOF-HAFERKAMP, R. VAN (1991), 'L'Interprétation de conférence à la Communauté Européenne', *Sociolinguistica*, 5: 64–9.

JESPERSEN, O. (1926), *Growth and Structure of the English Language*, 5/7th edns. (Leipzig: Teubner).

KALVERKÄMPER, H., and WEINRICH, H. (eds.) (1986), *Deutsch als Wissenschaftssprache. 25. Konstanzer Literaturgespräch*, Forum für Fremdsprachenforschung, 3 (Tübingen: Narr).

MULLER, S. H. (1964), *The World's Living Languages: Basic Facts of their Structure, Kinship, Location and Number of Speakers* (New York: F. Ungar).

NELDE, P. H. (ed.) (1990), *Language Conflict and Minorities/Sprachkonflikte und Minderheiten* (Bonn: Dümmler).

OSTROWER, A. (1965), *Language, Law and Diplomacy*, 2 vols. (Philadelphia: University of Pennsylvania Press).

PEI, M. (1966), *Glossary of Linguistic Terminology* (Garden City, NY: Anchor Books).

SKUDLIK, S. (1990), *Sprachen in den Wissenschaften. Deutsch und Englisch in der internationalen Kommunikation*, Forum für Fachsprachen, 10 (Tübingen: Narr).

STURM, D. (ed.) (1987), *Deutsch als Fremdsprache weltweit* (Munich: Hueber).

THOGMARTIN, C. (1980), 'Which Language for Students in Social Sciences? A Survey to Help Academics', *Anthropological Newsletter*, 21/2: 6.

TSUNODA, M. (1983), 'Les Langues internationales dans les publications scientifiques et techniques', *Sophia Linguistica*, 144–55.

ULIJN, J., and GORTER, T. R., (1989), 'Language, Culture and Technical-Commercial Negotiating', in Coleman (1989), 479–506.

UNESCO, *Statistical Yearbook/Annuaire Statistique* (Paris: Unesco Press).

WEIß, R. (ed.) (1992), *Fremdsprachen in der Wirtschaft. Ein Beitrag zu interkultureller Kompetenz* (Cologne: Deutscher Institutsverlag).

3 Germanness: Language and Nation

FLORIAN COULMAS

I   THE NATION-STATE

'It distinguishes the Germans', Nietzsche (1978: 131) once sighed, 'that they never get tired of asking who they are.' Although the Germans are hardly alone in indulging in the pastime of searching for an identity, Nietzsche pinpointed a noteworthy aspect of the spiritual make-up of one of Europe's belated nations. Before the Napoleonic wars Germany, as well as Italy, had not existed as a nation-state in even rudimentary form. By the end of the nineteenth century, however, Germany had not only joined the ranks of the more traditional nation-states, but become a nation-state superpower. A great deal of intellectual energy during that century was spent accompanying and digesting this process of nation-building. As a result, nation and state, which are usually identified in the Anglo-Saxon world as well as in France, are conceptually very distinct in Germany.

In spite of Germany's special status as a latecomer among Europe's major nation-states, it must not be forgotten that the idea of national self-determination is a modern one. As Hobsbawm (1990) has shown, the nation-state, in any variety that would be recognizable to us today, is no older than the American Constitution and the French Revolution. In the two centuries since then, the nation-state became the most important principle of political organization the world over. Two hundred years after the French, in the name of universalism, had ignited the fire of nationalism that would spread across Europe, at a time when the emergence of global problems made some believe, or hope, that the age of nations was past, Germany has reaffirmed the principle of the nation-state by bringing to an end what not a few of its neighbours also found 'unnatural': the coexistence of two states of one nation or, as it was more commonly put, the division of one nation into two polities.

Clearly, the nation-state is as much alive at the end of the twentieth century as it was at its beginning. Unwittingly, perhaps, Germany has once again epitomized important socio-political developments that have changed the political landscape in Europe. It will be up to historians to assess their full significance. In the present context we shall pursue a more modest aim. Since the catastrophe of the Nazi regime, nationalism has not been a respectable sentiment in Germany. Yet the recreation of a German nation-state in 1990 did not meet with much resistance. Was this because the nation-state ideology was so firmly entrenched among Germany's neighbours that it was impossible to follow another path, or is the resurrected German nation-state a 'natural' outcome of the 'Germanness' of its people, what in modern parlance has come to be called 'national identity'? Both have probably contributed to some extent and, if this is so, it may be of some interest to know what German identity consists in or is believed to consist in. Even this is too ambitious a question, however. We shall, therefore, concentrate on what is usually considered a key element of national identity and undoubtedly has played a crucial role in the creation of the German nation-state, the national language.

2   THE NATIONAL LANGUAGE AS A SOURCE OF LEGITIMACY

When Europe's most successful multilingual empire, the Habsburg monarchy, disintegrated, Max Weber (1956: 242) commented on the relationship between nation, state, and language as follows: 'Indeed, "nation state" has become conceptually identical with "state" based on a common language.' Prior to the French Revolution this had not been so: the identity of language, state, and nation was not an important issue. Why then did it develop into a powerful political motive in nineteenth-century Europe? The answer lies in both political and economic developments.

Kelman (1971) has identified two sources of legitimacy for a national system. The first is the population's 'sentimental attachment toward the nation'. The second is that the national system must meet the material needs of the population: this he calls 'instrumental attachment toward the nation'. As for his first criterion, it must be noted that in eighteenth-century Europe the sentimental attachment of the power élite to the nation was underdeveloped. The social barriers separating the populations of the many kingdoms and princely states from the dominant class were more pronounced than the national commonalities that united rulers and ruled. The most conspicuous evidence of this state of affairs was the élite's affinity to the French language (recall Voltaire's sarcastic remark that in Potsdam German was used only towards horses). In this sense there was no national language, because large parts of the power élite refused to treat German as such. This is what Myers-Scotton (1990: 25), referring to post-colonial patterns of language choice in African states, calls 'élite closure', an attitude she defines as 'a strategy by which those persons in power maintain their powers and privileges via language choices'. In Germany élite closure stood in the way of the nation-state. This is one reason why language became a critically important item in the ideological tool-box of nation-building in the nineteenth century. Overcoming élite closure by establishing a common language that ensured communication across all social strata was one way of creating a kind of unity and identity that superseded the personal allegiances of feudal social structure. Such unity and sense of belonging had been achieved, in some cases, by focusing on other features: a common patrimony, history, culture, and political idea. In Germany, however, because of its history of political fragmentation, the notion that it was in some sense natural for linguistic attachment and political allegiance to coincide became particularly important.

Meinecke's well-known distinction between *Kulturnation* and *Staatsnation* comes to bear here. A *Staatsnation* in his definition is one 'based on the unifying power of a common political history and constitution' (Meinecke 1962: 10). Germany lacked both the tradition of political unity which had moulded the French nation and the voluntaristic element of a new constitution which had created the American. Hence, in the German context the most promising path open to nation-building, as a reaction first to French cultural domination and later to military occupation, was that of a *Kulturnation*, one based on a common cultural heritage (ibid.). To this end the language was an indispensable tool. Although the process of linguistic unification, which had been given a focus and pushed forward by the Reformation, was hardly complete at the time of the French Revolution, the German language was more likely than any other social feature to lend itself to the creation of a national myth which could serve as a catalyst of 'sentimental attachment toward the nation'.

3   THE NATIONAL LANGUAGE AS AN ECONOMIC NECESSITY

What about Kelman's second source of legitimacy, the instrumental attachment toward the nation? As mentioned above, this has to do with material benefits the population derives from belonging to a national system. In what sense can a national system meet the material needs of a population better than, say, a feudal fief or a largely self-sufficient rural community? In a subsistence economy there is no obvious answer to this question. Only where production is for a market rather than for consumption are there obvious benefits to be derived from belonging to a national system, a national market, that is. As I have argued elsewhere in some detail (Coulmas 1993), language and trade or, more generally, language and economic activity are intimately linked with each other (see also Ammon, this volume). Wherever there is trade, there is communication, and the more extensive the trade, the wider the range of the language that serves to mediate it. While there are exceptions to this rule, it is a general tendency for which Europe's transformation from feudalistic agrarian self-sufficiency to industrial capitalism provides several examples.

Deutsch (1953) convincingly analysed nineteenth-century European nationalism in terms of communication theory, as a response, that is, to the need for more extensive and more tightly knit networks of social communication. The surge of linguistic nationalism can thus be understood as an economic necessity. This point has also been made by Gellner (1983), who sees nationalism primarily as a by-product of socio-economic developments rather than an ideological outgrowth of Romanticism or the French Revolution. Like Deutsch, he highlights the role of communication, while at the same time stressing the importance of industrialization. In modern industrial society, he maintains, for the first time in human history, explicit and reasonably precise communication becomes generally, pervasively used and important (Gellner 1983: 33).

For this kind of communication a standard language is needed, where dialects and social variation are sufficiently levelled not to obstruct mutual understanding across larger regions and social strata. It is important, therefore, that such a language is used universally by all members of the society and in all functional domains, including those formerly reserved to the 'prestige language' (Kahane 1986). Thus social communication had to be reorientated from primarily horizontal directions — the élite throughout Europe conversed in French — to include more vertical patterns creating a greater sense of civility and common belongingness up and down the social hierarchy.

Industrial production requires standardized and orderly procedures, as well as a mobile, homogeneous, and more highly educated population. These requirements imply the need for using a single standard language to reach all members of society who are drawn into the economic process. It is thus in the wake of the emergence of national economies worthy of that name that national languages come into existence. A national language in this sense is a standardized common language used by a large majority of the population. By affording its speakers access to the national market, including the labour market, the national language helps to increase the population's instrumental attachment towards the national system. Accordingly, one aspect of the national language is that it is the language of the national economy, including in particular the national labour market.

4 THE SUCCESS OF LINGUISTIC NATIONALISM

Linguistic nationalism in Germany has often been explained as a reaction to French supremacy. The common adversary, so the argument goes, triggered an ideology which emphasized the commonality of the many German states. Clearly, the Revolutionary wars were conducive to any intellectual current with anti-French overtones. It seems questionable, however, whether this could have been enough to sweep away the thoroughly cosmopolitan orientation of German intellectuals during the age of Enlightenment. Goethe, according to his secretary (Eckermann, 31 January 1827), held that *national* literature did not mean much because the epoch of *world* literature had arrived, and Schiller (n.d.: 429) stated with grand emotionalism: 'Early on I lost my fatherland, to exchange it for the wide world.' Such sentiments, one would imagine, could be silenced by war for a while, but hardly eliminated from public discourse altogether. A more credible explanation of the success of nationalism, linguistic or otherwise, is to be found in the socio-economic needs first on the social level, to break down élite closure, and secondly on the economic level, to establish a means of communication suitable for integrating a national market. Of course, these were not the motives of the proponents of linguistic nationalism in Germany, but the case can be made that it was thanks to these objective needs that the ideology of linguistic nationalism caught on and was able to ignite more than a flash in the pan of momentary enthusiasm.

Ironically, the Enlightenment had in a sense paved the way for linguistic nationalism, because for the liberating power of knowledge to be released beyond the educated élite it was essential that it be expressed in the vernacular language. However, while for the universalistic purposes of the Enlightenment any language was suitable as long as it was sufficiently elaborate to express the ideas that needed to be expressed (Coulmas 1989*a*), linguistic nationalism had to stress the particular qualities of one language distinguishing it from all others. It was Herder's Romantic notion of *Volksgeist* (national spirit) which provided the intellectual foundation for such claims to singularity. A language, in his view, incorporates a national spirit, preserving, as it does, a link with the past since time immemorial and hence with the mythical roots of a nation. Thus, although the vernacular as opposed to the prestige language was assigned a crucial role by both the Enlightenment and Romanticism, the ways in which it was instrumentalized were quite different. The Enlightenment enlisted the vernaculars to gain access to the realm of knowledge, emphasizing their instrumental qualities, their openness and universal expressive power. Romanticism, by contrast, treated these same languages as unique manifestations of national identities, emphasizing their sentimental qualities, their closedness and parochial peculiarity.

Ever since, both of these orientations have been part of the ideological baggage attached to the German language (as well as many other European standard languages), at times receding into the background to re-emerge later in somewhat modified forms. The above-mentioned socio-economic developments which were conducive to the establishment of standard German as a common language with nation-wide validity pertain to the instrumental qualities of the language. Linguistic nationalism, however, was all about its sentimental qualities. To put it in sociological terms, linguistic nationalism emphasized language as a medium of creating *Gemeinschaft* (community), while socio-economic developments at the same time required the language to take on qualities making it a suitable medium for creating *Gesellschaft* (society). Although this seems paradoxical, it may have been just this contradiction that caught the imagination and made many embrace linguistic nationalism. Through its look backward in search of imaginary roots and its emphasis on uniqueness, authenticity, heritage, and intimacy, the ideology provided an antidote to the threatening horrors of modernity, uprootedness, anonymity, and arbitrary replaceability.

5   FROM AUTHENTICITY TO PREJUDICE

Linguistic nationalism in Germany took on different faces. Herder had no political designs, nor was his idea of a national spirit enshrined in a national language accompanied by an intent to put one language above others. He thought that a multiplicity of distinct and authentic languages could best give expression to the multi-facetedness of humanity. It is not without reason, therefore, that the father of linguistic nationalism is sometimes celebrated as a champion of linguistic pluralism (Fishman 1985).

However, others less tolerantly inclined erected an edifice of militant nationalism on the foundations Herder had laid. In his *Addresses to the German Nation* Johann Gottlieb Fichte, a major proponent of

German idealism, first gave philosophical authority to the claim of the identity of nation, state and language: 'It is incontestably true that, wherever a particular language is found, a separate nation exists which is entitled independently to take charge of its own affairs and govern itself' (Fichte 1808, 12th Address).

Proclaimed in French-occupied Berlin, this proved to be a powerful concept. But Fichte was not content merely to establish a link between language and national self-determination. Like other nationalists after him (see Fishman 1972 for examples), he insisted that the language of his choice was superior to others. In the face of the high prestige and obvious refinement of French, this was not an easy point to argue, but Fichte's solution was to draw a distinction between 'living' and 'dead' languages. 'A living language,' he wrote, 'if compared with another one, can well be highly cultivated, but it can never in itself achieve the same perfection and formation that a dead language so easily obtains' (Fichte 1808). Living languages were those with an unbroken and 'pure' tradition, such as German. Dead languages, on the other hand, were languages with mixed and broken-off traditions like Latin–Celtic French or English. True, German was not as polished as French, but then it was not degenerate either. Such ideological acrobatics allowed Fichte to turn vice into virtue: German was not only distinct from other languages, but alive and more authentic, incorporating, as it did, the spirit of the *Urvolk*. The Germans, he said, speak 'a language which is shaped to express the truth' (ibid.).

That languages are different only makes a difference if there is a significant connection between language and thought, especially if one language is better equipped than others to 'express the truth' or, in more relativistic terms, if different languages express different truths. It was Wilhelm von Humboldt who took it upon himself to provide this idea with some philosophical credibility (see also Dittmar, in this volume). A key notion in his linguistic thinking is that of the 'genius' or 'character' of a language. Although Humboldt's work testifies to the same *Zeitgeist* as Fichte's, it is to his credit that he submitted the question of the differences between languages and their relationship with nations to systematic investigation. He carried further Herder's notion of the inseparability of language and thought, and thus also came to the conclusion that differences between languages involve differences in the understanding and interpretation of the world.

Language for Humboldt was a medium that unites individuals while separating groups. These groups are nations. They cannot be thought of without languages and vice versa, since 'our historiography nowhere justifies the assumption that a nation ever existed prior to its language'

(Humboldt 1823/1963: 69). Hence, 'the concept of a nation must be based especially upon language . . . Language by its own force proclaims the national character' (1830/1963: 561).

Rather than being at the origin of national identity, linguistic differentiation and language formation are often an outgrowth of national will (consider, for example, Dutch vs. German, Croatian vs. Serbian, Czech vs. Slovak, Hindi vs. Urdu). But in spite of this obvious problem, the notion of the identity of language and nation was an attractive point of reference for many nationalists who wanted to exploit language for their own purposes. During the nineteenth century the German language, along with many others, was glorified as a national monument. Jacob and Wilhelm Grimm planned to offer their encyclopaedic historical dictionary 'with joy and pride on the altar of the fatherland'. The Franco-Prussian War of 1870–1 provoked a wave of linguistic purism (the linguistic equivalent of xenophobia and racism), and the First World War saw the publication of many patriotic works about the German language, such as Eduard Engel's *Sprich Deutsch! Zum Hilfsdienst am Vaterland* (1916). By this time, notwithstanding the existence of several predominantly German-speaking states, the German language had become a firm component of Germany's 'national identity'.

6 RACE VS. LANGUAGE

Enter the movement of National Socialism. It might be expected that linguistic nationalism would have accorded well with the Nazis' expansionist pleas on behalf of a *Volk ohne Raum* (a people without space) and their call of *heim ins Reich!* addressed to Austrians and German minorities in eastern Europe. As a matter of fact it did not. While Italian irredentism used language as the focal point of its legitimacy, the Nazi ideology assigned language a decidedly secondary role. Language, after all, can be acquired; race, as understood by the Nazis, cannot. Considering the fact that conquered peoples have often adopted their conquerors' language, Hitler contemplated:

Since nationhood, or rather race, lies not in language but in the blood, one can speak of Germanization only if and when the blood of the subdued had been transformed. This, however, is impossible even if the resulting mixed product were to speak the language of the formerly superior race a thousand times (Hitler 1939: 428–9).

Hitler also realized that the concept of nationality based on language would force him to accept as Germans many whom he despised and intended to exterminate: 'Clearly, no one would dream of recognizing the purely superficial fact that many of the lice-infested exodus from the east speak German as proof of their patrimony and peoplehood' (ibid. 430).

For a *Volk* whose identity is rooted in *Blut und Boden* (blood and soil) language is too fickle and intangible a feature to serve as a reliable criterion of belonging. Language does not preclude a subjective, voluntaristic affiliation with a nation as a 'plébiscite de tous les jours', to quote Renan's (1882) influential definition. Indeed, linguistic nationalism would have saved millions had it been the ideology of the Third Reich, for most of those who perished in the gas chambers spoke German.

7 DIVISION OF STATE AND LANGUAGE

After the fall of the Third Reich it became evident that the essence of linguistic nationalism had survived intact. Fichte's idea of a natural identity of language, nation, and state was deeply ingrained and influenced even linguists in their dealings with language. This time, however, it was turned on its head. As a result of the lost war, Germany's integrity as a political entity was destroyed. As of 1949, the identity of language, nation, and state plainly ceased to exist. Switzerland and Austria had always been special cases, but the creation of two states on German territory was irreconcilable with the idea of the nation-state with its national language. 'A former country in central Europe' is what the 1966 edition of the *Random House Dictionary* stated under 'Germany'. Political integrity can be suspended by force and quickly, but what about the language?

As a reaction to Germany's division many linguists and journalists concluded that a split in the language would inevitably follow the establishment of two separate states. After all, a state had to have its national language. Just as if they were hunting for inside scoops instead of taking time to investigate the facts, and just as if languages were objects that can change virtually overnight, some professional observers of the German language started publishing articles about West German and East German barely a decade after the two German states had been founded (for a critical appraisal of such writings, see Schlosser 1981). This may have been the result of an authoritarian tradition of respect for the state which, somewhat perversely, turned the original historical idea that a speech community was entitled to political self-determination around to imply that political autonomy would bring about linguistic autonomy.

This idea also had a role to play in the Cold War. The communist

dictatorship tirelessly tried to secure as many of the traditional para-
phernalia of the nation-state as it could, since it was denied recognition
in the Western world as a respectable state. That the socialist nation
of German workers and peasants should have its own language was a
clearly stated policy aim which trickled down to dictionary definitions
by obliging lexicographers. At the same time claims were laid, much
in keeping with the authenticity maxim of linguistic nationalism, to
the linguistic heritage, or at least to the good parts of it. To quote
but one example, Walter Ulbricht, who dominated GDR politics for
two decades, declared: 'There is a big difference between the traditional German language of Goethe, Schiller, Lessing, Marx, and
Engels, which is replete with humanism, and the language as it is used
in certain circles of the West German Federal Republic, which is
defiled by the spirit of imperialism' (Ulbricht 1970).

This proclamation, and many similar ones, rang rather hollow, not
least because public discourse in the GDR right down to its very end
was a close cognate, if not the unbroken continuation, of Nazi-speak.
As Schlosser (1991: 15) has pointed out, the leadership of the GDR
in their official publications appealed to the collective solidarity of the
Volk and, in spite of the obvious ideological differences, their linguistic
self-representation was shockingly similar to that of the Third Reich.
The *Volksgerichtshof* had just ceased to execute its terror justice, when
the first *Volksrichter* (People's judges) were installed in the Soviet-
occupied zone; the *Volkssturm*, Hitler's last-ditch stand, had been dis-
mantled only a few years before, when the GDR created its *Volksarmee*.
Countless other new institutions were adorned with names that made
reference to the *Volk*, having been taken, one might be led to think,
straight out of the dictionary of the *Völkischer Beobachter* (the National
Socialists' newspaper). Proletarians went to work in *Volkseigene Be-
triebe*; the legislative body of the GDR was called *Volkskammer*, the
police force *Volkspolizei*, and the entire polity was a *Volksdemokratie*.
That *Volk* was a key term of Nazi ideology did not prevent the commu-
nists from appealing to the *Volk* in their attempt to provide the new state
with some kind of legitimacy. Although most of the population did not
actively support the regime, the constant calls for collective solidarity
and reference to East Germans as the *Volk* of the GDR did not remain
without effect. This became appallingly obvious when the East German
population was finally able to cast off the yoke of their repressive regime.
The war-cry resounding through the streets of East German cities as
an apotheosis of the downfall of the communist government, 'Wir sind
das Volk' (we are the people), later slightly but significantly modified to
'Wir sind ein Volk' (we are one people), took many West Germans by
surprise, because they were no longer used to thinking of themselves

and referring to themselves as a or the *Volk*. Arguably, the West German collectivity, thanks partly at least to American re-education programmes after the war, were much less of a *Volk*. In fact, *Volk* was discarded from public discourse west of the demarcation line. Even the *Volkswagen* was coyly transmuted into a *VW*.

In this sense there really was a certain difference in the language use of East and West. Other lexical items could be cited in support of this claim, especially words susceptible to ideological loading, many of which were defined in East German dictionaries in a way acceptable to the communist masters of meaning. However, to conclude on the basis of some diverging connotations and even outright semantic differences that the German language was in the process of splitting into two was merely an absurd misunderstanding of linguistic variation (see Barbour and Stevenson 1990: 174–9) coupled with the uncritical acceptance of the dogmas of linguistic nationalism in reverse.

8   SUMMARY AND CONCLUSIONS

As the twentieth century draws to a close, the 'German question' is once again on the agenda, and once again many commentators, Germans and others, are concerned with the question of what constitutes 'Germanness'. In this short chapter we have concentrated on one aspect of German identity which is relevant to this book, the German language.

As a catalyst of nationalism, the German language fulfilled in the past a variety of functions, three of which were particularly significant. First, on the economic level linguistic nationalism coincided roughly with industrialization and, as has been argued, was a necessary by-product of this process. The national market required a national language, and its promotion became a matter of economic necessity. Increasing vertical and horizontal communication in a more mobile society called for a higher level of standardization in many domains of social life, notably in language. Thus, secondly, on the social level, unifying the nation linguistically was a means of breaking down élite closure and softening social class barriers. Finally, on the political level language was recruited as a symbolic boundary marker, which provided the rationale for a claim to self-determination. Language was an easily recognized feature for distinguishing 'us' from 'them'.

A number of influential concepts have been reviewed on the preceding pages (Herder's *Volksgeist*, Fichte's *Urvolk*, and Humboldt's *Nationalcharakter der Sprachen*), and I have shown how these notions, despite fundamental differences in the intellectual orientations of their authors, combined to single out language as the key element of national distinctiveness in the German tradition since Romanticism. For historical reasons, other issues around which nationalism centred in nineteenth-century Europe, such as territory, political institutions, or religion, were less suitable for developing an ideology of national individuality in the German context. But that the German language was an expression of the profound depth of the German mind was an idea that inspired many.

Yet language is not a 'natural' vehicle for nationalism: nationalists have usurped language only when it suited their purposes. During the most excessive period of German nationalism the German language was relegated to a subordinate role in the ideological system, because the Nazis stressed primordial ties rooted in the fictitious concept of an eternal and unalterable (superior) race, which is hereditary rather than acquired.

After the War, I have argued, linguistic nationalism resurfaced in Germany in a very implicit and domesticated form. Barely a decade after two states had been founded on German territory, linguists began to look for the inevitable division of German into two distinct national languages: inevitable, that is, according to the doctrine of linguistic nationalism which postulates the identity of nation, state, and language. In the wake of the disintegration of the East German state, these concerns were quickly deposited in their proper place, in the waste basket of history. Which brings us to the present.

While these lines were being prepared for printing, Germany has been going through a critical period of self-assessment, rethinking its place in the world. Emotional discussions about who belongs to German society and who does not remain the order of the day. Nationalist sentiments have come to the fore in a most unpalatable way. This is not the place (or the author) to make predictions about the outcome of these developments. However, the observations put forward in this chapter suggest that at a time when the Germans are once again preoccupied with defining boundaries between 'us' and 'them', it would be useful and prudent to keep a watchful eye on what intellectuals and the mass media have to say about the national language and the spirit it enshrines.

Further Reading

Barnard (1969)
Berlin (1976)
Kohn (1960)
Reiss (1955)

References

BARBOUR, S., and STEVENSON, P. (1990), *Variation in German: A Critical Approach to German Sociolinguistics* (Cambridge: Cambridge University Press).

BARNARD, F. M. (ed.) (1969), *J. G. Herder on Social and Political Culture* (Cambridge: Cambridge University Press).

BERLIN, I. (1976), *Vico and Herder* (London: Hogarth).

COULMAS, F. (1989*a*), 'Language Adaptation', in Coulmas (1989*b*), 1–25.

—— (ed.) (1989*b*), *Language Adaptation* (Cambridge: Cambridge University Press).

—— (1993), *Language and Economy* (Oxford: Blackwell).

DEUTSCH, K. W. (1953), *Nationalism and Social Communication*; 2nd edn. 1966 (Cambridge, Mass.: MIT Press).

FICHTE, J. G. (1808), *Reden an die deutsche Nation*, in *Johann Gottlieb Fichte sämtliche Werke*, vii, ed. J. H. Fichte (Berlin, 1846), 257–516 [reprint 1965, Berlin: de Gruyter].

FISHMAN, J. A. (1972), *Language and Nationalism* (Rowley, Mass.: Newbury House).

—— (1985), 'Positive Bilingualism: Some Overlooked Rationales and Forefathers', in Fishman *et al.* (1985), 445–55.

—— Gertner, M. H., Lowy, E. G., and Milan, W. G. (eds.) (1985), *The Rise and Fall of the Ethnic Revival* (Berlin: Mouton).

GELLNER, E. (1983), *Nations and Nationalism* (Oxford: Blackwell).

HITLER, A. (1939), *Mein Kampf*, 2 vols.; 1-vol. unabridged edn. (Munich: Zentralverlag der NSDAP).

HOBSBAWM, E. J. (1990), *Nations and Nationalism since 1780: Programme, Myth, Reality* (Cambridge: Cambridge University Press).

HUMBOLDT, W. VON (1823), *Über den Nationalcharakter der Sprachen*, in *Werke in fünf Bänden*, iii. *Schriften zur Sprachphilosophie*, ed. A. Flitner and K. Geil (1963; Stuttgart: Cotta), 64–81.

—— (1830), *Über die Verschiedenheiten des menschlichen Sprachbaus und ihren Einfluß auf die geistige Entwicklung des Menschengeschlechts*, in *Werke in fünf Bänden*, iii. *Schriften zur Sprachphilosophie*, ed. A. Flitner and K. Geil (1963; Stuttgart: Cotta), 368–756.

KAHANE, H. (1986), 'A Typology of the Prestige Language', *Language*, 62: 495–508.

KELMAN, H. (1971), 'Language as an Aid and Barrier to Involvement in the National System', in Rubin and Jernudd (1971), 21–51.

KOHN, H. (1960), *The Mind of Germany: The Education of a Nation* (New York: Scribners).

MEINECKE, F. (1962), *Weltbürgertum und Nationalstaat*, in F. Meinecke, *Werke*, v, ed. Hans Herzfeld *et al.* (Munich: Oldenbourg).

MYERS-SCOTTON, C. (1990), 'Elite Closure as Boundary Maintenance: The Case of Africa,' in Weinstein (1990), 25–42.

NIETZSCHE, F. (1978), *Jenseits von Gut und Böse*, in *Werke in zwei Bänden* (Munich: Carl Hanser Verlag), ii. 11–173.

REISS, H. S. (ed.) (1955), *The Political Thought of the German Romantics 1793–1815* (London: Blackwell).

RENAN, E. (1882), 'Qu'est-ce qu'une nation?' in *Œuvres complètes*, ed. H. Psichari (Paris: Presses universitaires de France), i. 887–907.

RUBIN, J., and JERNUDD, B. (eds.) (1971), *Can Language be Planned?* (Honolulu: University of Hawaii Press).

SCHILLER, F. (n.d.), *Sämtliche Schriften. Horen* edn., vol. ii (Munich).

SCHLOSSER, H. D. (1981), 'Die Verwechslung der deutschen Nationalsprache mit einer lexikalischen Teilmenge', *Muttersprache*, 91/3–4: 145–56.

—— (1991), 'Deutsche Teilung, deutsche Einheit und die Sprache der Deutschen', *Aus Politik und Zeitgeschehen*, B 17: 13–21.

ULBRICHT, W. (1970), Interview, *Neues Deutschland*, 17 June.

WEBER, M. (1956), *Wirtschaft und Gesellschaft: Grundrisse der verstehenden Soziologie*, 4th edn. prepared by H. Winckelman (Tübingen: J. C. B. Mohr). [*Economy and Society*, ed. G. Roth and C. Wittich (Berkeley: University of California Press, 1978).]

WEINSTEIN, B. (ed.) (1990), *Language Policy and Political Development* (Norwood, NJ: Ablex).

4 Norms and Reforms: Fixing the Form of the Language

WOLFGANG WERNER SAUER and HELMUT GLÜCK

I   INTRODUCTION

Orthography is boring. It is a subject for elderly folk who love order, vote Conservative, and always keep their dog on a lead. Spelling is a burden for most schoolchildren, students, teachers, secretaries, and office workers. In more and more professions (in anglophone as in German-speaking countries) you have to master orthographic rules and conventions without gaining recognition for it: it is simply something which is taken for granted. If you are not sufficiently competent, you are considered uneducated and unsuited for promotion. At the same time, mastering the art of spelling is a particularly thankless task, as you can only be said to have achieved it if you are able to write a text without the reader finding anything remarkable about it. Should we therefore only discuss orthographical issues from socio-political perspectives? Why do we write the way we write? Who determines that we should write in this way and why do we conform to it? Why do we not write differently: why do members of the left and right, employers and trade unionists, feminists and male chauvinists, church-goers and free thinkers, all follow the same orthographical rules?

Orthography does not have to be boring. It does not have to be a subject just for elderly folk. The aim of our chapter is to show that orthographical issues can be of interest from several different perspectives: linguistic, historical, socio-political, and cultural, and we shall focus our attention on German spelling, which has taken over a thousand years to reach the form which is taught in schools and universities today, and which even now is a constant source of public controversy.

The debate in Germany on what constitutes the best form of spelling has long been of a rather political nature; there is a conflict between those who wish to preserve tradition and a sensitivity for the language through a conservative orthographical approach and those who seek to democratize spelling, to create a form of spelling for the entire population and not just for intellectuals. This debate is bound up with cultural considerations; many people fear, for example, that abolishing the capitalization of nouns (a speciality of which Germans have become sole practitioners since the Danes reformed their spelling in this respect in 1948) would result in a cultural collapse, that children would no longer be able to read the 'classics'. Many had similar fears when the old German form of handwriting (the 'Sütterlin-Schrift') and 'Gothic' print were abolished in German schools. No one protested very loudly, however, as this took place in 1941 under Hitler's regime. Afterwards it was too late.

We thus want to try to show that orthographical issues can be considered from a variety of different viewpoints, illustrating this by various examples. We shall begin (in Section 2) with an outline of the historical development of German spelling. The next section will examine the extent to which it became nationalized in the last third of the nineteenth century and how the *Duden* established itself as the standard reference work and the primary influence on standardization. In the subsequent sections, we shall show how hesitantly and tentatively (in the face of often massive attacks) the efforts at reform which many see as necessary are being advanced today. Finally, we shall raise some points concerning current problems.

2   THE FORMATION AND DEVELOPMENT OF GERMAN SPELLING

Huius enim linguae barbaries, ut est inculta et indisciplinabilis, atque insueta capi regulari freno grammaticae artis, sic etiam in multis dictis scriptu est propter literarum aut congeriem aut incognitam sonoritatem difficilis.

(This barbaric language is inelegant and crude and unaccustomed to obeying grammatical rules. It is also hard to write, due to the accumulation of characters in many words and their strange sound.)

The language being described in this way is German. The author of these lines is Otfrid, a monk who wrote a version of the Gospel in the town of Weissenburg in Alsace around 865, depicting the life and sufferings of Jesus. The Latin-educated monk had very good reasons for making use of such a crude language: his little flock were to have the Bible brought to them in their own tongue.

Otfrid's Gospel is one of the oldest works written in German. He had few if any models available to show him how the vernacular (*theotisce*) was to be written. In a letter written in Latin to Luitbert, archbishop of Mainz, from which the above quotation is taken, Otfrid reflects on the difficulty of transcribing this language. He used the Latin alphabet as his character source for the written representation of his Rhine Franconian dialect, a task to which it is little suited. Otfrid further objects that German 'sometimes demands three ⟨uuu⟩' of which the first two represent a consonant, that some vowels do not catch quite the right tone, and that in order to write in Franconian it is necessary to use the (Greek) characters y, k, and z, which are scarcely used in Latin.

These issues, discussed in the context of this earliest encounter with the written form of German, sound remarkably modern. The first transcription of every language to be written in an alphabetical writing system must involve the regulation of grapheme–phoneme correspondences. The Latin alphabet, however, was only partially suited to the task of transcribing Germanic languages such as Franconian or Old English, so that ambiguities and variations manifested themselves from the beginning. Otfrid, naturally enough, followed the principle of depicting the spoken (that is, what was heard) in the written. The principle 'Schreibe, wie du sprichst' (write as you speak) still has its supporters in the debates on German orthography today (Müller 1990, Günther 1985). For some 1,100 years it has been discussed as the problem of 'Laut–Buchstaben-Beziehung' (the relationship between sound and letter). Another feature of contemporary German was not yet relevant in Otfrid's day: the use of capital and small initial letters. For hundreds of years the majuscule (the capital letter) was used as a graphic device in the formulation of texts to highlight passages of particular significance, for example, the beginnings of chapters or, as we see in Otfrid's work, the beginnings of verses. The chapter beginnings of many handwritten manuscripts of the Middle Ages show how greatly the vividly fashioned majuscule, the initial, was bound to the aesthetic enjoyment of ornate images. Even after the invention of printing, the form of the foregrounded capital was determined by ornamental considerations: even long after Gutenberg it was a decorative feature and not the expression of a grammatical principle.

After the practice of emphasizing proper names, especially sacred ones, through the use of initial capitals had been established, a fundamental change in the function of the capital letter was introduced. Nouns (or more accurately: the heads of noun phrases) were specially marked with an initial capital. This characteristic survives today only in German, but in previous centuries it was also a widespread custom in other languages. In English, the rules stipulate the capitalization of book titles, headings, and so on: a remnant of the former practice.

In German, the syntactically motivated capitalization of nouns became customary around 1800 and later became codified, although it was implemented (as in English) in a rather casual way until the eighteenth century: words which the author felt to be particularly important could be capitalized, but this was not regarded as compulsory.

Johann Christoph Gottsched (1700–66), in § 46 of his '*Grundlegung einer deutschen Sprachkunst*' gives a vivid description of the gradual increase in the use of the majuscule:

46. Man hat nåmlich, um der Zierde halber, schon in alten Zeiten, den Anfang jeder Schrift mit einem so genannten großen Buchstaben gemacht; und dadurch der ersten Zeile eines jeden Buches ein Ansehen zu machen gesucht. Man gieng hernach weiter, und gab auch jedem neuen Capitel, jedem neuen Absatze, und endlich jeder neuen Periode eben dergleichen Zierrath. Endlich gaben die Poeten, die Würde ihrer Arbeiten anzuzeigen, jeder Zeile ihrer Gedichte, oder jedem Verse, einen größern und zierlichern Anfangsbuchstaben. (Gottsched 1748: 57)

(In former times each text was already begun with a so-called capital letter by way of decoration; this was to give the first line of every book a striking appearance. This was then taken further, so that each chapter, each paragraph, and finally each sentence was given the same decoration. Eventually, in order to demonstrate the dignity of their work, poets gave each line of their poems or each verse a larger and more decorative initial letter.)

In the following paragraph Gottsched goes on to trace how God's name, the names of famous people, of countries and towns and 'in the end all people without exception' gradually became capitalized. And as capitals were so clearly legible, one also granted 'certain important nouns this privilege'. This had been the case in every European nation and so it had remained. And yet:

Wir Deutschen aber sind noch weiter gegangen, und haben wegen der, bey der letzten Art der Wörter vorkommenden vielen Unrichtigkeiten, darein sich viele nicht finden können, alle Nennwörter, davor man **ein**, oder **der**, **die**, **das** setzen kann, mit großen Buchstaben zu schreiben angefangen. (Gottsched 1748: 58)

(However, we Germans went even further and, because of the many errors which occur in the latter type of words and which many people have difficulty with, started to write all nouns which can be preceded by 'a' or 'the' with a capital letter.]

Then he articulates what many still think today: 'daß unsere Sprache einen so merklichen Vorzug der Grundrichtigkeit vor anderen erhält' (our language has the notable advantage over all others of being fundamentally correct). And (also very modern) he rails against several innovators who sought to abolish capitalization again, because its rules were too difficult. 'To abandon such a well established custom': that is something we Germans simply do not do, even today!

Of the two rules Gottsched formulated with regard to capitalization (XXII and XXIII), the first is still partially applicable and the second remains completely valid.

XXII. Regel: Man setze im Anfange jeder Periode, und in Gedichten vor jedem Verse, einen so genannten großen Buchstab.

(Rule XXII. At the beginning of every sentence and in poems at the beginning of every verse there should be a so-called capital letter.)

This applies without exception to the beginning of sentences, but the rule is no longer binding for lyric poetry.

XXIII. Regel: Man schreibe nicht nur alle eigene Namen, sondern auch alle selbståndige Nennwörter mit großen Anfangsbuchstaben.

(Rule XXIII. Not only all proper names but also all independent nouns should be written with capital letters.)

The dilemma of whether or not to capitalize in German stems initially from the difficulty (and in some areas impossibility) of clearly categorizing all instances of capitalization and of formulating general rules accordingly. This is evident in the work of one of Gottsched's contemporaries, Chrysostomus Erdmann Schröter, a Saxon bureaucrat and author of a best seller of the time whose title-page is reproduced in Figure 4.1. This book with its baroque title went through several editions (in a continually changing form) in the space of a few years. It is about a thousand pages in length. The third edition, which appeared in the same year as Gottsched's *Sprachkunst* (1748), is quoted here. In the third section, 'Orthographia oder Rechtschreibekunst', Schröter formulates the rules for capitalization. According to him the following should be capitalized:

Alle Substantiva, oder solche Worte, die eine Sache, ohne Zuthuung eines andern, verståndlich ausdrücken; da ich **der**, **die** oder **das**, vorsetzen, und die Sache sehen, hören und betasten kan; ingleichen die Namen der Månner, Weiber, Ståpte, Dörfer, und dergleichen. (Schröter 1748: 6)

(All nouns, or such words which comprehensibly express an object without the addition of another word; before which I can place 'the' and if I can see, hear, or touch the object; likewise the names of men, women, towns, villages, and so on.)

Schröter's formulation of rules is not as rigid as Gottsched's. He states that although other languages may be different, one should none the less proceed with capitalization in German, given that every language has its own peculiarities. In German, after all, one does not write '*Vater* (father) with an F, just because the English, Scots, Danes, Swedes, and Norwegians do' (Schröter 1748: 36–7).

Allzeitfertiger

und auf allerley Fälle gerichteter

Briefsteller,

Welcher der Jugend, nach zuförderst festgesetzter

Orthographie und Stilographie,

So wohl allerley Arten Briefe, Wechsel,
Obligationen, Contracte, Abschiede 2c. und was sonst
im gemeinen Leben, und insonderheit bey der löblichen Kauf-
manschafft erforderlich ist,

als auch, durch in Kupfer gestochene

Vorschriften/

Die Lateinisch-und Teutschen Buchstaben,
nach den Grundstrichen und Wörtern, deutlich vor Augen leget;

FIG. 4.1. Title-page of Schröter's *Briefsteller* of 1748

Instances when one should write a word with a small initial letter, only implicitly mentioned in Gottsched's capitalization rules, are broadly dealt with by Schröter: adjectives and words that 'ein Seyn, Thun oder Leiden, bemerken' (signify being, doing, or suffering) are to be written with a small letter. His examples: *alt* (old), *schön* (beautiful), *ich liebe* (I love), *du leidest* (you suffer), *ihr seyd arm* (you are poor), etc. (p. 37).

What did the educated contemporary of around 250 years ago discover if he wanted to know what should be capitalized? If (to use today's concepts) he had consulted the theory (Gottsched) and the popular words of advice (Schröter), they would have explained to him that nouns or substantives in front of which an article can be placed and that are 'concrete', signifying something that can be seen or touched, as well as proper names, all have a capital letter at the beginning of the word. All other words are written with a small initial letter. By and large this principle, which actually captures the essence of the current norms, remains reasonably satisfactory. It is not, however, completely watertight. Only independent concrete nouns which are able to take an article are supposed to have a capital letter. However, as the criterion of a word being able to take an article outweighs the criterion of its being concrete, the words *Zeitgeist* (spirit of the age) and *Weltanschauung* (philosophy of life) are written with a capital exactly like words such as *Tisch* (table) and *Stuhl* (chair). But *der einzelne* (the individual) and *im folgenden* (in the following) are both written with a small initial letter, although I can touch an individual person and can see the following section.

The principle would be very effective, though, if it were to be applied so that all words capable of taking an article and/or all concrete words in Schröter's sense were capitalized, but all others not. 'Use a capital letter in the context just mentioned, but a small one in cases of doubt': if such a rule were used, one of the major obstacles of German orthography would be easier to negotiate. That capitalization communicates syntactic information is indisputable; whether this is indispensable for the reading and writing of German, though, is a debatable issue. At all events, it is clear that its formation and historical development had nothing to do with grammatical considerations, as is occasionally suggested.

From the point of view of those mid-eighteenth-century authors who did have something to say about the subject, there would have been no reason for the 'orthography question' to become an issue in German studies during the following century. The style of the German classical writers tended primarily to have a stabilizing influence: the spelling practices associated with Goethe, Schiller, and their contemporaries can be regarded as being widely established by the beginning of the nineteenth century. Frequent reference is made to this by, amongst others, the linguist and expert on orthographical questions Dieter Nerius (see e.g. Nerius *et al.* 1987: 14). The writings of Johann Christoph Adelung show the state of the language at the time. Adelung, who is frequently mentioned in the same breath as Gottsched, merely summarized the contemporary state of language usage and unlike his predecessors formulated little that was new. He did, however, give a more detailed exposition of the prevailing rules. In his *Deutsche Sprachlehre für Schulen* (German Grammar for Schools), of which the 4th edition of 1801 is quoted here, he examines the capitalization of 'proper names and the adjectives derived therefrom') (e.g. Europe > European) and pronouns of address (1801: 487–8). Adelung refines the rules, expanding them to related areas, and there is a tendency towards greater standardization.

It is also clear that another area, namely the written reproduction of speech, was not yet very significant at this time. There were variations, for example, in the length of the vowel (*Name* (name) in Gottsched and Schröter, *Nahme* in Adelung), but these did not result in any interference in communication.

In summary, one can see that at the turn of the nineteenth century a standard practice of spelling had developed, which made it possible to extract information from written texts without difficulty. The remaining orthographical variations were not a serious obstacle for the reading public. That 'more recently formulated written language' which 'went under the name of *Hochdeutsch* [High German]' (Adelung 1801: 5), had achieved the widespread circulation which was a prerequisite for effective communication in the German-speaking area. This 'pure German' or 'good German' (as Adelung calls it) was, however, still in the first instance the property of the 'upper classes of the nation', dependent upon their 'culture and taste': but in the course of further developments its norms were accepted in all German-speaking areas (at least as far as the form of the written language that we are dealing with here is concerned). The principle on which these norms are based is often summarized in histories of the language by the maxim which had already been so dear to Otfrid a thousand years earlier: write as you speak. In an inverted form this sentence was to have a great effect upon the spoken language.

Closely linked with the political movement towards unification in Germany after 1815 (the temporary achievement of which was represented by the founding of the German Empire by Bismarck in 1871) was the rapid establishment of a German 'national science' (*Germanistik*). The names Jacob and Wilhelm Grimm may be seen as an embodiment of this complicated process. Their apparently meticulous historical and philological work and their preoccupation with the monuments of early German history (equal attention being given to the areas of the law, literature, and language) served primarily to strengthen national consciousness and were, however esoteric they may have seemed, in every sense political. 'Was haben wir denn gemeinsames als unsere sprache und literatur?' (What else do we have in common other than our language and our literature?) asks Jacob Grimm not without pathos in the Preface to the first volume of his German dictionary (1854: 9).

In the process of this examination of the past, early German history was transformed by the Romantic vision of the researcher into an idyll, a standard against which to measure an imperfect present. Even the language of the time appeared to the philologist as a pale shadow of its former self in long past better days.

Vor sechshundert Jahren hat jeder gemeine Bauer Vollkommenheiten und Feinheiten der deutschen Sprache gewußt, d. h. täglich ausgeübt, von denen sich die besten heutigen Sprachlehrer nichts mehr träumen lassen.

(Six hundred years ago every common peasant was aware of the perfections and nuances of the German language, by which I mean he used them daily, nuances that the best language teachers today could not even dream of.) (Grimm 1819: 2)

And it is precisely these language teachers, in Jacob Grimm's opinion, that have brought about the wretched condition of the language through their 'woolly and erroneous rules' (Grimm 1819: 2). Even their names are not withheld: it is Adelung who stands accused.

In this Preface to the German Grammar of 1819 it is still the form of the German language which is of primary concern. Thirty-five years later Jacob Grimm examines orthography. From today's perspective he seems to be a radical reformer. He consistently uses a small initial letter in his writings (capitals only appear at the beginning of a paragraph and with proper names) and grumbles about the inadequacies of the Latin alphabet in the context of its 'application to German sounds', about the fluctuations and shameful inconsistency' of German spelling, about the 'accumulation of vowels and of consonants, which gives German writing the impression of being sprawling, stiff, and sluggish', and about representations of vowels:

wenn man nahm, lahm, zahm schreibt, warum nicht auch kahm? oder umgedreht, wenn kam, scham, name gilt, warum nicht nam, lam, zam? (Grimm 1854: 70)

(If one writes *nahm, lahm, zahm*, why not also *kahm*? Or conversely, if *kam, scham, name* are correct, why not *nam, lam, zam*?)

Furthermore, he criticizes the shortcomings of the ⟨ch⟩ (*Sicht* (sight), *Flucht* (flight)), and the ⟨th⟩ (*Thal* (valley), *Theil* (part)), which he finds unnecessary (Grimm 1819: 73–4) and which was actually abolished in the reform of 1901. He also points out that structurally, one of the letters ⟨f⟩, ⟨v⟩, or ⟨w⟩ is 'completely dispensable' and suggests discarding one and 'then redetermining the relationship between the others' (1819: 78).

Jacob Grimm derives his critique, as he does the justifications for his suggestions for change, from examples of the history of the German language. As modern as his explanations may sound, to make him the father of the reform movement would be to misinterpret his intentions. He does not want to simplify orthography for pedagogical reasons, like all reformers since Konrad Duden. To act as language teacher or schoolmaster is far from his mind. In a speech to the Berlin Academy in 1847 he spoke with derision about the move towards standardization in the language and made fun of the tendency 'to pedantry in language':

In der sprache aber heiszt pedantisch, sich wie ein schulmeister auf die gelehrte, wie ein schulknabe auf die gelernte regel alles einbilden und vor lauter bäumen den wald nicht sehn. (Grimm 1847: 328)

(But being pedantic about language means proudly holding on to one's own fixed views on everything, the way a schoolmaster does to the rules he teaches or a schoolboy to the rules he learns, and therefore not seeing the wood for the trees.)

This tendency of Jacob Grimm's to adopt a casual attitude towards linguistic matters is usually overlooked. Generations of philologists have been glad to cite Grimm the philologist, but they have preferred to ignore Grimm the satirist and cynic. His pedant, who through sheer linguistic proficiency advised 'his consumptive wife not to drink *eselsmilch* (ass's milk) but only *eselinnenmilch* (the milk of a female ass)' (1847: 329) has regrettably been forgotten. It is interesting to note that there is a sense of continuity with regard to this specific point: a short time ago it was publicly suggested that one should say *Kuckuck-innenei* (the egg of a female cuckoo) rather than *Kuckucksei* (cuckoo's egg) as only female cuckoos lay eggs, and *Amselmännchengesang* (the song of the male blackbird) rather than *Amselgesang* (the black-bird's song) as female blackbirds are unable to sing (Vogt 1992). This is presumably meant to be sarcastic. In Grimm's time, how-ever, it was the pedants (especially Germanists) who prepared the way for the final codification of orthographical, lexical, and grammatical norms.

3   THE NATIONALIZATION OF GERMAN SPELLING AND THE ESTABLISHMENT OF THE *DUDEN* AS A STANDARD REFERENCE

There is general agreement that the state of German spelling at the beginning of the last third of the nineteenth century was deplorable. Before 1871, every large German state and many of the smaller ones had their own rule-books for spelling. Orthography was as disunited as the country. In fact there were not that many rule-books and their differences focused primarily on details. It is true that the school-teachers who taught using these individual state rule-books had a lot of freedom in their actual teaching, as there was no definitive orthographical dictionary. The Prussian Minister for Culture, Raumer, complained in 1857 that:

Es ist ein unerträglicher Zustand, wenn in einer Anstalt der Lehrer der einen Classe die Schreibweise für falsch erklärt und mit allen Mitteln auszutreiben versucht, die der Lehrer der hervorgehenden Classe mit ebensolchem Eifer den Schülern eingeprügelt hatte. (Raumer 1863: 301)

(It is an unacceptable situation for a teacher in an institution to tell a class that their spelling is wrong and then attempt by every means to eradicate precisely what the previous teacher had drilled into the pupils with an equal amount of fervour.)

This assessment, however, refers primarily to the numerous words with two alternative spellings, which differed consistently from one another, in small ways (e.g. *Classe/Klasse* (class) and the ending *-iren/ ieren*). This did not affect the communicative capability of the orthographical system: the degree of variation might even have been less than in the German language today, where, for example, *Zentrum* competes with *Centrum* and *Center* (centre).

The dominant trend at the time was towards standardization and the breakthrough in all areas of life came after the founding of the German Empire. With the realization at Versailles of the so-called *kleindeutsch* solution to the unification of the empire by Chancellor Bismarck in 1871 (the unification of the German states without German-speaking Austria) after the military victory over France, it was finally possible to achieve unity and standardization. Currency, measurements, weights, railways, postal services, and industrial production norms all had to be standardized in order to facilitate trade within the new national territory, and legal systems were to be harmonized. This process has highlighted the current parallels in many areas following the merging of the former GDR with the Federal Republic of Germany in 1990.

However, one area hardly affected the population of the new empire: that of the still unstandardized orthography. In fact, it took five years for the minister responsible in the dominant state of the empire, Prussia, to concern himself with orthographical questions. The *Verhandlungen der zur Herstellung größerer Einigung in der deutschen Rechtschreibung berufenen Konferenz* (Conference on the Establishment of Greater Standardization of German Orthography) took place in Berlin in 1876. Amongst those present was the 45-year-old headmaster Konrad Duden, who some years previously had published a slim orthographical dictionary in the small town of Schleiz where he worked. After ten days of tough negotiations the conference made a number of remarkable resolutions. The most notable of these was 'that length should only be graphically represented for the vowels e and i, as these occur in stressed as well as unstressed syllables' (quoted from Nerius 1975: 63). Following this ruling one would write

Kan (boat), *Färe* (ferry), *Son* (son), *Höle* (cave), *Hun* (chicken), *Gefül* (feeling). Only the need for differentiation between homophones allowed a continued distinction to be made between *Ruhm* (glory) and *Rum* (rum). The issue of capitalization was not on the agenda in 1876.

The proposals of the conference (which also included the replacement of ⟨c⟩ by ⟨k⟩ and ⟨z⟩) although published with an explanation, were never put into force. According to contemporary accounts, Bismarck worked himself up into a fury over the new orthography, calling it *Sprachkonfusion*, which would only 'bewilder' the people, who were having to get used to all sorts of reforms, and banned it without further ado (see Sauer 1988: 87).

Still in the same year, Konrad Duden, who was a passionate advocate of the 'new orthography', published a short work with the title *Die Zukunftsorthographie nach Vorschlägen der zur Herstellung größerer Einigung in der deutschen Rechtschreibung berufenen Konferenz erläutert und mit Verbesserungsvorschlägen versehen von Konrad Duden'* (The Orthography of the Future in accordance with Suggestions Made at the Conference for the Establishment of Greater Standardization within German Spelling, with Suggestions for Improvement by Konrad Duden).

In this work he puts forward the case for reform with verve. He justifies his position in a radically democratic fashion:

Der Reiz, welcher für den sachlich Hochgebildeten darin liegt, daß er durch die Gestalt des Wortes an die Herkunft desselben, an die Begriffsentwicklung, die es durchlaufen hat, vielleicht selbst an die Wurzel, aus der es entsprossen ist, erinnert wird, ja daß ihm hie und da eine, freilich meist sehr schwankende, Anung aufdämmert, warum diese Lautgruppe mit der Funktion betraut ist, gerade diesen Begriff zu bezeichnen — dieser Reiz, sage ich, ist dem Volke im großen und ganzen, zu dessen Gebrauch die Schrift da ist, völlig unverständlich, und er hat mit dem Zweck der Schrift nichts zu schaffen. . . . Reinliche Beschränkung auf den Zweck ist überall gut, darum ist diejenige Orthographie die beste, welche, das historische Studium der Sprache den Gelehrten überlassend, nichts weiter will als treu und sonder Müh' das gesprochene Wort widergeben. (Duden 1876: 11–12)

(There is a certain attraction for the highly educated man in the fact that the form of a word reminds him of its origin, of its conceptual development, perhaps of the root from which it sprang, in the fact that it sometimes dawns on him, albeit mostly very hazily, why this group of sounds is accorded the function of designating this particular concept. But in my view this attraction is by and large completely incomprehensible to the mass of the people for whose use writing exists, and it has nothing to do with the purpose of writing. . . . It is always a good thing to limit oneself strictly to the purpose; therefore, the best orthography is the one which, leaving the historical study of the language to the scholars, seeks to do nothing more than to reproduce the spoken word faithfully and without excessive effort.)

These words are clearly directed against the historicization of linguistics and it is equally clear that Konrad Duden is concerned with the establishment of an 'orthography for the people'. The linguist Wolfgang Ullrich Wurzel, author of the only biography of Konrad Duden, describes the failure of the 1876 attempts at reform as a 'fiasco'. 'By means of a bureaucratic decree the results of a commission of renowned experts were declared null and void and relegated to the archives — nothing changed' (Wurzel 1985: 66). Wurzel comes to the conclusion that 'the consequences of this defeat . . . have not been overcome even today' (1985: 67).

Konrad Duden came to terms with the defeat remarkably quickly. In 1880 he issued the *Vollständiges Orthographisches Wörterbuch der deutschen Sprache* (Complete Orthographical Dictionary of the German Language), which was published by the Bibliographical Institute in Leipzig. Its modest vocabulary gave the spelling of approximately 28,000 keywords, contained hardly any definitions and only a few grammatical comments on articles and the endings of both the genitive singular and nominative plural. It is not only the publishing house that to this day regards this slim volume as the first *Duden*: the German Post Office celebrated its centenary with a special issue. In linguistics, this edition is regarded as the first in a series of *Duden* spelling dictionaries (see Sauer 1988: 13 ff.).

This first *Duden* was based on the Prussian rules in operation at the time; some variations of spelling in other states, for example Bavaria, were also taken into account. The book became a best seller and was rapidly followed by new editions. These subsequent *Duden* dictionaries gradually increased the number of words covered from edition to edition and also included additional details on words that were 'better avoided' and the origin of many words. The fact that the 'Complete Orthographical Dictionary' conformed to the dominant principles of spelling in Prussia and included etymological elements, that is to say precisely the kind of historical component he had earlier rejected, shows with what ease Konrad Duden accepted the prevalent attitudes and abandoned his radical 'Orthography of the Future'. The reward for this was not long in coming. When the orthographical question was brought up once again by the authorities responsible for cultural policy after the death of Bismarck in 1898, Konrad Duden was invited to take part in the deliberations. In 1901 the *Beratungen über die Einheitlichkeit der deutschen Rechtschreibung* (Discussions on the Standardization of German Spelling), known today as the Second Orthographical Conference, took place in Berlin.

The host was the Reichsamt des Inneren (Ministry of the Interior). Those participating included not only the most important government officials, but also representatives of the individual German states and of the Austrian monarchy. Switzerland had declined to take part, as the *Duden* had already been adopted as the definitive spelling dictionary there some years previously. Konrad Duden was one of the two participants at the conference who had also been present in 1876. This time, however, he was more than just a face in the crowd: through his 'Orthographical Dictionary' he had become an authority. The conference progressed at a remarkably rapid pace. After only three days of negotiation the rules for German spelling had been reworked. The minutes documenting the course of the discussions have been preserved and are reprinted in the appendix of Nerius and Scharnhorst (1980).

The Berlin conference could not be regarded as representing a reform of German orthography. Broadly speaking the rules were brought into line with the norms which had been established a good twenty years previously in the *Duden*. Of those suggestions made in 1876 only an amended version of the rule concerning the use of ⟨th⟩ and the replacement of ⟨c⟩ by ⟨k⟩ or ⟨z⟩ in foreign words remained. The question of capitalization continued to be ignored. All those present were able to approve the new framework of rules and in 1902 the 'Rules of German Spelling' were published in the form of a pamphlet. Since then there have been countless reprints. Originally published 'on the command of the Royal Prussian Ministry for Religious Instruction and Medical Affairs', it was also published long after the demise of the Prussian Empire without official sanction by the same publishing house, Verlag Weidemann, which moved to Berlin in 1902.

The currently available edition (without the imprint of the publisher) is published in both Dublin and Zurich and was last printed in 1964. This set of rules was once again given official approval in 1955:

Die in der Rechtschreibreform von 1901 und den späteren Verfügungen festgelegten Schreibweisen und Regeln für die Rechtschreibungen sind auch heute noch verbindlich für die deutsche Rechtschreibung. (*Bundesanzeiger* (The Federal Legal Gazette), no. 242, 15 Dec. 1955: 4)

(The orthographical conventions and rules laid down by the Orthographical Reform of 1901 and later decrees are to this day still binding for German orthography.)

This ruling was agreed by the various education and arts ministers of the Federal Republic of Germany's *Länder*. The *Bundesanzeiger* went on to state that 'in Zweifelsfällen die im *Duden* gebrauchten Schreibweisen und Regeln verbindlich sind' (in cases of uncertainty the orthographical conventions and rules used in the *Duden* are binding). And so it is that the current editions of *Duden* are regarded as being semi-official. Teachers make decisions about marks on the basis of the *Duden*, judges base their verdicts in relevant cases upon its rulings, the majority of Germans have faith in their *Duden*. And the compilers of the *Duden* in Mannheim actively encourage this ideology: since 1986 they have incorporated the additional line 'based on the official orthographical code' into the title of their 'Orthography'.

It was already clear, however, immediately after the second Orthographical Conference, that the framework of rules was unsatisfactory. The principal witness for the prosecution is Konrad Duden. Already in 1902 he writes in the preface to the seventh edition of his Orthographical Dictionary,

daß die so entstandene 'deutsche Rechtschreibung' weit davon entfernt ist, ein Meisterwerk zu sein. . . . Ihr Hauptvorzug besteht darin, daß sie überhaupt da ist und allgemeine Gültigkeit hat. (pp. iv–v)

(that the 'German orthography' which came into being in this way is far from being a masterpiece. . . . Its main advantage lies in the fact that it exists at all and that it has general validity.)

At the same time, however, he also urges that 'jetzt keineswegs für alle Zeiten ein Stillstand eintreten soll' (on no account should there now be a permanent standstill). Ironically, a standstill has been in effect now for over ninety years!

4   CHALLENGES TO THE AUTHORITY OF THE STANDARD NORMS

The process of the codification of German orthography had reached a conclusion in 1901, but the nature of this conclusion provided sufficient cause for further suggestions for reform. Already in the same year as the first appearance of the pamphlet of rules, a comprehensive reform concept entitled *Die lautlichen und geschichtlichen Grundlagen unserer Rechtschreibung* (The Phonetic and Historical Foundations of our Orthography) came on the market. The author was Oskar Brenner, a professor at Würzburg University who had taken part in the orthographic conference the previous year as a Bavarian representative. In the opening meeting there, Brenner had already voiced the need 'for strong intervention in the status quo' and had pleaded for 'radical simplification' (minutes according to Nerius and Scharnhorst 1980: 33). His proposal, published in the work cited above, dealt in particular with the clearest way to classify sounds and letters in relation to one another (these relations were later known as 'grapheme–phoneme correspondences'), with the simplification of the marking of vowel quantity (length), and the strict limitation of capitalization to a minimum of cases. These form the main issues of the reform debate to the present day. Other areas of possible orthographical reform addressed by Brenner are: the writing of foreign words, hyphenation (syllabification), the formation of compounds, and questions of punctuation.

The history of this century's numerous suggestions for reform shows that in recent times there has been a shift in the importance and value accorded to individual points. Since at least the 1970s educationalists in particular have openly encouraged *gemäßigte Kleinschreibung* (a modification of spelling which would dispense with capital letters for most nouns), as it is in the use of capital letters that the majority of mistakes occur. Figures for the proportion of all writing errors (excluding punctuation) accounted for by these mistakes vary from 30 per cent to 50 per cent, with some authors setting the figure even higher (see Drewitz and Reuter 1974: 85). A current review of this area is to be found in *Die Rechtschreibung des Deutschen und ihre Neuregelung* (The Orthography of the German Language and its Revision; 1985: 21 ff.).

The alteration of phoneme–grapheme correspondences has ceased to be an issue since the so-called Wiesbaden Recommendations of 1958. The proposals put forward by the *Arbeitskreis für Rechtschreibregelung* provoked a lively discussion, but contained no suggestions for changes in this area.

To summarize, however, we can say that the two areas which played a predominant role from the beginning of the history of orthography are still those determining today's debates on what consitutes an ideal orthography. Owing to the structural limitations of the Latin alphabet used as the foundation of German orthography the relationship between sounds and spelling has been a bone of contention since the time of Otfrid, and the curious rulings affecting capital letters have caused controversy since their provisional regulation in the eighteenth century. Both areas were largely ignored in the modification of the orthographical system in 1901, but both have become dominant preoccupations of most reform proposals and both have become favourite targets for critics of reform.

The inadequacies of the current orthographical system can be seen most clearly in schools, in the classroom. As early as 1912, just ten years after the official introduction of the regulations, a teacher from Breslau called Kosog comments that

Die deutsche Rechtschreibung nämlich ist nichts weiter als ein wahres Schul-
kreuz; denn wenn man die Zeit, die dafür aufgewendet wird, die Tränen, die
um ihretwillen von den Schülern alljährlich vergossen werden, summieren
könnte, man würde erschrecken über das Unheil, das dieser Unterrichtsge-
genstand Jahr für Jahr anrichtet. (Kosog 1912:3)

(German orthography is actually nothing more than a cross schools have to
bear: if one could calculate the time expended on it, the tears spilt on its
account by pupils all year round, one would be shocked at the damage this
subject inflicts year after year.)

Kosog takes issue with two points: the alphabet and capitalization.
Of the former he remarks (1912: 4): 'Unser Alphabet bietet nämlich
auf der einen Seite zu wenig, auf der anderen zuviel . . . Es gibt kaum
etwas Regelloseres und Willkürlicheres als dieses Alphabet' (On the
one hand our alphabet offers too little, on the other hand too
much. . . . There is hardly anything as disorganized and arbitrary as
this alphabet). If one really were to follow the rule 'write as you speak'
(or as you hear), then even a little word like *Fuchs* (fox) alone could
be represented in sixty-three different ways: Kosog's examples range
from *Fuks* to *Phucks*, *Vugs*, *Fux*, and *Phux* (p. 5).

To highlight the issue of capitalization Kosog gave seventy people of
both sexes a dictation of thirty lines: none of his guinea-pigs was able
to complete it without mistakes. This text with its multiple difficulties
created a furore in the press at the time, resulting in many letters which
confirmed 'the impossibility of reproducing the dictation without mistakes'. Kosog's conclusion (p. 24): 'eine Rechtschreibung, die selbst von
den Gebildetsten im Volke nicht beherrscht wird, hat ihr Daseinsrecht
verwirkt, und je eher sie verschwindet, desto besser' (an orthography
which cannot even be mastered by the most educated in the land has
forfeited its right to exist: the sooner it disappears the better).

There are two possible ways to bring about orthographical change.
One is from below. If people no longer write in the way the rule-books
demand, then the rules will alter at some point; the development of
American orthography is an example of this process. There are other
normative systems which have also done this in the last few decades:
dress codes, table manners, forms of greeting and address, stylistic
rules in general have changed or survived through the power of the
people. As we shall show, examples of this can be given for some
aspects of German orthography.

The other way is from above, when through laws or decrees the
state changes the norms and thereby standard writing practices. This
seems to be what the orthographical reformers are waiting for, since
it is very rare for one of them to dare to publish a text which contains
no capital letters (e.g. Drewitz and Reuter 1974). No one, to our

knowledge, has tried to alter the marking of vowels. A democratic state, it appears, does not dare to undertake orthographical reform. Only the National Socialist government produced a revised orthography with simplification in mind, but it was abandoned towards the end of the Second World War. Although the Nazi era had a profound effect on the German language as a whole (see Section 5 below), the *'grossdeutsch'* (Pan-German) orthography (it was to be 'clear, simple and strong') disappeared with the Reich.

After the consolidation of the two German states in 1949, the orthographical question became caught up for a while in the conflicts of the Cold War. East and West Germany were not prepared to enter into negotiation with one another, but nor did either of the two states venture into orthographical waters alone. Attempts at reform in conjunction with Austria and/or Switzerland would have been possible: this would have resulted in other areas being pressurized into action.

5   CURRENT DEBATES ON ORTHOGRAPHICAL REFORM

Since the beginning of the 1980s, experts from German-speaking countries have at last been discussing the idea of reform at successive conferences. In 1988 they published a proposal for the restructuring of German orthography. The directive given by the state was to 'present suggestions for a reform of the framework of rules governing the areas of hyphenation, the writing of foreign words, and relations between sounds and spelling'. The issue of capitalization had already been excluded through the wise foresight of the authorities. They were clearly aware of the emotions which suggestions of change in this area would arouse. Even the moderate proposals for reform in the other areas resulted in media uproar. Every commentator, feature writer, and talk-show host was suddenly an expert on orthography. They united under the banner of 'preservation'. 'Die Muttersprache ist wie die Landschaft, in die man hineingeboren wird, etwas Angestammtes, eine Heimat, aus der niemand vertrieben werden darf' (The mother tongue is like the country in which one is born, an inheritance, a *Heimat* (home) out of which no one should be driven), stated the national newspaper the *Frankfurter Allgemeine Zeitung* on 12 August 1988. The worthy gentlemen from the commission, the 'reformers' (in quotation marks), were suddenly 'communist pigs', 'idiots', 'bearded revolutionaries' (*Sprachreport*, 4 (1988)).

A few trivial aspects of the proposals regarding issues of sound-to-spelling relations were used to ridicule the suggested reforms. The authors had wanted to alter some cases of so-called *Dehnungs-schreibungen* (ways of marking vowel lengthening): *Aal* (eel) was to become *Al*, *Moor* (moor) *Mor*, and the *Kaiser* (emperor) was to change its ⟨ai⟩ to ⟨ei⟩. The battle against the reform as a whole was waged with ludicrous examples such as 'der Keiser im Mor'. The brave reformers had actually only dared to tackle individual words, and had not intervened structurally in the writing of vowels: even the word *Mai* (May) was left unscathed. It was obvious that hardly any of the commentators in the media had read the reformers' extensive explanations, whose phraseology was not exactly crystal clear.

For some time now virtually no one writing texts for public consumption had observed any version of the beautifully precise rules governing the writing of two or more words separately or as a compound. The alteration of hyphenation patterns is done by the computer. No PC or typesetting computer is able to cope with the existing rules of hyphenation. The following examples of incorrect hyphenation are all taken from daily newspapers. The errors result from fundamental limitations of the software:

1. It is unable to recognize morpheme boundaries: for example, *Aben-dessen* (supper), *al-tägyptisch* (ancient Egyptian), *Alteisenmann* (scrap-metal dealer), *al-tehrwürdig* (time-honoured), *Amo-kläufer* (madman), *ange-blich* (supposedly), *An-tarktis* (Antarctic), *Anzeigent-eil* (advertisement section), *Armeer-eserve* (army reserve), *Artike-lende* (end of an article), *Artisch-ocke* (artichoke), *Atlanti-krouten* (Atlantic sea-routes), *Atomzei-talter* (atomic era).
2. It is unable to recognize fixed combinations of letters (bound graphemes): for example *alp-habetisch* (alphabetical), *Amp-hitheater* (amphitheatre), *ät-hiopisch* (Ethiopian).
3. It takes too literally the rule of thumb that the combination ⟨st⟩ should not be split: for example *Abga-stest* (test of level of exhaust emissions), *Abschied-stournee* (farewell tournament), *Ausdruck-stypen* (types of expression).
4. It fails to deal effectively with the so-called *Fugen-s* (-*s*- linking two components of a compound): for example *Amt-seid* (oath of office), *Anfang-schor* (opening chorus), *Aufstieg-schancen* (promotion prospects), *Ausdruck-stypen* (forms of expression).

The use of either ⟨ph⟩ or ⟨f⟩ in frequently used foreign words, for example *Photographie, Fotographie* (photography), is not just a recent phenomenon. ⟨c⟩ and ⟨z⟩ are used interchangeably, not only in advertising (*Zigarette* is less chic than *Cigarette*). In some words ⟨k⟩ and ⟨z⟩ are restricted in their usage on historical grounds: *Kongress-Zentrum* (congress centre) cannot be abbreviated (as KZ can only possibly be understood as an abbreviation for *Konzentrationslager* (concentration camp), whereas *Congress Centrum* (CC) is permissible. Even the *-eur* ending can hardly ever be replaced with *-ör*: *Koifför* (for *Coiffeur*) would barely be decipherable, *Likör* sounds 'tackier' than *Liqueur*. Punctuation only irritates pedants these days, and to a large extent its importance depends on the situation and social context in which it is used. For example, it is expected to be more or less correct in published texts and in Germanists' seminar papers, but the comma before 'und' and the semicolon after certain complex sentence structures are no longer an educational issue.

So the efforts of the last reform proposal were in vain. The reformers were not officially allowed to comment on capitalization; on the issue of sound-to-spelling relations they were far less radical in their suggestions than their colleagues in 1876; and the rest nobody wanted to hear. Admittedly, after the incorporation of the GDR, the German state did have more important tasks than to formulate rules on hyphenation in compound words ... And the language community, or more accurately, those of its members who work with the language, are putting pressure on the rules from below. East Berlin called its television channel *Deutscher Fernseh Funk*, which was described rather sourly by the 1991 *Duden* as 'a deviation from orthographical rules' (p. 204), according to which only *Deutscher Fernsehfunk* is acceptable. Ironically, they have been conspicuously violating this principle themselves since 1989: the words *Deutsches Universal Wörterbuch* are emblazoned on the spine of their dictionary.

A further interesting development in the writing of compound nouns is the use of capital letters for each of the components of the compound (*PostGiro*, *SchülerFerienTicket*). This use of capitals within a word is a device often used by creative advertising copywriters to emphasize words within words and typically occurs at morpheme boundaries in compounds. However, this technique has another important function in contemporary German, which is to form gender-neutral human nouns (see Hellinger, in this volume). *LeserInnen* (readers), *BerlinerInnen* (Berliners) are perfectly normal terms for the majority of *Leser und Leserinnen* (male and female readers) of *taz*, the Berlin newspaper which is not just read by *Berlinerinnen und Berlinern* (female and male Berliners). The *PolitikerInnen* (politicians) of the former Red–Green coalition government of the city introduced this spelling innovation. This isolated ruling, since withdrawn, has not yet been adopted by the rest of the country, but it may only be a matter of time before it becomes more widely accepted.

From the viewpoint of the 'official' orthographical rules, all these forms are regarded as incorrect. The authoritative book of norms, the *Duden* spelling dictionary, has deliberately ignored language development in this area for years. The chief editor of the *Duden* published in Mannheim, Günther Drosdowski, defined the *Duden*'s purpose as follows: 'to establish the spelling of words according to official rules and try to ensure that the number of written variations is kept to a minimum, so that the communicative function of the written language is not impaired' (1987: 26). What he overlooks is that standard writing practices change for precisely those communicative reasons. Women do not just want their presence to be linguistically implicit, they want to be explicitly acknowledged by the language. Furthermore, in many cases the use of the capital within a word is an expression of creative pleasure in manipulating the written language.

Admittedly, the imperial German preceptors of the official rules could not have foreseen developments like these; nevertheless, in a democratic society orthography cannot remain an eternal monument to past eras. Closing your eyes to changes will not make them go away. It is not enough for the new 1991 *Duden* coyly to change a section heading from 'Notes for the User' to 'Notes on Dictionary Usage' (p. 9) and for the reader to find in the same edition that the most 'radical' alteration deep in the jungle of rules is that now, within names written in capital letters 'the ß may also be used for reasons of clarity' (Rule 187). Thus, HEINZ GROSZE is now promoted to HEINZ GROßE. It is almost moving to read in the Preface that 'the presentation of the guidelines for orthography and punctuation has been improved' (*Duden* 1991: 5), until one finds 212 rules following on barely 50 pages. Spellings like *PostSparen* and *EuroCard* may still be regarded as a fad, but the large variety of gender-specific descriptions in written form cannot simply be dismissed: the number of people who have searched through the rules section of the *Duden* in vain to find them must run into millions.

Another discrepancy in the noble claim of the *Duden* editorship to 'ensure the standardization of orthography in German-speaking areas' becomes more and more obvious from edition to edition: the German language is being infiltrated by English and American words and expressions at an increasingly rapid rate. Anglicisms are adapted to the morphological-syntactic structure of German to varying degrees and some more rapidly than others (see Glück and Sauer, in this volume). However, full integration, including the adaptation of loan forms to the German orthographical system, is only rarely achieved. At the turn of the last century 'cakes' became *Keks* (which actually means 'biscuit'!) and 'strike' *Streik*; the development of the plural forms *Kekse* and *Streiks* completed the integration progress. It will be interesting to see if *cornflakes* are transformed into what one already sees described on the packaging as *Kornfleks*. Konrad Duden went into battle against linguistic 'intruders who have not earned the right to German citizenship' (Duden 1887: p. iv). His descendants simply record them and make do without the rhetoric. There is still enough room, however, for a few tentative exceptions. Why, according to *Duden*, one can call two very young children *Babys* but two elegant women either *Ladys* or *Ladies* remains the secret of the Mannheim editors.

The problem of the spelling of words integrated into German from another language is a complex one because (as in English and French) the spelling of the original language is normally adopted, even if it does not correspond to the rules of German orthography. One writes *Joystick* [tʃɔɪstɪk] and not *Tschojstick* in German, just as in English one writes *weltschmerz* [wɛltʃmɜːts] and not *weltshmerts*. However, whereas such foreign words are phonologically integrated into the English language and therefore made to conform to the structural pattern of the English system, foreign words in German, at least those borrowed from English and French, retain the pronunciation of the language they originate from. One not only writes *Balkon*, *Teint*, and *Grand* but also pronounces them [balkõ], [tẽ]; and [gʁã] as in French, although (standard) German possesses no nasal vowels. The 'German' pronunciation of such words ([balkoŋ], [tɛŋ], [gʁaŋ]) would be regarded as uncultured. This leads to the problem that every German must be familiar with the phonological and orthographical systems of both English and French (and increasingly of Italian), in order to be able to read and write their own language; words like *hors d'œuvre*, *gratin*, *fettucini*, *gnocchi*, *ice cream*, and *cheddar cheese* find their way on to many German menus, but also *ćevapčići* and *ražnjići*, *giros* and *souvlaki*, *vinho verde* and *köfte kebab*. Whereas, however, the first group has to be mastered phonetically (and, if necessary, orthographically), this is hardly ever the case in the second group.

6   CONCLUSIONS

There is no doubt that the degree to which orthographical norms are regarded as absolutely binding has changed in many cases since 1901. The norms have been considerably relaxed, influenced amongst other things by the almost playful nature of their use in advertising campaigns. The *säzzer* and *säzzerin* (compositor; normally spelt *Setzer*, *Setzerin*) of the Berlin *tageszeitung* have become almost legendary, enlivening articles with their individual commentary. The use of the capital I within words (*LehrerInnen* (teachers), *SchülerInnen* (pupils)) will probably become established, despite the entrenched opposition of the *Duden* editorial staff and the reservations of linguists, representing as it does a socio-political demand. 'Incorrect' spellings, used millions of times, will gradually become accepted as 'correct', for in the end it is actual usage which is the determining factor.

After all, the prevailing orthographical rules are themselves only the product of the particular social structures that existed at the turn of the century, when norms were established on the basis of the writing conventions current at that time. As history has shown, the *Sprachkultur* (language culture) of German does not simply mean the extent to which the written form of the language has been standardized. On the one hand, we continue to uphold an orthographical system that Konrad Duden's successors consider to be fixed in perpetuity. On the other hand, the set of rules embodied in the current *Duden* has expanded to such an extent that virtually no one can cope with it in its entirety. Some way must therefore be found to liberalize the system, if a mastery of 'correct writing' is not to become the exclusive property of a minority.

Admittedly, writing is no longer the primary means of communicating over long distances. In Konrad Duden's time there were few telephones, no radios, no televisions. Today, however, information is transmitted to a much greater extent by means of the spoken language than in writing. But this does not mean that writing is unimportant, and it is a matter for some concern that in spite of universal education many people still feel inhibited in expressing themselves in writing. Precisely in order to combat these inhibitions and to make writing a cultural asset accessible to all, we should discuss ways of changing its norms to make them more user-friendly. After all, for centuries these norms were not fixed but were in a state of constant flux. The 'classical' writers of German literature managed without a codified norm, and even today no one has difficulty in reading, for example, Goethe texts in the *Vollständige Ausgabe letzter Hand* (complete final edition) of 1830.

The editors of the *Duden* may continue to determine what form German orthography should take in order to enable efficient written communication without sacrificing tradition. Linguists may deliberate on possible future spelling systems. But attempts to achieve absolute standardization should be resisted. In particular, aspects which do not affect the physical form of words, the evolved relationship between phoneme and grapheme, could be changed straight away: for example, hyphenation and some aspects of writing loanwords or of capitalization lend themselves to more liberal treatment. There is also room for a greater degree of tolerance towards errors: much of what is considered to be 'wrong' is anyway no more than a creative attempt to manipulate spelling in a playful way. Advertisements in particular show how a deliberate violation of the rules can grab a reader's attention. This cheerful anarchy, which is practised even by state organizations such as the Federal Post Office and the Federal Railways, will not lead to a collapse of the orthographical system. Hopefully it will lead to orthographical issues being taken a little less seriously and allow them to be dealt with in a more liberal fashion. The elderly folk who find these developments alarming will, no doubt, see things differently.

Further reading

Duden Rechtschreibung der deutschen Sprache (1991)
Grimm (1847)
Nerius *et al.* (1987)
Die Rechtschreibung des Deutschen und ihre Neuregelung (1985)
Sauer (1988)

References

ADELUNG J. CH. (1801), *Deutsche Sprachlehre für Schulen* (Berlin).
BRENNER, O. (1902), *Die lautlichen und geschichtlichen Grundlagen unserer Rechtschreibung* (Leipzig).
DREWITZ, I., and REUTER, E. (eds.) (1974), *Vernünftiger schreiben/reform der Rechtschreibung* (Frankfurt: Fischer).
DROSDOWSKI, G. (1987), *Rechtschreibung und Rechtschreibreform aus der Sicht des Dudens* (Mannheim: Dudenverlag).
DUDEN, K. (1876), *Die Zukunftsorthographie nach den Vorschlägen der zur Herstellung größerer Einigung in der deutschen Rechtschreibung berufenen Konferenz erläutert und mit Verbesserungsvorschlägen versehen von Konrad Duden* (Leipzig: Bibliographisches Institut).
—— (1880), *Vollständiges Orthographisches Wörterbuch der deutschen Sprache* (= Duden[1]; Leipzig: Bibliographisches Institut).
—— *Vollständiges Orthographisches Wörterbuch der deutschen Sprache mit etymologischen Angaben, kurzen Sacherklärungen und Verdeutschungen der Stichwörter* (= Duden[3]; Leipzig: Bibliographisches Institut).
—— (1902), *Orthographisches Wörterbuch der deutschen Sprache* (= Duden[7]; Leipzig: Bibliographisches Institut).
DUDENREDAKTION (ed.) (1991), *Rechtschreibung der deutschen Sprache Auf der Grundlage der amtlichen Rechtschreibregeln* (= Duden[20]; Mannheim: Dudenverlag).
GLÜCK, H., and SAUER, W. W. (1990), *Gegenwartsdeutsch* (Stuttgart: Metzler).

GOTTSCHED, J. CH. (1748), *Grundlegung einer deutschen Sprachkunst nach den Mustern der besten Schriftsteller des vorigen und jetzigen Jahrhunderts* (Leipzig).

GRIMM, J. (1819), *Vorreden zur Deutschen Grammatik von 1819 und 1822* (Darmstadt).

—— (1847), '*Über das Pedantische in der deutschen Sprache*', in *Kleine Schriften* (Berlin, 1864).

—— (1854), *Vorreden zum deutschen Wörterbuch* (Darmstadt, 1961).

GÜNTHER, H. (1985), 'Probleme beim Verschriften der Muttersprache. Otfrid von Weissenburg und die lingua theotisca', *Zeitschrift für Literaturwissenschaft und Linguistik*, 15/59: 36–54.

KOSOG, O. (1912), *Unsere Rechtschreibung und die Notwendigkeit ihrer gründlichen Reform* (Leipzig).

MÜLLER, K. (1990), '*Schreibe, wie du sprichst*'. *Eine Maxime im Spannungsfeld von Mündlichkeit und Schriftlichkeit* (Frankfurt: Lang).

NERIUS, D. (1975), *Untersuchungen zu einer Reform der deutschen Orthographie* (Berlin: Akademie-Verlag).

—— and SCHARNHORST, I. (eds.) (1980), *Theoretische Probleme der deutschen Orthographie* (Berlin: Akademie-Verlag).

—— et al. (1987), *Deutsche Orthographie* (Leipzig: Bibliographisches Institut).

OTFRID, *Evangelienbuch* (1905), ed. Oskar Erdmann, Altdeutsche Textbibliothek, 49 (Tübingen: Niemeyer).

RAUMER, R. VON (1863), *Gesammelte sprachwissenschaftliche Schriften* (Frankfurt, Erlangen).

Die Rechtschreibung des Deutschen und ihre Neuregelung (1985), ed. Kommission für Rechtschreibfragen des Instituts für deutsche Sprache (Düsseldorf: Schwann).

Regeln für die deutsche Rechtschreibung (1902), im Auftrage des königlich-preußischen Ministeriums der geistlichen, Unterrichts- und Medizinal-Angelegenheiten (Berlin). Most recent ed. Dublin, Zurich, 1969.

SAUER, W. W. (1988), *Der 'Duden', Geschichte und Aktualität eines Volkswörterbuchs* (Stuttgart: Metzler).

SCHRÖTER, CH. E. (1748), *Allzeitfertiger und auf allerley Fälle gerichteter Briefsteller* (Leipzig).

VIERECK, W. (ed.) (1980), *Studien zum Einfluß der englischen Sprache auf das Deutsche*. (Tübingen: Narr).

VOGT, W. (1992), 'Amselmännchengesang', *Frankfurter Allgemeine Zeitung*, 18 Jan. (letter to the editor).

WURZEL, W. U. (1985), *Konrad Duden* (Leipzig: Bibliographisches Institut).

5 Directions of Change in Contemporary German

HELMUT GLÜCK and
WOLFGANG WERNER SAUER

I INTRODUCTION

The German language is currently changing rapidly, perhaps more rapidly than at any other time in its history. Innumerable radio programmes and almost twenty television channels are available in almost every household and extensive use is made of them. Most people own several radios and at least one television set. The sales figures show that the publishing houses are in a state of crisis, while the figures for newspaper sales are also sinking. Listening comes before reading, and speech affects linguistic usage infinitely more than written texts. For centuries, however, standard norms were derived from the written language. Until recently, dictionaries basked in reflected glory by taking their corpus predominantly from literary texts, and illustrating it with passages by distinguished authors of the last two hundred years. There were admittedly a few selected quotations from some of the higher-quality newspapers, but this was merely paying lip service to modern developments, and colloquial language was taboo, even in its spoken form.

This situation has changed slightly, as is shown by newer editions of current monolingual dictionaries (such as the *Deutsches Universalwörterbuch* (1989), published by the Dudenverlag. But how can this fact be reconciled with the existence of a *Wörterbuch der deutschen Umgangssprache* (Dictionary of Colloquial German), boasting almost a thousand pages? This work contains the vocabulary which is used above all in oral communication and which is often at variance with what is considered 'good' German. We do not wish to give the impression that colloquial speech should be the primary yardstick for descriptions of current German usage. To describe current trends on this basis alone would mean rejecting substantial parts of the prevailing norms. We shall merely try to depict certain linguistic changes which indicate trends away from accepted grammatical conventions, and which seem to us to be representative of the ways in which the language is changing.

In our discussion of trends in contemporary German, we shall deal with some of the topics which are normally described in grammars, but our examples will be drawn from observation of German as it is actually used. We do not claim to portray all the changes currently taking place in standard German, merely some of the most significant developments. We have already dealt with this subject, with different emphases, on several occasions (see Glück and Sauer 1985, 1987, 1990) and believe we can now distinguish between the accidental and the systematic in this respect. We shall not, therefore, document each individual example; references come from written texts (fiction and non-fiction), periodicals, newspapers, advertising brochures, etc., while examples drawn from spoken sources (radio and television) are indicated as such.

Our system is not set out in a way comparable to any particular grammar. Grammar writers may complain that what follows is not their concern, or that what is central for them has been omitted: in the space available here we have necessarily adopted a selective approach, and in these circumstances it is impossible to please everyone. Nevertheless, it would be gratifying if writers of future grammars found our discussion stimulating.

2   WORD-FORMATION PATTERNS

2.1   *Nouns*

Processes of truncating polysyllabic words in German are as productive today as they ever were. In most cases, this creates bisyllabic abbreviations in which the first syllable carries meaning while the second functions as a derivative element, indicating an emotional or gender-specific characteristic. The first syllable thus refers back to the first morpheme of the original word, without necessarily reproducing it identically. In this way, semantic references are more likely to be obscured than emphasized. Thus *der Ostdeutsche* (East German) mutates into *Ossi*, *der Westdeutsche* (West German) becomes *Wessi*, and *der Bundesbürger* (federal citizen) becomes *Bundi*. The use of these abbreviations is widespread and they are even to be found in the columns of serious newspapers. *Ossi* and *Wessi* appear in the 1991 Duden dictionary.

The description *Wossi* has only recently been coined, and is a blend of *Wessi* and *Ossi*, used to describe those people who come from the

West, but are not *Besserwessis* (a 'blend' of *Besserwisser*, know-all, and *Wessi*); rather, they concern themselves with the difficulties and problems of the East Germans, and are thus accepted by them. Other words following the same pattern include *Ziggi* (for *Zigarette*), and it even extends to proper names, such as the once popular *Gorbi* (for Gorbachov, the former Soviet leader) and *Ötzi* (for *Ötztaler*, the recently discovered mummified man from the Stone Age). Bisyllabic words can also be abbreviated in this fashion: for example, *Putzi* (for *Putzfrau*, cleaning-woman) or *Touri* (for *Tourist*).

This morphological model has, amongst other things, long been used to produce diminutive forms of forenames: for example, *Siegfried* becomes *Siggi*, *Gabriele* becomes *Gabi*. The *-i* morpheme thus expresses a sense of informality and familiarity, forming what Fleischer (1983: 201) calls 'expressiv-kosende Formen' (expressive, affectionate forms), that are consciously contrasted with the source form. This popular pastime of creating new diminutive forms continues unabated. For several years, for example, manufacturers have been calling the simple match *Zündis* (from *Zündholz*), and a lavish advertising campaign transformed heavily polluting juggernauts into harmless *Brummis* (from *brummen*, to rumble or drone). Peter Tomuscheit (1992: 24) talks of an 'infantilen Trend zum Kosewort' (infantile trend towards terms of affection), going as far as to accuse advertising psychologists of launching a nation-wide *'Verschnullerungskampagne'* (literally 'a campaign to flood the country with dummies/pacifiers (*Schnuller*)', here meaning to reduce everyday language to 'baby-talk').

However, the *-i* has no positive emotional connotations in abbreviations such as *Ami* (for *Amerikaner*), *Nazi* (for *Nationalsozialist*) or the recent *Stasi* (for *Staatssicherheitsdienst*, the secret security force of the former GDR). Bisyllabic abbreviations with specific sex indicators include *Fascho* (for *Faschist*) or *Macha* as the female equivalent to *Macho*. *Reala* and *Realo* are descriptions of female and male members respectively of the pragmatic wing of the German Green Party. Even comparative and superlative forms can be derived from them: for example *Realissima/o*. Monosyllabic abbreviations following the pattern *Prof* (from *Professor*) and *Kat* (from *Katalysator*, catalytic converter) are less common than those with the *-i* ending. A new word in this category is *Rep* (from *Republikaner*, a member of the right-wing Republican Party in Germany), which is now included in the *Duden*.

All of these abbreviations form their plurals, where necessary, with the suffix *-s*. The suffix *-e* is also common in the formation of feminine nouns which are derived principally from verbs. The bisyllabic words formed in this way are of a very informal, colloquial character. For instance, *Leihe* (from *leihen*, to lend) is the name of a car-hire company; *Putze* is an alternative to *Putzi* (from *Putzfrau*, cleaning-woman); and *Denke* (from *denken*, to think) was used in a magazine interview by the head of the German rail service in the context 'die Denke muß sich ändern' (the thinking must change). Formations of this kind have long been a distinctive feature of the Berlin dialect in particular: *Sause* (from *sausen*, to charge, race) for a pub-crawl, *Plätte* (from *plätten*, to iron, press) instead of *Bügeleisen* (for iron), for example, have in turn influenced forms such as *Glotze* (from *glotzen*, to stare), a derogatory term for the television. They are often considered brash and vulgar.

A further type of abbreviation is the contraction of syllables, which has always been a popular way of creating company names, such as *Hertie*, a chain of department stores named after its founder Hermann Tietz. *Azubi* (from *Auszubildende/r*, trainee) has recently gained acceptance as an alternative term for *Lehrling* (apprentice), which is no longer used officially. As these syllable contractions often occur in terms used by state security agencies, they have uncomfortable associations for some people: for example, *Kripo* (for *Kriminalpolizei*, police), *Schupo* (for *Schutzpolizist*, policeman), *Gestapo* (for *Geheime Staatspolizei*, secret state police under National Socialism) or *Stasi* (see above). The term *stino* (from *stinknormal*, utterly/boringly ordinary), used by young people to describe the petit bourgeois, has equally negative connotations.

Formations ending in *-itis* also tend to have a predominantly negative aspect. In standard German the suffix is commonly used in connection with diseases (for example *Gastritis*, *Hepatitis*), and in medical terms indicates some form of inflammation. The words thus formed are often unique, but the model is very productive, as its judgemental undertones are widely understood. For instance, someone who has a manic craving to play the clarinet could be said to have *Clarinetitis*, whilst *Betonitis* (from *Beton*, concrete) is an apt description of the traffic minister's predilection for smothering the country in roads (and the development itself might be called *Autobahnisierung*). The *-isierung* suffix is used in a similar fashion to *-itis*, and also has negative associations. It is used to warn against those trends specified in the root of the word: for example 'Keine Verboutiquisierung unserer Wohnviertel' is how a local paper opposes the proliferation of boutiques in its area. The 'Europäisierung der nationalen Währungen' (Europeanization of national currencies) is also understood in this same negative light.

Even the common *-ismus* suffix, which originally had no negative connotations, is now used mostly in a denigratory fashion. The connotations attached to words such as *Bürokratismus* and *Militarismus* are transferred to new constructions such as *Kohlismus*, analagous to Thatcherism, a policy of redistribution of wealth in favour of the already wealthy. On the other hand, *Genscherismus* is seen as positive, as the West German foreign secretary from 1974 to 1992 was very popular; but then hardly anyone can really explain what Genscherism actually entails.

Some morphemes are intended to create the opposite effect, especially those which are used increasingly in advertising and politics. The pseudo-prefixes *Euro-*, *Öko-*, and *Bio-* invite approval, as do the suffixes *-team*, *-partner*, and *-mat* and those magic words which can be used in either position: *Service*, *City*, and *System*. They may also be combined in various ways: for example *Europartner*, *Ökosystem*, and *Serviceteam*. These forms generally have virtually no denotative meaning, but have connotations ranging from chic to extremely positive. *Biosocken* and *Ökoküchen* (eco-kitchens) are the glorious fetishes of the modern German lifestyle, *Cityhopper* and *Intercitys* its indispensable components. The cashpoint is called a *Bancomat* and a condommachine is the *Condomat* (but, as the *Süddeutsche Zeitung* reported, the products themselves cannot be called *McCondoms*, as McDonalds has banned this particular neologism).

2.2 *Adjectives*

The creation of new adjectives, using both straightforward compounding and certain derivative processes (such as using affixes, or 'free' morphemes used as affixes), is very popular in contemporary German. The root is either an adjective or a participle, and the modifier may be an adjective, a verb, a noun, or an abbreviation. This type of formation has a long history, and already occurred in Gothic: for example *gastigoþs* (hospitable). Examples of this type of compounding, which again occur frequently in advertising, are:

Adjective	+ Adjective	leichtflockig, vollschmeckend (light and fluffy, rich-tasting)
Verb	+ Adjective	pflegeleicht, knitterfrei (easy-care, crease-resistant)
Noun	+ Adjective	hautverträglich, babyzart (easy on the skin, baby-soft)
Abbreviation	+ Adjective	pH-neutral, PVC-beschichtet (pH-neutral, PVC-coated)

Compounds whose first element reinforces and intensifies the second are particularly common: for example *supergut*, *megagut*, and *spitzentoll* (incredibly good). These intensifiers are also found as free morphemes in colloquial speech:

> Das finde ich mega/super/spitze
> (I think that's mega/fantastic/brilliant)

Some of the root words in this category occur so often that they almost assume the character of derivative suffixes (Römer 1980: 47). This group includes examples such as *-fertig* (-ready), *-frisch* (-fresh), and *-sicher* (-proof).

As far as adjectival derivatives are concerned, the productivity of individual suffixes is very variable. Thus the *-sam* suffix has gone entirely out of fashion, with words like *achtsam* (attentive, careful) and *ehrsam* (respectable, honourable) sounding archaic to many ears. The *-bar* suffix, however, although not far removed semantically from *-sam*, is enjoying a boom because it can be used with almost every transitive verb. It is even used nonsensically for the sake of effect: for example, a soft-drinks firm advertised its unbreakable plastic bottle as *unkaputtbar* (unbustable). Two other suffixes, *-haft* and *-lich*, are facing strong competition from *-mäßig*, with well-known forms such as *schulmeister-haft* (schoolmasterly) and *amtlich* (official) being replaced, particularly in spoken German, by new forms. For example, one hears 'amtsmäßig brauch' ich noch einen Stempel' (officially I still need a stamp) or 'sei nicht so schülermäßig' (don't be such a schoolkid). Another common feature of the spoken language is the positioning of adverbs formed with *-mäßig* at the end of a clause, normally together with the particle *so*, in a sense that is similar to (American) English forms with -wise:

> Hamburg ist prima, so stadtmäßig
> (Hamburg is great, as towns go)
> Die Leute müssen noch 'ne Menge lernen, so abfallmäßig
> (People still have a lot to learn, rubbish-wise)
> Ich hab' viel zu tun, (so) unimäßig
> (I've got a lot to do, university-wise)

This construction may also be used at the beginning of clauses:

> Urlaubsmäßig hab' ich Bock auf England
> (I quite fancy England for a holiday)
> Essensmäßig gibt's bei Toni die beste Pizza
> (Foodwise, the best pizzas are at Toni's)

The rampant advance of *-mäßig* is viewed negatively by many Germans, but it is none the less not entirely confined to youth or particularly informal usages.

The *-ig* suffix is one of the most productive adjective suffixes of modern German (Fleischer 1983: 259). The words formed with it are informal and modern-sounding, with examples such as *poppig* (trendy), *fetzig* (mind-blowing), and *flippig* (cool), and are often used in advertising as expressive neologisms: examples are *schokoschmackig* (chocolatey), *kartoffelig* (potatoey), and *pfandig* (used for a returnable bottle, the word is reminiscent of *pfundig*, a once popular word for 'great, fantastic'). Finally, there is a large group which may be regarded as examples of neutral denominalization, for example:

> trendige Blusen (trendy blouses)
> formatiger Mann (a man of stature)

Formations of this sort usually replace expressions using prepositions or relative clauses, showing the tendency towards abbreviation. So, instead of being sold *zum alten Preis* (at the old price) before a price rise, cigarettes are offered as *altpreisig*, and the characteristics of punks, freaks, and hooligans are embodied in the adjectives *punkig*, *freakig*, and *hoolig*.

2.3 Verbs

The creation of new verbs is less common than that of nouns. It occurs overwhelmingly in attempts to integrate words of English origin into German, and not only those which are derived from nouns. Considerable effort may be expended in making these English verbs manageable within the German syntactic system. They follow the same conjugation pattern as weak verbs, and the orthographical form of the English stem is retained. The degree to which these forms are adapted to the German system can be measured by such features as the form of prefixes and participles or personal/temporal inflections.

The model is not new (forms such as *du hast gejazzt* have appeared in dictionaries since the war), but the number of these verb forms has grown dramatically in recent years: *coachen*, *talken*, *leasen*, *joggen*, and *scannen*, for instance, are frequently used and fully conjugable. The 'importation' of some forms may lead to complications, as for example in the case of English verb forms ending in *-le*, which may appear to be similar to the German verbal infix *-el-* (for example *lächeln*, to smile): this sometimes results in duplicate forms such as *recyclen* and *recyceln*, and may introduce uncertainties of inflexion, as in the participles *recycled*, *recycelt*, and *gerecycelt*. In other examples, however, the English structure has disappeared, and the word is fully adapted to the German paradigm, as in *geleast*, *gescannt*, and *gebootet*.

The extent to which words can be integrated appears to depend on how alien the original English orthography is to German. For example, *recyclen* and *designen* are more foreign-looking than *joggen* or *pushen*. *Hübsch designed* (nicely designed) is more common than *hübsch designt*; we have no attested examples of *hübsch gedesignt* or *heftig pushed* (violently pushed): *gepusht* would be normal for the latter. The *Deutsches Universalwörterbuch* (1989) records a Germanized form *puschen* in the sense of 'to drive, set in motion', but it is used less often than the competing form. Other duplications include *surfen/sörfen* and *anturnen/antörnen* (to turn on in the sense of 'their music really turns me on').

Forming verbs by adding prefixes to English root forms always follows the German pattern: for example *anpowern* (to warm up, get going — as of an audience), *aufstylen* (to make more chic), *ausflippen* (to freak out), *reinmoven* (to visit, check out), and *vertrusten* (to form into a trust). English progressive forms ending in -*ing* replace the nominalized form of the German infinitive with some verbs, for example *Recycling, Relaxing, Sharing,* and *Sponsoring*. In some cases, newly formed verbs derived from English replace older verbs (as in *sponsern* instead of *sponsieren*), or provide some difference in meaning from the extant form, for example *promovieren* (to promote, now only used in the sense of achieving a doctorate) and *promoten* (to promote in the sense of 'advertise'). *Faksimilieren* has the same meaning as *faxen* (to make a facsimile), which is now used as an abbreviation of *telefaxen* (to fax) instead of the Germanized *fernkopieren*.

3   THE MORPHOLOGY OF NOUNS

3.1   *Genitive*

The genitive is the case used to indicate dependent relationships within complex nominal expressions, often with possessive or partitive meaning. In some instances the genitive is the object case, that is, verbs or adjectives govern the genitive, for example 'Bernd bedarf eines Helfers' (Bernd needs a helper), 'Eva erinnert sich des Vorfalls' (Eva remembers the incident), 'Annette ist des Lateinischen kundig' (Annette has a good knowledge of Latin), 'Elisabeth war des Lobes voll' (Elizabeth was full of praise). The genitive also indicates adverbial and predicative relationships, for example 'erhobenen Hauptes' (with one's head raised), 'eines schönen Tages' (one fine day), or 'er ist des Todes' (he is doomed). Some prepositions govern the genitive, for example *wegen* (because of), *trotz* (in spite of), *dank* (thanks to). The 'subject genitive' (that is, the expression of a subject in the genitive) is very rare, although it is occasionally demanded by certain idioms or phrasal verbs, as in 'Der Worte sind genug gewechselt' (Enough words have been exchanged).

Peter Braun points out a sociolinguistic peculiarity of the genitive, namely its close relationship with the written language. For that reason 'sind Volkssprache und Mundarten zu keiner Zeit ein Feld für Genitivobjekte gewesen' (everyday speech and dialects have never been an area for genitive objects) (Braun 1987: 111). From that it is easy to conclude that the genitive is not the most popular feature of contemporary German, because the strict writing-based norms of the standard language from the first half of the twentieth century are no longer fully applicable to today's colloquial language.

The decline of the genitive is lamented by the representatives of these strict norms. Ludwig Reiners, the author of several widely distributed works on 'good style' and self-appointed 'specialist in medical diagnoses of the German language' (Braun 1987: 143), believes the genitive is actually dying out (Reiners 1949: 213). This is certainly not the case. The object genitive is undoubtedly in decline, but this can be explained mainly by the fact that the verbs concerned, such as *jemandes harren* (to await someone) and *sich jemandes schämen* (to be ashamed of someone), are obsolescent. In other cases, the genitive forms compete with prepositional objects:

> sich jemandes erinnern/sich erinnern an jemanden
> (to remember someone)

The same applies to some adjectives which are governed by the genitive:

> einer Sache begierig/begierig nach
> (eager for something)
> einer Sache voll/voll von
> (full of something)

Accumulations of genitive attributes in noun groups remain common, particularly in administrative, legal, and academic texts, but expressions like 'im Zuge der Folgen dieser Zeit des Umbruchs . . .' (in the course of the consequences of this period of radical change . . .) are seen as unwieldy and unattractive.

In both written and spoken language, the possessive genitive (in the form of the so-called Saxon genitive) is used in the initial position: for example, 'Hannovers Kassen sind leer' (Hanover's coffers are empty). In second position the use of the possessive genitive alternates with prepositional expressions using *von*: for example, 'Die Kassen Hannovers sind leer'/'die Kassen von Hannover sind leer'; however, the pre-positioned dative/genitive 'der Stadt ihre Kassen sind leer' would be considered decidedly colloquial or dialectal (see Heringer *et al*. 1980: 66 ff.).

The prepositions *wegen* and *trotz*, which can take either the genitive or the dative, are further evidence that the genitive is not disappearing altogether. Which case is used depends on the textual or pragmatic context. With *trotz* the genitive has become the rule (see Schröder 1986: 189), although it originally took the dative. *Wegen*, however, is now used with genitive only in formal texts, such as seminar papers, radio news reports, or official announcements. A sentence like 'wegen eines Bieres brauchen wir uns doch nicht zu streiten' (let's not argue over a beer) is clearly dysfunctional: indeed, if you were to utter it in the pub it might actually cause an argument!

One last observation about the genitive's orthographical realizations: until recently, if she opened a fast-food shop, Gabi would have been quite happy calling it *Gabis Imbiß*. Now it would be *Gabi's Imbiß*. In the new *Länder* in particular, the apostrophe is seen as proof of a modern lifestyle: out of forty newly registered businesses in eastern Germany in 1990, all but two clearly felt it was *de rigueur*. Furthermore, it seems that the adoption of the apostrophe following the *-s*, also taken from English, is even more stylish: for example, *Abramskis' Farbenshop* or *Berlins' nettester Biergarten*. These trends thus run counter to current English practice, where the possessive apostrophe is rapidly disappearing in this context.

3.2 *Accusative*

The accusative is the case of the direct object, the accusative object. It is the basis for classifying verbs as transitive or intransitive. The accusative object in active sentences corresponds to the subject in passive sentences, as long as the verb concerned allows a passive construction: for example 'Maya begrüßt Jenny' (Maya greets Jenny) corresponds to the passive construction 'Jenny wird von Maya begrüßt' (Jenny is greeted by Maya), but 'Bernd bekommt einen Brief' (Bernd receives a letter) cannot become *'Ein Brief wird von Bernd bekommen'. Many verbs and prepositions govern the accusative, sometimes even two accusatives: thus 'Das Spiel in Rostock kostete die Eintracht 1992 die Meisterschaft' (the game in Rostock cost Eintracht the championship in 1992); or in addition to another case, for example the dative (*bieten*, to offer someone something), the genitive (*berauben*, to rob someone of something), or various prepositional cases. In some instances, adjectives used predicatively require the object to be in the accusative: for example 'er ist solche Schwierigkeiten gewohnt' (he's used to such difficulties).

In some contexts, the accusative object reflects or complements the meaning of the verb: for example 'sie kämpfte einen schweren Kampf' (she fought a hard fight). The accusative can also be used as the adverbial case: 'er arbeitete die ganze Nacht' (he worked all night long), or 'das kostet 100 Mark' (that costs 100 Marks). With some verbs, the accusative still competes with the dative object, and duplicate forms are common: for example 'es ekelt mir/mich vor dem Geruch' (the smell of it disgusts me). However, with several verbs, the accusative has become so dominant that the dative has been displaced, and the accusative is now used exclusively: thus 'mich friert/freut/hungert' (I'm freezing/pleased/hungry). In a few instances, the accusative also competes with other forms: 'ich erinnere das Buch' instead of 'ich erinnere mich des Buches' or 'ich erinnere mich an das Buch' (I remember the book).

The accusative is the case least affected by the trend for grammatical features to lose their distinctive form, and it is generally correctly marked. With some indefinite pronouns, it is becoming increasingly common for the ending to be omitted in the accusative: for example 'Der Oberst kennt niemand(en)/jemand(en)' (the colonel knows no one/someone). There are some uncertainties in agreement between elements, with either the marking disappearing in the second marked element, for example

> einen stark illustrier*ter* Bericht erhalten unsere Kunden
> (our customers will receive a well-illustrated report)

or false agreement being established between the two, as in this headline from the *Süddeutsche Zeitung*:

> Den Künstler braucht heute niemanden
> No one needs an artist nowadays)

However, it is worth pointing out that in colloquial speech certain articles (in the sense of the classification in Helbig and Buscha 1989) may be used without an ending: 'hast du mal ein Groschen?' (have you got a penny?), 'Kein Schutt abladen' (no tipping), and 'gib mir mal mein Mantel' (give me my coat, will you?). This practice is sufficiently widespread and systematic not to be dismissed as 'careless pronunciation'.

3.3 *Dative*

The dative case marks indirect objects (dative objects). Many verbs and prepositions govern an obligatory (*begegnen*, *vertrauen*) or an optional object in the dative (*helfen*, *versprechen*), often in addition to the accusative (*bieten, rauben*) and various prepositional objects. In some cases adjectives used predicatively require the dative as the object case: for example 'das sieht ihm ähnlich' (that's just like him) or 'die Dame ist mir nicht bekannt' (I don't know the lady, literally 'the lady is not known to me').

Like the accusative, the dative has no specific case-meaning. There are none the less various special functions, such as the possessive dative, which expresses belonging, scope, or possession: thus 'dem Opa fallen die Zähne aus' (Grandpa's teeth are falling out), 'ihm ist der Vater gestorben' (his father has died), 'ihnen wurde das dritte Kind geboren' (their third child was born); and the *Trägerdativ* (dative objects used in descriptions concerned with the wearing of some form of clothing): for example 'der Tante rutschen die Strümpfe' (auntie's stockings are slipping). The dative also competes with prepositional objects to express the beneficiary of certain actions: 'er holt dem Vater ein Bier' (he fetches his father a beer) instead of 'er holt ein Bier für den Vater'. Optional elements that indicate, for example, the addressee of the action expressed by the verb are represented by the so-called free dative: for example 'der Einserkandidat trägt seinem Professor die Tasche' (the star pupil carries his teacher's case) or 'der Kellner schüttet dem Gast die Suppe auf die Hose' (the waiter spills soup on the customer's trousers).

The decline of the dative -*e* ending, which began in the early nineteenth century, is now almost complete. This ending occurred very often with masculine or neuter nouns and is historically the case marker of the so-called strong declension of these two genders. The -*e* still occurred in polysyllabic words in the middle of the nineteenth century: for example *vor Gerichte* (in/before the court), *im Zusammenhange mit* (in connection with), *beim Abschiede* (on departure). However, although it no longer appears in these contexts, it is sometimes still used with monosyllabic words, mainly for stylistic reasons: thus 'meinem Kinde soll es besser gehen' (my child is said to be getting better). On the whole, though, the use of the dative -*e* is limited to certain fixed expressions, where it either always occurs (as in *im Falle eines Falles*, 'if it comes to it'; or *zu Hause*, 'at home') or occurs more frequently than the form without the ending (as in *im weitesten Sinn(e) des Wortes*, 'in the broadest sense of the word'; or *zum Wohl(e)*, 'cheers').

Like the genitive object, the dative object faces competition from prepositional objects: for example *jemandem schreiben* versus *an jemanden schreiben* (to write to someone). This tendency is sometimes reinforced when the verb itself is obsolescent: for example *jemandem entfliehen* becomes *vor jemandem (ent)fliehen* (to run away from someone). However, the number of verbs which take an obligatory dative object is larger and more stable than those which require the genitive object. Peter Braun (1987: 113) lists some 220 verbs which either take a dative object exclusively or can take one in addition to an accusative or prepositional object: for example 'Die Maid begegnet dem Wolf. Der Wolf bietet dem Mädchen seine Freundschaft an. Rotkäppchen verhilft dem Wolf zu einer Mahlzeit' (The girl meets the wolf. The wolf offers her his friendship. Little Red Riding Hood helps him to a meal). The possessive dative ('dem Wolf seine Freundin', the wolf's friend) is often used in colloquial speech, but according to Jung (1984: 271) should be avoided in the written language.

3.4 *Plurals*

The formation of plurals is the 'poor relation' of grammars and guide-books on style (see Glück and Sauer 1990: 60 ff.). Admittedly, the former do cover the regular patterns of plural formation of nouns as well as some doublets, such as *der Strauß > die Strauße* (ostriches)/ *Sträuße* (bunches). Both, however, ignore the uncertainties and viol-ations of the normal forms, which are so extensive that they even have an impact on the Duden orthographical dictionary, which has the headword *Visa* (the plural of *Visum*). This suggests that as a result of English influence the form *Visa* has actually become established in popular usage as a singular, and there are many other examples of this: for example 'du mußt das Errata korrigieren' (you must correct the error; *Errata* is in fact the plural form of *Erratum*), 'die Rarissima ist . . .' (the extremely rare item is . . . ; again, *Rarissima* is actually the plural form of *Rarissimum*). Confusion between the plural -*a* suffix of Latin neuter nouns with the -*a* suffix of the Latin feminine singular is probably responsible for the incorrect gender assignment in these examples. The plural of these incorrectly constructed words is formed by adding an -*s* (thus *die Visas*).

With other loanwords there are many duplicate forms to be found where the derivative in the source language does not correspond to the German paradigm. In this case, German falls back on more familiar endings:

Atlas   Atlanten/Atlasse
Kaktus   Kakteen/Kaktusse

The Graeco-Roman plural in these and the following examples is seen by some people as an indication of superior education:

Thema   Themata/Themen
Komma   Kommata/Kommas

However, this ignores the functional aspect: the *-ta* forms are correct in technical registers, but they seem somewhat out of place in everyday texts. None the less, forms such as *die Genusse und Modusse des Deutschen* (the genders and moods of the German language) are scarcely more acceptable, despite being created by analogy to other words, than are hypercorrections (*Boni* instead of *Bonusse*). Uncertainties that occur, for example, with 'exotic forms' such as Italian food and drink, which are now no longer confined to Italian restaurants, are covered by the trusty *-s* plural: even the Duden records *Pizzas*, *Espressos*, and *Cappuccinos*. Indeed, Duden editors have become increasingly liberal with the use of these *-s* plurals in recent years (Glück and Sauer 1990: 62). Some South Germans complain about the growing popularity of this form, seen as 'North German', but the process cannot be stopped. The plural *-s* is the rule with the many abbreviated words in contemporary German: for example *Promis/Prominente* (VIPs), *Nudos/Nudisten*, *Profs/Professoren* all follow the old paradigms established by words like *Pullis/Pullover*, *Autos/Automobile*, and *Loks/Lokomotiven*. The normally unmarked plurals of the many new words ending in *-er* (such as *Computer*, *Composer*, and *Scanner*) are sometimes given the *-s* ending, again because of influence from English.

Those plural forms which we have elsewhere ascribed to a *Bewegungssprache* (difficult to translate as it is intentionally ambiguous, referring both to 'the movement' (of the left) and to 'the emotions'; Glück and Sauer 1990: 64) are still popular: for example 'ich habe Ängste (literally, 'I have anxieties'; the normal expression using the singular 'ich habe Angst' means 'I am afraid'), 'Boris muß mentale Widerstände überwinden' (Boris has to overcome mental obstacles, literally 'resistances'). New examples of this include: 'Literatur ist ein Ensemble vergangener Zukünfte' (literature is an ensemble of past futures) and 'X—ein Paradies von Einkaufswelten' (X—a paradise of shopping worlds).

4   COMPARISON OF ADJECTIVES

Comparative forms of adjectives are changing visibly, with the so-called regular form of the comparative and superlative ending in *-er* and *-st* respectively being applied more generally. In this way semantically nonsensical comparatives and superlatives are created (e.g. *das weißeste Weiß*, the whitest white), words that in themselves express a superlative are made 'even more superlative' (e.g. *die optimalste Lösung*, the most optimal solution), and English adjectives become declinable (e.g. *der softere John Major*). Other adjectives previously classified as indeclinable may also now be declined and have comparative forms (e.g. *der superste Held*, the most super hero). Conversely, comparatives are sometimes formed for the sake of emphasis by using the particle *mehr*: thus *mehr leicht als andere* (lighter than others) instead of *leichter als andere*. As this example shows, some neologisms and adjectives derived from English follow this pattern even where the English itself would use an *-er* comparative.

Other recent trends include the use of pseudo-prefixes to give adjectives a superlative colouring: *megagut, superbillig, spitzentoll* (mega good, super cheap, incredibly fantastic); and the use of adverbs as 'comparative particles' to indicate degrees of intensity: *total gut, echt gut* (really good), *durchaus erfreulich* (entirely gratifying). The pattern itself is old (cf. *recht, sehr*), but the number of comparative particles is increasing.

5 SYNTAX

There are three positions for the finite verb in German sentences. The possible sentence patterns depend to a large extent on the position of the finite verb, and its displacement may alter the mode of the sentence, as in:

'Peter lernt Englisch'/'Lernt Peter Englisch?'
(Peter is learning English/Is Peter learning English?)

where the inversion creates a question, or it may lead to constructions which without hesitation would be deemed totally incorrect by German speakers: for example *'Peter Englisch lernt'. The finite verb can only appear in the first or second position in most main or dependent clauses. An exception is the subordinate clause introduced by certain conjunctions, where the finite verb moves to the end.

The verb occurs most commonly in second position, as in 'normal' main clauses, in yes/no questions requiring confirmation (e.g. 'Peter lernt doch Englisch?', Peter's learning English, isn't he?), in complementary questions ('Was lernt Peter?', what is Peter learning?), and in subordinate clauses which are not introduced by a conjunction and function either as a subject or as an object ('Ich glaube, Peter lernt Englisch', I think Peter is learning English). The finite verb occurs in first position in all other yes/no questions, imperatives, postpositional main clauses, optative clauses not introduced by a conjunction (e.g. 'Lernte Peter doch Englisch!', if only Peter learned English!; the particle *doch* is obligatory here), and in conditional or concessive clauses not introduced by a conjunction ('Lernt Peter Englisch, darf er nach Neuseeland fahren', if Peter learns English, he can go to New Zealand).

In all the above cases the position of the verb is strictly adhered to. The final position of the finite verb in a subordinate clause introduced by a conjunction is also rigidly adhered to in standard written German. However, for many years now there have been signs that this practice is changing. In some subordinate clauses introduced by a conjunction the finite verb appears in the same place as it would in a main clause, that is, in the second position: 'Peter hat keine Zeit, weil er *lernt* Englisch' (Peter has no time because he's learning English). This change has spread throughout the spoken language, and is now heard increasingly even in situations where the standard language might be expected, such as in radio and television broadcasts. It occurs mainly after the conjunction *weil*, but can also be observed after *obwohl* (although) and *während* (while). Recently, this change in verb position has also begun to affect other conjunctions, and the verb may now occasionally be found in the second position after *falls* (in case), *ob* (whether), *sobald* (as soon as), and *wenn* (when, if).

Actual examples of this subordinate clause pattern in print are largely restricted to the reproduction of speech, especially in advertisements. A detailed explanation of this phenomenon may be found in Ulrike Gaumann's study 'Weil die machen jetzt bald zu' (1983). Her comment that both possibilities — finite verb in second or final position in the subordinate clause — were possible until the sixteenth century does not entirely stand up under close scrutiny. An examination of three sixteenth-century texts (*Historia von D. Johan Fausten* of 1587, the *Lalebuch* of 1597, and the *Fortunatus* of 1509) shows that clauses using *weil* occur only very rarely anyway, and that in subordinate clauses introduced by a conjunction the traditional final position of the finite verb is kept to throughout. Only Hans Sachs uses the conjunction *weil* with a causal meaning. Thus in *Ein Kampfgespräch* (1532) both 'weil Ihr Euch bübisch stellt' and 'weil Ihr seid bübisch' appear. In the latter instance the position of the verb may be determined by the demands of metre, but other examples occur in Hans Sachs.

Peter Eisenberg (1989) treats these 'incorrect' *weil* clauses as a major problem for any attempt to establish fixed linguistic norms, stressing the discrepancy between the concept of a norm and structural descriptions of actual utterances. His explanation that the speaker uses *weil* 'dann, wenn er eine Begründung eher zögerlich vorbringt oder sie gar nicht erst sucht, so daß nach *weil* leicht eine Pause entsteht' (when he expresses a reason only tentatively or is not even looking for one, so that it is quite likely that a pause will follow) is only plausible at first sight. This argument cannot be extended to the other conjunctions mentioned above, and must therefore be modified. Speakers tend to introduce a pause if they are about to begin a complex subordinate clause construction in which they wish to express something which is not closely related to the content of the main clause. However, there remains the question of why this variation in the position of the verb has not occurred after conjunctions such as *bevor* (before), *nachdem* (after), and *soweit* (as far as). Other grammars either do not deal with this change (Duden, Erben, Helbig-Buscha) or simply dismiss it as incorrect (Engel 1988: 730). Should grammarians not instead acknowledge that different constructions exist in written and spoken German, and accept their use in appropriate contexts? (For a detailed discussion of *weil* clauses in functional and pragmatic terms, see Günthner 1993 and Schlobinski 1992: 315–44.)

Violations of the norm will always occur as long as the standard form of the language is based on the written language. The sentence unit so beloved of grammars is, whatever the descriptive model may be, still a structure that primarily applies only to the written language. Spoken forms are characterized by what the grammarians term anacoluthon or ellipsis. The 'sins' of those who use the language horrify prescriptive grammarians, whose maxim is 'speak as we write'. We prefer to take Fritz Mauthner's view:

Man sollte nie vergessen, daß die Sprache nicht der Grammatiker wegen da ist. Das scheinen aber die Grammatiker zu glauben, trotzdem nicht einmal die bescheidene Umkehrung berechtigt wäre. (1913: 208)

(We should never forget that language is not there for the grammarians. They, however, seem to think that it is, despite the fact that not even the converse would be entirely justified.)

Mauthner's *Beiträge zu einer Kritik der Sprache* (Contributions to a Critique of Language) offers a wealth of stimuli to the non-normative but nevertheless judgemental writers of more modern grammars. As Mauthner (1913: 207) also says: 'Es ist gar nicht auszudenken, wie langweilig eine vollständige Sprache nach dem Herzen der Grammatiker wäre' (It is almost inconceivable how boring a complete language designed to suit the grammarians would be). He spells out what the grammarians think

wenn ich in der Kneipe auf mein Glas klopfe, anstatt zu sagen 'Ein Bier'. Sage ich aber ausdrücklich 'Ein Bier', so nennt das der Grammatiker wirklich eine Ellipse. (1913: 207)

(if I knock on my glass in the pub, instead of saying 'a beer'. If I actually explicitly ask for 'a beer', then the grammarians would say that really is an ellipsis.)

How his conception of order can be fulfilled can be gleaned from the rest of Mauthner's treatise. The fundamental lesson is that if you are really thirsty, you must recognize the difference between pragmatic and grammatical utterances.

6 CONCLUSIONS

We would like to conclude our exposition of the changing grammatical norms in contemporary German with a point which may be generalized as 'the influence of English'. There are two distinct strands to this, one of which is based more on non-linguistic reasons (gender-marking), while the other is more an internal linguistic matter (borrowing or adaptation). Both are hardly dealt with in current grammars as they are considered marginal to the subjects with which such works are concerned.

We have dealt at length with gender-marking in 'Welfengarten EINS' (1990). The issue here is the problem of how to indicate natural and grammatical gender. There is considerable popular demand for some consistency between these two disparate entities. This trend started in the USA and reached Germany several years ago (see also the chapters by Sauer and Glück and by Hellinger in this volume). The practical consequences have been most obvious in the designation of human nouns, where the unmarked form, indicating both men and women, is disappearing in many fields. Until recently, *der Grammatiker* was used generically for both female and male authors of grammars. If this chapter were written in German, we would in fact use this form intentionally, as none of the grammars we have mentioned had a female editor. However, even if there had been authors of both sexes, we would not have chosen the popular form *VerfasserInnen*. The use of this economical device (using the feminine plural form but with a capital I at the beginning of the suffix to mark it as a special form) may be convenient in writing, but in the spoken language it obviously has to be dissolved into *Verfasserinnen und Verfasser*. The bakery which advertises for a 'nette Verkäuferin für unser Team' (nice sales assistant — the form is marked as feminine — for our team) does not want to employ a male assistant and can thus omit the capital 'I'. Even if a job is not advertised gender-specifically, companies rarely choose to use the form with the capital 'I', usually preferring to use forms with parentheses or a stroke:

> Wir suchen ein(e) Ingenieur(in)
> Wir suchen ein/e Ingenieur/in
> (We are looking for an engineer)

However, in both of these alternatives the masculine form of the article is in the wrong case: it should be *einen*, but the only way to incorporate the correct form (other than by using the passive) would be to use a more long-winded formulation such as:

Wir suchen eine/einen Ingenieurin/Ingenieur

None of these versions is particularly easy to read, and all of these solutions to the problem of explicit reference become rather unwieldy in longer texts, whether the nouns themselves are repeated or they are replaced by pronouns: for example 'Ihr/ihm wird . . . geboten' (she/he will be offered . . .), 'der Nachweis ihres/seines Abschlusses' (evidence of her/his qualifications).

The introduction of the neuter form as a solution to this problem (for example *das Student* as a means of combining *die Studentin* (fem.) and *der Student* (masc.) in one form) has found little favour outside the works of feminist authors (see especially Pusch 1984: 46–68). The baker round the corner persists obstinately with the masculine noun *der Lehrling* (apprentice) even if the person concerned is actually female. Universities and left-wing/Green ministers prefer, even in official documents, to use clearly unmarked terms deriving from participles such as *die Studierenden/Lehrenden/Gelehrten* (students/teachers/scholars). Official forms often have two alternatives, of which the one that does not apply is to be crossed out, as in English with Mr/Mrs/Ms.

This desire to make gender-marking explicit is significantly affected by social and age factors. Older graduates tend to do so consistently, whilst people in 'practical' occupations tend to show as little interest in the issue as 20-year olds. The same is true of *frau*, the 'feminine alternative' to the indefinite pronoun *man* (one); the alternation between *man* and *frau* is nowadays more likely to be found in advertisements for large banks than in young students' seminar papers.

These students are much more inclined to colour their language by borrowing or adapting words and patterns from English. English word formation patterns and inflectional forms derived from English, as described above (especially in Section 2) are found most frequently in the language of younger people: for example 'Welche sweet mouse kann Basic von Unix unterscheiden?' (which sweet mouse can distinguish between Basic and Unix?) asked a 'Lonely Heart (18)' in a personal column, and 'genug gesingelt' (single for long enough) appeared on the same page.

Technical terminology is particularly affected by the use of loan words. Advertising copy-writers, who seem to shrink from hardly any verbal barbarisms, are immensely influential and have produced a multitude of English, pseudo-English, and hybrid descriptions of consumer goods. Many women wear *bodies, leggins, leisure-shirts,* and *french knickers* in *Trendcolors,* many cyclists have become *Mountainbike-Piloten,* riding a *Checker Pig* or a *Mountain-Goat* with *Upside-down-Teleskopgabeln* and *Aerospace Rahmendämpfung* and wearing an *Airtech Radhelm* on their heads (these cycling terms are taken from the *Streiflicht* column of the *Süddeutsche Zeitung* of 11 June 1991). Electrical shops advertise *CD-Players, Tuner, Hifi-Midi-Systeme, Super-VHS-Video-Movies,* and *Portable-Komponenten-Systeme.* Occasionally, daily newspapers even publish glossaries to help their readers understand editorial articles; the *Frankfurter Allgemeine Zeitung,* for example, compiled a vocabulary for *Terminmarktgeschäfte* (26 January 1990), so that one could find out what expressions such as *Arbitrage, at the market, at the money, in the money, out of the money, Basistitel, call, future, hedge, margin, Optionsfrist, put, switching, Volatilität,* and *Zeitwert* meant in this context.

These latter phenomena will inevitably have an effect on grammatical norms. And apart from individual matters of personal judgement, these norms are changeable. Mauthner realized eighty years ago:

daß die Worte oder Sprachbewegung sich zwar vom Menschen auf Menschen übertrugen, nicht durch Fortpflanzung, sondern durch Nachahmung, und daß eine konservative Tendenz vorhanden ist, daß jedoch neben dieser Tatsache ... das Anpassungsvermögen der Sprachen, das heißt die Willkür des Menschen, seine Bewegungen zu ändern, unbegrenzt ist. (Mauthner 1912: 113)

(that words or linguistic changes are passed on from person to person not by reproduction but rather by imitation, and that there is a tendency to be conservative, but that despite this fact ... the adaptability of languages, that is, the arbitrary tendency of human beings to change their habits, is unlimited.)

Perhaps it is only the grammars that still have to change?

Further Reading

Braun (1987)
Eisenberg (1989)
Engel (1988)
Glück and Sauer (1990)
Keller (1990)
Mauthner (1913)

References

BASLER, O. (n.d.), *Grammatik der deutschen Sprache* (Geneva, New York).
BRAUN, P. (1987), *Tendenzen in der deutschen Gegenwartssprache*, 2nd edn. (Stuttgart: Kohlhammer).
BÜNTING, K. D., and ADER, D. (1991), *Grammatik auf einen Blick* (Chur: Isis).
DUDEN (1973), *Grammatik der deutschen Gegenwartssprache* (Mannheim: Dudenverlag).
—— (1989), *Deutsches Universalwörterbuch*, 2nd edn. (Mannheim: Dudenverlag).
—— (1991), *Rechtschreibung der deutschen Sprache auf der Grundlage der amtlichen Rechtschreibregeln* (Mannheim: Dudenverlag).
EISENBERG, P. (1989), *Grundriß der deutschen Grammatik*, 2nd edn. (Stuttgart: Metzler).
ENGEL, U. (1988), *Deutsche Grammatik* (Heidelberg: Julius Groos).
FLEISCHER, W. (1983), *Wortbildung der deutschen Gegenwartssprache*, 5th edn. (Leipzig: Bibliographisches Institut).
GAISER, K. (1950), 'Wieviel Grammatik braucht der Mensch?', in Rötzer (1973).
GAUMANN, U. (1983), *'Weil die machen jetzt bald zu'. Angabe- und Junktivsatz in der deutschen Gegenwartssprache* (Göppingen: Kümmerle).
GLÜCK, H., and SAUER, W. W. (1985), 'La crise de l'allemand', in Maurais (1985), 219–79.
—— —— (1987), 'Neue Wortbildungsmuster im Deutschen als Problem der Auslandsgermanistik', in Radwan (1987), 73–103.
—— —— (1990), *Gegenwartsdeutsch* (Stuttgart: Metzler).
Grundzüge einer deutschen Grammatik (1981), ed. Autorenkollektiv (Berlin: Akademie-Verlag).
GÜNTHNER, S. (1993), ' "... Weil — man kann es ja wissenschaftlich untersuchen" — Diskurspragmatische Aspekte der Wortstellung in WEIL-Sätzen', *Linguistische Berichte*, 143: 37–59.
HELBIG, G., and BUSCHA, J. (1989), *Deutsche Grammatik. Ein Handbuch für den Ausländerunterricht*, 12th edn. (Leipzig: Verlag Enzyklopädie).
HERINGER, H. J., STRECKER, B., and WIMMER, R. (1980), *Syntax. Fragen—Lösungen—Alternativen* (Munich: UTB Verlag).
JUNG, W. (1984), *Grammatik der deutschen Sprache*, rev. G. Starke (Leipzig: Bibliographisches Institut).
KELLER, R. (1990), *Sprachwandel. Von der unsichtbaren Hand in der Sprache* (Tübingen: Francke).
KÜPPER, H. (1990), *Wörterbuch der deutschen Umgangssprache* (Stuttgart: Klett).
MAURAIS, J. (ed.) (1985) *La Crise des langues* (Quebec, Paris: Robert).
MAUTHNER, F. (1912), *Zur Sprachwissenschaft* (= *Beiträge zu einer Kritik der Sprache*, ii; Leipzig).
—— (1913), *Zur Grammatik und Logik* (= *Beiträge zu einer Kritik der Sprache*, iii; Leipzig).

Pusch, L. F. (1984), *Das Deutsche als Männersprache. Aufsätze und Glossen zur feministischen Linguistik* (Frankfurt: Suhrkamp).

Radwan, K. (ed.) (1987), *Kairoer Germanistische Studien*, ii (Cairo).

Reiners, L. (1949), *Stilkunst. Ein Lehrbuch deutscher Prosa*, 2nd edn. (Munich: Biederstein).

Römer, R. (1980), *Die Sprache der Anzeigenwerbung*, 6th edn. (Düsseldorf: Schwann).

Rötzer, H. G. (1973), *Zur Didaktik der deutschen Grammatik* (Darmstadt: Wissenschaftliche Buchgesellschaft).

Schlobinski, P. (1992), *Funktionale Grammatik und Sprachbeschreibung* (Opladen: Westdeutscher Verlag).

Schröder, J. (1986), *Lexikon deutscher Präpositionen* (Leipzig: Verlag Enzyklopädie).

Tomuscheit, P. (1992), 'Wenn der Gorbi mit dem Mölli . . . ,' *Die Welt*, 9 May.

Wahrig, G. (1986), *Deutsches Wörterbuch. Mit einem 'Lexikon der deutschen Sprachlehre'* (Gütersloh: Bertelsmann).

6 After the Wall: Social Change and Linguistic Variation in Berlin

HELMUT SCHÖNFELD and
PETER SCHLOBINSKI

I INTRODUCTION

Many of the features of sociolinguistic variation discussed in this chapter apply more or less generally in the eastern and western parts of Germany. However, we have chosen to focus on Berlin, as it was for so long the symbol of division (a kind of microcosm of the political situation in post-war Germany) and after the fall of the Wall it was where contact was most immediately (and painfully) re-established, so that both contrasts and emerging changes are most clearly to be seen here. This divided speech community is now reunified: what does this mean for language development in the new capital city and for the people who for over a generation lived in two adjacent but increasingly distinct communicative worlds?

The vernacular spoken in Berlin is certainly the most thoroughly investigated urban vernacular in Germany and reflects like no other the variety of different developments in East and West. The popularity of 'Berlinish' is evident not only from the numerous academic publications on the subject but also from the fact that the major weekly newspaper *Die Zeit* recently published a full-page article on it (Zimmer 1992). In fact, the academic interest in Berlinish goes back a long way: the work of Agathe Lasch (1928) was fundamental in pointing the way for further research.

In more recent work regional and social variations of Berlinish have been investigated (see especially Schlobinski 1987, Dittmar, Schlobinski, and Wachs 1986, Dittmar and Schlobinski 1988*a*, Schönfeld 1986*a*, 1986*b*, 1989*a*, and 1989*b*, and Johnson 1991). The division of the city and the linguistic differences this gave rise to were of particular interest for social dialectology. As early as 1945 strong trends in the use of the different variations could be observed and, cemented by the construction of the Wall, these finally resulted in differing evaluations of the dialect and its use.

While in West Berlin Berlinish developed into a sociolect which was stigmatized by the middle and upper classes, the development in East Berlin was quite different. There Berlinish developed into a regional variety which spread rapidly in the direction of Brandenburg and eventually pushed Low German completely out of the picture. But interestingly Berlinish was seen as positive in East Berlin. It was to be heard on East German radio, and even teachers and politicians *berlinerten*. For many East Berliners, Berlinish developed a double function: as a familiar language it gave them a sense of identity, and it also distinguished them from the Saxons, whose language was identified with functionaries and political parties.

The diverse developments of Berlinish must be seen in the context of the emergence of two very dissimilar societies and the general effects of this on the German language. The removal of two disparate societies resulting from the unification of Germany therefore also brought these divergent linguistic processes to a halt, channelling them into a one-way street. The dominant system defines not only the social and political rules but also the linguistic and communicative values and norms which the new citizens have to abide by. The catchphrase *Besserwessi* (see Glück and Sauer, this volume) aptly encapsulates a whole series of problems and prejudices in the relationships between East and West Germans, which is obvious to those who follow the daily newspapers. We hope to clarify in this chapter what it actually means to subject one society, with its own linguistic forms and rules, to another.

The specific vocabulary of Berlinish has been covered in dictionaries since the end of the nineteenth century: from the glossary by Trachsel (1873) and *Der richtige Berliner* by Meyer (first edition 1878, tenth edition 1965) to the *Berliner Wörterbuch* for Berlin (West) by Schlobinski (1986) and the *Berliner Wörter und Wendungen* for Berlin (East) by Wiese (1987); moreover, the language's peculiarities in word formation and syntax have also been described, for example in *Der richtige Berliner* (Meyer 1878). In 1928 Agatha Lasch published a comprehensive investigation of the origins and development of Berlinish, concentrating on a structural inventory but also dealing with historical and sociolinguistic questions.

Following a lull in the post-war years, West German linguists again turned their attention to Berlinish in the 1980s. In East Berlin, however, a group of sociolinguists at the Akademie der Wissenschaften had been working on the subject since the end of the 1960s. Many of these linguists felt strong personal associations with the dialect and much of the early work was firmly in the dialectological tradition (see also Dittmar, in this volume). Their research focused on the heterogeneity of the language inventory (such as speech varieties, different levels in the dialect/standard continuum), the social distribution of speech varieties (in terms of the knowledge and use of the dialect), and the social evaluation of speech varieties. They also emphasized the importance of studying speech varieties in specific communicative situations and the meaning of linguistic norms for the individual speaker. Not all of these research areas were covered in the same depth. For example, group-specific speech patterns (such as *Jugendsprache*) and the linguistic integration of outsiders into the Berlin speech community were given pride of place, and the influence of Berlinish on neighbouring areas was also covered in detail. For obvious reasons, their research on the city as a whole was almost entirely limited to the linguistic situation in Berlin in the nineteenth and early twentieth centuries: from 1961 West Berlin was no longer accessible to East German academics.

In West Berlin a project on the so-called Berlin Urban Vernacular was carried out between 1982 and 1984. The main results are published in Dittmar and Schlobinski (1988*b*: 1–144), and a summary is given in Barbour and Stevenson (1990: 112–25). Active co-operation and common research projects between research groups in East and West finally became possible in 1990. The prime objective of this new collaboration was to compare existing results and to analyse them from a unified perspective, and specific aspects have been identified for further analysis. The key questions now are what differences there are between East and West Berlin, and what social meanings these differences have (first results are given in Schlobinski and Schönfeld 1993).

2   THE LINGUISTIC SITUATION IN BERLIN BEFORE 1945

Like all big cities Berlin has not developed as a uniform entity in economic, social, or indeed linguistic terms. This is especially true of the knowledge and use of the different language varieties and their social evaluations. Berlin expanded through the coalescence of many different settlements. The city of Greater Berlin came into being in 1920 as a result of the fusion of seven independent towns and fifty nine local districts; this has been called 'the unification of Berlin'. There are very considerable differences between these areas in their economic and social structures and also in their regional associations, and many of them have become home to numerous immigrants.

These developments had a considerable impact on the existing language varieties used in the Berlin area. The varieties of the localities

which became part of greater Berlin (around 1920) have been investigated by the surveys of German linguistic atlases. The survey conducted in 1880 comprised forty sentences to be translated or converted
into the local dialect. In thirty-one areas of what later became Greater
Berlin the sentences were converted into a Low German dialect, in
eighteen areas into Berlinish, in four areas into a mixture of colloquial
speech and Low German dialect, and in two areas into a dialect that
originates in the Palatinate. Low German dialects are linguistically
quite distinct from the Berlin vernacular. Berlinish, a mixture of Low
German dialect, High German written language, and elements of
Upper Saxon vernacular, developed in the sixteenth century in 'Old
Berlin' and appeared in all social groups (although not immediately
and not all to the same extent) in the ninteenth and twentieth centuries. As a result, the Low German dialect survived in some areas at
least to some extent. In 1960 Helmut Schönfeld was still able to detect
individual speakers of this dialect in outer areas of Berlin, but this was
no longer the case in the 1980s.

Since the end of the eighteenth century Berlinish has been subject
to a wide variety of social and situational influences, and in recent
times these have become more pronounced. By the 1940s Berlinish
was characterized by considerable internal variation and strong interference from other varieties, and it embraced a number of distinct
'levels'. The level furthest from the standard variety shows to this day
the linguistic structure of a *Halbmundart* (semi-dialect), while the level
that is closest to the standard variety does not include all the most
familiar elements of Berlinish. Some of the most striking elements of
Berlinish (which have very different social variations) are:

[e:] instead of /ai/ (Middle High German /ei/, as in *keen* for *kein*)
[o:] instead of /au/ (MHG /ou/, as in *ooch* for *auch*)
[u] instead of /au/ (restricted to *uff* for *auf*)
[p] instead of /pf/ (as in *Appel* for *Apfel*)
[k] instead of /x/ and /ç/ (as in *ick* for *ich*, or *-ken* for *-chen*)
[t] instead of /s/ (as in *wat* for *was*, *dat* for *das*)
[j] instead of /g/ (as in *jut* for *gut*, *liejen* for *liegen*)
[ɣ] instead of /i/ (as in *ümmer* for *immer*, *Mülch* for *Milch*)

Other common features are the 'post-stress -*e*' (as in *ville* for *viel* and
icke for *ich*) and the uniform case for dative and accusative (e.g. *ma*
instead of *mir* and *mich*; see Schlobinski 1988). Some phonological
characteristics which were very common in 1940 are now considered
old-fashioned in both East and West Berlin or are no longer used:

• unrounding of /y:, ø:/ to [i:, e:] (as in *miede* for *müde*, *scheen* for
  schön)

- [i] instead of /ai/ (in the prefix *ein-*: e.g. *infalln* for *einfallen*)
- [j] instead of /g/ before consonants (as in *jroß* for *groß*)

The following short text exemplifies differences between the standard variety, Berlinish, and Low German (Central Brandenburg) dialects:

Standard: Ich weiß nicht, was ich dazu sagen soll. Das mache ich mir allein. Du hast da nichts zu suchen.

Berlinish: Ick weeß nich, wat ick daßu saren soll. Dit mach ick mir alleene. Du hast da nüscht ßu suchen.

Low German dialect: Ick wett nich, wat ick doa sall tue seien. Dät moak ick mei alleene. Dau hes doa nüsch tue süekene.

An interesting aspect of the internal differentiation of the Berlin vernacular and its major variants is the variation within the different 'communication communities' in the city. In the 1940s there were local differences in terms of the inventory and structure of Berlinish and its usage and evaluation between inhabitants of individual districts of Berlin and even within the old town itself. Each individual quarter of the old town had its special characteristics: for example the *Mietska-sernenviertel*, the *Bankenviertel*, the *Villenviertel*, and the Jewish quarter. Many parts of the city and dormitory areas had a highly heterogeneous social structure, indeed in some cases there were even differences within one building between the part of the house facing the street and that facing on to the backyard. These social differences frequently affected the language, and this has often been highlighted in the local literature. But even in the 1960s and 1980s older inhabitants of East Berlin reported that there had been differences in the structure and usage of the standard within the city and that two varieties of Berlinish had existed even before 1945. Indeed they often claimed that in the 1930s you could recognize which area of the city someone came from merely by their speech.

The stronger position of Berlinish as against that of the Low German dialect in other city areas and a greater degree of uniformity in the Berlin vernacular in the whole of the city resulted from better transport conditions, the general process of industrialization, the employment of many Berliners throughout the city, and through the work of schools, mass media, and political organizations between 1933 and 1945. Considerable changes in the relationship between the various language forms occurred during the war years. The distribution of inhabitants according to social class in particular areas and their relative isolation was greatly reduced: many men became soldiers, many women worked in industry, and large numbers of women, children, and older people were evacuated from the city. As a result, they came into prolonged contact with people who did not speak Berlinish. Whole areas of the city were severely damaged, and some were completely devastated. An endless flood of refugees, many of whom stayed for a long time, introduced other regional languages to the city.

3   THE DEVELOPMENT OF THE LANGUAGE BETWEEN 1945 AND 1989 IN A DIVIDED CITY

After the Second World War the same linguistic starting-point existed in individual areas of Berlin. From 11 July 1945 Berlin was divided into four sectors and put under the control of the Allies. Social, economic, and political developments in the Soviet sector were very different from those in the Western sectors. The Soviet sector was linked to the Soviet occupied zone and the Western sectors to those zones controlled by the Western powers. The separation of West Berlin resulted from a long process affecting all aspects of public and private life.

In 1948 the city was divided administratively under two city councils, and the introduction of different currencies had serious consequences for the economy and the population. With the founding of the Federal Republic and the GDR in the autumn of 1949, the division between the two parts of the city was deepened. The GDR called Berlin its capital city, while West Berlin became more strongly integrated into the legal, financial, and economic system of the Federal Republic. Many in West Berlin sought to reconstruct the city as a leading cultural metropolis and wished to make it a 'window on to the free world'. Despite increasing difficulties, thousands of Berliners were able to travel daily between the two parts of the city: until the construction of the Wall around 53,000 people from the GDR and East Berlin worked in West Berlin and around 12,000 from West Berlin worked in the East of the city. Almost a quarter of the students at West Berlin universities came from the East. However, as the years went by West Berliners lost interest in visiting the Eastern part, while throughout the 1950s numerous refugees from the East arrived in West Berlin every day.

On 13 August 1961 work began on blocking the borders to West Berlin. From then on it was only in exceptional circumstances that citizens from East Berlin and the GDR had access to West Berlin, and for West Berliners access to East Berlin was also restricted. After that both parts of the city underwent separate developments. From the point of view of the language, the main result was that the Berlin dialect was stigmatized in the West and largely restricted to working-class areas, while in East Berlin, by contrast, the dialect was spoken by most of the people and it even became the prestige variety, spreading out into the surrounding areas.

4   AFTER THE FALL OF THE BERLIN WALL

Between 1945 and 1989 many differences arose between the standard language and Berlinish in both parts of the city, especially in terms of vocabulary. For West Berliners much of the vocabulary which was developed in the GDR seemed incomprehensible: according to some estimates, almost 2,000 words and phrases specific to the GDR were in common use there. Of course, most of these terms belonged to the terminology of the socialist state and its economy, but many were not simply part of the Party jargon or journalese (e.g. *Exposition* for *Ausstellung*, exhibition; *Jahresendflügelpuppe* for *Weihnachtsengel*, Christmas angel). Many were also commonly used in professional and everyday life (e.g. *Aktivist*). However, most of these words disappeared with the disappearance of their source and others were simply replaced by corresponding West German words. Some sections of the media continued to call for 'the cleansing of the old burdens of the language', and it remains to be seen which words and phrases will survive this discontinuity. One specific casualty of the process was the East Berlin concept *Stadtbezirk* (city district), which was banned after public discussion and replaced by the West Berlin term *Bezirk*.[1]

However, since the fall of the Wall the inhabitants of East Berlin have been bombarded with a bewildering stream of new developments, new concepts, words, and names which have to be mastered. East Berliners must now know and use many new words simply in order to be able to cope with everyday life, as they are constantly confronted with this new vocabulary, in both official and informal contexts. At the same time, much of their old stock of language has simply gone out of use. Especially in the first months after unification (1990–1) this led to many problems and considerable social tension. Even well-educated people complain that they cannot understand 'this new bureaucratic language'.

This inadequate command of West German officialese constitutes a serious language barrier for most East Berliners. However, civil servants and officials in West Berlin do not really understand the

---

[1] There is a further category of words, which were either coined or acquired new meanings after the fall of the Wall: e.g. *Abwicklung* (a euphemism for the liquidation of an institution or sacking of staff) and *Seilschaft* (old boy's network).

extent of these difficulties. For example, on 9 December 1991 an article in the *Berliner Zeitung* referred to the situation in job centres: 'In addition to the long waiting times, job centre visitors criticize the bureaucratic jargon. The advisers constantly receive instructions to speak in everyday language. None the less, official dealings remain firmly anchored in the language of the law.' It is therefore not surprising that many East Berliners came to talk of a new 'linguistic wall' and complained that they had 'lost their language'. For example, a woman from East Berlin wrote in the *Berliner Zeitung* on 25 July 1991:

We need a new dictionary. Who invented these childish and ridiculous abbreviations like Trabbi, Ossi, Wessi, Kita and now also Zivi? When I read Kita, I thought it was a new girl's name. The best thing would be to publish a manual with these words and their explanations. Foreigners couldn't possibly understand what they're supposed to mean.

East Berliners in general are affected by these changes in many ways, at work and in their private life as well as in their dealings with local authorities and institutions. Many of the new words they confront relate to what for them are new situations, others replace old concepts. In some cases, not only old GDR terms were replaced but even words that had been in common use before 1949, and this was seen by older Berliners as a step backwards, at least at the beginning. New words relating to employment and social security included terms such as *Arbeitslosengeld*, *Arbeitslosenhilfe* (both are forms of unemployment benefit), *Arbeitsbeschaffungsmaßnahme* (job creation schemes), *Kurzarbeit* (short-time working), and *Altersübergangsgeld* (early retirement allowance); *Arbeitnehmer* (employee) replaced *Arbeiter* (worker), *Team* replaced *Brigade* or *Kollektiv*, *Personalchef* (personnel manager) replaced *Kaderleiter*. There were far-reaching changes too in education and training: for example, the old uniform school system was replaced by a variety of types of institutions such as *Grund-*, *Haupt-*, *Mittel-*, and *Realschule*. Officially, *Kindertagesstätte* (Kita) is supposed to be used instead of *Kinderkrippe* and *Kindergarten*, but many parents continued to use the old expressions. Day-care institutions were previously called a *Hort* both officially and in everyday speech, but are now known officially as *offene Ganztagsbetreuung* or (in an East Berlin newspaper) *ganztägige Hortbetreuung* or *Schulhort*, and are run by an *Erzieher/in* (previously *Hortner/in*). The school-leaving certificate, previously known in East Berlin as the *Abschlußbescheinigung* is now officially called the *Zertifikat*, the subject that used to be known as *Kunsterziehung* (art education) is now *bildende Kunst* (BK), and the *Polylux* is now the *Overhead-Projektor*.

West and East German words and concepts from other areas of life meet head on in Berlin too: for example *Plastik* and *Plast*, *Renovierung* and *Rekonstruktion*, *KFZ-Schein* and *Zulassung* (driving licence). Numerous abbreviations often cause difficulties: *KOB* (*Kontaktbereichsbeamter*, police community liaison officer) is used instead of the East Berlin version *ABV* (*Abschnittsbevollmächtigter*), *WIP* (*Wohnungsbaugesellschaft in Prenzlauer Berg*, housing association in Prenzlauer Berg, a district in East Berlin) has taken the place of *KWV* (*Kommunale Wohnungsverwaltung*), *Azubi* (*Auszubildende/r*, trainee) is now preferred to *Lehrling* (apprentice).

Many narrower areas of everyday life are also deeply affected by these linguistic changes. For example, various new foods, dishes, and drinks with new labels have arrived from West Berlin and West Germany. Many have become especially well known as they are seen and heard a great deal: for example Turkish dishes like *Döner* and *Kebab*. In other cases the East German word has been replaced, and the things themselves also seem to have changed: for example *Hamburger*, *Hämbörger* instead of *Grilletta*, *Hot Dog* instead of *Ketwurst*, *Hähnchen* is now used for *Broiler* (roast chicken). Young people have adopted these words and their pronunciation particularly quickly: thus *Pommes frites* is now *Pommes* instead of *Pomm fritts*. The greater variety of products now available has also led to confusion: for example East Berliners were used to differentiating between minced pork (*Hackepeter*) and minced beef (*Schabefleisch*), but now a whole range of new products and terms have arrived from West Berlin, such as *Gehacktes vom Rind und Schwein*, *Rindergehacktes* and *Gehacktes gemischt*, and to add to the confusion many shops display the same products under different names.

In the GDR, the general official attitude towards foreign words was very negative and many campaigns were launched against English words and pop songs with English lyrics. However, the number of foreign words used in East Berlin has now risen considerably: for example *Timing, out, Crash-Kurs, Count-down, Discounter*. The frequent use of particular modish foreign words by East Berliners is still often considered undesirable and results in a negative evaluation of the speaker, even though they are often used by West Berliners: among these are *Outfit, Feeling, Promotion, Touch*. Other fashionable expressions, such as *lecker* (tasty) and the particles *eh* (in the sense of *sowieso*, anyway: 'der is eh nich da') and *halt* (in the sense of *eben*, just or well: 'das war halt gestern so') were already used by individual East Berliners before the fall of the Wall, but have now been widely adopted, especially by those who work in West Berlin or who are consciously assimilating the West Berliners' language use. They are even used in very pronounced Berlinish, and are used both to integrate speakers into the 'new' Western society and to distinguish them from the 'old' Eastern one.

The conscious distancing effect can be seen very clearly in the changes in professional and commercial terms which have been introduced in the East especially since 1991. Frequently, East Berliners do not even understand these terms and rarely use them. Instead of the words *Geschäft* or *Laden* (for shop), words such as *Salon* (e.g. *Hundesalon*), *Studio* (e.g. *Haarstudio*), *Zentrum* (e.g. *Treppenberatungszentrum*), and *Shop* (e.g. *Videoshop*), have been borrowed from West Berlin and appear quite frequently now in East Berlin. The *Wickelraum* (changing room) for babies is now called in one children's swimming pool in East Berlin *Baby-Kosmetik Centre*. Early in 1992 the name of a small dry cleaner's (*Reinigungsgeschäft*) was changed to *Kleenothek*. It is even common to see the word *Pub* on some East Berlin *Gaststätten* (although, as the results of a survey of 300 East Berliners revealed, the common West Berlin word *Pinte* has not been adopted). The previously common *Schlächter* (butcher) had already been replaced by *Fleischer* or *Fleischerei*, and as early as 1991 the very uncommon *Metzger* (a South German word that had migrated via West Berlin) was to be seen on one shop. The change is particularly clear from the word *Friseur* or *Frisör*: the terms *Salon, Haarsalon, Haarpflege-Salon, Haar-Dressing, Hair-dressing, Coiffeur*, and even *Inter-Coiffeur* are becoming increasingly common, and not only in the more up-market salons. However, Berliners still usually prefer to say *Frisör*.

Dictionaries of Berlinish (such as for East Berlin Wiese 1987, and for West Berlin Schlobinski 1986) document the extent of the potential lexical differences between East and West. However, they can give a misleading impression of the actual differences in everyday use. Many of the words listed in these dictionaries are no longer used, or were used only for limited periods or only by certain groups.

The complexity of the vocabulary of Berlinish in terms of actual use was shown clearly in a very extensive empirical investigation carried out in the 1980s by Helmut Schönfeld in East Berlin and Peter Schlobinski in West Berlin. The extent to which seven supposedly common Berlin words were recognized was studied by means of a survey of 500 inhabitants of Berlin. The results showed that many old variants were well known in both East and West: for example *Jören* (for *Kinder*, children) and *schwofen* (for *tanzen*, to dance). However, the use of these older forms frequently varied according to which part of the city the informant came from. For example, West Berliners made more use of *schnieke* (great, terrific: WB 35 per cent; EB 4 per cent), whereas *Piepel* (young boy: WB 11 per cent; EB 27 per cent) was heard more often in East Berlin. Recent colloquial forms coined in one part of the city did not seem to have been adopted in the other: for example, *fesch* (great, terrific) was only encountered in West Berlin, while *fetzig*, *gefetzt*, and *urst* were used with the same meaning exclusively in East Berlin (an exception to this rule is the popularity in East Berlin of the words *geil* and *supergeil*, again meaning terrific, super in Western *Jugendsprache*).

Certain labels, terms, and words from everyday Berlinish are commonly used in specific situations in one part of the city and are by and large not known in the other part. Some of these differences are cited in the dictionaries mentioned above and have also been checked by means of non-representative sampling. Examples for West Berlin include: *Großer Gelber* (for double-decker bus), *Türkenkoffer* (for plastic bag), *Mafiatorte* and *Pappdiskus* (for pizza), *Bonnys Ranch* (for the Bonnhöfer psychiatric clinic), *Kulturclubs* (for the fountain beside the *Gedächtniskirche*); and for East Berlin: *Schlenki* (for articulated bus), *Blutblase* (for the red caps of the platform staff in East Berlin train stations), *Neu Deli* (for a block of flats with delicatessen shops on the Alexanderplatz), *Sankt Walter* (for the East Berlin television tower; the name alludes to former GDR leader Walter Ulbricht), *Tränenpavillon* (for the former border crossing between East and West Berlin at Friedrichstraße). Other recently coined terms remained restricted to the particular district or group in which they originated or were current only for a short time: for example in East Berlin *Putenrennen* (for Women's Day celebrations), *Kulturpickel* (for the *Kongresshalle* in East Berlin), *Pinkelklötzchen* (for kidney), and *Suppenschmiede* (for kitchen) (Schönfeld 1989*b*).

On the other hand, greater media exposure and individual mobility has meant that many of these words have become known in other areas: for example *Pallazo Prozzo* (for the *Palast der Republik* in East Berlin) and *Hungerharke* (for the monument to the Berlin airlift in West Berlin). Colloquial expressions for city districts, which were previously common only in specific groups, have now also become more widely known through the media: for example *Prenzelberg* for Prenzlauer Berg, *Höhenschöngrünkohl* for Hohenschönhausen.

Throughout the city, the workplace has now become a great linguistic melting-pot. This applies both at the level of everyday language and in terms of the specific registers of given occupations, such as the army, the police, and the railways. Our knowledge of these speech forms and the communication problems and other developments resulting from unification is still very limited, but this area too should be a fruitful one for future research, as distinct vocabularies had developed in each part of the city. A common feature of the language of waiters in the dining cars of West Berlin trains, for example, was to use abbreviated forms to avoid confusion (e.g. *Wald* for Schwarzwälder Kirschtorte), while their counterparts in the East Berlin Reichsbahn also developed their own jargon (e.g. *Anna* for Tasse Kaffee).

Many of the new words and abbreviations from West Berlin are essential for quick and simple communication and have been adopted immediately by the East Berliners. Frequently, however, the East Berlin variant is still used in speech through force of habit, even if the thing it designates has changed (e.g. *Altersheim* instead of *Seniorenheim* for old people's home, *Kaufhalle* instead of *Supermarkt*). In some cases, where two variants exist, some East Berliners adopt the West Berlin form only very hesitantly or even reject it out of protest. This is the case, for example, with the words *Kita* and *Tram*. As an East Berlin woman wrote in the *Berliner Zeitung* on 25 February 1991:

Mein Enkel, stolz darauf dem Krippenalter entwachsen zu sein, weigert sich störrisch, in die Kita zu gehen. Bei dem Wort Kindergarten strahlende Augen, bei Kita Protest und Widerwillen. Ich will alles dafür tun, daß dem kleinen die schöne altdeutsche Bezeichnung Kindergarten nicht abhanden kommt.

(My grandson, proud of having outgrown the creche, refuses point blank to go to the *Kita* (day nursery). At the mention of the kindergarten he's all smiles, but say *Kita* and he protests. I intend to do all I can to ensure that the little fellow doesn't lose the fine old German word kindergarten.)

With similar strength of conviction, many East Berlin readers of the *Berliner Zeitung* rejected the short form *Tram* in favour of *Straßenbahn* in the summer of 1991, although this preferred form (introduced in East Berlin by the Senate) had been used in the media and also in compound forms such as *Trambahn, Tramschiene*.

Many individual variants coming from West Berlin or West Germany in general are understood by East Berliners. However, they have still met with considerable hostility, especially when widely used (e.g. *lecker* rather than *wohlschmeckend* for tasty). Many are also misunderstood and used incorrectly (e.g. *Flieger* for *Flugzeug* (plane), as *Flieger* in East Berlin was used commonly instead of the word *Pilot*). Curiously enough, though, many old East German forms are already being reused, as are many East German products, after a period of rejection by East Berliners, a fact which is exploited to the full in advertising campaigns: for example *Berliner Pils—Bier von hier*. This even applies to the archetypal East German word *Broiler* (roast chicken). After the fall of the Wall, many had adopted the West German version *Hähnchen*, but many restaurants and fast food outlets continued to be run under the name *Zum Goldbroiler*. The products themselves are sold under both names, and at one particular East Berlin outlet a sign appeared in 1992 offering *Hähnchen* but accompanied by the following notice: 'Hier dürfen Sie noch Broiler sagen' (you may still say *Broiler* here).

Finally, in addition to the more striking lexical contrasts between East and West, it is worth mentioning differences in idiomatic and stylistic features of speech. In the West, for example, it was common to say 'etwas rechnet sich' (something pays off, is worth while), 'ich würde mal sagen' (I'd say), 'das ist out' (that is out of date). Other phrases equally distinctively marked the speaker as East German: for example 'die Frage steht' instead of 'die Frage stellt sich' (the question arises), 'sich keinen Kopf machen' (not to let something worry you), 'das ist Fakt' (that is a fact), 'orientieren auf (ein Problem)' (to focus on, turn your attention to). (For an extensive discussion of this and other contrasts, see Schlosser 1990.) There were also differences in terms of the constitution of specific text types. A trivial example of this is the contact columns in newspapers: until 1990, East Berlin advertisements were factual and brief, simply stating age, occupation, physical features, hobbies, and so forth, while West Berliners described their appearance in more forthright terms in order to show themselves in the best possible light.

5   CONFRONTING THE NEW REALITY: CHANGING PERCEPTIONS AND ATTITUDES TOWARDS VARIATION

Differences in social norms and values, in social behaviour and language use, in the awareness of language norms and the confrontation with situations which had previously been encountered only indirectly through radio and television: all this was now experienced directly. This applied also to habits and patterns of language use, both within Berlinish and in the use of the standard variety, and although similar experiences confronted Germans in all parts of the country Berliners were obviously in the front line.

Surveys conducted in the summer of 1990 showed which differences in language use were recognized at that time and how they were evaluated. Teenagers in East Berlin believed that the linguistic flexibility and facility in using the standard variety correctly, which they associated with West Berliners, was something positive and worth emulating. More recent surveys suggest that this opinion is still held by many East Berliners, but others increasingly reject what they see as the sterile and artificial use of the standard variety and condemn the pompous and vacuous speech behaviour they associate with it.

According to respondents, West Berliners were 'more professional' in their use of language and often used it as a means of showing off: this applied to style, choice of words, language strategies, and the use of different speech forms. The East Berliners also believed that West Berliners generally use the standard variety and speak Berlinish rarely if at all: when they did use Berlinish, then it was almost only in private everyday situations, where a form of the vernacular very close to the standard was used.

West Berliners questioned in the surveys declared that East Berliners still often used the very strong 'old-fashioned' forms of Berlinish and that they used the vernacular in situations where it would generally not be used in West Berlin. This applied even to East Berliners who, to judge from their clothing and general manner, belonged to the middle and upper classes: this speech behaviour would be considered very uncommon among members of the corresponding social classes in West Berlin.

The process of reunifying the city since 1989 has brought about many very wide-ranging social changes, which in turn have led to changes in the use of language varieties, especially among East Berliners. Many inhabitants of East Berlin have been obliged to adapt to greater use of the standard variety in official and public situations, and for many people this has been far from easy. Even those who appear on radio and television still often use the version of Berlinish that is furthest from the standard or other variants of the vernacular. They may try to adapt to the perceived demands of the new situation, but speech habits cannot be changed as easily as clothes. This was evident in 1989 not only in public interviews with artists and sports personalities, but also in the public speech of leading figures from the fields of business, science, and politics. It is still particularly noticeable in the speech behaviour of many politicians who since November 1989 have held prominent positions in political parties and public office.

However, new linguistic requirements have been imposed on 'ordinary' East Berliners too, not only those who are in the public eye. The reasons for this are the increased regional and social mobility, private and professional contacts with inhabitants of the old *Länder*, and ever increasing occupational requirements. This has become particularly clear in job advertisements which have appeared in East Berlin newspapers since the middle of 1989. Many employers explicitly look for applicants who among other things 'should possess a good knowledge of German, be persuasive, and have good verbal skills'. This applies not only to jobs in, say, advertising or banking or to secretarial posts: 'self-confident nurses with good communicational skills' and 'persuasive and enthusiastic sales people' are also in great demand. It is often made clear that there will be written tests and an initial interview, and in many cases applicants who are strong Berlinish speakers will not be considered. Indeed, within a few months of the Wall coming down, newspapers were advertising courses offering training in language skills, such as 'rhetorical skills', 'conversation practice', and 'conflict resolution', and public-sector employees in particular are encouraged to take part in seminars and courses on topics such as 'using language that is easy to understand' and 'dealing with the public' (from an advertisement in the *Berliner Zeitung* on 28 May 1991).

By visiting public agencies, department stores, discos, and so forth in West Berlin, many East Berliners are continually confronted with what for them are the new norms for language use. Increasingly, as in the days before the Wall was built, many people from the East are working with West Berlin colleagues. Some West Berliners also work in East Berlin but mostly as managers and advisers. Surprisingly enough, however, very few of those questioned even in the early surveys spent any time in the 'other' part of the city, and indeed many say they visit the other part less often now that the initial euphoria has died away. You stay in your own area, with the result that direct contact with the actual speech behaviour of the 'others' remains limited. Old stereotypes survive.

West Berliners are often asked how they actually recognize an East Berliner. Common responses in the early days in particular included references to clothes that were 'obviously from the East' and shopping bags made of nylon or cotton, and other aspects of 'visible' behaviour, such as going around in groups. Some people, however, claimed that you could also distinguish the inhabitants of one part of the city from the other after a short conversation. The evidence offered is often rather vague but is generally based on value judgements, which relate to speakers rather than their speech. Apart from obvious individual features, such as the use of words like *urst* (for great, terrific) or the pronunciation of *auch* (also) as *ooch*, they would also identify now redundant terms such as *Kollektiv* instead of *Team* or *Gruppe*, or *Lernvermögen* instead of *Intelligenz* (Liebe Reséndiz 1992: 130). It is also common to ridicule the continued use of the more 'extreme' (non-standard) forms of Berlinish, especially in public contexts such as at supermarket checkouts or in university seminars. Even when East Berliners attempt to accommodate to the new norms, they are still criticized for failing to abandon some of the older speech habits: for example, the emphatic use of titles (e.g. 'Herr Professor') in exaggerated politeness, and the deferential and almost apologetic manner towards strangers (e.g. asking for information in institutions: 'Guten Tag! Entschuldigen Sie bitte . . . Vielen Dank!'). The curt response of one West Berliner interviewed on television in 1992 is representative of the views of many: 'die (Ostberliner) ha'm die Einheit jewollt un müssen sich nun unsren Jargon aneignen' (the East Berliners wanted unity, so now they must learn to speak like us).

6 CONCLUSIONS

For most West Berliners very little has changed since the opening up of the Wall, other than of course the removal of the insular nature of the city and the loss of some financial benefits and other perks. For the East Berliners, however, their whole world and way of life has changed dramatically. Many fitted in very quickly to the new order, but for many others a deep loss of purpose and security, inferiority complexes, and identity crises were the result. Their self-confidence and self-esteem plummeted, and the situation was often exacerbated by private problems such as unemployment. This in turn resulted partly in feelings of helplessness and a growing feeling that Berlin was still divided in two. A survey carried out in 1992 showed that many believed that this 'inner wall' was getting higher all the time.

The social and linguistic facts of life in the whole of Berlin have affected the East Berliners in many different ways, including their norms of language use. Many East Berliners want to become like their neighbours in the West, but others use the standard variety only in very few and quite specific situations. Some even deliberately speak Berlinish when talking to West Berliners. Living together continues to bring more changes and new problems, and the painful process of improving communication between East and West is far from over.

It will be interesting to see how the linguistic systems converge at the level of individual forms and how the outcome of this process is evaluated. Writing in *Die Zeit* in 1992, Dieter Zimmer claimed that Berlinish was on the advance, and it is certainly true that the Berlin dialect is increasingly evident in the media. It remains to be seen whether Berlinish will now be seen as a prestige variety, as in East Berlin, or stigmatized, as in West Berlin. However, we believe that many East Berlin dialect speakers will come under pressure to adapt and will therefore move in the direction of colloquial varieties closer to the standard. Whatever happens, Berlin offers a unique opportunity to document and study language change taking place against the background of dramatic social and political change.

Further Reading

Burkhardt and Fritzsche (1992)
Dittmar and Schlobinski (1988*a*)
Lasch (1928)
Muttersprache 3 (1993), *Sprache nach der Wende* (special issue)
Schildt and Schmidt (1986)
Welke *et al.* (1992)

References

BARBOUR, J. S., and STEVENSON, P. (1990), *Variation in German: A Critical Approach to German Sociolinguistics* (Cambridge: Cambridge University Press).
BURKHARDT, A., and FRITZSCHE, P. (eds.) (1992), *Sprache im Umbruch. Politischer Sprachwandel im Zeichen von 'Wende' und 'Vereinigung'* (Berlin, New York: de Gruyter).
DITTMAR, N., and SCHLOBINSKI, P. (eds.) (1988*a*), *Wandlungen einer Stadtsprache. Berlinisch in Vergangenheit und Gegenwart* (Berlin: Colloquium Verlag).
————— (eds.) (1988*b*), *The Sociolinguistics of Urban Vernaculars* (Berlin, New York: de Gruyter).
————— and WACHS, I. (1986), *Berlinisch. Studien zum Lexikon, zur Spracheinstellung und zum Stilrepertoire* (Berlin: Arno Spitz).
JOHNSON, S. (1991), 'Gender, Group Identity and Variation in Usage of the Berlin Urban Vernacular'. Ph. D. thesis, University of Salford.
Kommunikation und Sprachvariation (1981), 'Autorenkollektiv under the leadership of W. Hartung and H. Schönfeld' (Berlin: Akademie-Verlag).
LASCH, A. (1928), *Berlinisch. Eine berlinische Sprachgeschichte* (Berlin: Verlag von Reimar Hobbing).
LIEBE RESÉNDIZ, J. (1992), 'Woran erkennen sich Ost- und Westdeutsche? Eine Spracheinstellungsstudie am Beispiel von Rundfunksendungen', in Welke *et al.* (1992), 127–39.
MEYER, H. (1878), *Der richtige Berliner in Wörtern und Redensarten* (Berlin).
SCHILDT, J., and SCHMIDT, H. (eds.) (1986), *Berlinish. Geschichtliche Einführung in die Sprache einer Stadt* (Berlin: Akademie-Verlag).
SCHLOBINSKI, P. (1984), *Berlinisch für Berliner und alle, die es werden wollen* (Berlin: arani).
—— (1986), *Berliner Wörterbuch* (Berlin: Marhold); 2nd edn. 1993 *Berliner Wörterbuch. Der aktuelle Sprachschatz des Berliners* (Berlin: Spiess Verlag).
—— (1987), *Stadtsprache Berlin. Eine soziolinguistische Untersuchung* (Berlin, New York: de Gruyter).
—— (1988), 'Über den "Akkudativ" im Berlinischen', *Muttersprache*, 98/3: 214–25.
—— and SCHÖNFELD, H. (1993), 'DFG-Bericht zum Projekt "Berlinisch".' Unpublished final report.

134   *Helmut Schönfeld and Peter Schlobinski*

Schlosser, H. D. (1990), *Die deutsche Sprache in der DDR zwischen Stalinismus und Demokratie* (Cologne: Verlag Wissenschaft und Politik).

Schönfeld, H. (1981), 'Regional definierte Sprechergruppen', in *Kommunikation und Sprachvariation* (1981), 130–68.

—— (1986*a*), 'Die berlinische Umgangssprache im 19. und 20. Jahrhundert', in Schildt and Schmidt (1986), 214–98.

—— (1986*b*), 'Prozesse bei der Herausbildung regionaler Umgangssprachen im 19. und 20. Jahrhundert', in *Umgangssprachen und Dialekte in der DDR. Wiss. Beiträge der Friedrich-Schiller-Universität Jena*, 162–75.

—— (1989*a*), 'Ziele und Methoden der Erfassung und Beschreibung regionaler Umgangssprachen in der DDR', in *Theoretische Positionen der Beschreibung gesprochener und geschriebener Sprache. Protokollband* (Leipzig: Pädagogische Hochschule), 94–101.

—— (1989*b*), *Sprache und Sprachvariationen in der Stadt. Zu sprachlichen Entwicklungen und zu Sprachvariation in Berlin und anderen Städten im Nordteil der DDR*, Linguistische Studien des Zentralinstituts für Sprachwissenschaft der Akademie der Wissenschaften, Reihe A, 197 (Berlin).

Trachsel, C. F. (1873), *Glossarium der berlinischen Wörter und Redensarten, dem Volke abgelauscht und gesammelt* (Berlin).

Welke, K. *et al.* (eds.) (1992) *Die deutsche Sprache nach der Wende*, special issue of *Germanistische Linguistik*, 110–11 (Hildesheim: Georg Olms Verlag).

Wiese, J. (1987), *Berliner Wörter und Wendungen* (Berlin: Akademie-Verlag).

Zimmer, D. (1992), 'Balina Schnauze', *Die Zeit*, 18 Sept., 106.

7 Theories of Sociolinguistic Variation in the German Context

NORBERT DITTMAR

I ORIGINS AND DEVELOPMENT OF SOCIOLINGUISTICS IN GERMAN-SPEAKING COUNTRIES

1.1 *Humboldt as the Forerunner of the Modern Study of Language in Society*

Sociolinguistics in German-speaking countries started with language-philosophical reflections on the expressive power and logical capacity of German as one of Europe's national languages. The claim that French was superior to German in terms of logic and clarity of expression produced many philosophical and philological dissertations on the 'Reichthum und Armut der deutschen Sprache' (wealth and poverty of the German language: see Dieckmann 1989; 'poor' and 'rich' in this context can be seen in global and historical terms). But do today's sociolinguists have better instruments and taxonomies than their sociolinguistic ancestors more than one and a half centuries ago?

The competition between nations over linguistic prestige was at the same time a competition about the mental capabilities of European national languages, which brings us to the 'German' cradle of socio-linguistics: Wilhelm von Humboldt's thesis of 'linguistic relativity', formulated in the nineteenth century. In Humboldt's view there is no doubt that language and thought are interdependent:

Das Denken ist aber nicht bloss abhängig von der Sprache überhaupt, sond-ern, bis auf einen gewissen Grad, auch von jeder einzelnen bestimmten. Man hat zwar die Wörter der verschiedenen Sprachen mit allgemein gültigen Zeichen vertauschen wollen, wie dieselben die Mathematik in den Linien, Zahlen, und der Buchstabenrechnung besitzt. Allein es lässt sich damit nur ein kleiner Theil der Masse des Denkbaren erschöpfen, da diese Zeichen, ihrer Natur nach, nur auf solche Begriffe passen, welche durch blosse

I would like to express my gratitude to Wolfdietrich Hartung and Patrick Stevenson for many helpful comments and suggestions.

Construction erzeugt werden können, oder sonst rein durch den Verstand gebildet sind. Wo aber der Stoff innerer Wahrnehmung und Empfindung zu Begriffen gestempelt werden soll, da kommt es auf das individuelle Vorstellungsvermögen des Menschen an, von dem seine Sprache unzertrennlich ist. (Humboldt 1904: iv. 21–2, cited in Werlen 1989: 56)

(However, thinking is not merely dependent on language in general, but to a certain extent on each individual specific language. Admittedly, there have been attempts to replace the words of the various languages with generally valid symbols, in the same way as lines, numbers, and letters are used in mathematics. However, only a small part of what is thinkable can be captured in this way, as these symbols by their very nature only fit those concepts which can be produced by mere construction or else are formed purely by reasoning. But when the substance of inner perception and feeling is to be formed into concepts, then it is a question of the individual person's powers of imagination, which cannot be separated from his language.)

Ever since Humboldt, many German linguists have been preoccupied with the question of whether and to what extent differences in language use imply differences in mental abilities, and whether speakers have social advantages or disadvantages because of their 'relative language use'. The thesis of linguistic relativity has removed the social innocence from reflections on language. Today, the causal interrelationship between the 'individual powers of imagination' and varying language use can still be seen as the original driving force of sociolinguistics.

1.2   German Contributions to the Sociology of Language

As an outstanding representative of a sociological orientation in linguistics, Hugo Schuchardt in his essay *Über die Lautgesetze: Gegen die Junggrammatiker* (On Sound Laws: Against the Neogrammarians) investigated the Neogrammarians' claim that the laws of sound changes operated without exceptions. This dissertation, written in 1885, was very important for the development of the theory of variation. Schuchardt formulated some insights which remain relevant for today's sociolinguistics. I shall pick out just two:

1. 'I assume there to be a mixture of languages even within the most homogeneous linguistic community.'
2. 'Every stage of language is a transition stage, every one just as ordinary as any other . . . and thus [one can see] plainly the fluid transitions of its spatial and temporal differences.' (1885/1972: 20)

Unlike other linguists before them, Humboldt and Schuchardt in their dialectic of form and function showed the relevance of language and variety contact for the theory of language.

What is the significance of the theorizing about the sociology of language in German writings towards the end of the nineteenth and beginning of the twentieth centuries? We can explain it by taking an example from Georg von Gabelentz. Gabelentz thinks that 'customs and rules' (meaning approximately culture, religion, and law) are guiding factors in the social use of language that are revealed in the forms and functions of linguistic expressions and structures and that they should be investigated comparatively (Gabelentz 1891/1984: 245). His cross-linguistic approach to comparing, for example, Japanese, Korean, Polynesian, and Kri-specific forms of politeness would today be called ethnographic. The correctness of this interpretation is borne out by Gabelentz's observation of culture-specific speech acts of swearing, respect, and politeness in different societies (ibid. 248).

Gabelentz calls 'classes' the 'driving force' of social change (thus 'the needs of upper classes to take precedence over lower classes', ibid. 249). The so-called mechanism of 'hypercorrection' (the lower middle class overgeneralizes the prestigious pronunciation of the upper middle class and upper class), a notion that was introduced by Labov in the twentieth century, is described by Gabelentz as follows: 'Ever since clothing regulations disappeared, lower-class people can dress like gentlemen. But they cannot follow them; because what has been adopted by them is not valid any more for the upper classes' (ibid. 250). Gabelentz sees the 'language community' as the site of socially conditioned linguistic change: it extends as far as the possibility of linguistic intercourse. Everybody has his own individual language (idiolect), his circle of linguistic intercourse, which he informs, and from which he receives. His friends' circles intersect with his own and with other people's circles. And so it continues, from A to B, B to C, and so forth, to the point where the community of dialect speakers or even of speakers of the language A has ceased to exist (ibid. 275).

Gabelentz also recognizes 'men's language' and 'women's language', whose differences he illustrates as follows:

Among the Chiquitos in the province of Santa Cruz, there also exists a sharp difference between men's and women's language. Women have to use specific words for certain terms; or they do not pronounce the initial sounds of words or the last syllable of the suffixes, or they use completely different forms (ibid. 284).

The contribution of these historical writings to the theory of sociolinguistics is to document the evolution of sociolinguistic thinking.

As there is, in my opinion, no technical progress in social science, they help us to realize when research is on the wrong track and to find new beginnings. In doing so, it is necessary to reconstruct the 'heuristic' considerations explicitly and theoretically in order to make the linguistic and sociological terms defined in this way usable in empirical research: this is a task for German sociolinguistics that should offer many new insights.

1.3 *'Subversive' and 'Authoritarian' Sociolinguistics: Developments in West and East*

As in other national and cultural contexts, German sociolinguistics developed as a consequence of social and political conflicts. However, things are much more colourful and much stranger in German socio-linguistics than in other disciplines. Between 1960 and 1989 we had two kinds of sociolinguistics in Germany: a Western one (FRG) and an Eastern one (GDR), as reflected by the following keywords:

FRG	GDR
appeared 'subversively'/linguistic differences cause social ones;	'official linguistics' ordered from above/language is investigated in terms of social needs;
'unequal language' = 'social inequality': this hypothesis has to be proved by the best possible descriptive instruments;	with support from sociolinguistics, linguistic norms have to be developed in order to achieve a socialist 'language culture';
the researcher's drive to prove that unequal language entails social inequality results in the development of superior instruments for the description of variation and theories.	sociolinguistics seeks to document the favourable conditions of communication in socialist countries and to formulate and improve norms and theories; neglect of empirical methods.

Thus, a paradox has arisen that only sociologists, psychologists, and historians can solve: in the GDR, the development of sociolinguistics was officially supported as a science to form a theory and develop instruments in the service of society, but this resulted in largely trivial theoretical findings; by contrast, West German sociolinguistics, unwanted by the dominant classes of society, developed a subversively active, strongly empirical approach to research that had to operate with more and more refined theoretical and descriptive methods in its fight to gain respect. Constantly required to justify its existence, WSL (Western sociolinguistics) has done just that by high-quality research guided by theory, while the highly official legitimation accorded to ESL (Eastern sociolinguistics) seems to have stripped it of all creativity.

Of course, the academic market must be taken into consideration: Western sociolinguists had better access to data (above all the freedom to record spoken language) and relevant literature than their Eastern counterparts. But on closer inspection, some interesting facts emerge. WSL is orientated predominantly towards the description of behaviour (following the maxim: describe what is and not what could be), that is, it ignores psychological factors concerning the relationship between language and thought and provides us instead with superior 'surface descriptions' or formulates theoretical suggestions; but the 'inner perception', which was such an important dimension for Humboldt, has been left aside as a 'disruptive factor'. The immunization against 'internal counterparts' to social variation has led to highly developed descriptive models of variety differences in WSL. ESL, on the other hand, never tires of stressing the fact that research on variation based purely on behavioural descriptions neglects the different mental activity of speakers associated with the use of different varieties, the question of language norms (appropriate vs. inappropriate language use), and the ethical–practical side of 'language culture' that is essential to linguistic communities (see Section 4 below for a more detailed consideration of this point).

The subtly differentiated methods of description in the West have improved the technology of data analysis but have provided only limited insights into questions of norms and explanations. ESL has attached more weight to these issues by operating on a coherent theoretical basis, but at the same time it has been rather premature in proclaiming its findings without testing them empirically. Their joint task should now be to bring together WSL's descriptive instruments and the concepts about the relationships between language, thought, and social structure developed by ESL in a new constructive perspective.

2   DESCRIPTIONS OF VARIETIES: PROBLEM SPACE

In order to describe variation in German, we have to consider sociopolitical, territorial, social, gender-, class-, group- and age-specific parameters. They determine the 'variety space' (see Section 3.2 below) and establish the constraints that delimit linguistic characteristics of varieties as a continuum of features. By way of illustration, each of the following examples represents a specific parameter of the variety space.

'National' Variants of German
1 (*a*) Wir befetzen uns (Austrian)
⟨FRG Standard: streiten⟩ (argue)

  (*b*) Der ist verunfällt und wurde verbüßt (Swiss)
⟨FRG Standard: der hat einen Unfall gehabt und Bußgeld zahlen müssen⟩
(he had an accident and had to pay a fine)

'Territorial' (Dialectal) Variants: (a) Franconian (b) Berlin vernacular
2 (*a*) Wir haben sie runter laß muß fahren
⟨Wir haben sie herunterfahren lassen müssen⟩
(we had to have them driven down)

  (*b*) Die kriecht so'n Ölfmeter dat die ja nich aus de Oogn kieken kann
⟨die kriegt so einen Schlag ins Gesicht, daß sie nicht mehr aus den Augen gucken kann⟩
(she'll get such a smack in the face that she won't be able to see)

Supra-Regional Colloquial Language
3 Ich krieg die nicht immer weggeschmissen
⟨ich tue mich schwer/ich bringe es nicht fertig, sie wegzuwerfen⟩
(I can't bring myself to throw them away)

'Group-Specific' Varieties/Registers
4 Es wird nicht geschehen, daß ein von der CDU aufgestellter Kandidat abgenickt wird
(*abnicken* is similar to 'go through on the nod': the speaker is saying that his party, the FDP (Liberal Democrats), will not automatically give their support to any candidate proposed by their coalition partners the CDU (Christian Democrats))

'Age-Specific' Varieties/Registers
5 Er hat da wieder so 'ne tussi ausgegraben, so'n geilen fisch
(youth slang: he picked up a girl)

'East-German, GDR-Specific' Variants
6 (*a*) die Wohnraumlenkerin orientierte auf die Lösung, den Kredit
abzukindern
[The municipal housing officer proposed the solution of paying
off the debt by getting extra child allowances (that is, having
more children).]

(*b*) A: ich will jetzt nach dem 5. Diplom machen
B: So schnell? Mußt ja'n intellektueller Superstar sein!
A (an East German): I'm going to do my diploma exam at the
end of the 5th [meaning 5th year of studies]
B (a West German): So soon? You must be an intellectual superstar! [She interprets '5th' as 5th semester, the normal unit of
study in West Germany.]

'Learner' Varieties (a), 'Foreigner Talk' (b)
7 (*a*) die samstag, ich, wann ich komme feierabend und komme
zuhause, sage zu mir: 'ich niks ganz gut, habe fieber'
⟨als ich am Samstag Feierabend hatte und nach Hause kam,
sagte meine Frau zu mir: 'mir geht's schlecht, ich habe Fieber'⟩

(When I came home after work on Saturday my wife said to
me: 'I'm not well, I've got a fever'.)

(*b*) ich hab nichts gegen Türkischmann Türkischfrau ... nur ihr
anders sprechen wie wir ... andere Sprache ... oben meine
Schrank is eine Türkischmann, immer beten machen
⟨ich habe nichts gegen türkische Männer oder Frauen ... nur
Ihr redet anders Deutsch als wir ... eine andere Sprache ist
das ... da oben, wo mein Kleiderschrank ist, habe ich einen
türkischen Nachbarn, der betet immerzu⟩

(I've nothing against Turkish men and women, it's just that you
speak differently from us, it's another language, upstairs where
my wardrobe is there's a Turkish man, he just prays all the
time)

Examples (1) and (2) represent the 'diatopic' or spatial dimension of
variation; Austrian and Swiss German (1) each have nationally recognized and codified standards of linguistic norms. As far as the tolerance
of different norms is concerned, therefore, the diversity of German
clearly differs from, say, French. Example (2*a*) is Franconian and

shows a marked syntactic difference from the standard variety, while
the Berlin dialect, example (2*b*), is marked only by lexical (*kieken*) and
phonetic (*dat, oogn*) variation, and this variety is generally closer to
standard German.

Example (3) can be assigned to the 'diasituative' dimension of the
variety space: colloquial language, in this case marked by the use of
the 'recipient passive' (*Rezipientenpassiv*), is used in informal situations
depending on the interlocutor. Example (4) illustrates 'diastratic' variation, that is, the group-specific varieties (including social class):
abnicken is an expression used by politicians in parliament, and its use
therefore implies a special group-specific knowledge in the interlocutor.

Youth language, as in example (5), belongs to the 'diachronic'
dimension of variation: these are varieties that show special lexical
characteristics and routine formulas that are typical of a certain age
group only. They are adopted for a while and given up later.

'Diatopic', 'diastratic', 'diasituative', and 'diachronic' are central
dimensions of variation that Nabrings (1981) elaborated in her work
on the theory of variation. Thanks to (West) German sociolinguistics,
the parameters of the variety space that govern linguistic constraints
of particular varieties have been determined explicitly. In contrast
to anglophone sociolinguistics, which more or less limits the study
of variation to the empirical description of low-level grammatical
variants, German sociolinguistics attaches great importance to a
theoretical foundation from which specific linguistic characteristics
of varieties can be derived according to the parameters of the variety
space.

I would like to add a fifth parameter to Nabrings's four, to account
for the continuum of 'interlanguages' in German. This dimension is
illustrated by example (7): (*a*) is the utterance of an Italian after a
five-year stay in Germany; (*b*) is commonly called 'foreigner talk': a
native speaker talks to a Turk in broken German (see also Rost-Roth,
this volume). German sociolinguistics has also formulated the
theoretical insight that one form can correspond to more than one
function: for example, 'youth language' (example 5) can also be seen
as 'diastratic' (group-specific), 'foreigner talk' (example 7*b*) as 'diasituative'. It is important to understand the interaction of form and function as a continuum; this is the only way of interpreting the difference
between East and West after reunification, as illustrated in (6), at the
same time as 'diatopically', 'diastratically', 'diasituatively', and 'dia-
chronically' conditioned variation. As none of the four terms describes
the specific particularity of this variation, I suggest the introduction
of the term 'diacommunicative' (specific to a communication com-

munity). It takes into account the different socio-political history of communication communities within a linguistic community.

3   THEORIES OF SOCIOLINGUISTIC VARIATION I: WEST GERMAN AND AUSTRIAN PERSPECTIVES

In what particular ways does German sociolinguistics contrast with sociolinguistics in other (for example anglophone) countries? Perhaps the main contrast is that in the German tradition, to a greater extent than in other contexts, theoretical considerations are subjected to strict empirical testing and can thus be developed further. In this section, I shall illustrate this contention by discussing three important aspects of German sociolinguistics. First, in (West) Germany the theory of socio-linguistic codes has been validated and theoretically reinforced by ran-dom samples (see Section 3.1). Secondly, 'variety grammar' (VG), a comprehensive alternative grammar, has been developed; in contrast to the variable rules developed by Labov and others, which describe only single linguistic features defined and isolated as variables, VG provides a complete grammatical description of a variety that is documented by a linguistic corpus incorporating extra-linguistic factors like age, social class, social network, etc. defined precisely in the variety space (see Sec-tion 3.2). Thirdly, the recognition that the description of variation as a surface phenomenon grasps its symptoms but conceals causal, prag-matic factors, that many forms of variation are connected with discur-sive processes, led to the elaboration of Gumperz's anthropological concept of 'contextualization' (see Section 3.3).

West German sociolinguistics took its starting-point from class-specific language research. Nowadays, key sociolinguistic concepts like 'deficit' (the thesis of the linguistic poverty of the lower classes) and 'difference' (the thesis of the functional equivalence of culturally dif-ferent ways of speaking; see also Dittmar 1976) are also applied in the debate on 'men's' vs 'women's' language: some female sociolinguists consider the female way of speaking to some extent as a form of verbal deficit, while others interpret 'the more controlled, more dialogical and more polite speaking of women' as powerful in the sense of a 'difference' (Günthner and Kotthoff 1991*b*: 22). Similar theoretical questions are discussed in connection with 'youth language' (does it 'lack' verbal and grammatical means when compared to adults' lan-guage? See Schlobinski, this volume) and 'semilingualism' (linguistic deficits of migrant children in two languages: see also Rost-Roth, this volume). Linguistic difference as a consequence of social inequality is therefore an important theoretical focus of German sociolinguistics.

How is it that in West Germany methodological reflections on sociolinguistic facts quickly break with certain current theories, hypotheses, and empirical methods, introduce critical and fundamental reflections, and achieve unorthodox theoretical soundness and further methodological developments? To give an answer à la Kuhn would take a whole book; I tend to adopt the admittedly rather sweeping hypothesis that this particular kind of research was (and is) so fruitful because unlike American sociolinguistics it was not organized in orthodox schools (see the dominant and authoritarian roles of Labov and Fishman, who run their schools like a business). In the (former) West Germany, there were (and still are) no strictly organized schools of sociolinguistics. Unorthodox ways of thinking encourage critical tightrope-walking, individual approaches to problem solving, and theoretical analysis of specific questions. However, the absence of schools also has disadvantages: sociolinguistic activities are not supported by centralized institutions (as was the case in East Germany) but often dissolve into individual, noncommittal reflections.

Modern sociolinguistic theorizing in German sociolinguistics is moving towards stimulating 'cross-cultural' and 'cross-linguistic' comparisons: on the one hand towards the comparison of different varieties of German, on the other hand towards the study of a particular type of variety like 'urban vernacular' or 'genderlect' (a term coined by Günthner and Kotthoff 1991*b*: 17) in communication communities with different languages.

3.1   *The Theory of 'Linguistic Codes'*

In the late 1960s and early 1970s, the autonomy of linguistics was undermined by various types of 'linguistics plus a hyphen' (as in text- and sociolinguistics). The main focus of sociolinguistics in the FRG was the hypothesis formulated by the British educational sociologist Basil Bernstein that in Western democracies the lower classes are at a social disadvantage as against the middle class because of inadequate linguistic socialization and, as a consequence, inferior academic capability. Overnight, because of this theory, linguistics became an exciting and popular subject of interdisciplinary study. Bernstein's dichotomous relativistic world-view, the result of a dualistic perception of society seen through the distorting spectacles of language, was explicitly reconstructed by Dittmar and Klein (1972) and Dittmar (1976: chaps. 1–3).

According to this view, two different social structures, LC and MC (lower class and middle class) determine two different mental strategies of acquisition and applications of procedural knowledge (speech strategies) that produce two different class-specific language varieties (codes). The thesis is as follows: LC speakers have fewer opportunities for social advancement than MC speakers because of the linguistic characteristics of their utterances. This is a 'relativistic' view because it relates language to social experience and assumes a close relationship between speaking and thinking as its basis. This thesis, once carefully formulated by Humboldt and taken up again and stated more precisely by Whorf, Sapir, and others (see Werlen 1989), has passed through several stages of research but has not yet been proved satisfactorily (there is no evidence to prove or disprove it). Nevertheless, 'poverty' vs. 'richness' of language (in this context 'restricted' vs. 'elaborated', or 'simple' vs. 'complex') is a metaphorical formulation with almost magical effects, a phenomenon for which I have coined the term 'dichotomistic fallacies'.

'Linguistic codes' à la Bernstein are, according to modern sociolinguistic terminology, a kind of register. In order to understand the concept of 'code', it is helpful to conceive of it from the perspective of the hearer's expectations: that is, many or few words and flexible or rigid grammatical style signal, in relation to the type of speech situation, whether the speaker comes from a MC or a LC background. In other words: in a formal speech situation, such as a job interview, one would anticipate that a MC speaker will use a broader vocabulary and repertoire of 'legitimate' rules belonging to the standard variety than a LC speaker. Therefore, the language of lower classes is to a certain extent simpler and more predictable because of a higher recurrence of specific language patterns. Conversely, it is more difficult to know in advance what MC speakers are going to say because they have more lexical and syntactic options at their disposal. The 'typical' LC speaker, because of his/her authoritarian and rigid role expectations, applies speech strategies which in actual discourse result, for example, in a typical intonation pattern, breaks while speaking, repetitions of certain words and verbal stereotypes. Thus there is the assumption of a close relationship between social roles, thought, and ways of speaking.

Now, is the thesis that the grammatically and lexically less differentiated speech styles of the lower classes are inferior to the linguistic styles of the middle and upper classes relevant today? It certainly is, although it has become increasingly obvious that codes can be studied only in terms of pragmatic parameters, and not formal-grammatical ones (Klann 1975). Auwärter (1988), for example, shows that language use is much more dependent on speech situation and particular speaker characteristics than has been assumed.

However, the sociology of the 'legitimate' language advocated by Pierre Bourdieu (1982) has provided the class-specific description of language with a postmodern theoretical framework that can be applied to the problem of 'language and social inequality'. Bourdieu's popular thesis that speakers of a speech community are evaluated on the 'linguistic market' according to the social value of the variety they use (a continuum between the prestigious and the stigmatized use of varieties) changes Bernstein's 'sociogenetic' parts of the theory into socio-political ones: it is no longer 'richness' or 'poverty' of linguistic varieties that is important (linguistically they are seen as equivalent) but their social and normative credibility, their 'legitimacy'.

Are the 1990s the age of 'post-stratification', when language barriers have been overcome? Unfortunately, this is not the case for Germany: sociolinguistic codes are topical again, but the research community has not yet realized it. If one accepts the idealization that West Germans and East Germans live in the same speech community but in different communication communities, one can compare the linguistic behaviour of West Germans ('*Wessis*') with that of East Germans ('*Ossis*') in terms of linguistic codes – but only with considerable abstractions. Cut off from the international leisure scenes and markets of the Western world, '*Ossis*' were controlled in their social perceptions and needs, living under a kind of bell-jar of socialist monoculture, and were also constrained by the 'corset' of a state-run language culture. Compared with the differentiated role expectations of a prestige- and market-orientated Western society, their repertoire of roles was somehow less flexible, less prepared to adjust to complex, market-specific situations requiring a strategic use of language that can exploit or avoid disadvantages or advantages in interactional situations.

In 1989, as soon as *Ossis* appeared in supermarkets and among the West German public, and held speeches in parliament or other institutions, it was possible to study their rather rigid register. Instead of lower and middle class one could speak simply of *Ossi-* and *Wessi-milieu*. Again, one can find a complex relationship between language, social structure, and thought. Many Germans were able to see for themselves that West Germans and East Germans have different registers at their disposal, which they use in concrete situations with varied success. However, this observable fact would be hard to prove scientifically, as tests do not always succeed in checking a hypothesis (see Dittmar 1976: chap. 3). The most promising approach would be to document natural situations of communication between *Ossis* and *Wessis*, but this requires a lot of technology and research staff.

In the light of experience over the last forty years, one can predict various contrasting speech patterns in German–German dialogues, for example in buying and selling situations, in debates, negotiations, and speeches in public situations. In everyday life, both stylistic versatility and even differences in vocabulary and everyday idioms lead many to dismiss East German language behaviour as a restricted knowledge of the 'legitimate' language in Bourdieu's sense of the term. East German expressions like *urst stark* (very strong), *demokratische Kneipe* (democratic pub), *Reinigungsbrigade* (cleaning brigade), *Stomatologe* (dentist), *Fit-Wasser* (washing-up liquid) have to be replaced by their West German equivalents *sehr stark, populäre/alternative Kneipe, Straßenkehrer, Zahnarzt,* and *Geschirrspülmittel*. East German expressions are not important any more in reunited Germany, they are not 'legitimate' any more: the old norm is ridiculed just like other East German things. On the other hand, one can now find more indirect uses of language: for example, events that are paraphrased verbosely and euphemistically by *Wessis* are expressed by *Ossis* in a more straightforward and simple manner. *Ossis* do not use as many particles like *wohl, überhaupt, eigentlich, sowieso,* to modify or tone down their utterances, and they say what they think, not always regarding speaking as a kind of '*Imagepflege*' (image building).

Can we diagnose from these communicative restrictions a verbal or communicative 'deficit' that is the result of a discrepancy in relation to the knowledge of West Germans? Is the '*Ossi*-code' the successor of the 'restricted code', does the East–West contrast correspond to class differences? Differences certainly exist, old norms are useless in the prestigious Western society, new norms have to be acquired. Younger people and those who have the necessary intellectual capacities may be able to adjust to the 'new' society after a period of transition and change, but for older generations it is too late to overcome this linguistic relativity.

The language use of *Ossis* is more a question of social evaluation than of linguistic 'deficit' (see also my argumentation in Dittmar 1976, chap. 3). East German language usage, once a fixed, legitimate norm, is not a psychological handicap but an expression of differing communicative (and social) values for different purposes. Therefore we do not need linguistic compensation programmes for *Ossis* (that is, courses in rhetorical skills), but their irrevocable and binding integration in all areas of life. From this, I derive the 'principle of social and linguistic exposure': if *Ossis* are exposed to new norms, an appropriate and flexible language will develop that can cope appropriately with communicative tasks. In sociolinguistic terms, this means that language use adjusts functionally and dynamically to the social context. Sociolinguistic codes belong to the topic of language and social inequality, but the dynamics of such codes is far more significant than the dichotomous theory discussed above.

3.2   *The Study of Variation: Variety Grammar vs. Variable Rules*

The reconstruction of the theory of sociolinguistic codes led first to the description of the 'very restricted' German codes of Spanish and Italian migrants, and then to the ambitious formulation of a 'variety grammar' (VG). The VG was conceived as an instrument for describing linguistic behaviour by means of an explicit, probabilistically evaluated grammar, dependent on the basic parameters 'space', 'speech situation', 'social group', and 'historical periods'. This represented considerable progress over the earlier work on sociolinguistic codes: it was no longer important to check a hypothesis on the relationship between social structures (lower classes, upper classes) and linguistic variables. Instead, one tried to grasp heuristically what kind of correlations exist between varieties and extra-linguistic parameters. Therefore, it was considered practical to use a context-free grammar in order to describe varieties against the background of 'diachronic', 'diasituative', 'diatopic', and 'diastratic' dimensions of the variety space. The VG was successfully applied to the description of geographical varieties (Palatinate dialect: see Senft 1982) and learner varieties (Heidelberger Forschungsprojekt 'Pidgin-Deutsch', 1975, Klein and Dittmar 1979). Thus it represents a sociolinguistic alternative to Labov's 'variable rules'.

The basic concepts of a VG are the variety space, a reference grammar, and the probabilistic weighting of rules. The variety space is an ordered set of varieties that is to be described. Let us assume linguists want to describe the syntactic differences in

- war letters as against travel letters ('registers' that we may call R1 and R2)
- around 1917, 1942, 1967 (the factor 'time' with the values T1, T2, T3)
- by peasants as against intellectuals of the bourgeoisie (the factor 'social class' with the values S1 and S2).

Here, we have a three-dimensional variety space with $2 \times 3 \times 2 = 12$ varieties. According to this variety space, (R1, T2, S2) is the variety of war letters written around 1942 by intellectuals of the bourgeoisie.

For a syntactic description we need a reference grammar. This is a type of grammar (for example generative, functional, or relational) that has to be formulated explicitly, so that it becomes possible to describe the relevant linguistic characteristics of the varieties in question. It isolates individual varieties by applying certain rules with a certain probabilistic weighting. The concept of probabilistic rule-weighting can be explained by so-called context-free phrase structure grammars: consider the example from Klein (1988: 1000) in Figure 7.1. The example shows how the noun phrase (NP) develops from an elementary to a complex pattern of use through the learner varieties V1 to V6, which may be used by six different informants. The 'figures' show the relative frequency of rule application for each variety. If we add up the scores for each variety separately, we always end up with 1; but the distribution of the application of the five rules for each variety is different. In V1, for example, virtually only simple nouns are produced, whereas in V2 nouns with articles and the incorrect rule 'noun followed by adjective' also occur. In V3, the post-noun position of the adjective is used more frequently, and for the first time the attributive adverb is added to the noun. The tendencies of V3 are developed further in V4. The 'wrong' rule (Det N Adj) is reduced in

			V1	V2	V3	V4	V5	V6
NP	- - - ➤	N	0.9	0.6	0.3	0.2	0.2	0.2
NP	- - - ➤	Det N	0.1	0.3	0.3	0.3	0.3	0.3
NP	- - - ➤	Det Adj N	0.0	0.0	0.0	0.0	0.4	0.4
NP	- - - ➤	Det N Adj	0.0	0.1	0.3	0.4	0.0	0.0
NP	- - - ➤	Det N Adv	0.0	0.0	0.1	0.1	0.1	0.1

FIG. 7.1.   Variety grammar: hypothetical example of a 'rule block'

V5, and in V6 a use of rules appears that comes close to the standard.

The VG makes this development obvious in a precise and flexible way (and in principle it can do this on all linguistic levels). In VG, we call the NP rule in the example a 'rule block': it is defined by the identity of the category 'NP', with the total of the rule applications for a particular variety adding up to 1. This particular point gives it an advantage over Labov's variable rules as they have certain problems with statistics.

In order to work fruitfully with a VG, it is necessary to adopt the following procedure:

1. the variety space has to be determined (extra-linguistic variables like social status, gender, age, social network);
2. the variables that are to be investigated must be specified, for example certain morphological or syntactic rules, lexical variation, or alternative realizations of phonological rules;
3. a corpus must be gathered containing a certain number of utter-

ances that allow representative descriptions of the variables in question;

4. the quantitative distribution of the types of utterances in question has to be determined;
5. a suitable type of grammar for the description has to be chosen;
6. on the basis of the frequencies of rule occurrences found in the corpus probabilistic weightings have to be calculated.

However, the VG is a purely descriptive instrument: it describes the data in terms of the whole corpus, but it does not explain anything. It simply describes precisely certain regularities and leaves the question of interpretation open, which some linguists and sociolinguists find unsatisfactory. The model has been formulated precisely with respect to variable rules, which have generally been accepted as an instrument of description in the USA and are widely used nowadays. In Germany, however, both variable rules and VG have been used only occasionally by linguists, and the development of VG did not lead to the founding of a 'school'.

Variable rule analysis has been applied to the system of Berlin variants in a technically optimal way in Schlobinski (1987). On the basis of selected phonological variables, differences in the realization of Berlin urban vernacular were shown in relation to various parameters: district, age, gender, East and West. In the area of dialectology, VG has been used only to describe the dialect of Kaiserslautern (see Senft 1982). The most productive area of application of VG has been in Second Language Acquisition. The developments from elementary to more complex learner varieties were described very successfully by means of context-free, probabilistically weighted rules. An overview of the results of the so-called Heidelberg Project can be found in Klein and Dittmar (1979).

Clearly, descriptive adequacy (of a given corpus) is not a sufficient condition for a sociolinguistic theory. The question therefore is how VG can serve certain explanatory intentions. As we have already seen, one of the basic questions of sociolinguistics is the relativity of language use. To what extent does language use influence thought? To what extent is thought influenced by social context? We cannot answer these questions merely by using the technical tool of variety grammar. I take a rather defensive position with respect to this theoretical issue: what we should look for are the most relevant non-linguistic factors which best explain the variation under investigation. Thus sociolinguists should be guided in a very pragmatic way by those explanations which benefit from fruitful correlations between linguistic and non-linguistic parameters.

3.3   *Ethnographic Approaches*

It is thanks to Peter Auer that the concept of 'contextualization' introduced by J. J. Gumperz has been improved theoretically through his work on German–Italian bilingual language contact. In Auer's view, Gumperz's notion of 'contextualization' comprises in most general terms 'all activities by participants which make relevant, maintain, revise, cancel . . . any aspect of context which, in turn, is responsible for the interpretation of an utterance in its particular locus of occurrence' (Auer 1992: 4). This determination of context can be called flexible and reflexive.

While Gumperz assigns prosodic and gestural features to contexts of utterances, Auer emphasizes that the concept of contextualization is a dynamic one. It is basically a concept applicable to data that flexibly experiences changes in context. Above all, one has to assume a multidimensional relationship between context and text or discourse instead of a unidirectional one. Context is a constantly changing continuum that proceeds in time and forms different backgrounds to utterances. Crucial to Gumperz's concept of 'contextualization' is the view that the relationship between linguistic utterances and context is reflexive, that is, they influence each other; it is important to see that language is not only context-determined but itself contributes to the construction of context in a fundamental way. Context in this view is not just a given facet of interaction but the result of the participants' common social effort to share essential properties of the social context. Therefore it is not a collection of social or material facts but a collection of cognitive schemes or models of what is relevant to interaction at any time. Context comprises:

1. certain material and social facts of interaction that can be described by objective observers;
2. information that is not known by the participants before the interaction begins and that is independent of it.

The emergent context parameters refer to types of linguistic activity that cannot be predicted by the social or material circumstances of the interaction. They also refer to facets of social knowledge that all participants share but that have to be converted from the participants' invisible cognitive dispositions or resources into commonly available backgrounds.

Auer makes a major contribution to a typology of 'contextualization cues'. Such contextualization cues cannot be predetermined by a theory. However, it emerged from practical research that the class of non-referential, non-lexical contextualization cues, particularly

prosody, gestures, bodily movements, eye contact, and linguistic variation (including speech styles), are important elements of context. In
this way, contextualization can be defined as a relationship between
a speaker, a context, an utterance, and a (non-referential) contextualization cue: 'Contextualization cues are used by speakers in order to
enact a context for the interpretation of a particular utterance' (Auer
1992: 31).

Taking into account the process by which participants create a
certain context, Auer distinguishes between 'brought about' and
'brought along' aspects of contextualization: certain aspects are
'brought along' into interaction, others are 'brought about' in interaction. According to Auer, three groups of 'context' schemes can be
distinguished. The first group of context schemes is determined only
by the participants' intra-episodic contextualization work. This
includes such things as knowledge about how to tell a story in a
specific cultural context, how to refer deictically, what to tell first and
what follows. The second group of contextual schemes are the so-
called default assignments, which imply certain fixed assumptions
about the world. They are based on the 'default knowledge' that
represents knowledge about typical objects and situations. 'Default
knowledge' is a fundamental part of our common-sense knowledge
and is used to make cognitive systems ready for use by closing gaps in
our knowledge with standard assumptions about the world. Relevant
knowledge is used when anaphorical relations have to be discovered
in order to understand a text. 'Default assignments' also incorporate
certain expectations about a topic, such as when talking about crimi-
nals and the police in the framework of the institutions of speech.
Therefore, 'default knowledge' is part of the 'brought along' knowledge (for example knowledge about the social roles of parents and
children).

A third group of context parameters can be even less 'brought about'
than the assigned default norms. It comprises the physical environment of the interaction as well as the period of time and visible features
of the participants (gender, ethnic group, etc.). Deixis and gestures/
body movements refer to this environment. Certain aspects of the
environment are made transparent for the participants and therefore
serve as contextualization cues. If these environmental factors are missing in contextualization work, this can mean in some cases that they
are irrelevant to the situational context of a specific interaction. This
is also important if we want to estimate the significance of parameters
like gender or ethnic group.

Context and contextualization are a broad field for a kind of linguistics that follows symbolic-systematic forms of analysis. Contextualiz-

ation is of great significance for microethnographic sociolinguistics as it connects linguistic and non-linguistic behaviour, variation, prosody, gestures, and body movements. Only these multifunctional relationships give us a key to social meaning that cannot be understood appropriately without context. These concepts undoubtedly lead to progress in the area of sociolinguistic interactional analysis within a theory of social-cognitive ethnography.

3.4 *Sociolinguistic Styles*

In German sociolinguistics we distinguish between 'narrow' and 'broad' concepts of style. The former is connected with the attempt to integrate the 'variation' of linguistic styles into an (explicit) model of grammar (see, for example, the work of the Austrian linguists Dressler and Wodak). A broader and more pragmatically based concept of 'style' has been developed by ethnographically orientated sociolinguists who are particularly interested in the description of structures of dialogues (a representative publication is Hinnenkamp and Selting 1989). Thus the notion 'style' is a good indicator of the so-called pragmatic secession: linguists who pursue a pragmatic approach seek to distinguish themselves from 'system linguists'. However, in order to give a satisfactory definition of style, both approaches have to be taken into consideration. The key question for 'system linguists' is as follows: how can styles of linguistic activity be taken into account appropriately? For their part, the 'secessionists' must answer the following question: how do you explain the stylistic variation of linguistic expressions and meanings?

Stylistic Variation in Grammar: Input-Conditioned Variety Switches
Taking the example of standard German in Austria and the Vienna dialect, Dressler and Wodak have shown in many essays that coexisting dialectal varieties can be described only by unidirectional optional rules or bidirectional rules of linguistic change. Unlike Labov, who regards the variation of a variable as a continuum from a standard form to a dialectal form and does not take the intention of the speaker into account, Dressler and Wodak argue that the input of a rule must be the form intended by the speaker. In other words, it is presupposed that the speaker consciously intends to speak either Austrian standard or Vienna dialect. This means that attitudes and psychological factors govern the use of the different varieties. Dressler and Wodak postulate bidirectional input-switches between forms of standard and dialect because we do not know whether the underlying intention of the speaker was dialect or standard.

Now what is new in Dressler's and Wodak's approach to socio-linguistic variation? Unlike Labov and Klein with their models of variable rules and variety grammar, which separate the description of behaviour from the psychological reality, the Vienna authors think that psychological reality of linguistic relativity is a *conditio sine qua non*. They are not concerned with formal correlations between linguistic and sociological data: their argument is rather that psychological mechanisms intervene between social categories and the characteristics of individual speakers. This means that there is a psychologically conditioned free variation that can be assigned to non-social factors. Seen in this way, 'style' is the individual dimension of this variation, which is filtered by psychological processes. This view is strengthened by concrete examples of the behaviour of defendants in the courtroom, where socio-psychological categories like authority, dominance, and intimacy appear to have a greater significance than the social speech situation.

Pragmatic Speech and Communication Styles
The book *Stil und Stilisierung. Arbeiten zur interpretativen Soziolinguistik* (1989) edited by Hinnenkamp and Selting gives a representative survey of new work on 'sociolinguistic styles'. The contributors understand 'style' in accordance with Dittmar (1989) as *Gestalt* (form, product) and *Gestaltung* (formation, dynamic process), that is, the context-dependent and recurrent choice of linguistic expressions is bound to an interactive process of communication and understanding, which is necessarily based on interpretation. With the help of the notion of style, these for the most part ethnographically and ethno-methodologically orientated authors seek to grasp the characteristics that co-occur as linguistic expressions/meanings and discursive sequences in typical formulations. Thus the concept of sociolinguistic styles has been rediscovered by these linguists because it offers them a chance to integrate the semantic role of linguistic expressions in the analysis of sequences of interactional patterns.

From this perspective, according to Selting and Hinnenkamp (1989: 5), most investigations aim to establish 'what for the participants is the meaningful use of co-occurrent linguistic means of structure and expression ... in the developing situation of interaction'. In fact, these are programmatic formulations that have not yet been put into practice very much in discourse analysis; that is, the co-variance of interactive sequences and linguistic means that constitutes style in this sense is not yet sufficiently understood. Auer (1989) shows plainly that 'style' is more than a superordinate concept for individual observations of conversational events. His theoretical paper tries to explicate the notion of style through its proximity to the category of the 'naturalness' of language; on prosodic, phonological, morphological, syntactic, and textual levels, 'naturalness' must be included in the notion of style as a grammatical characteristic. While such grammatical features appear on these levels as 'linguistic variation', Auer understands style 'only in relation to one interpreting participant of the culture in question in relation to another' (Auer 1989: 29). In his view, style is the 'set of interpreted, co-occurring linguistic and/or non-linguistic characteristics that are ascribed to (groups/roles of) persons, kinds of text-types, media etc.' (ibid. 30), and he illustrates his concept with some examples of discourse styles.

A Compromise between Traditionalists and Secessionists

My concept of style (see Dittmar 1994) is systemic (-*emic* in contrast to -*etic*) and ecological. 'Style' is an ordering power: it imposes systematic and measurable constraints on a seemingly chaotic variation (in linguistics often called 'free' variation). While grammar aims at an abstract description of a priori knowledge independent of time and place, problems relevant to concrete communication, such as factual representation, exchange, and type of relationship, are solved by the use of 'style' as selected from within a set of alternatives in time and place. Stylistic registers translate communicative goals and purposes into adequate verbal representations and expressions. 'Styles' have many facets and serve a variety of needs. They form human relationships in a substantive way and provide us with the necessary and subtle means to express the need for (1) exchange and balance, (2) belonging (to a group) and distance/separation (from another group), and (3) the projection of actions and events (saliency, contrastiveness, and selection).

Underlying every conversation is a drive for exchange: I call this 'conversational energy' (CE). This CE is transformed into conversational activities (routine formulas, openings, closings, summaries, consequences of speeches, repairs, anticipation, turn-taking rules, etc.) and thematic guidelines (communicative goals and purposes). The processing of (discourse) topics and communicative activities represents a burden for both speaker and hearer, from which they need to be relieved. The process of exchange (informing, factual representation, requests/demands, etc.) brings both burden and its relief. Proximity and distance result from the creation of the exchange. Speakers have to find their balance between both poles. The accommodation to a specific state can be undertaken with stylistic register.

'Style' is simultaneously both form (*Gestalt*) and formation (*Gestaltung*). 'Stylistic form' is habitual: its expression is a 'static' trace of formal realizations in the verbal encoding of morpho-syntactic meaning in the message (which corresponds to the Labovian linguistic variable and its hypothetical 'social significance'). 'Stylistic formation' is the process of putting speech in a 'form', that is, in an expressive discourse which continuously accommodates individual intentions to topical relevance, situational context, and pragmatic conditions.

In the traditional study of style and in the more recent variationist-orientated sociolinguistics, 'style' is understood as a variable set of relatively context-free alternating forms (variants) for functional semantic equivalents (cf. Thibault 1982). In contrast to Labov's notion of style as a form of acquired habitual behaviour, I view 'style' in an 'ecological' and 'systematic' sense as an ordered system of preferential tendencies (preferences) of language use (of speakers). Forms of expression, which are context-bound and filtered through discursive frameworks, are selected from the various levels of the individual linguistic variety space and combined by means of co-occurrence restrictions (grammatical and semantic limitations) into a specific stylistic level. 'Style' is therefore an expression of goals and purposes and partially strategic, partially habitual in nature. Discursive frameworks indicate the type of exchange and the communicative burden which must be balanced or relieved. In this process modality of communication plays an important role, namely proximity and distance within the relationship, representational projection, etc. Styles represent the accommodation to the intentions and roles of the interactional partners.

There are, then, four levels between which we have to differentiate when we speak of 'style':

1. the level of social action (intentions, action goals and purposes);
2. the frameworks which interlocutors must set as a prerequisite to taking part in common activities, so that these activities are orientated towards and focused on the goals and purposes;
3. the conversational activities (exchange, projection of action and events) which move along a topical guideline (loading of the communicative task) according to the framework and which constitute conversational patterns that control the validity of the frame and the closure of the topical guideline and, at the same time, repair breaks in the frame and realign them with the context; and
4. the level of linguistic expressions and their form (address forms, epistemic expressions, routine formulaic expressions, active–passive, connectors, etc.), which are filtered through the framework, conversational activities, and topical guidelines, and which

constitute a corresponding stylistic situation via co-occurrence restrictions.

Figure 7.2 illustrates this relationship. In contrast to Auer's descriptive concept of style, mine is a functional one: the functional determination of stylistic means of expression is taken into account in a social-psychological model (cf. for example the work by Sandig (1986) and Schlobinski, this volume).

4 THEORIES OF SOCIOLINGUISTIC VARIATION II: EAST GERMAN PERSPECTIVES

4.1 *Early Work on Social Dialectology in the GDR*

In the former GDR, dialectologists were the pacemakers in studies on the sociology of language and sociolinguistic thinking. The same was true in the FRG in the 1950s and 1960s. Both East and West German dialectology draw on many observations on the sociology of language in the second half of the nineteenth and the beginning of the twentieth centuries. In the early years of this century and also in the 1920s, macro-sociological categories like social class, social status, profession, group, and education were already discussed in connection with the preservation and decline of dialects and linguistic change.

The book *Sprachsoziologische Studien in Thüringen* by Rosenkranz and Spangenberg (1963) was fundamental to the development of GDR-specific sociolinguistics: it deals with 'linguistic change in the Thuringian dialectal area' and 'tendencies of national linguistic developments in Thuringia'. The empirical foundation of both sections of the book is remarkable. There are statistical arguments and results of questionnaires, and dialect accounts from Thuringia are given showing evidence of the decreasing use of dialect and diagnosing the emergence of an 'educated colloquial language'.

Rosenkranz, for example, emphasizes the particular significance of written language in the age of industrialization, spreading more and more and driving the dialect out of industrial centres. He sees colloquial language as a kind of compromise variety between local dialects and written language. Its form is largely determined by its purpose (to achieve optimal understanding), and it stays closer to written language but integrates parts of the dialectal system. Moser's differentiation between 'dialect', 'colloquial language', and 'common language' (*Gemeinsprache*) on the horizontal level, and 'vernacular' (*Volkssprache*), 'refined vernacular' (*erhöhte Volkssprache*) and 'standard'

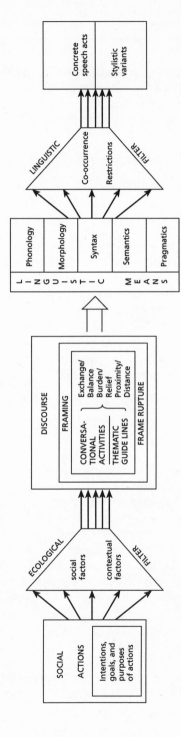

Fig. 7.2. Linguistic and extra-linguistic dimensions of the concept of 'style'

(*Hochsprache*) in terms of 'vertical' social stratification seems to have only limited application to conditions in Thuringia. Rosenkranz sees dialects only as local dialects, and he questions the existence of 'Thuringian', 'Bavarian', etc. as superordinate regional dialects. Urban centres are of great importance for him because they serve as melting-pots between dialectal and written varieties. In short, the decisive development in the second half of the twentieth century is seen by Rosenkranz in the emergence of 'colloquial language'. It is an 'extensive lingua franca (*Verkehrssprache*) without strict norms' (1963: 19). Rosenkranz names seven essential features of this *Verkehrssprache* (ibid. 21–2):

1. Stylistically, it is a vernacular: 'a simple enumerative way of speaking through the connection of many main clauses by "and then . . ."', hardly any abstract terms, strong reference to emotions, metaphorical expressions as in the dialect.
2. It is a 'lingua franca' in the sense that it avoids highly localized vocabulary and sound forms; more generally it avoids everything that makes understanding more difficult.
3. From a sociological point of view, it is a language of urban origin, that is, it is clearly related to written language.
4. It avoids 'vulgarity'; the crudeness of gutter language ('Rosenkranz's expression) is paraphrased, as seen in the following examples: *Armleuchter* (euphemistic expression for *Arschloch*/arsehole) or *Scheibenhonig* (euphemistic expression for *Scheiße*/shit). Using Brown and Levinson's terminology of politeness, one could say: urban expressions are more indirect and less 'face-threatening'.
5. The vernacular is subdivided into communication communities that are connected with each other mainly by social areas; but it is not confined to specific geographical areas as is the case with local dialects. 'In this connection, several layers of colloquial language become congruent, i.e. cross-sections of several speech communities' (1963: 22).
6. The present state of the sound system of colloquial language is marked by a rather wide range of varieties. Its function is to make it possible for the speakers to adjust to their interlocutors.
7. The vocabulary of colloquial language is drawn from various different varieties. Words and expressions of different origin are used at the same time, so that there is no single choice of word or norm that is considered correct. Taking the village of Mansfeld as an example, Rosenkranz gives an impressive illustration of the influence of industrialization and shows how industry,

changes in agriculture, commuting, and urbanization trigger off a dynamics of linguistic change between influences of written language and local dialects.

4.2 *The Role of Sociolinguistic Research in the Socialist Society of the GDR*

The theses on Marxist sociolinguistics by Grosse and Neubert (1974*b*) have much more to do with the sociology of language than with dialectological theorizing. Both authors developed the model of the 'Sociolinguistic Differential' (1974*b*: 13) that was often referred to in the former GDR. The sociolinguistic differential of communication is a complex unity comprising the following four factors: linguistic 'codes', 'transmitter', 'receiver', and the prevailing 'type of speech situation'. The variants of language use that are relevant to society (described as grammatical or semantic phenomena) are formed by 'transmitter', 'receiver', and 'speech situation'. The variations in language use according to transmitter and receiver are described as 'sociolinguistic layers', the variants resulting from different speech situations as 'sociolinguistic spheres'. Both layers and spheres can be combined in 'sociolinguistic systems'. With respect to varieties, Grosse and Neubert distinguish 'dialect', 'sociolect', and 'functional style' (the last reflecting requirements of the speech situation or communicative functions). Examples of the terminological delimitation of dialect, sociolect, and functional style are given, but there are no practical illustrations or instructions on how to describe variation within a corpus.

The 'Sociolinguistic Differential' did not in fact have much impact on empirically orientated research in the 1970s and 1980s. However, the theoretical considerations did serve to determine the directions research should take. The major point of theoretical interest is Grosse and Neubert's statement that a correlation between linguistic and social data in the sense of a co-variance or co-variation is theoretically unsatisfactory because it presupposes a far-reaching and unacceptable separation of social and linguistic factors. They argue that there is no simple causal relationship between language and society because both are related dialectically to each other and the effects of their relationship are mutual.

This argument is developed further in Hartung and Schönfeld (1981), based on the thesis that the relationship between language and society is mediated by 'linguistic-communicative activity'. The starting-point is that 'mental and linguistic-communicative activity are first of all essential components of work' (Hartung 1981: 47). Following the concept of 'activity' developed by Leontiev in Soviet psychology, Hartung defines linguistic-communicative activity as a 'specific mediation of the permanent alternation of externalization and internalization of mental activity. Its autonomy is based on the fact that there is no other possibility of mediation [between these two processes] which could have anything like the same effectiveness' (Hartung 1981: 52). Thus linguistic variation appears to be based on conditions of activity: 'from this point of view, linguistic variation does not reflect social and situational differentiations directly but models certain conditions of communicative activity which can be related to such differentiations'.

These considerations had a major influence on the approach to discourse analysis which was initiated (largely by Hartung) in the last years of the former GDR from the perspective of the sociology of language. The main focus of this work was on communication in work processes and the resolution of group conflicts. Since 1992 these investigations on discourse have been continued at the Institut für deutsche Sprache in Mannheim.

4.3   *Communicative Norms*

Hartung (1977*b*) introduces the distinction between 'speech community' and 'communication community', which allows him to distinguish communication norms in the former GDR that changed under socialist conditions from those in the FRG without questioning either nation's affiliation to the 'German speech community'. According to Hartung, 'norms' are 'intellectual benchmarks that serve as a basis for people who enter mutual communicative relationships as speakers/ writers and hearers/readers, showing them how they can or should proceed in realizing their communicative relationship' (1977*b*: 12). While Lerchner (1973), following Coseriu and Semenyuk, uses the double character of linguistic norms (internal-linguistic vs. external-social) as a 'central link' between the autonomy of the language system (= possibilities/potential rules) and the regularities of linguistic varieties (= reality) for the benefit of linguistic theory, Hartung (1977*b*: 29 ff. and 50 ff.) introduces a new definition of the sociolinguistic notion of norms on the basis of activity theory.

The relationship between 'system' and 'norms' is rather different as regards the level of language use (= social level) than as regards the level of social meaning (= level of the reflection of social reality). On the level of linguistic-communicative activity, system and norms are more or less congruent, while on the level of social meaning, the notion of norms has a broader scope than the notion of system because it includes not only the linguistic form of the product of activity but also the basic pragmatic conditions of communication as such (politeness, openings and closings of verbal interaction, principles of exchange, semantic and communicative appropriateness, etc.). These considerations lead Hartung (1977*b*: 39 ff.) to the differentiation of various types of norms: grammatical-semantic (text and discourse) and situative (types of speech situations) ones on the qualitative level, pragmatic and interaction-specific ones with reference to the basic conditions of communication.

The 'norm' is one of the key notions of sociolinguistic theory: choice and alternation of varieties are governed by norms. Consequently, norms are the crucial feature of a sociolinguistic theory in which social structures and language system are in a reciprocal relationship. Following Hartung's attempt to redefine the notion of norms, a number of empirical studies appeared dealing with the linguistic manifestations and differentiated realities of the newly created communicative norms in the GDR and drawing their data mainly from workers in socialist enterprises.

Particular emphasis was placed on the analysis of lexis in the evaluation of the results of these studies. It was used to show how and to what extent the language had changed under socialist conditions and how its new norms corresponded better to the communicative needs of society. Considerable linguistic change was demonstrated by the example of 'neologisms' that were used with different frequencies by different groups (Herrmann-Winter 1979: 132 ff., Donath and Schönfeld 1978: 42–3). Another important result was that 'the concept of a North German vernacular with its own subsystems on all linguistic levels and its own norms cannot be supported by the results of our analysis' (Herrmann-Winter 1979: 256). In other words, dialects were being replaced by a vernacular which was influenced to a large extent by the standard variety.

Finally, it seems important to me that different levels of linguistic variation were understood as being controlled by various extralinguistic factors: syntactic characteristics (elliptical clauses, occurrences of assimilation and reduction) were interpreted as features of spoken language in the context of informal 'speech situations', and phonetic variants were understood as 'spatial' markers, while the GDR-specific neologisms reflected the socio-economic, cultural, and political conditions (Herrmann-Winter 1979: 254–5).

4.4   *The Concept of 'Language Culture'* (Sprachkultur)

The sociolinguistic concept of language culture that had been developed in the former GDR at the end of the 1970s can be called the practical application of the sociolinguistic notion of norms to the task of maintaining the 'quality' of the language. Ising (1977) and Scharnhorst and Ising (1978 and 1982) provide us with a theoretical documentation of the concept, and the latter (1978: 332) give a more precise account of the tasks of 'language culture':

(*a*) the culture of the language is the condition of the system of the literary language, its degree of stability, its richness of meaning, its stylistic layers, and its ability to serve in all areas of linguistic communication, particularly public communication;

(*b*) the cultivation of language, i.e. efforts to improve the culture of the language. This activity is directed towards the development of a literary language in accordance with known regularities and expressive needs of society;

(*c*) the culture of speech, i.e. the condition of the sum total of linguistic utterances, the level of social communication through language;

(*d*) the cultivation of speech, i.e. efforts to improve the culture of speech. This activity is directed towards the improvement of the level of linguistic communication.

According to Hartung (1984), *Sprachkultur* is the practical application of the concept of language norms: it is central to 'the evaluation of language and linguistic interaction and to the promotion of an unhindered and confident command of language' (Hartung 1984: 70). To establish a rational notion of language culture, it is not enough to distinguish between 'narrow' and 'broad' concepts of culture. Hartung (1984: 72) concedes that 'culture only begins at a certain above-average level', for we distinguish between 'culture' and 'non-culture' in negative terms (e.g. 'they have no gastronomic culture'), but the qualitative evaluation of behaviour in terms of positions in a hierarchy should not be based on élitist principles. However, if we replace this 'narrow' conception of culture with a 'broad' one, we may arrive at a concept that embraces all areas of human social activity, but it still remains to a large extent a 'relativistic' notion, because various ways of dealing with language are seen as functionally equivalent as long as the relevant communicative tasks are fulfilled.

Hartung gets round this dilemma by trying 'to make a trivial broad concept of culture more interesting by means of qualitative restrictions' (1984: 73). This can be achieved by incorporating norms, as culture is not only 'the result of current critical analysis of living conditions, but also of a *tradition of critical analysis*, which has

accumulated experience and fixed it in the form of norms' (ibid.).
From the point of view of the historical development of normative
processes in communication communities, 'not all of the various forms
are functionally equivalent; the norms of the standard language occupy
a superior position' (1984: 74). The standard language owes this
superior position to its function of recording social knowledge in
written form and thereby making it widely available: access to the
standard language provides access to this social knowledge. Admittedly, it is not just a question of lexical knowledge, but rather much
more fundamentally of 'linguistic processes of fixing and presenting
social knowledge, which are not restricted to the written form but
also have a considerable impact on oral communication' (ibid.). This
widely effective mediation of language culture and the practice of
fostering 'appropriate' language use presuppose that various kinds of
norms (such as grammatical and communicative, oral and written) can
be understood sociolinguistically: 'the notion of language culture then
relates to the level on which such communicative norms are used'
(1984: 80).

5   CONCLUSIONS: INTEGRATING PERSPECTIVES FROM WEST AND EAST

In a retrospective review of research on grammar in the former GDR,
Manfred Bierwisch talks pejoratively of the 'endless insistence on the
social character of language with the result that it remained unclear
what it was actually supposed to mean' (Bierwisch 1992: 171). I cannot
agree with this rhetorical oversimplification of the actual achievements
of sociolinguistic research in the GDR. One of its main concerns was
a broad, comprehensive linguistic education of all social classes that
was to be based on the knowledge of communicative norms and the
functionally appropriate use of dialect, standard, and vernacular. This
objective was the principal justification for theoretical and (to a lesser
extent) empirical sociolinguistic research in the GDR.

However, as long as expensive research in the West on 'restricted'
and 'elaborated' codes, or on undirected second language acquisition
and learner varieties only results in an improvement of linguistic techniques for the description of variation, because research projects lose
their funding after only five years as a result of new trends, or practical
efforts to apply the results of empirical research are not supported, the
future 'United German' sociolinguistics urgently needs East German
knowledge on the effectiveness of communicative norms and the principles of language culture. Studies on youth language and on discourse

with and about foreigners (Jäger 1992) show that relying on linguistic relativity, on a 'natural order' in the sense of the 'difference' (as opposed to 'deficit') conception of this society does not provide equality of opportunity for free. The way forward now is for West German technical–empirical know-how to join forces with the East German ethic of language culture in order to create a sociolinguistics that does not tolerate a barrier between theory and practice.

Further Reading

Dittmar and Schlobinski (1988)
Hartung and Schönfeld (1981)
Hinnenkamp and Selting (1989)
Scharnhorst and Ising (1976, 1982)
Schlieben-Lange (1991)

References

Ammon, U., Dittmar, N., and Mattheier, K. (eds.) (1987–8), *Socio-linguistics: An International Handbook of the Science of Language and Society*, 2 vols. (Berlin: de Gruyter).
Auer, P. (1989), 'Natürlichkeit und Stil', in Hinnenkamp and Selting (1989), 27–59.
—— (1992), 'Introduction: John Gumperz' approach to contextualization', in Auer and di Luzio (1992), 1–37.
—— and di Luzio, A. (eds.) (1992), *The Contextualization of Language* (Amsterdam: Benjamins).
Auwärter, M. (1988), 'Das Experiment in der Soziolinguistik', in Ammon *et al.* (1988), 922–31.
Bierwisch, M. (1992), 'Grammatikforschung in der DDR: Auch ein Rück-blick', *Linguistische Berichte*, 139: 1–13.
Bourdieu, P. (1982), *Ce que parler veut dire* (Paris: Éditions de Minuit).
Brown, P., and Levinson, S. (1988), *Politeness: Some Universals in Language Usage* (Cambridge: Cambridge University Press).
Dieckmann, W. (ed.) (1989), *Reichthum und Armut deutscher Sprache. Reflexionen über den Zustand der deutschen Sprache im 19. Jahrhundert* (Berlin, New York: de Gruyter).
Dittmar, N. (1976), *Sociolinguistics: A Critical Survey of Theory and Application* (London: Edward Arnold).
—— (1989), 'Soziolinguistischer Stilbegriff am Beispiel der Ethnographie einer Fußballmannschaft', *Zeitschrift für Germanistik*, 4/10, 414–23.
—— (1994), 'Sociolinguistic Style Revisited: The case of the Berlin Speech Community', in *Verbal Communication in the City*, proceedings of sym-posium on 'Verbale Kommunikation in der Stadt', Bern, 1990 (Tübingen: Narr).

DITTMAR, N., and KLEIN, W. (1972), 'Die Code-Theorie Basil Bernsteins,' in Klein and Wunderlich (1972), 15–35.

—— and SCHLIEBEN-LANGE, B. (eds.) (1982), *Die Soziolinguistik in romanischsprachigen Ländern* (Tübingen: Narr).

—— and SCHLOBINSKI, P. (eds.) (1988) *The Sociolinguistics of Urban Vernaculars* (Berlin: de Gruyter).

DONATH, J., and SCHÖNFELD, H. (1978), *Sprache im sozialisitschen Industriebetrieb* (Berlin: Akademie-Verlag).

DRESSLER, W., and WODAK, R. (1989), *Fachsprache und Kommunikation. Experten im sprachlichen Umgang mit Laien* (Vienna: Bundesverlag).

GABELENTZ, G. VON (1891/1984), *Die Sprachwissenschaft. Ihre Aufgaben, Methoden und bisherigen Ergebnisse* (Darmstadt: Wissenschaftliche Buchgesellschaft).

GROSSE, R., and NEUBERT, A. (eds.) (1974a), *Beiträge zur Soziolinguistik* (Halle: Niemeyer).

—————— (1974b), 'Thesen zur marxistisch-leninistischen Soziolinguistik', in Grosse and Neubert (1974a), 9–24.

GÜNTHNER, S., and KOTTHOFF, H. (1991a), *Von fremden Stimmen* (Frankfurt: Suhrkamp).

—————— (1991b), 'Von fremden Stimmen: Weibliches und männliches Sprechen im Kulturvergleich', in Günthner and Kotthoff (1991a), 1–51.

HARTUNG, W. (ed.) (1977a), *Normen in der sprachlichen Kommunikation* [= *Sprache und Gesellschaft*, 11] (Berlin: Akademie-Verlag).

—— (1977b), 'Zum Inhalt des Normbegriffs in der Linguistik', in Hartung (1977a), 9–69.

—— (1981), 'Differenziertheit der Sprache als Ausdruck ihrer Gesellschaftlichkeit', in Hartung and Schönfeld (1981), 26–72.

—— (1984), 'Sprachkultur als gesellschaftliches Problem und als linguistische Aufgabe', *Jahrbuch 1984 des Instituts für deutsche Sprache*, 70–81.

—— and SCHÖNFELD, H. (eds.) (1981), *Kommunikation und Sprachvariation* [= *Sprache und Gesellschaft*, 17] (Berlin: Akademie-Verlag).

HEIDELBERGER FORSCHUNGSPROJEKT 'PIDGIN-DEUTSCH' (1975), *Sprache und Kommunikation ausländischer Arbeiter. Analysen, Berichte, Materialien* (Kronberg: Scriptor).

HERMANN-WINTER, R. (1979), *Studien zur gesprochenen Sprache im Norden der DDR* (Berlin: Akademie-Verlag).

HINNENKAMP, V., and SELTING, M. (eds.) (1989), *Stil und Stilisierung. Arbeiten zur interpretativen Soziolinguistik* (Tübingen: Niemeyer).

HUMBOLDT, W. VON (1904), *Wilhelm von Humboldts Werke*, ed. Albert Leitzmann (Berlin: Königliche Akademie der Wissenschaften).

ISING, E. (ed.) (1977), *Sprachkultur—Warum, wozu?* (Leipzig: Akademie-Verlag).

JÄGER, S. (ed.) (1992), *BrandSätze. Rassismus im Alltag* (Duisburg: DISS).

KLANN, G. (1975), *Aspekte und Probleme der linguistischen Analyse schichtenspezifischen Sprachgebrauchs* (Berlin: Max Planck Institut für Bildungsforschung).

KLEIN, W. (1988), 'Varietätengrammatik', in Ammon *et al.* (1988), 997–1006.

—— and DITTMAR, N. (1979), *Developing Grammars: The Acquisition of German by Foreign Workers* (Heidelberg, New York: Springer).

—— and WUNDERLICH, D. (eds.) (1972), *Aspekte der Soziolinguistik* (Frankfurt: Athenäum).

LERCHNER, G. (1973), 'Sprachnorm als linguistische und soziologische Kategorie', *Linguistische Studien*, A/3: 9–31.

LÖFFLER, H. (1985), *Germanistische Soziolinguistik* (Berlin: Erich Schmidt Verlag).

MÖSER, H. (1964), *Das Aueler Protokoll. Deutsche Sprache im Spannungsfeld zwischen West und Ost* (Düsseldorf: Schwann).

NABRINGS, K. (1981), *Sprachliche Varietäten* (Tübingen: Narr).

ROSENKRANZ, H., and SPANGENBERG, K. (1963), *Sprachsoziologische Studien in Thüringen* (Berlin: Akademie-Verlag).

SANDIG, B. (1986), *Stilistik. Sprachpragmatische Grundlegung der Stilbeschreibung* (Berlin: de Gruyter).

SCHARNHORST, J., and ISING, E. (eds.) (1978, 1982), *Grundlagen der Sprachkultur. Beiträge der Prager Linguistik zur Sprachtheorie und Sprachpflege*, 2 vols. (Berlin: Akademie-Verlag).

SCHLIEBEN-LANGE, B. (1991), *Soziolinguistik*, 3rd edn. (Stuttgart: Kohlhammer).

SCHLOBINSKI, P. (1987), *Stadtsprache Berlin* (Berlin: de Gruyter).

SCHRÖDER, H. (1921), 'Hyperkorrekte (umgekehrte) Schreib- und Sprechformen besonders im Niederdeutschen', *Germanisch-Romanische Monatsschrift*, 9: 19–31.

SCHUCHARDT, H. (1885/1972), 'On Sound Laws: Against the Neogrammarians', in Vennemann and Wilbur (1972), 39–72; originally publ. 1885 as *Über die Lautgesetze: Gegen die Junggrammatiker* (Berlin: Oppenheim).

SELTING, M., and HINNENKAMP, V. (1989), 'Einleitung: Stil- und Stilisierung in der interpretativen Soziolinguistik', in Hinnenkamp and Selting (1989), 1–23.

SENFT, G. (1982), *Sprachliche Varietät und Variation im Sprachverhalten Kaiserslauterer Metallarbeiter. Untersuchungen zu ihrer Begrenzung, Beschreibung und Bewertung* (Bern: Lang).

THIBAULT, P. (1982), 'Style, Sense, Function', in Dittmar and Schlieben-Lange (1982), 73–85.

VENNEMANN, T., and WILBUR, T. (eds.) (1972), *Schuchardt: The Neogrammarians and the Transformational Theory of Phonological Change* (Frankfurt: Athenäum).

WERLEN, I. (1989), *Sprache, Mensch und Welt. Geschichte und Bedeutung des Prinzips der sprachlichen Relativität* (Darmstadt: Wissenschaftliche Buchgesellschaft).

8 Language in Intercultural Communication

MARTINA ROST-ROTH

I   INTRODUCTION

Many people living in Germany today speak a language other than German and consider themselves members of another culture. These people may be '*Gastarbeiter*' (guest workers), '*Asylsuchende*' (refugees seeking political asylum in Germany), or '*Aussiedler*' (former residents of eastern European countries who are of German descent). Not surprisingly, therefore, contact within the German context often entails problems of intercultural communication.

At the same time, Germany's particular political situation as formerly two separate states and now one unified country also makes the encounters between '*Ossis*' and '*Wessis*' (East Germans and West Germans) an issue of intercultural communication. This very special situation allows us to observe the effects of the different cultural backgrounds resulting from different political, economic, and educational systems in societies sharing basically the same language. However, language studies reveal that usage differs in the two parts of Germany as well, so that the *Ossis* and *Wessis* can still be said to regard each other to a certain extent as foreigners.

This chapter briefly surveys the research on intercultural communication and discusses the implicit assumption that most intercultural communication is problematic and likely to result in misunderstanding and miscommunication (Section 2). It then considers the proportion of the population of Germany who are non-Germans and the present political policies towards integration and the teaching of foreign languages (Section 3). The next two sections discuss linguistic investigations of foreigners' second language acquisition and the ways in which German native speakers adjust to non-native-speaker levels of German competence. Section 6 deals with the study of intercultural communication in institutional settings such as doctor–patient

I am much indebted to Patrick Stevenson for his friendly editorial advice.

consultations and other advisory and counselling situations – a 'classic area' of German intercultural communication research. Section 7 presents some findings in the area of contrastive pragmatics, and Section 8 offers some observations on the contact between East and West. Finally, the last section will discuss some of the different approaches to explaining miscommunication and misunderstanding in Germany – and beyond.

2   RESEARCH ON INTERCULTURAL COMMUNICATION

Most research on intercultural communication assumes an inherent potential for misunderstanding and conflict. According to Hinnenkamp (1989), there are two basic approaches towards an explanation for this, each with a different theoretical implication: he distinguishes between the approaches in which intercultural problems are considered the result of cultural differences in the way communication is conducted and those in which socio-psychological factors are thought to be responsible for miscommunication.

In the first case, there is a basic assumption that the common features of the interacting cultures will function adequately but that their differences may lead to 'transfer', resulting in misinterpretation and conflict (in this context, Hinnenkamp refers in particular to the work of the American linguist John Gumperz: see e.g. Gumperz 1982). The other approach explains intercultural communication problems in terms of in-group and out-group identification. This sociopsychologically orientated approach reveals that cultural identity and feelings of group membership are just some of many features that may have a bearing on social behaviour, that is, members of different cultures and language communities may also have common beliefs in their roles as women or their enthusiasm for sport and so on. Therefore conflicts are not considered to be an automatic consequence of cultural differences. Instead it becomes obvious that the effect of cultural differences may be quite unpredictable, and may be determined by the different interests of the parties involved (cf. Barth 1969*b* and Streeck 1985).

Most empirical research focuses on pragmatic contrasts in language use and in the resulting misunderstanding, and pioneering work in this area has been carried out by John Gumperz. A central concept in his theory is 'contextualization' (for a critical discussion, see Auer 1992, which develops this theory further; see also Dittmar, this volume):

Constellations of surface features of message form are the means by which speakers signal and listeners interpret what the activity is, how semantic content is to be understood and how each sentence relates to what precedes or follows. These features are referred to as contextualization cues. (Gumperz 1982: 131)

According to Gumperz *et al.* (1979: 21), misunderstandings in intercultural communication occur for three main reasons:

(1) Different cultural assumptions about the situation and about appropriate behaviour and intentions within it.
(2) Different ways of structuring information or an argument in a conversation.
(3) Different ways of speaking: the use of a different set of unconscious linguistic conventions (such as tone of voice) to emphasise, to signal connections and logic, and to imply the significance of what is being said in terms of overall meaning and attitudes.

For a critical discussion of Gumperz by German linguists, see Streeck 1985, Redder and Rehbein 1987, and Hinnenkamp 1989: 9 ff.)

Gumperz's work focuses mainly on intercultural communication with members of ethnic minorities which have distinctive characteristic language features, such as in Asian English. Intercultural communication in Germany, however, is generally characterized by the fact that immigrants and foreigners often have low competence in the German target language and/or are still at an intermediate stage of learning or acquiring the language (for a characterization of the different degrees of competence achieved by non-native speakers, see Section 4 below). It is only possible to talk of the emergence of an 'ethnic variety' of German when it comes to the second and third generation of immigrants (see Hinnenkamp 1989: 13).

Basic research into the use of German in intercultural communication has been done on communication in institutional settings, such as between doctors and patients and in the educational context (see also Wodak, this volume). Pragmatic analysis of discourse structures shows that problems and conflicts arise from culture-specific communication patterns, which may then lead to misinterpretations (cf. Rehbein 1985*a* and Redder and Rehbein 1987, which present a number of different studies).

Apart from these empirical studies, which look at the actual problems in documented communications, the term 'intercultural communication' is currently the subject of much discussion in various disciplines. Since the mid-1980s, there have been an increasing number of publications under this heading (after Asante *et al.* 1979, a traditional handbook for intercultural training, several collections of essays were published, especially in Germany: Rehbein 1985*a*,

Redder and Rehbein 1987, Gerighausen and Seel 1987, Knapp *et al.* 1987, Albrecht *et al.* 1987, Wierlacher 1987, Wierlacher *et al.* 1987 and 1988, Neuner 1988, Spillner 1990, Reuter 1992. See also Brunt and Enninger 1985 and Blommaert and Verschueren 1991).

Intercultural communication is also a major issue in the social and educational sciences: intercultural education and intercultural learning have become catchwords in educational studies and important strategies in attempts to tackle the problems caused by growing racism.

Intercultural communication has become a popular topic in foreign and second language teaching as well. Numerous articles and publications in Germany have appeared since the late 1980s in the area of German as a Foreign Language (*Deutsch als Fremdsprache*). However, these publications all seem to use the label 'intercultural communication' as a general term, even though they are rooted in such different traditions as social studies, intercultural learning, linguistic pragmatics, and the ethnography of speaking.

The 'intercultural boom' has also had an impact on the field of literature and Intercultural German Studies (*Interkulturelle Germanistik*; see Wierlacher 1987 and Zimmermann 1989), and intercultural problems have been a relevant issue in the political and international sphere too (see e.g. Council of Europe 1986 and the *Dokumentation über die VI. Konferenz der Europäischen Kulturminister* 1990, where the subject of the conference was 'Multicultural society and European cultural identity').

Finally, intercultural aspects have also been examined in the context of the East–West German question. For many years, contrastive studies dealt primarily with official or public discourse, especially as represented in the media (see especially Hellmann 1984). But after the unification of Germany the focus of studies in this area shifted to concentrate more on intercultural contact (see e.g. Gumperz 1991, Barden and Großkopf 1992, Ylönen 1992). These more recent studies on East–West contact will be discussed in more detail in Section 8.

3   MULTICULTURAL SOCIETY AND INTERCULTURAL ENCOUNTERS

Investigations of intercultural communication in Germany have mainly been concerned with the groups of migrant workers who have been coming to Germany since the 1960s from southern European countries (see especially Sections 4, 5, and 6 below). Refugees from other countries seeking asylum in Germany, arriving in increasing numbers since the early 1980s, have been relatively neglected in linguistic research.

Since the end of the 1980s, there has been an enormous increase in the number of people coming to Germany from eastern European countries and the former USSR. A considerable number of them belonged — or at least claimed to belong — to German minorities in the countries they came from, but this tells us nothing about their actual proficiency in German (for an overview, see Born and Dickgießer 1989, Bade 1990, and Rosenberg and Weydt forthcoming). Official German policy, as far as these *Aussiedler* or so-called *Rückkehrer* ('returning' immigrants) are concerned, has been to offer them a variety of measures to facilitate rapid integration or assimilation, and this group has therefore had the most extensive programme of language teaching.

In considering statistics on the size of the groups that constitute what is now multicultural Germany, one has to bear in mind that many foreigners live and work in Germany illegally, so that the official figures are almost certainly an underestimate. Nevertheless, they give some impression of the current situation. The official reports of the German Federal Bureau of Statistics, for example, show the annual percentage of foreign nationals in Germany to be between 6.4 per cent and 7.6 per cent for the period 1975 to 1987 (see Hamburger 1989: 7). That is to say, there was a relatively stable and quite substantial percentage of foreign nationals residing in Germany during that period: in absolute numbers, this means that in 1987 4.5 million foreigners lived in the Federal Republic. More than two-thirds of them had lived there for more than eight years. From 1988 to 1989 there was a rise in the number of foreign nationals, so that in 1990 they made up 8 per cent of the population, *Asylsuchende* and *Aussiedler* not included (!). The newspaper *Der Tagesspiegel* (15 July 1992: 1) reported the following numbers: 6 million foreigners living in Germany, with 1.5 million coming from the European Community. Furthermore, 12 per cent of all the children born in Germany in 1990 have parents who are not German nationals.

The life of immigrants in Germany, as in other European industrial societies, is characterized by a high degree of social isolation, and there is typically little contact between the different ethnic groups (see Barbour and Stevenson 1990: 195). This clearly offers very little opportunity or even incentive for language learning. Dittmar and Stutterheim (1985: 180) list the following domains for possible encounters between members of the indigenous and immigrant populations: institutions; shops; language tuition and schooling; medical care; workplace; public places, parks, and streets; leisure activities;

private places. Of these potential areas for contact, research has mainly focused on communication in institutional settings.

There are almost no statistics on the extent of immigrants' language knowledge, but some data on their participation in language courses may be a helpful indicator. Since 1975 the Sprachverband Deutsch für ausländische Arbeitnehmer in Mainz, e.V. has taken on the task of co-ordinating and subsidizing language instruction for immigrants employed in Germany. These language classes are explicitly designed for *ausländische Arbeitnehmer* (foreign workers) and their families. They are only open to people from countries of the European Union or the so-called *Anwerbeländer*, countries such as Morocco and the Phillippines, from which Germany recruited labour. Every year 70,000 to 90,000 people have registered for these language classes, with women representing 60 per cent of total enrolment (Paleit 1991).

Refugees granted political asylum, and especially the large number of asylum seekers who arrived in Germany and Berlin during the 1980s, have been excluded from these kinds of language classes. Sometimes they have been assisted by philanthropic church and political organizations. Incentives, like the offer of Bremen's Social Democrat government to subsidize tuition for special courses, are the exception rather than the rule. Since the legal status of the asylum-seeking refugees is extremely precarious, and because most of them live isolated from the rest of the community in *Asylantenheimen* (refugee hostels), their opportunities for contact with native speakers and untutored language acquisition are also very limited. Especially for people coming from Asia and Africa, it seems to be extremely difficult to have social contact with native Germans.

In contrast to other refugees, the group of *Aussiedler* from eastern Europe and the former USSR or the new Commonwealth of Independent States, have received considerable official support from various institutions. These people are even required to participate in language classes in order to facilitate their integration into German society. As a result, the language classes for the *Aussiedler* and those supported by the Sprachverband work under totally different conditions.

So the German government's policies for integrating foreign nationals vary according to the immigrant groups concerned. Measures to support language tuition and successful language acquisition seem to be considered a necessary means of implementing integration, but as the German government clearly has completely different interests regarding the integration of the various groups, educational measures differ accordingly. Therefore not all groups of foreigners have the same preconditions for intercultural encounters with native German speakers.

The children and young people of the second and third generations have special problems to face. In addition to the specific linguistic problems, some cultural groups (such as Muslims) have to deal with extremely difficult cultural contrasts in value systems and patterns of behaviour. Cultural values and even forms of communication encountered in school may also be diametrically opposed to their own cultural norms and socialization patterns. For this reason, differences in language use and cultural background fundamentally affect the second and third generations as far as school achievement is concerned. The support of the family and academic achievement are decisive factors in terms of these people's future options in the world of work. They often have an unsatisfactory bilingual education, which frequently results in 'semilingualism' (see Section 4 below). This is a particularly pressing problem because the children of immigrants who were born and raised in Germany are often faced with the threat of being sent back to a 'home' they have never lived in. This is a particularly sad irony for a generation which has serious difficulty writing and even speaking the language of the 'home' country, a country that has become even more foreign to them than their present country of residence, Germany, where they are (and always have been) considered foreigners, both legally and socially.

The legal insecurity of many foreign nationals is symptomatic of the problem that Germany's official policy at the beginning of the 1990s still refuses to recognize that Germany has already become an *Einwanderungsland*, that is a country open for immigration. This prevents the development of an effective immigration policy, which in spite of present restrictions would still be able to guarantee some legal security. A change in the official political position is necessary in order to provide a firm basis for integrative efforts and equal opportunities in education, specifically for measures to support language acquisition. The present policy towards foreign nationals can only exacerbate their already unstable living conditions.

The living conditions of the immigrant population have become even worse as a result of increasing hostility towards foreigners and a sharp increase in blatant public racism since the late 1980s. Although the German economy must be considered quite strong in comparison to other national economies, there is a growing tendency among the indigenous population to blame the foreigners for Germany's rising rate of unemployment, housing shortages, and the financial cuts being made in various areas. These trends may be exacerbated by the fact that many people are having difficulty adapting to the changes which have followed (re)unification. At the same time, reunification has also strengthened feelings of nationalism in Germany. Many intercultural encounters in the 'new' Germany have already resulted in the murder of foreign nationals and these tragic events demonstrate the far-reaching consequences of nationalism, isolationism, and xenophobia. Integration policy and efforts to develop a multicultural society have to confront these tendencies, which have also had an effect on the programmes of political parties and election results.

All of these factors should be taken into account in studies of intercultural communication in this context, as an analysis of the linguistic problems alone can only give an incomplete picture. Nevertheless, linguistic and pragmatic studies may be valuable in providing the groundwork for intercultural education and language teaching, thus forming a basis for intercultural encounters which might actually help to improve mutual understanding. At the same time, we still have only minimal knowledge about how intercultural communication actually works — or does not work — and how much conflict potential is due to different cultural backgrounds and expectations, and about what the impact of language problems might be.

4   IMMIGRATION AND SECOND LANGUAGE ACQUISITION

As we have already seen, conditions affecting the learning of the language of the host country vary according to the various groups of immigrants in Germany. Immigrant workers and their families are offered a wide selection of language classes, from ones held just for Turkish women, for example, to intensive courses for the unemployed, or *Alphabetisierungskurse* (reading and literacy courses). Course participation, however, is voluntary and there are still many people among the foreign workers and their families who learn German outside the classroom in what is called untutored language acquisition.

This so-called natural language acquisition by the immigrant workers has been of great interest for socio- and psycholinguistic research (see Klein 1984 for an overview). For example, the process of language acquisition by adults has been studied in comparison to first language acquisition, and a particularly interesting question here was when language acquisition was interrupted and 'fossilization' began to occur. However, since a detailed discussion of the various theoretical issues raised in this context has been given by Barbour and Stevenson (1990: 192 ff), only a few of the more important points need to be mentioned here:

- The study of *Gastarbeiterdeutsch* or GAD ('guest worker' German) initially concentrated on a structural description of the acquisition process, that is, the syntactic, morphological, lexical, and phonetic properties of the 'learner varieties'. Reduction, omission, and paraphrasing were found to be typical features of learner discourse. The basic question was to find out whether there are certain 'sequences' in the natural acquisition process.

- Another important question is how far the learners' varieties are influenced by their first languages and, if there is evidence of 'transfer', which features of the native language are transferred and which are not.
- A further question that has been raised is the extent to which GAD might be influenced by 'foreigner talk' (FT), the simplified speech style often used by native speakers when talking to foreigners (for a further discussion of FT, see Section 5 below).
- Learner varieties may also be influenced by regional varieties, most obviously in the rapid acquisition of tags at the end of utterances, such as *gell* in the south of Germany, *ne* in the north, and the use of *wa* in Berlin.
- Finally, GAD has many features which are typical of most pidgins, and there has been considerable discussion of whether it should be considered to represent the emergence of a new pidgin.

The first major sociolinguistic project on natural language acquisition in Germany was the Heidelberg Research Project on 'Pidgin German', which was carried out between 1974 and 1979 and which analysed forty-eight 'learner varieties'. In particular, it described the transition from elementary to more developed varieties, showing for example that semantically empty categories such as copulas were learned relatively late in the process (see Heidelberger Forschungsprojekt 'Pidgin-Deutsch' 1975, Klein and Dittmar 1979). Language acquisition of adult migrants was also investigated in a number of studies by Meisel, Clahsen, and Pienemann (see e.g. Clahsen *et al.* 1983), focusing on the acquisition of word order, negation, and inflectional morphology. Kuhberg (1987) and Stutterheim (1988) describe how temporal meanings are usually first expressed lexically and through discourse pragmatics, and only later by grammatical means (see Barbour and Stevenson 1990: 195 ff. for a more thorough review and examples of *Gastarbeiterdeutsch*).

In addition to these earlier studies, two more recent projects should also be mentioned here. In a detailed longitudinal study of the second language acquisition of Polish migrants, the P-Moll Project (Moll = Modalität im Längsschnitt von Lernervarietäten, 'Modality in a longitudinal study of learner varieties') examined modality in terms of discourse, semantics, and syntax (see Dittmar *et al.* 1990). The researchers found, for example, that modal features are acquired in the following order (Dittmar and Terborg 1991: 358):

1. *bitte* (please)
2. *müssen* (must) and *können* (can)
3. *denken* (think)
4. *möchten* (would like) and *wahrscheinlich* (probably)
5. *sicher* (certainly)
6. *wollen* (want to)
7. *vielleicht* (maybe)

and they conclude:

> It has occasionally been suggested that modality is a function which learners acquire late in the acquisition process of verbal expressions. . . . We have evidence that formal indicators of modality and modal expressions occur quite early. (Dittmar and Terborg 1991: 370–1)

The corpus of the P-Moll data allows comparisons to be made to a large-scale project of the European Science Foundation on 'second language acquisition by adult immigrants' (see Bremer *et al.* 1988). The basis of this ESF Project was that there would be speakers of two source languages acquiring the target language in each country studied. In Germany, for example, the project looked at Turkish- and Italian-speakers learning German. The acquisition process of these learners can then be compared to, say, Italian-speaking learners of Swedish in Sweden and to Turkish-speaking learners of Dutch in the Netherlands. These comparisons should eventually enable us to reach some conclusions about the influence of the individual first and target languages on the acquisition process itself.

Besides the more or less structural description of second language acquisition, there remains the basic question of how successful the process of language acquisition is. That is to say, is there early 'fossilization' or a development of 'interlanguages' which approach the norms of standard German? Or, at the very least, is the acquired language functional enough for the learners' communicative needs?

As the assessment of learners' language competence is a highly complicated procedure, there have been very few empirical studies, and these were only conducted on very limited populations. The Heidelberg Research Project on 'Pidgin German' (HPD 1975) did find that a range of social factors determines the relative success of language acquisition ('success' is defined as the degree to which learners approximate native-speaker competence in standard or regional varieties of German):

1. The main factor is the amount of contact the learners have with Germans in their leisure time.
2. A second important factor is the age of the learner at the beginning of the acquisition process.

Other factors are:

3. the amount of contact they have with Germans at work;
4. the kind of employment or occupation they have; and
5. the length of their stay in Germany

Even though this hierarchy of factors was only based on studies of a population of southern European immigrant workers, it also seems to apply to other groups. The nature and quality of interaction and intercultural contacts with 'native Germans' is more decisive than the length of stay. But once again taking into account the fact that different integration policies apply for different immigrant populations, which either get a variety of or even no support at all, it seems evident that a basic factor in successful language acquisition is also the individual expectations regarding a future life in Germany.

The same applies to the second and third generations, who were born in Germany. Even if they do not seem to have as many problems in learning the language of the 'host' country as their parents, they still face problems which arise from an often unsatisfactory bilingual education (see Ehlich 1981 and 1986, Romaine 1989). Furthermore, young people and children alike play a unique and particularly important part in intercultural communication, as they often take on the role of translators in communication between their parents and native Germans (for the linguistic aspects of this 'indirect exchange', which has been labelled *Sprachmitteln* (language mediation), see Section 6).

5   FOREIGNER TALK AND XENOLECTS: NATIVE SPEAKERS' ADAPTATION TO NON-NATIVE SPEAKERS' COMPETENCE

Not only the language learners' speech has been the subject of linguistic investigations, but also native speakers' speech or 'Foreigner Talk' (FT) in interaction with foreign language learners (for an overview see Hinnenkamp 1982, Roche 1989, and Rost 1989). The language varieties that native speakers use in such situations have been characterized by Ferguson (1977 and 1981) as a form of register variation and presumed to be universal (for a critical discussion of the concept of FT as a 'register' see Roche 1989).

Typical features of German FT were first described by Clyne (1968;

see also Bodemann and Ostow 1975, HPD 1975, Hinnenkamp 1982, and Roche 1989). In particular the following features have been observed:

- repetition of syllables, words, parts of utterances
- paraphrases and reformulations
- phonetic 'clarification'
- deletion of pronouns, articles, copula
- overgeneralization of infinitives
- negation by *nix*
- use of *du* instead of *Sie* for singular 'you'.

Examples of some of these features cited in the literature include:

> Türkischmann du? (Hinnenkamp 1989: 102)
> (Turkish man you?)
>
> hast du Arbeitserlaubnis? (Rost 1989: 265)
> (have you work permit?)
>
> das ist weil alles kommt weite Wege (Rost 1989: 265)
> (that is because everything comes long distances)
>
> ich hab nix gegen türkischmann, wirklich nich, auch gegen türkischfrau nix, aber, drohn mit messer und mit schere? (Roche 1988: 401)
>
> (I have nothing against Turkish man, really I don't, also nothing against Turkish woman, but, threaten with knife and with scissors?)

Significantly, the simplifying and clarifying features do not necessarily all operate in the same way: in his empirical investigation of German 'xenolects', Roche (1989) has shown that there are different degrees of simplification and 'relative distance' from the native speakers' normal variety. Furthermore, the last feature in the list above (*du* versus *Sie*) shows that FT as a register is not only characterized by simplification and deletion, but also includes social implications (see also the discussion of 'Türkischmann Du' in Hinnenkamp 1991 and Section 7 below on 'forms of address'). This raises the question of whether the use of FT as a whole should be considered as an expression of contempt towards the foreign interlocutor.

Interactive features of German FT, such as specific and frequent use of questions, repairs, and activities to confirm understanding, have been described by Hinnenkamp (1989) and Rost (1989). Hinnenkamp (1987) and Rost (1990) demonstrate that many of the features considered to be typical of FT register variation and xenolects occur (only) as a result of repair sequences, whose purpose is to achieve mutual understanding. Interactions between native and non-native speakers often need more repairs than native–native interactions, that is, there are more indications of non-understanding and more comprehension checks. 'Side sequences' of this kind considerably influence the native speaker's adjustment to non-fluent speakers. The following example, showing some typical reformulations, repetitions, and 'decomposition' of a question, is taken from Rost (1990: 35):

NS. was hast du bei der Polizei gemacht?
   (what did you [familiar singular] do at the police station?)
NNS. bitte?
   (sorry?)
NS. was habt ihr bei der Polizei gemacht?
   (what did you [familiar plural] do at the police station?)
NNS. eh nicht verstanden
   (er not understood)
NS. eh. . . eh ihr beide seid zur Polizei gegangen
   (er. . . er you two went to the police)
NNS. ja
   (yes)
NS. mit der Freundin. . . eh was habt ihr dort gemacht?
   (with your girlfriend. . . er what did you do there?)

Smith *et al.* (1991) found in their study that the extent to which native speakers are prepared to accommodate to non-fluent speakers is limited. They show that there are not only strategies to accommodate, but also strategies designed to restrict accommodation. In the end, it is a question of how much motivation there is for the participants either to make the necessary effort to ensure communication or to break it off completely. In some situations the participants tend to break off a conversation, and in others either the interest of the speakers or, in the case of institutional encounters, an obligation to complete the communication results in the continuation of the conversation until a specific communicative goal is achieved. The factors involved here include not only the individual situation and institutional conditions but also the participants' (social) appraisal of each other: Hinnenkamp (1991: 106 ff) argues that Bourdieu's concepts of 'symbolic capital' and 'symbolic profitability' may also result in an 'ethnic habitus' and 'ethnic capital', which condition linguistic exchanges in intercultural contexts.

6 INTERCULTURAL COMMUNICATION IN INSTITUTIONAL SETTINGS

The first comprehensive collections of studies on intercultural communication in Germany contain a considerable number of investigations on communication in institutions (cf. Rehbein 1985*a* and Redder and Rehbein 1987). This is because it was especially within the tradition of linguistic pragmatics, which has dealt intensively with language use in institutions (see also Wodak, this volume), that intercultural communication first received significant consideration.

In intercultural contact in institutions, the parties involved encounter each other in role relationships. Generally speaking, the native speaker has the role of the institutional representative (doctors, counsellors, administrative staff, etc.) and the non-native speaker takes on the role of the client (patients, those seeking advice, applicants in public agencies, etc.). A crucial aspect of communication in institutions is the specific expectations in terms of behaviour associated with the various roles (cf. Ehlich and Rehbein 1980 and Rehbein 1985*b*: 18). Problems can then arise in intercultural communication because the interactants, often without being aware of cultural differences, base their interaction on very different expectations. The clarification and resolution of such miscommunication will then be further hindered or even blocked if there are problems at the language level too, even in cases where the misunderstanding is consciously recognized.

Gumperz's initial studies on communication in institutional settings referred to conversations at bank counters, in counselling sessions, and in job interviews (see Gumperz *et al.* 1979 and Gumperz 1982). They showed how prosodic features (such as intonation or the accentuation of certain utterance segments), which serve as contextualization cues, can lead to misunderstanding: for example, prosodic elements often determine whether an utterance is to be understood as a polite request, a command, or an expression of annoyance.

It is precisely such details of linguistic realization which are often decisive in institutional settings in terms of the successful completion of tasks which may have far-reaching consequences. This is especially true in the case of job interviews, such as those Gumperz examined, or student advisory sessions, studied by Erickson and Schultz (1982) and which they refer to as 'gate-keeping situations'. Significant situations of this kind have not yet been investigated in Germany to any great extent, but Grießhaber (1990) conducted a study with an exemplary analysis of job interviews. These interviews revealed how transfer from Turkish to German at the discourse level can lead to disadvantages for foreign applicants (see also Grießhaber 1987). Consider, for example, the response of a Turkish woman when asked whether the job for which she was applying might not be too difficult:

INTERVIEWER. Ist das nicht zu schwer für Sie?
(Isn't that too difficult for you?)

APPLICANT. Aber man muß sch [schwer], ja? Das kann schwer sagen, ah man muß das versuchen. Man. Man muß... wie soll ich dat sagen, wenn man nicht macht, was gibt[s] [dann]?
(But you have to difficult, don't you? That can say difficult, er but you must try. You. You must... how should I say, if you don't do it, what is there then?)

Grießhaber (1990: 405 ff) argues that the Turkish woman's reaction to the German interviewer's question is not convincing in the German context. The conventional response would be to counter the clerk's doubts by emphasizing one's abilities and previous experience. The Turkish applicant's strategy appears to be to adopt an approach that would be considered successful in Turkish contexts: she tries to establish her suitability for the job by showing that she is aware of the difficulty of the task involved and that she is prepared to make a serious attempt to tackle it.

The rest of this section will deal with intercultural communication in various institutional settings.

6.1 *Public Agencies*

Hinnenkamp (1985) based his observations regarding 'Zwangskommunikation auf der Behörde' (obligatory communication in public agencies) on tape-recorded conversations at an immigration office. The *Ausländerbehörde* represents a significant, decisive, central point in the experience of (im)migrants, as this is where decisions regarding residence and work permits are made. One of many prerequisites for some of these permits is adequate knowledge of the German language. For this reason, this communication situation represents a double burden for foreign nationals. It is not only important to be able to communicate in German well enough to be understood effectively, but at the same time the communication constitutes a kind of test.

Furthermore, these situations are especially difficult in a linguistic sense, as very specific demands in terms of bilingual competence are made. Hinnenkamp describes the daily scenario of most 'guest workers' as 'more or less stable diglossic situations' which are divided into a German half, including the production situation (namely work), and a reproductive sphere (namely leisure), in which other familiar cultural values and another language dominate. Even for those who have acquired a level of competence in German that is adequate for these day-to-day requirements, the exceptional nature of the situation in a public agency places demands on their communicative capacity which may be overwhelming. In fact, even many German native speakers have difficulty in coping with the type of discourse which is required in institutional communication.

Compensatory measures are being developed in some 'German for Foreigners' classes, where particular attention is paid to pragmatic aspects of language use (rather than the more conventional emphasis on grammatical structures, vocabulary, etc). It is generally acknowledged that it is not enough to teach the language itself: it is also necessary to provide specific preparation for dealing with concrete situations (for a comprehensive concept of language instruction with practical survival strategies for foreign nationals, see Barkowski *et al.* 1979 and 1980). However, as Grießhaber (1987) shows, using role plays to practise 'making complaints at the employment office' is not at all easy, and even well-meaning attempts to help prepare people for these situations can only approximate the actual institutional conditions.

6.2   *Making a Complaint in a Shop*

To be successful in performing speech acts such as 'making complaints', non-native speakers need to be aware of the particular conceptions of such notions as 'rights' and 'obligations' which apply in the relevant cultural setting. For example, in his analysis of classroom simulation using role plays to practise the speech act 'making a complaint in a shop', Grießhaber (1985) shows how ineffective this kind of instruction can be if what Rehbein calls 'complementary knowledge' of the foreign students (in this case Turkish women) is not adequately taken into account.

A good example of how misunderstanding can arise as a result of different conceptions of reciprocal rights and obligations is Ohama's (1987) study of the particular problems facing Japanese learners of German. Language difficulties do indeed complicate the situation, but they are not in the end seen as the fundamental reason for the misunderstanding. The problem arises when a Japanese customer tries to return a purchase without a receipt. The assistant does not know what to do and says: 'wat mach ich 'n da nun?' (what should I do?). The customer does not recognize that there is a problem and interprets the remark as an indication that she is going to be reimbursed. This initial misunderstanding then leads to further complications (Ohama 1987: 28). The basic conflict here is that in Japan the proof of purchase is generally supplied by means of argumentative discourse procedures, whereas in Germany a 'material' proof of purchase in the form of a receipt is generally required. (Complaints have also been examined in studies of second language acquisition: see e.g. Bremer *et al.* 1988 and Dittmar *et al.* 1990).

6.3   Legal Consultations

Legal discourse is highly conventionalized, and here too lawyers and their clients may have very different conceptions about rights and options. Mattel-Pegam (1985) analysed a consultation between a German lawyer and an Italian prisoner, showing how this conversation goes awry. The Italian client has already been convicted and now wants to suggest steps to the lawyer in the hope of getting an appeal, in view of the unusually heavy sentence. In the lawyer's opinion, however, these suggestions are not viable options for a representative of the German legal system. Moreover, the lawyer views the consultation with the client as a moan or whinge (*lamento*), since he considers him to be guilty. Language problems are also taken into account, since both communicate through an interpreter.

Becker and Perdue (1982) analysed a consultation between a German lawyer and a Turkish client. They found that the client's knowledge of German did not allow him to give the relevant information in a concise and orderly way. They also argue that both interactants (mis)interpreted each other's utterances according to rigid patterns of expectations about what was relevant and what was not.

6.4   Medical Institutions

Special problems can arise for foreigners and non-native speakers in an area of vital importance: medical care. Studies on communication in the medical field have clearly shown that the particular institutional conditions and the characteristic linguistic behaviour of the staff can result in difficulty, even for patients who are native speakers, in articulating their needs within the context of a doctor's examination (see also Wodak, this volume). Problems can arise concerning the understanding of medical terminology and scientific knowledge, and these problems are exacerbated by the constant time pressure in such situations. For non-native speakers and people not familiar with German institutions and forms of health care, the problems are magnified.

Rehbein (1986) clearly illustrated these problems through a direct comparison of the treatment of two patients, a native German speaker and a non-native speaker (a Turk), in a general practitioner's surgery.

The analytical and methodological basis for this comparison is provided by an analytical/descriptive model of behaviour and procedures ('Handlungsmuster', cf. Rehbein 1977) considered typical for this institution, including categories such as 'complaint description', 'symptom identification', 'diagnosis', and 'proposed therapy'. The analysis of dialogues showed that in the interaction with the Turkish patient the doctor relied more on his professional knowledge, requesting less information from the patient himself. In describing the complaint, which normally provides the physician with the information needed for further decisions, there is not so much scope for this. The knowledge of the non-German patient and his personal experience receive (even) less consideration than those of the native German-speaker. Furthermore, the diagnosis and proposed therapy (even if expressed well-meaningly in simplified German) are not always easy for the patient to understand. The following extract shows how the use of colloquial language does not necessarily prevent (and indeed might even cause) difficulty in understanding:

ARZT. Dann hatten Sie ja Schmerzen am Po, nicht?
PATIENT. [*Schweigt*]
ARZT. [*2 sec.*] Hier hatten Sie ja Schmerzen, nicht?
PATIENT. Ja, das, die die . . . Jes (= jetzt) Moment. . .
ARZT. Is weg jetzt, nicht?

(DOCTOR. And then you had a pain in your bottom, didn't you?
PATIENT. [*silence*]
DOCTOR. [*2 sec.*] You had a pain here, didn't you?
PATIENT. Yes, that, the the (. . .) now, wait . . .
DOCTOR. It's gone now, isn't it?)

In this case, it is simple to repair the misunderstanding by gestures and facial expressions. It is obviously more difficult to make such repairs in situations where the part of the body is not visible and the problem is more abstract. This example also shows how the interaction is typically shaped by information offered by the doctor, which then merely has to be confirmed or denied by the patient.

Rehbein also investigated communication in counselling sessions (see also the following section of this chapter). It has become commonplace to employ 'language mediators' in such situations. Unlike interpreters, these mediators are not professionally trained, but are usually volunteers who have sufficient command of the languages involved. Normally, the mediators are members of the same minority group as the client who have already lived in Germany for quite some time.

Rehbein's (1985c) analysis of 'language mediation' in a medical counselling session tried to determine the principles used in transmitting linguistic information in this type of situation. An essential aspect of this mediation procedure is that a variety of changes are made to the original statements. The aim is to bridge the gap between professional knowledge and everyday lay knowledge, making the utterances mutually understandable. This also entails making adjustments to take account of different cultural background knowledge. This can involve adding or omitting certain points, supplementary explanations, or a shift in focus. These German–Turkish comparisons in a medical context show that other differences exist besides those of conventional institutional behaviour. These other fundamental differences include conflicting perceptions of the course of an illness, what the relevant symptoms are considered to be, and even the importance attached to consultations with doctors.

6.5   Counselling Sessions

The problems analysed by Rehbein (1985c) are by no means confined to medical counselling sessions: the role of culture-specific frames of reference has also been investigated in other types of counselling (cf. also the study of conflicting expectations in Gumperz *et al.* 1979, Gumperz 1982).

A study conducted by Backa (1987), for example, analysed a very typical situation in the area of counselling for foreign nationals. Counsellors in a community centre for foreigners attempted to create an informal atmosphere for their sessions, in the hope of reducing any possible anxieties connected with such counselling situations. However, Backa showed that the informality of the situation actually created additional problems, as there were no clearly defined institutional regulations regarding who should perform the counselling and when. It was therefore necessary to negotiate these things during the session itself. In fact, this particular session ends with the client being referred to another counselling centre! This is not an isolated example: the same thing was observed in Hinnenkamp's study of a counselling session in a public agency (Hinnenkamp 1985: see above).

6.6   Conclusions

A characteristic feature of all these studies is the fact that problems in understanding are attributed primarily to culture-specific structures of knowledge. These structures lead to differing expectations in terms of the patterns of action in the various institutions. The degree of linguistic competence (in this case, in German) is assumed to play a less significant role in communication in certain institutions than the knowledge of the conventional behaviour required in these interactions. Nevertheless, language difficulties represent an additional burden, making it even more difficult to resolve misunderstandings.

7 CONTRASTIVE PRAGMATICS

In the field of contrastive pragmatics, and related areas such as contrastive discourse and text analysis, misunderstanding in intercultural communication is generally attributed to the fact that 'habits' in the verbalization of speech acts and text forms are transferred from the first language to the target language. This can lead to behaviour which is regarded by members of the target language society as inappropriate or even impolite (for an overview, see Hinnenkamp 1989: 6 ff, and House 1985; see also Coulmas 1978, which outlines the scope of contrastive pragmatics in a description of the contrast between two languages and possible instances of interference). For example, Coulmas (1981*b*) examines the problems that arise from the different functions of expressing thanks and apologizing in German, English, and Japanese.

The following overview offers an account of studies in contrastive pragmatics which reveal differences between German and other languages or cultures. This is in itself a very large field, and in the space available here it would be difficult to give a genuinely representative selection. On the one hand, there are many 'smaller' publications which deal with German but do not provide any major findings. On the other hand, there are also many studies dealing with other languages (such as English and Japanese) which might help to shed some light on contrasts with German. The discussion here will therefore necessarily be highly selective.

7.1 *Listening Behaviour*

To gain a complete picture of linguistic interaction, both speaking and listening behaviour need to be taken into account. Listening behaviour was first considered in the field of intercultural communication in the USA by Erickson and Schultz (1982), who revealed differences in listening response between black and white Americans. The major differences were that the white subjects tended to give more acoustic signs of acknowledgement than the black subjects. The white Americans nodded their heads more often and they had more eye contact for longer periods of time with the speaker. Erickson and Schultz clearly show that such differences can lead to uneasiness and to instances of major misunderstanding and misjudgement. For example, white speakers, who often take on the higher-status position of gate-keeper, tend to interpret the listening behaviour of blacks as inattentive or even as an expression of lack of understanding.

Another cross-cultural study which deals with the role of the listener and is based on a highly standardized experiment includes an analysis of the interpretations of German-speakers. The intention of the so-called PONS Test (cf. Rosenthal *et al.* 1979) was to verify hypotheses on the 'decodability' of non-verbal behaviour in a cross-cultural comparison. Certain national groups fared better than others in comparison to the American (native-native) control group, and cultural and linguistic proximity were seen as decisive factors. German, together with other Germanic and central Indo-European language groups, are relatively 'close' to American English. White US-American non-verbal behaviour is therefore assumed to be understood 'more easily' or 'better' by Germans than by speakers of Chinese, for example.

A study conducted by Quasthoff-Hartmann (1987), which dealt specifically with German, analysed listening behaviour between native and non-native speakers in conversational narration. Quasthoff-Hartmann concluded that if non-native listeners fail to perform the expected 'back-channel behaviour' (brief utterances such as 'I see', 'right', 'really?', which show that they are following what the speaker is saying) simply because they do not understand what is being said, this might be misinterpreted as lack of interest, which in turn might inhibit the continuation of the narrative. Friction may also arise if listeners either do not know what the appropriate listening signals in the target language are or utter the right signals in the wrong place.

7.2   *Narrative Style*

Cultural differences may also be manifested in narrative style. Redder (1985) compared the verbalization of a picture-book story by Turkish and German children. She found that the Turkish children stuck more closely to the sequence of the pictures than the German children, probably because it facilitated the process of language production. The major difference between the versions, however, was that the narratives of the Turkish children were characterized by much stronger value judgements and a moralizing tone. Redder felt that this could be explained at least partially by the differing approaches to education in the two countries.

Rehbein (1987) also compared narratives of Turkish and German schoolchildren. His study found that the German children used more referential features in retelling the story than the Turkish children, that is, they summarized and made comments more than the Turkish children did. The Turkish children's retelling of the story showed that they were less capable of this more distanced form of narration only in their second language, German. Paradoxically, it seemed to be easier for them to approximate the 'German' narrative style in their first language, Turkish, even though the retelling of the story in Turkish by a control group of Turkish children in Turkey clearly demonstrated the high value attached there to an exact and literal form of memorization.

Fienemann (1987) compared the different ways in which a French woman related a certain event first to a German and then to a French interlocutor. This cross-cultural comparison shows that the presentation of an event may vary in an 'addressee-specific' manner. Differences can be observed in the presentation of the protagonist's picture of herself, which is achieved through the selection of different types of narrative. The event concerns an argument between a French holiday-maker and a lifeguard:

Der Schwerpunkt der deutschen Version liegt . . . auf einer anderen Stelle als der der französischen. In der deutschen Version werden mehr von S's [der Protagonistin, M.R.] Gedanken, Argumentationen, Angsten u.s.w. mitgeteilt, es geht mehr um sie selbst als Person, während in der französischen Version die Auseinandersetzung zweier Protagonisten im Vordergrund steht. . . . Der entscheidende Unterschied zwischen beiden Versionen ist die unterschiedliche Bewertung der Lösung des Konflikts . . . Während es in der deutschen Geschichte als Normverstoß gerechtfertigt werden muß, wird es in der französischen Erzählung als Sieg über den Kontrahenten dargestellt. (Fienemann 1987: 164)

(There is a difference of emphasis between the German and the French versions. In the German version more is said about the protagonist's feelings, arguments, fears, etc: it's more about her as a person. In the French version the focus is on the dispute between the two protagonists. . . . The crucial difference between the two versions is the different evaluation of the way the conflict is resolved. In the German story it has to be justified as a violation of conventional behaviour, while in the French narrative it is depicted as a victory over the opponent.)

7.3 Written Texts

Intercultural contrasts are not confined to the spoken language: there is often considerable variation in the construction and the style of written texts. The most comprehensive studies in this field to date have been conducted by Clyne (1981 and 1991).

Clyne (1991) gives a detailed comparison of academic writing in German and English. English texts are generally structured in a more linear fashion, while German texts display much more excursiveness, with a greater number of digressions. German texts also provide fewer definitions, and data and references are less integrated into the text.

Not surprisingly, this variation in text construction often leads to difficulties in comprehension. German — and English (!) — texts written by German-speaking academics are often difficult for English native speakers to understand. This is not only due to the language but also because of the different text construction (see also Sachtleber 1991 for a comparison of German and French scientific texts; or Marui and Reinelt 1985 on the paragraphing of texts by Japanese students of German).

Another aspect of written texts which is especially significant for intercultural contacts is that of correspondence. The conventions involved in letter-writing vary from country to country. For example, there are different rules for openings and closings, all of which have a direct implication for the relationship between sender and addressee (see e.g. Stolt 1992 for German–Swedish and Büchle 1991 for German–Spanish variation; see also 'forms of address' in Section 7.6).

7.4 General Intercultural Contrasts

Some studies offer insights into more general intercultural contrasts. For example, in contrasting German and Swedish communicative behaviour, the previously cited article by Stolt (1992) includes aspects of letter-writing, variation in greeting and departure rituals, differences in the ways of talking about emotions, and variation in forms of address.

Kotthoff (1989*b*) deals specifically with German–American contrasts, collecting so-called 'impressions'. According to her study, Germans tend to make compliments less often and are more direct in making criticisms than US-Americans. Differences are also observed in aspects of non-verbal behaviour such as spatial distance (according to Hall and Hall 1983, Americans tend to maintain less distance between themselves and their interlocutors than Germans do).

In the academic field, significant differences can also be seen in styles of speaking. The ability to speak without notes is not as important to Germans as it is to Americans, and Germans are also less concerned with amusing their audience. A particularly significant contrast in communicative terms (and one that may seem remarkable, to say the least, to the anglophone reader) is the fact that incomprehensibility is not necessarily considered a failing in German academic texts;

on the contrary, it may be valued as a marker of a 'serious' scientific style (see Kotthoff 1989*b*)!

German–Turkish differences are demonstrated by Ilkhan (1987). He shows that the behaviour of Turkish learners of German can be misinterpreted when they transfer the conventional Turkish behaviour of initially rejecting offers. This observation also applies to many other cultures. A further example is the direct translation of fixed expressions in situations where they would be inappropriate for the German context. For instance, in Turkish the phrase '*Geçmis olsun*' (literally, let it be over) is often translated into German as *Gute Besserung* (get well soon) but is used not only when someone is ill but also in the context of car accidents or before examinations.

Similar forms of transfer are mentioned by Uhlisch (1991), regarding learners of German who speak Russian as their native language. For example, the Russian *posdraviat* (wishing health) is converted to *gratuliere!* (congratulations) and used in situations in which Germans do not normally offer congratulations, such as on May Day (1 May) or at Christmas.

Günthner and Rothenhäusler (1986) describe 'deficits in the area of socio-pragmatic competence' in interethnic communication between Germans and Chinese learners of German in China. They show that the use of *vielleicht* (maybe) by Chinese learners of German should not be interpreted as uncertainty or doubt about what they are saying, but rather as a politeness signal. Other sources of misunderstanding in this context are the level of tolerance in dealing with lulls in the conversation (in this regard, cf. also Enninger 1987) and the literal translation of certain set forms of greeting, such as *Qu na-li?* (where are you going?) and *Chi guo-le ma?* (have you already eaten?), which should not be taken literally (see also Günthner 1989).

7.5   *Conversational and Argumentative Styles*

In a comprehensive empirical study of linguistic differences in argumentative behaviour between Germans and US-Americans, Kotthoff (1989*a*) shows that it is characteristic of German native speakers to emphasize aspects of dissent more strongly. This finding reinforces the results of other contrastive studies on language and culture comparing German-speakers with speakers of other languages (see also Byrnes 1986 for German–American contrasts in conversational styles). For example, Reuter *et al.* (1989) describe problems in German–Finnish business communication, which are reflected in statements such as 'Germans have an aggressive style of argumentation' and 'Finns cannot take criticism' (see Reuter *et al.* 1989: 260). Torres and Wolff (1983) describe the same differences in dealing with criticism with regard to German–Spanish contact.

Kotthoff also investigated conversational closings. The closings in American conversations are generally longer than in German ones, and they often include explicit expressions of appreciation ('It was nice meeting you') and extensive thanking sequences. By contrast, Germans usually end conversations much more quickly. Kotthoff's analysis of the behaviour of US-American learners of German is especially informative in terms of intercultural contact. The conversational closings in such encounters tend to be initiated abruptly and then completed relatively quickly. Summarizing statements, which usually appear quite often in conversations in English amongst American speakers, hardly occur at all in conversations involving American learners of German. These characteristics clearly cannot be attributed to cross-cultural transfer: they are the result of the learners' inadequate competence in the target language. Therefore, these observations may also be relevant for learners of German with native languages other than English.

7.6 *Forms of Address*

For many foreigners and non-native speakers, the differentiation between and proper usage of *du* (informal 'you') and *Sie* (formal 'you') involves considerable difficulty. This is especially true for native speakers of languages such as English, which do not make such a distinction. This problem can lead to rather curious forms of behaviour in encounters between English- and German-speakers involving changes in situation and language. Speicher (1985) illustrates this very clearly: at a social gathering, he observed two couples, one English and the other German, who used *Sie* plus *Herr/Frau* and last name when speaking German but switched to first names when speaking English (cf. also Kuglin 1977 for German–Turkish contrasts in the same vein).

However, Torres and Wolff (1983) show that problems do not arise only when speakers have no such distinction in their native language. Since the scope of Spanish *tu* is much greater than that of *du* in German, Spaniards often inappropriately use *du* when speaking German, and Germans use *Usted* in Spanish in cases where *tu* would be the expected form.

Transfer problems like this can be particularly problematic, since the cause of the inappropriate behaviour is often not recognized by the interlocutors, and they tend to jump to (false) conclusions about each other's intentions. However, even for native speakers of German the choice between *du* and *Sie* is not always clear-cut. It varies according to social group and is even subject to temporary changes in trends and styles (see Clyne 1984: 124–8). It is interesting to note that there were also differences in the usage of *du* and *Sie* in the former East and West Germany; in the following ironic passage Monika Maron, a writer from the GDR, tries to distinguish between the different meanings of *du* as used in the two German states:

Ihr Du ist das Boutiquendu: kann ich dir helfen; das therapeutische Alternativdu: das Gespräch mit dir war mir sehr wichtig, du; das Englischdu: how do you do. Mein Du ist das allgegenwärtige Gewerkschaftsdu: Kollege Meier, hast du dich schon mit dem Kollegen Müller beraten; und natürlich das Minderheitendu, eine konspirative Herzenssache. (quoted from Kretzenbacher 1991, who has dealt in depth with the German-German *Du*)

(Your Du is the Boutique-Du: Can I help you?; the Therapeutic-Alternative Du: That talk with you really meant a lot to me; the English-Du: How do you do. My Du is the ubiquitous Union-Du: Colleague Meier, have you already talked to Colleague Müller?; and of course, the Minority-Du, a conspiratorial affair of the heart.)

7.7 *Conclusions*

It is typical of much of the literature referred to here that impressions are often represented which are of a more or less intuitive nature. Even though the evidence provided is often very weak, these comparisons are quite revealing as they deal with the mutual perception of cultural differences, pointing out relatively directly how possible mutual (mis-)interpretations and the stereotyping of individual cultural groups may arise (see also Hall and Hall 1983, Bausinger 1988 and n.d.). However, if we are to gain a more accurate picture of how miscommunication results from contrasting patterns of linguistic and cultural behaviour, these intuitive observations will have to be reinforced by detailed empirical analyses (such as Kotthoff 1989*a*).

8 INTERCULTURAL CONTRASTS BETWEEN EAST AND WEST GERMANY

Of course, differences in language use between East and West Germany were by no means confined to different usage of personal pronouns. Earlier studies tended to concentrate on the written language and in particular on lexical contrasts in public language (see e.g. Hellmann 1984 and in more satirical vein Röhl 1991), but with the advent of the *Wende* attention focused on more general problems (Hellmann 1989, Schlosser 1990).

More recently still, however, studies have been undertaken on interactive aspects of encounters between *Ossis* and *Wessis*. Barden and Großkopf (1992), for instance, investigated the tendency to shift away from the characteristic Saxon dialect amongst East Germans who have moved to the West, basing their study on the analysis of sociolinguistic variables. Gumperz (1991) has also conducted fieldwork in the new *Länder* to see whether his concept of 'minorization' can be applied to the relationship between East and West Germans. His hypothesis that East Germans tend to downgrade themselves and that West Germans tend to be rather arrogant corresponds to widely held public perceptions, epitomized in the term *Besserwessi* (see Glück and Sauer, this volume), which was twice nominated as the 'word of the year' for 1991 (see Strecker 1992).

These tendencies have been studied in detail by Ylönen (1992). In her analysis of a conversation between two businessmen at the Leipzig trade fair, one from the East and the other from the West, she shows how status differences and power relationships are articulated and how comprehension difficulties arise. For example, a single word may give rise to a misunderstanding, with potentially serious consequences:

Mit 'Partner' möchte der Dresdner offensichtlich den gemeinsamen Charakter des Geschäftsvorhabens unterstreichen und die Grundlage für gute Arbeitsverhältnisse schaffen. Er meint damit 'Teilnehmer' am Geschäft in einem (positiv konnotierten) kollektiven Sinn.... Das Mißverständnis entsteht, weil der Westdeutsche eine derart vorsichtige Redeweise in Geschäftsanbahnungen nicht gewohnt ist. Er interpretiert 'Partner' als 'Teilhaber' und glaubt aufgrund dieser Wortwahl, daß die ... Firma als potentieller Investor für ein Gemeinschaftsunternehmen angesprochen wird. (Ylönen 1992: 18)

(By using the word 'partner', the man from Dresden clearly wants to emphasize the joint nature of the business proposal and establish the basis for a good working relationship. What he means is 'participants' in the business in a collective sense (with positive connotations).... The misunderstanding arises because the West German is not used to such a cautious manner of speaking when beginning business negotiations. He interprets 'partner' in the technical commercial sense and takes the choice of this word to imply that his firm is being approached as a potential investor in a joint venture.)

Other differences between East and West Germans are also identified here:

- The West German adopts a higher rhetorical tone in order to assert his dominant position: 'Das heißt, das sähe so aus, daß äh Sie hier vor Ort das Produkt benötigen, unser Know-how' (So it looks as if er you need the product here on the spot, our know-how).

- There are differences in fluency: the West German in this situation appears to be an eloquent and practised speaker, while the East German expresses himself increasingly awkwardly.
- Many of the West German's utterances imply speech acts such as 'instructing' and 'imposing conditions', which again serves to demonstrate his dominant position.

It is difficult to predict how relationships between East and West Germans will develop in future. It may be that the differences between them will rapidly even themselves out, but experience suggests that, now the initial euphoria over unification has worn off, the tendency to maintain demarcation lines between groups that has been observed in other international and interethnic contexts will apply in Germany too. Indeed, as the term 'Kultur des Mißverständnisses' (culture of misunderstanding) suggests, the differences may be more extensive and more ingrained than had previously been thought (see Good 1993). So for the time being we can only speculate what effect increasing mobility and economic and social policies will have on the interaction between East and West Germans and the development of mutual understanding.

Ylönen concludes her article with a quotation by Jens Reich (a prominent citizens' rights activist from the former GDR), in which he comments on the mental division of Germany:

'Es ist durchaus möglich, daß in einem Land zwei Gesellschaftsstrukturen nebeneinander bestehen ... mit zwei Gesellschaften kann man auch zwei Kulturen und zwei Sprachen haben. Ich meine "Sprachen" in einem gehobenen Sinne, nicht die erbsenzählende Untersuchung darüber, ob in einem Gebiet Brathendl heißt, was sie im anderen Broiler nennen. Ich meine Sprachmuster, Denkschablonen, ja sogar die verfeinerten Elemente der Körpersprache, an denen man die Herkunft eines Sprechers ausmachen kann.' Und in bezug auf die osteuropäische Orientierung der ehemaligen DDR kommt er zu dem Schluß: 'sie hat doch zwei Generationen so tief geprägt, daß wir uns jetzt mit einem Tschechen oder Polen über existentielle Probleme leichter verständigen können als mit vielen Westdeutschen.' (Ylönen 1992: 20)

('It is perfectly possible for two social structures to exist side by side in one country ... with two societies you can also have two cultures and two languages. I mean "languages" in a sophisticated sense, not the nit-picking study of whether a roast chicken is called a "Brathendl" in one area and a "Broiler" in the other. I mean linguistic patterns, ways of thinking, even the refined elements of body language which reveal the speaker's origin.' And as regards the orientation of the former GDR towards eastern Europe he concludes:

'it had such a profound influence on two generations that it is easier for us now to discuss existential problems with a Czech or a Pole than with many West Germans.')

I too shall let this quotation speak for itself, although it is important to note Reich's careful formulation 'with many' West Germans. In my opinion, sweeping generalizations are dangerous both in relation to intercultural contact and in the analysis of intercultural communication, as specific and possibly diverging features and interests are then all too easily marginalized as 'subcultural' factors.

9　CONCLUSIONS

The study of intercultural communication in the German context shows that there may be many different ways of explaining miscommunication and misunderstanding. In investigations of communication in institutions, for example, culture-specific expectations and norms have been found to be the main reasons for the problems that arise. In the field of contrastive studies it is usually assumed that transfer leads to inappropriate behaviour. Studies in discourse analysis have shown that in many cases inadequate foreign language competence hinders communication, which in turn leads to the misinterpretation of speakers' intentions. Conversely, the study of East–West German contact shows that even using the 'same' language does not guarantee mutual understanding.

What is common to all of these studies is that they confirm the assumption that intercultural communication can be viewed as 'communication under difficult conditions'. However, the examples of 'misunderstanding' and 'miscommunication' in intercultural encounters described in this chapter can affect the relationships between those involved in very different ways. Some inappropriate behaviour may be regarded merely as a *faux pas* or even as amusing (of course, it is not so amusing if it is not intended or if you are the butt of the joke). But inappropriate behaviour often leads to far-reaching misunderstanding which may inhibit the development of a social relationship and which can be very detrimental to those involved. More seriously still, language problems, miscommunication, and misinterpretations place an added burden on intercultural contact, which in turn fosters prejudice and reinforces stereotypes.

Yet in the end, deeper-seated conflicts of interest and sociopsychological processes are responsible for forming the basis of contact between members of different groups. The crucial factors here are the level of tolerance shown towards others and the amount of effort made towards achieving mutual understanding. This means that the role of language problems in intercultural contexts needs to be qualified. On the one hand, there are many examples which show that language competence and language problems are insignificant and that it is stereotypes and prejudices, or other (assumed or actual) differences in interests, that define and determine intercultural conflict. On the other hand, numerous studies demonstrate that even if the interaction partners make well-meaning efforts, intercultural communication remains very difficult precisely because of conflicting linguistic and cultural norms. This can turn out to be particularly problematic, as many forms of linguistic behaviour, expectations, and the possibility of culture-specific variation are not consciously recognized. Furthermore, language in intercultural communication can be problematic (for both interactants and investigators) because 'language' and 'culture' are ultimately inseparable concepts.

Further reading

Rehbein (1985*a*)
Redder and Rehbein (1987)
Knapp *et al.* (1987)
Kotthoff (1989*a*)
Hinnenkamp (1989)
Blommaert and Verschueren (1991)

References

ALBRECHT, J., DRESCHER, H. W., GÖHRING, H., and SALNIKOW, N. (eds.) (1987), *Translation und interkulturelle Kommunikation. 40 Jahre Fachbereich Angewandte Sprachwissenschaft der Johannes Gutenberg-Universität Mainz in Germersheim* (Frankfurt: Lang).
ALTHAUS, H.-P., HENNE, H., and WIEGAND, H. E. (eds.) (1980), *Lexikon der germanistischen Linguistik*, 2nd edn. (Tübingen: Niemeyer).
APELTAUER, E. (ed.) (1987), *Gesteuerter Zweitspracherwerb*, (Munich: Hueber)
ASANTE, M., NEWMARK, E., and BLAKE, C. (1979), *Handbook of Intercultural Communication* (Beverly Hills: Sage).
AUER, P. (1992), 'Introduction: John Gumperz' Approach to Contextualization', in Auer and di Luzio (1992), 1–37.
—— and DI LUZIO, A. (eds.) (1985), *Interpretive Sociolinguistics* (Tübingen: Narr).
—— —— (eds.) (1992), *The Contextualization of Language* (Amsterdam: Benjamins).
BACKA, S. (1987), 'Interkulturelle Probleme in der Beratung — eine Fallstudie', in Redder and Rehbein (1987), 53–68.

BADE, K. (ed.) (1990), *Neue Heimat im Westen. Vertriebene, Flüchtlinge, Aussiedler* (Münster: Westfälischer Heimatbund).

BARBOUR, S., and STEVENSON, P. (1990), *Variation in German: A Critical Approach to German Sociolinguistics* (Cambridge: Cambridge University Press).

BARDEN, B., and GROßKOPF, B. (1992), ' "Ossi Meets Wessi": Social and Linguistic Integration of Newcomers from Saxony', Paper presented at the LAUD Symposium 'Intercultural Communication', Duisburg, 23–7 Mar.

BARKOWSKI, H., HARNISCH, U., and KUMM, H. J. (1979), 'Sprachlernen mit Arbeitsmigranten im Wohnbezirk', *Deutsch lernen* 1: 5–17.

—— —— —— (1980), *Handbuch für den Deutschunterricht mit ausländischen Arbeitern* (Kronberg: Scriptor).

BARTH, F. (ed.) (1969*a*), *Ethnic Groups and Boundaries* (Bergen: Universitatsverlaget).

—— (1969*b*), 'Introduction', in Barth (1969*a*), 9–38.

BAUSINGER, H. (1988), 'Stereotypie und Wirklichkeit', *Jahrbuch Deutsch als Fremdsprache*, 14: 157–70.

—— (n.d.), *Fremde Deutsche. Alltagskultur aus der Sicht ausländischer Studenten* (Tübingen: Ludwig-Uhland-Institut für empirische Kulturwissenschaft).

BECKER, A., and PERDUE, C. (1982), 'Ein einziges Mißverständnis. Wie die Kommunikation schieflaufen kann und weshalb', *Osnabrücker Beiträge zur Sprachtheorie*, 22: 96–121.

BLOMMAERT, J., and VERSCHUEREN, J. (eds.) (1991), *The Pragmatics of International and Intercultural Communication* (Amsterdam: Benjamins).

BODEMANN, M., and OSTOW, R. (1975), 'Lingua Franca und Pseudo-Pidgin in der Bundesrepublik', *Zeitschrift für Literaturwissenschaft und Linguistik*, 18: 122–46.

BORN, J., and DICKGIEßER, S. (1989), *Deutschsprachige Minderheiten. Ein Überblick über den Stand der Forschung für 27 Länder* (Mannheim: Institut für deutsche Sprache im Auftrag des Auswärtigen Amtes).

BREMER, K., BROEDER, P., ROBERTS, C., SIMONOT, M., and VASSEUR, M. T. (1988), *Procedures Used to Achieve Understanding in a Second Language.* Final report to the steering committee of the European Science Foundation additional activity 'second language acquisition by adult immigrants' (London).

BRUNT, R., and ENNINGER, W. (eds.) (1985), *Interdisciplinary Perspectives on Crosscultural Communication* (Aachen: Rader Verlag).

BÜCHLE, K. (1991), ' "Briefkontakte" — Aspekte eines interkulturellen/interlingualen Textvergleichs', *Kultur und Sprache, Berliner Beiträge zu Deutsch als Fremdsprache*, 1: 19–30.

BYRNES, H. (1986), 'Interactional Style in German and American Conversations', *Text*, 6/2: 189–207.

CLAHSEN, H., MEISEL, J., and PIENEMANN, M. (1983), *Deutsch als Zweitsprache* (Tübingen: Narr).

CLYNE, M. (1968), 'Zum Pidgin-Deutsch der Gastarbeiter', *Zeitschrift für Mundartforschung*, 35: 130–9.

200   *Martina Rost-Roth*

CLYNE, M. (1981), 'Culture and Discourse Structure', *Journal of Pragmatics*, 5: 61–6.

—— (1984), *Language and Society in the German-speaking Countries* (Cambridge: Cambridge University Press).

—— (1991), 'Zu kulturellen Unterschieden in der Produktion und Wahrnehmung englischer und deutscher wissenschaftlicher Texte', *Info DaF*, 18/4: 376–83.

COULMAS, F. (1978), '*Kontrastive Pragmatik*', in Faber *et al.* (1978), 53–60.

—— (ed.) (1981*a*), *Conversational Routines: Explorations in Standardized Communication Situations and Prepatterned Speech* (The Hague: Mouton).

—— (1981*b*), ' "Poison to your Soul!" Thanks and Apologies Contrastively Viewed', in Coulmas (1981*a*), 69–91.

COUNCIL OF EUROPE (1986), *Intercultural Education: Concept, Context, Curriculum, Practice* (Strasbourg).

DITTMAR, N., and STUTTERHEIM, C. VON (1985), 'Communication Strategies of Migrants in Interethnic Communication', in Auer and Di Luzio (1985), 179–214.

—— and TERBORG, H. (1991), 'Modality and Second Language Learning. A Challenge for Linguistic Theory', in Huebner and Ferguson (1991), 347–84.

—— REICH, A., SCHUMACHER, M., SKIBA, R., and TERBORG, H. (1990), 'Die Erlernung modaler Konzepte des Deutschen durch erwachsene polnische Migranten. Eine empirische Längsschnittstudie', *Info DaF*, 17/2: 125–72.

DOKUMENTATION (1990), 'Folgen der Arbeitsmigration für Bildung und Erziehung', *Deutsch lernen*, 69–87.

EHLICH, K. (1981), 'Spracherfahrungen. Zur Sprechhandlungssituation von Kindern ausländischer Arbeiter', in Nelde *et al.* (1981), 23–40.

—— (1986), 'Xenismen und die bleibende Fremdheit des Fremdsprachensprechers', in Hess-Lüttich (1986), 43–54.

—— and REHBEIN, J. (1980), 'Sprache in Institutionen', in Althaus *et al*, (1980), 338–45.

ENGEL-BRAUNSCHMIDT, A., and SCHMÜCKER, A. (eds.) (1977), *Korrespondenzen. Festschrift für D. Gerhardt* (Gießen: W. Schütz).

ENNINGER, W. (1987), 'What Interactants Do with Non-Talk across Cultures', in Knapp *et al.* (1987), 269–302.

ERICKSON, F., and SCHULTZ, J. (1982), *The Counselor as Gatekeeper* (New York: Academic Press).

FABER, H. VON, KREIFELTS, B., and SIEGRIST, L. (eds.) (1978), *Technologie und Medienverbund. Sprachtests, kontrastive Linguistik und Fehleranalyse*, special issue of *International Review of Applied Linguistics*, on Gesellschaft für Angewandte Linguistik conference, 1978 (Heidelberg: Julius Groos).

FERGUSON, C. A. (1977), 'Simplified Registers, Broken Language and Gastarbeiterdeutsch', in Molony *et al.* (1977), 26–39.

—— (1981), ' "Foreigner Talk" as the Name of a Simplified Register', *International Journal of the Sociology of Language*, 28, 9–18.

FIENEMANN, J. (1987), 'Ein und dieselbe Geschichte? Erzählen auf deutsch und französisch', in Redder and Rehbein (1987), 151–72.

GERIGHAUSEN J., and SEEL, P. C. (eds.) (1987), *Aspekte einer interkulturellen Didaktik* (Munich: Iudicium).

GOOD, C. (1993), 'Über die "Kultur des Mißverständnisses" im vereinten Deutschland', *Muttersprache*, 103/3: 249–66.

GÖTZE, L. (ed.) (1987), *Deutsch als Fremdsprache. Situation eines Fachs* (Bonn: Verlag Dürrsche Buchhandlung).

GRIEßHABER, W. (1985), 'Zitieren von Handlungsmustern – "Recht im Alltag" im Unterricht für ausländische Arbeiter', in Rehbein (1985*a*), 257–75.

—— (1987), 'Personalisierung von Sachkonflikten. Beschwerden auf dem Arbeitsamt beim interkulturellen Rollenspiel', Osnabrücker Beiträge zur Sprachtheorie, 38: 69–84.

—— (1990), 'Transfer, diskursanalytisch betrachtet', *Linguistische Berichte*, 130: 386–414.

GUMPERZ, J. (1982), *Discourse Strategies* (Cambridge: Cambridge University Press).

—— (1991), 'Intercultural Communication: East Germany and West Germany', lecture at the Free University of Berlin, 8 July.

—— JUPP, T., and ROBERTS, C. (1979), *Crosstalk: A Study of Crosscultural Communication* (Southall: NCILT).

GÜNTHNER, S. (1989), 'Interkulturelle Kommunikation und Fremdsprachenunterricht', *Info DaF*, 16/4: 431–47.

—— and ROTHENHÄUSLER, R. (1986), 'Interethnische Kommunikation zwischen Deutschen und Chinesen', *Info DaF*, 13/4: 304–9.

HALL, E. T., and HALL, M. R. (1983), *Hidden Differences: How to Communicate with the Germans* (Hamburg: Stern Verlag).

HAMBURGER, F. (1989), 'Auf dem Weg zur Wanderungsgesellschaft. Migrationsprozeß und politische Reaktion in der Bundesrepublik Deutschland', *Deutsch lernen*, 14: 3–33.

HEIDELBERGER FORSCHUNGSPROJEKT 'PIDGIN-DEUTSCH' [= HPD] (1975), *Sprache und Kommunikation ausländischer Arbeiter. Analysen, Berichte, Materialien* (Kronberg: Scriptor).

HELLMANN, M. (1984), *Ost–West-Wortschatzvergleiche. Maschinell gestützte Untersuchungen zum Vokabular von Zeitungstexten aus der BRD und der DDR* (Tübingen: Narr).

—— (1989), Zwei Gesellschaften – zwei Sprachkulturen? Acht Thesen zur öffentlichen Sprache in der BRD und in der DDR, *Forum für interdisziplinäre Forschung*, 11: 27–38.

HESS-LÜTTICH, E. W. B. (ed.) (1986), *Integration und Identität* (Tübingen: Narr).

HINNENKAMP, V. (1982), *Foreigner Talk und Tarzanisch. Eine vergleichende Studie über die Sprechweise gegenüber Ausländern am Beispiel des Türkischen* (Hamburg: Helmut Buske Verlag).

—— (1985), 'Zwangskommunikative Interaktion zwischen Gastarbeitern und deutscher Behörde', in Rehbein (1985*a*), 276–98.

202   Martina Rost-Roth

HINNENKAMP, V. (1987), 'Foreigner Talk, Code-Switching and the Concept of Trouble', in Knapp *et al.* (1987), 143–80.

—— (1989), *Interaktionale Soziolinguistik und Interkulturelle Kommunikation. Gesprächsmanagement zwischen Deutschen und Türken* (Tübingen: Niemeyer).

—— (1991), 'Talking a Person into Interethnic Distinction', in Blommaert and Verschueren (1991), 91–110.

HOUSE, J. (1985), 'Contrastive Discourse Analysis and Universals in Language Use', *Papers and Studies in Contrastive Linguistics*, 20: 5–14.

HUEBNER, T., and FERGUSON, C. (eds.) (1991), *Second Language Learning and Second Language Theory* (Amsterdam: Benjamins).

ILKHAN, I. (1987), 'Kontrastive Überlegung zu einigen unterschiedlichen Kommunikationsmustern im Deutschen und Türkischen', in Götze (1987), 183–92.

KLEIN, W. (1984), *Zweitspracherwerb. Eine Einführung* (Frankfurt: Athenäum).

—— and DITTMAR, N. (1979), *Developing Grammars: The Acquisition of German by Foreign Workers* (Heidelberg, New York: Springer).

KNAPP, K., ENNINGER, W., and KNAPP-POTTHOFF, A. (eds.) (1987), *Analyzing Intercultural Communication* (Berlin, New York: Mouton de Gruyter).

KOTTHOFF, H. (1989*a*), *Pro und Contra in der Fremdsprache. Pragmatische Defizite in interkulturellen Argumentationen* (Frankfurt: Lang).

—— (1989*b*), 'So nah und doch so fern. Deutsch–amerikanische pragmatische Unterschiede im universitären Milieu', *Info DaF*, 18/4: 448–59.

KRETZENBACHER, H. (1991), 'Das deutschdeutsche Du,' *Deutsch als Fremdsprache*, 3: 181–3.

KUGLIN, J. (1977), 'Einige Bemerkungen zur Anrede im Deutschen und Türkischen', in Engel-Braunschmidt and Schmücker (1977), 261–78.

KUHBERG, H. (1987), *Der Erwerb der Temporalität des Deutschen durch zwei elfjährige Kinder mit Ausgangssprache Türkisch und Polnisch* (Frankfurt: Lang).

MARUI, I., and REINELT, R. (1985), ' "Ich will dem Mensch werde . . ." '— Modalisierungen in deutschen Aufsätzen japanischer Studenten', in Rehbein (1985*a*), 190–221.

MATTEL-PEGAM, G. (1985), 'Ein italienischer Strafgefangener konsultiert einen deutschen Rechtsanwalt,' in Rehbein (1985*a*), 299–323.

MEISSNER, B., NEUBAUER, H., and EISFELD, A. (eds.) (forthcoming), *Die Rußlanddeutschen — gestern und heute* (Cologne: Markus).

MOLONY, C., ZOBL, H., and STÖLTING, W. (eds.) (1977), *Deutsch im Kontakt mit anderen Sprachen, German in Contact with Other Languages* (Kronberg: Scriptor).

NELDE, P., EXTRA, G., HARTIG, M., and VRIENDT, U. J. (eds.) (1981), *Sprachprobleme bei Gastarbeiterkindern* (Tübingen: Narr).

NEUNER, G. (1988), *Kulturkontraste im DaF-Unterricht, Studium Deutsch als Fremdsprache, Sprachdidaktik*, 5 (Munich: Iudicium).

OHAMA, R. (1987), 'Eine Reklamation', in Redder and Rehbein (1987), 27–52.

PALEIT, D. (1991), 'Deutsch für ausländische Arbeitnehmer – die Situation 1991', *Deutsch lernen*, 3: 226–41.

QUASTHOFF-HARTMANN, U. (1987), 'Zuhöreraktivitäten in der interkulturellen Kommunikation', in Gerighausen and Seel (1987), 104–24.

REDDER, A. (1985), 'Beschreibungsverfahren türkischer Kinder auf Deutsch: Eine einfache Bilderfolge', in Rehbein (1985), 222–46.

—— and REHBEIN, J. (eds.) (1987), *Arbeiten zur interkulturellen Kommunikation* [= *Osnabrücker Beiträge zur Sprachtheorie*, 38] (Universität Osnabrück).

REHBEIN, J. (1977), *Komplexes Handeln. Elemente zur Handlungstheorie der Sprache* (Stuttgart: Metzler).

—— (ed.) (1985*a*), *Interkulturelle Kommunikation* (Tübingen: Narr).

—— (1985*b*), 'Einführung in die interkulturelle Kommunikation', in Rehbein (1985*a*), 7–39.

—— (1985*c*), 'Medizinische Beratung türkischer Eltern', in Rehbein (1985*a*), 349–419.

—— (1986), 'Institutioneller Ablauf und interkulturelle Mißverständnisse in der Allgemeinpraxis. Diskursanalytische Aspekte der Arzt–Patient-Kommunikation', *Curare*, 9: 297–328.

—— (1987), 'Diskurs und Verstehen. Zur Rolle der Muttersprache bei der Textverarbeitung in der Zweitsprache', in Apeltauer (1987), 113–72.

REUTER, E. (ed.), (1992), *Wege der Erforschung deutsch–finnischer Kulturunterschiede in der Wirschaftskommunikation* (Tampere: Publikationsreihe des Sprachzentrums der Universität Tampere).

—— SCHRÖDER, H., and TIITTULA, L. (1989), 'Deutsch–Finnische Kulturunterschiede in der Wirtschaftskommunikation. Fragestellungen, Methoden und Ergebnisse eines Forschungsprojektes', *Jahrbuch Deutsch als Fremdsprache*, 15: 237–69.

ROCHE, J. M. (1988), 'Codewechsel und Textkonstitution', *Linguistische Berichte*, 117: 397–409.

—— (1989), *Deutsche Xenolekte. Struktur und Variation der Äußerungen deutscher Muttersprachler in der Kommunikation mit Ausländern* (Berlin: de Gruyter).

RÖHL, E. (1991), *'Deutsch-Deutsch'. Satirisches Wörterbuch* (Berlin: Eulenspiegel).

ROMAINE, S. (1989), *Bilingualism* (Oxford: Blackwell).

ROSENBERG, P., and WEYDT, H. (forthcoming), 'Sprache und Identität. Neues zur Sprachentwicklung der Deutschen in der Sowjetunion', in Meissner *et al.* (forthcoming).

ROSENTHAL, R., HALL, J., ARCHER, D., DI MATTEO, M., and ROGERS, D. (1979), 'Measuring Sensitivity to Nonverbal Communication: The PONS Test', in Wolfgang (1979), 67–97.

ROST, M. (1989), *Sprechstrategien in 'freien Konversationen'. Eine linguistische Untersuchung zu Interaktionen im zweitsprachlichen Unterricht* (Tübingen: Narr).

Rost, M. (1990), 'Reparaturen und Foreigner Talk—Verständnisschwierigkeiten in Interaktionen zwischen Muttersprachlern und Nichtmuttersprachlern', *Linguistische Berichte*, 125: 24–45.

Sachtleber, S. (1991), 'Texthandlungen in der Wissenschaftssprache. Ein Vergleich deutscher und französischer Texte', in Spillner (1991).

Schlosser, H. D. (1990), *Die deutsche Sprache in der DDR zwischen Stalinismus und Demokratie. Historische, politische und kommunikative Bedingungen* (Cologne: Verlag Wissenschaft und Politik).

Smith, S., Scholnick, N., Crutcher, A., Simeone, M., and Smith, W. (1991), 'Foreigner Talk Revisited: Limits on Accommodation to Nonfluent Speakers', in Blommaert and Verschueren (1991), 173–85.

Speicher, J. K. (1985), 'The (mal)functioning of address forms in intercultural situations', in Brunt and Enninger (1985), 94–102.

Spillner, B. (ed.) (1990), *Interkulturelle Kommunikation*, Forum Angewandte Linguistik, GAL (Frankfurt: Lang).

—— (ed.) (1991), *Ein Europa—Viele Sprachen* (Frankfurt: Lang).

Stolt, B. (1992), 'Kulturbarrieren als Verständnisproblem', in Reuter (1992), 28–46.

Strecker, B. (1992), 'Besserwessi—Wort des Jahres 1991', *Sprachreport*, 1: 5.

Streeck, J. (1985), 'Kulturelle Kodes und ethnische Grenzen. Drei Theorien über Fehlschläge in der interethnischen Kommunikation', in Rehbein (1985*a*), 103–20.

Stutterheim, C. von (1988), *Temporalität in der Zweitsprache: Eine Untersuchung zum Erwerb des Deutschen durch türkische Gastarbeiter* (Berlin: de Gruyter).

Torres, G. M., and Wolff, J. (1983), 'Interkulturelle Kommunikationsprobleme beim Sprachenlernen, dargestellt an Mißverständnissen zwischen Spaniern und Deutschen', *Neusprachliche Mitteilungen*, 4: 209–16.

Uhlisch, G. (1991), ' "Wir gratulieren Ihnen zu Weihnachten!"—Sprach- und kulturkontrastive Analysen—als Beitrag zum interkulturellen Lernen, *Kultur und Sprache, Berliner Beiträge zu Deutsch als Fremdsprache*, 1: 11–18.

Wierlacher, A. (ed.) (1987), *Perspektiven und Verfahren interkultureller Germanistik* (Munich: Iudicium).

—— Eggers, D., Engel, U., Krumm, H. J., Krusche, D., Picht, R., and Bohrer, K. F. (eds.) (1987), *Erkenntnisgewinn von den Rändern her* [= *Jahrbuch Deutsch als Fremdsprache 1987*] (Munich: Hueber).

—— —— —— —— —— —— —— (eds.) (1988), *Kulturspezifische Aspekte der Sprachvermittlung Deutsch als Fremdsprache* [= *Jahrbuch Deutsch als Fremdsprache 1988*] (Munich: Hueber).

Wolfgang, A. (ed.) (1979), *Nonverbal Behavior: Applications and Cultural Implications* (New York: Academic Press).

Ylönen, S. (1992), 'Probleme deutsch-deutscher Kommunikation', *Sprachreport*, 2–3: 17–20.

Zimmermann, P. (ed.) (1989), *Interkulturelle Germanistik. Dialog der Kulturen auf Deutsch* (Frankfurt: Lang).

9 Critical Linguistics and the Study of Institutional Communication

RUTH WODAK

I INTRODUCTION

Many important changes have taken place in the study of language in social context since the first sociolinguistic studies were carried out in the USA and Great Britain in the early 1960s. In this chapter, I shall concentrate on the consequences of two particular developments: first, the shift of emphasis from the sentence or utterance to 'text in context', in other words to the study of discourse, which in turn led to the linking of sociolinguistics and discourse analysis; secondly, the more systematic approach to the study of context, which entailed both a more interdisciplinary perspective and a greater emphasis on qualitative (as opposed to quantitative) research. The results of these developments are that sociological, psychological, social-psychological, and historical perspectives have been incorporated into sociolinguistic study, and that the analysis of context has become an important object of study in its own right. In particular, a whole area of research has developed focusing on the study of language use in institutional settings.

I shall begin here by outlining some general areas of research within the field of institutional communication, and then go on to discuss in more detail the approach developed by the Vienna Applied Language Studies research group, which we refer to as *discourse sociolinguistics*. As this approach is based on a holistic analysis of a given context, rather than just the specific linguistic interaction of the participants, it will be necessary to give a fairly lengthy illustration. This will be based on a recent study of institutional communication in an out-patient ward of a hospital in Vienna (see also Lalouschek *et al.* 1990).

2 COMMUNICATION IN INSTITUTIONAL CONTEXTS

2.1 *Institutional Discourse*

Dieter Wunderlich offers the following definition of 'institutions':

Institutionen sind historisch gestaltete Systeme von — unter Umständen kodi-
fizierten — Regeln, die Tätigkeiten von Personen in mehr oder weniger
präziser Festlegung aufeinander abstimmen sollen, und zwar in Ausrichtung
auf Zwecke, die jeweils im Zusammenhang der gesellschaftlichen Produktion
und Reproduktion stehen. (Wunderlich 1976: 312)

(Institutions are historically constituted systems of (possibly codified) rules,
which are supposed to co-ordinate the activities of people in a more or less
precisely determined manner. This should be geared to purposes which are
related to social production and reproduction.)

This definition incorporates several important aspects of the organization and structure of institutions. First, Wunderlich emphasizes the
historical development of institutions. On the one hand, they are
constantly subject to change, in the sense that their rules are not
immutable and the communicative patterns that characterize them
differ according to social system and historical circumstances. On
the other hand, however, certain traditional elements are often well
preserved but lose their original functions and therefore become
rituals (for example, the wearing of traditional clothing at 'dissertation
defences' in many European universities or certain ritual formulations
in legal language: see Dànet 1980).

Secondly, both explicit and implicit rules that operate in institutions
have the function of organizing hierarchies: by regulating (communicative) behaviour, they orchestrate the relationships between the
people who work within institutions. Hierarchies are an important
characteristic of institutions, and they reveal the nature of internal
power structures. Finally, Wunderlich also mentions 'production' and
'reproduction' functions, which may often be contradictory: for
example, doctors in hospitals are supposed to cure their patients (this
is their 'explicit' function), but they are also supposed to teach new
doctors and socialize them in the institutional world of the hospital
(this is their 'implicit' function). This can lead to communicative misunderstandings if two doctors interview a patient while at the same
time (in the same discourse) one of the doctors tries to explain important medical phenomena to the other (see Menz 1991).

If we take this definition of institutions as a starting-point, it is easy
to see that the sociolinguistic study of institutional communication
can serve a variety of interests and has a number of practical advantages. For instance, institutional settings can be clearly delimited and

defined, and interaction between participants in institutional discourse is fairly simple to record and study; the correlation of linguistic and sociological variables is relatively straightforward; and this kind of research may have practical benefits, such as contributing to training programmes for lawyers or doctors (see Wodak 1984). It is also one of the principal objectives of 'critical linguistics' as defined by Wodak (1989).

This approach to the study of language in context aims to make both explicit and implicit rules and power structures in socially important domains transparent. In modern societies these domains are embodied in institutions, which are structured in terms of social power relationships and characterized by specific divisions of labour (cf. Mumby 1988, Weick 1985, Menz 1991). Within institutions, élites (typically consisting of white males) occupy the dominant positions and therefore possess power. They determine what Bourdieu (1979) calls the 'symbolic market', the value and prestige of symbolic capital: in other words, certain forms of communicative behaviour and certain registers of language are considered to be more prestigious than others. This is most evident in the technical registers used by all professional groups (for example, what in lay terms is often referred to as 'legal language' or 'legalese': see Dressler and Wodak 1989), but it also manifests itself less obviously in the form of preferred styles and certain communicative strategies (see Section 2.2 below).

Institutions have their own value systems, which are crystallized in the form of specific ideologies. However, it is important to distinguish between the explicit demands and expectations of the official institutional ideology and the implicit rules underlying everyday behaviour. These two sets of norms often lead to contradictions: for example, Van Dijk (1992) discusses a study of managers in large companies, who claimed in interviews that foreigners had equal chances of employment and that they explicitly supported 'affirmative action', whereas in reality foreigners are actively discriminated against. However, these contradictions are disguised by what Barthes (1974) calls 'myths', and in this way they are legitimized. Myths in this sense of the word are secondary semiotic systems, which both insiders and outsiders are supposed to believe and which mystify reality. A particularly striking example of this in medical institutions is the great knowledge that doctors are supposed to possess and their infallibility: the *Götter in weiß* (gods in white) is a common expression in Austrian German (see Menz 1991 for the description of other important myths in medical organizations).

One further aspect of the analysis of institutions should also be mentioned here. Investigations of the status quo should make it possible to devise and propose different communicative practices for those working in the institution (contributing to training programmes, for example), as well as different approaches from the clients' point of view (see Lalouschek *et al.* 1987). In this respect, sociolinguistic analysis is particularly important, as a purely sociological analysis tends not to make the individual dynamic processes clear enough. Using the discourse analysis approach, we can see the effect of both explicit and implicit institutional rules and norms in virtually every specific discourse, and we can demonstrate how structures are constantly being reproduced in each specific interaction (see Ehlich and Rehbein 1986, Strong 1979).

2.2   *Some Important Areas of Research*

The study of institutional discourse in German-speaking countries has been applied to many different areas. It is impossible to survey the whole field here, so I shall restrict myself to the major issues (for more detail, see Wodak 1987*a*).

Schools

Communication in schools was the first main focus of interest. The first studies were mainly influenced by the British sociologist Basil Bernstein (see e.g. Bernstein 1960), and analysed the supposed discrimination against working-class children on the grounds of their limited linguistic repertoire or 'restricted code' (see e.g. Oevermann 1972; for a discussion in English of research in German-speaking contexts, see Barbour and Stevenson 1990: chap. 6, and Dittmar, this volume). Much of the subsequent discussion revolved around the concepts of 'compensatory' or 'emancipatory' language-teaching programmes: that is to say, should we try to 'adapt' working-class children to middle-class standards, or should we adopt a more functional approach, stressing the functional adequacy of specific speech forms in given situations (Dittmar 1973)?

In these early studies of communication in schools, written texts were used to test children's linguistic behaviour in much the same way as Bernstein had done. Quantification and correlation of linguistic data with sociological variables were the main analytical methods used, and the so-called outsider perspective dominated (Dittmar 1987, Habermas 1981). More recent work has focused on three different perspectives. First, as the notion of discourse became more important, classroom interaction was analysed with respect to certain communicative patterns, especially 'teachers' questions' (Ehlich and Rehbein 1976, Redder 1982, Wodak *et al.* 1989). Ehlich and Rehbein also studied the different *Aktantenwissen* (worlds of knowledge) which determine the interaction between teachers and children. They demonstrate how teachers seek to convey knowledge in very specific ways, and identify six 'types' of knowledge that teachers typically transmit. The most important of these are *Musterwissen* (pattern knowledge) which means collectively known and internalized forms of action and behaviour, and *Routinewissen* (routine knowledge), which refers to automatic schemas of structured information. What this means is that on the one hand children are supposed to adapt to certain behavioural patterns, and on the other hand specific cognitive information should also be acquired. As the authors show, when different structures and schemas of knowledge coincide in the classroom, this frequently leads to misunderstanding.

Secondly, many researchers have turned their attention to immigrant children and their language acquisition (see Barbour and Stevenson 1990: chap. 7). In the 1970s and 1980s, this work was concerned primarily with migrant families from southern Europe and Turkey, but the fall of the Iron Curtain in 1989 added a new dimension to this issue, as large waves of immigrants from eastern Europe entered Germany and Austria (see Rost-Roth, this volume). Many Viennese schools, for example, now have up to 80 per cent foreign children, and new language teaching programmes will have to be developed.

Thirdly, one particular study (Wodak *et al.* 1992) has looked at communication in various settings throughout the whole institution of the school from the 'insider perspective': in the classroom, in staff meetings, in parent–teacher meetings. This research focused especially on gender differences and on hierarchical patterns of power. Three head teachers were analysed as case-studies, showing different forms of female leadership: the indirect style, the persuasive style, and the 'motherly' style. Among other things, it was possible to demonstrate in detail how institutional communicative patterns interact with gender-specific strategies: in the official meetings in the school institutional discourse clearly dominated, whereas in the more informal settings individual and gender differences became apparent.

The Law
A lot of research has also been conducted on legal language, both on courtroom discourse (e.g. Leodolter 1975, Hoffmann 1991) and on written texts (e.g. Pfeiffer *et al.* 1987). Here again we can distinguish between studies focusing on small sequences and small discourse units (specific forms of cross-examination, for example), and analyses of larger units or whole institutions. Significant findings of this research

include the impact of social class, gender, and experience on inter-action in the courtroom and the outcomes of trials: for instance, work-ing-class women fared less well in terms of opportunities to defend and present themselves and received more severe sentences than middle-class defendants. Another, rather different, aspect of this work is the fact that the comprehensibility of bureaucratic and legal lan-guage has become a major social and political issue, and sociolinguists have become involved in the reformulation of legal texts (see e.g. Wassermann and Petersen 1983).

Counselling

Uta Quasthoff (1980) and more recently Reinhard Fiehler and Wolf-gang Sucharowski (1992) have investigated language use in coun-selling. Quasthoff's analysis focused on the stories which clients tell and on the difficulties the 'experts' have in relating to their clients' experiences (see also Ehlich 1980, and Rost-Roth, this volume). Fiehler and Sucharowski have developed training programmes for specialists in the 'caring' professions on the basis of discourse studies in various institutions (hospitals, bureaucracies, universities).

Features of Institutional Communication

In many institutions, similar characteristics and phenomena have been observed. We can summarize the most important features as follows:

Anonymity: the names of the clients are known but the names of the experts are often concealed, so that the clients do not know who they are talking to; written discourse in laws, formulas, etc. is characterized by vagueness and by the use of the passive.

Ritualization: institutions have many rituals (for example in clothing, manners, language), which have developed historically but have lost their original function and are therefore difficult for the outsider to understand; they fulfil an important socio-psychological function in terms of group identity and integration.

Militarism: power and hierarchical relationships may be articulated either in open or in more subtle ways; many speech acts in institutions involve commands, but these are always one-way; the experts pose the questions, the clients must answer them; the experts may also interrupt, the clients may not; the clients are supposed to be able to 'tell their stories', but the experts usually occupy the floor longer than the clients.

The search for harmony: contradictions and conflicts are swept under the carpet; officially, those who work in the institution form closed groups and are either afraid to discuss their relationships with each other or incapable of doing so.

Sociological variables: social class, education, gender, and experience seem to be the most important static variables; middle-class clients and those with more experience typically receive better treatment and more time than others; (white) males usually occupy the positions of power.

Inexperienced experts: the most important points of contact between the inside and outside worlds (such as out-patient clinics, enquiry desks) are usually staffed by young, inexperienced experts; decisions are therefore often very slow, because the inexperienced staff have to wait for the opinions of older, more experienced colleagues; it is also often the case that more experienced staff occupy relatively low positions in the hierarchy (such as nurses), and they therefore have to be very careful in what they say so as not to pose a threat to the experts in higher positions (for example doctors).

2.3 Doctor–Patient Communication

Research into doctor–patient communication has generally been based on one of two approaches: the medical-sociological approach, which focuses on the institution, and the linguistic approach, which deals with the micro-structural aspects of communication. However, the two approaches have rarely been combined in a single study. Aaron Cicourel, one of the founding figures of doctor–patient research, has shown on the basis of selected interviews the advantages of a conversational analysis approach, in contrast to quantitative psychological investigations (see Cicourel 1981, 1984). He repeatedly insists that any analysis must take account of the structural framework as well as the different interests of the two main protagonists, the doctor and the patient.

Current American research is increasingly limited to the analysis of individual types of conversation, for example question–answer sequences and 'accounts', that is, narratives in which the speakers account for or explain their actions. Since these types of conversation are studied in isolation from the context of the complete discourse, they can only be interpreted up to a certain point (see e.g. Fisher and Todd 1983, Fisher and Groce 1990, Freeman 1987, Heller and Freeman 1987, Greene *et al.* 1986, West 1984, 1990, Todd 1983). However, two findings from recent work in the United States are relevant to the present context and are worth pointing out here. Richard Frankel (1983) describes the clash between the institutional world and the lay world as a 'frame conflict': value systems, the structuring of knowledge, and traditions all diverge and cause misunderstanding and conflict. One common concrete manifestation of this is the fact that doctors typically want to arrive as quickly as possible at a diagnosis, while patients often want to explain aspects of their biography and would also like to know the implications of their symptoms or illness. Elliot Mishler (1984) shows that even research on communication is typically conducted from the perspective of the medical paradigm, so that scientific interpretations are only made from the point of view of doctors and the signals coming from the patients are ignored in the analyses. I shall return to these approaches in my conclusions.

The medical-sociological approach dominated early work in German-speaking countries, and this lacked the necessary linguistic apparatus. At the centre of this research were case histories and conversations during doctors' rounds (see Köhle and Raspe 1982, Strotzka et al. 1984). Current research based on discourse analysis oscillates between two extremes: on the one hand, a mainly application-orientated approach, on the other hand, an approach which is concerned with the smallest units of discourse without an exact analysis of context (see Ehlich et al. 1989, Spranz-Fogaszy 1987 for examples of both perspectives.) Thomas Bliesener's study of doctor–patient conversations (1982) represents the first relevant attempt at defining a particular subsystem of everyday life in a hospital using discourse analysis. The doctors' rounds are broken down into their various features and phases, and individual problems in the patients' communication are pinpointed. Here again, however, the genuinely sociolinguistic aspect is not brought in: he does not investigate, for example, whether women and men or old and young patients are treated differently. He does make the important point, though, that communicative problems will not be solved simply by devoting more time to the patients, as the quality of the doctors' approach and of their conversations with patients also needs to be improved.

Within the framework of the Vienna Applied Language Studies programme, we have incorporated many ideas from both the medical-sociological and the linguistic approach and have recently begun research into case interviews (see Hein et al. 1985, Hein and Wodak 1987, Menz 1991; we have also done extensive research on therapeutic discourse: see Wodak 1986). In a study on general practitioners, Norbert Hein (1985) has shown clearly that there are language barriers between doctors and patients that are attributable to social class differences. Patients from a working-class background were treated condescendingly, and even the suggested treatment for the same symptom (broken sleep) was different for working-class and middle-class patients: the former were offered drug prescriptions, the latter were referred for psychotherapy. However, in these studies the definition of the discourse unit was again limited to the exploratory interview itself: everyday life at the institution was only touched upon (for further detail on this area, see Cicourel 1981, Wodak 1987*b*, Freeman and Heller 1987). In the next section, therefore, I shall look in some detail at a study that attempts to analyse precisely discourse in the context of the institution as a whole.

3   DISCOURSE AND CONTEXT: A STUDY OF EVERYDAY LIFE IN AN OUT-PATIENTS' WARD

3.1   *Setting and Methodology*

The Context of the Institution
All the studies mentioned so far were carried out from the 'outsider's perspective', that is, on the basis of certain preconceived assumptions specific discourses were extracted from the larger institutional setting and analysed in isolation. The sense of context in relation to the institution and to the daily life of the institution is therefore lost. Given this limited knowledge of the context, the interpretations that are made are only partially valid.

Therefore one of the key objectives of our study of an out-patients' ward at a Viennese hospital was to work from the 'insider's perspective' (this concept has been well defined by Habermas 1981, and has had a great impact on fieldwork and research design in the social sciences). This entailed observing and recording complete morning sessions as a single discourse unit, and having discussions with doctors and patients both before and after these sessions. On this basis, it was possible to define new, dynamic categories of analysis, which can be taken together with the familiar categories in interpreting the data (see below).

For an understanding of the context, it is important to realize that the out-patients' ward has very low status and prestige in relation to the rest of the institution. It is a kind of outpost and amongst other things serves as a training ground for young doctors, which results in the kinds of problems discussed earlier with inexperienced insiders working where experienced ones are arguably most necessary. Hierarchy, knowledge, experience, and gender are interlinked in a strange and unique way in the out-patients' ward. Inefficiency, bad organiz-ation, and bad training are disguised by the propagation of myths (see

Section 2.1 above), and stereotypes emerge: doctors never have enough time, they are never wrong, and there is simply no better way of doing things.

The Design of the Study

Seven morning sessions were covered in detail and eighty-three discrete conversations were recorded. The age of the patients ranged from 17 to 87 and there were roughly the same number of men and women. Five doctors were observed, three women and two men. The institution itself was very co-operative and supportive. However, some problems did arise, especially when doctors became nervous and felt overstretched by the many contradictory demands imposed on them.

The following sets of questions provided the basic framework for the study:

Context: What is the impact of the context, and which elements of the context are relevant in the interactive process? Is the setting, for example, more important than the experience and personality of the doctors, or is the influence of each element different? What typical discourse patterns can be identified?

If we assume the context to be a complex, multi-level phenomenon, the individual discourse unit of each morning session (the actual microtext) could be seen as the 'inner circle', and the remaining elements of the context could be seen as being grouped in concentric circles around it. The specific setting of the ward and the system of doctors and nurses then constitute the second circle, and the history and structures of the hospital form the outer circle. It is in the inner circle that variables concerning the people involved and interactive phenomena become particularly important (which doctor, which patient, what is the nature of the relationship between them, which other events take place?), but the complete context should always be taken into account.

Power: According to Pierre Bourdieu (1979: 355 ff), the élite define language and possess power, and the language of the élite is 'symbolic capital'. How is this capital 'invested', and how are differences in knowledge expressed in our specific case? Does a frame conflict (see Section 2.3 above) exist in the out-patients' ward? How do doctors convey their power? Which 'power registers' can be identified?

From other studies on the comprehensibility of news and legal texts (Lutz and Wodak 1987, Pfeiffer *et al.* 1987), we know that previous experience is an essential factor in dealing with institutions. Clients with more experience of the institution receive different and better treatment, they are regarded as (virtual) equals by the experts, and so the effects of the power structures are mitigated. Is this the case in the context of the hospital?

Myths and realities: How are the values and myths of the institution expressed? How are the contradictions between expectations and reality dealt with? Or, to put it more bluntly, how do the doctors cope with their everyday professional life?

Categories for Analysis

In dealing with these questions and assumptions, three dynamic categories were found to be important. First, the emergence of 'patient initiatives' (such as asking for information, relating stories, making complaints or judgements) and the ways in which doctors deal with these initiatives (for instance by answering, interrupting, or ignoring them) bring two conflicting elements into play at the same time: the doctors' exercise of power and the patients' voices. Secondly, we focused on doctors' problem-solving procedures: the specific discursive way in which doctors deal with each specific problem demonstrates the assumed contradiction between explicit rules, myths, and actual events. Thirdly, the verbal negotiation and formation of relationships was analysed: each time a relationship is established, fundamental factors of the management of relationships are applied. For example, if a personal relationship is established, then important rituals of politeness are observed, certain forms of address are used, and so on.

3.2 A Typical Morning Session

The Institutional Context

A male and a female doctor are working together during this specific morning session. Both have just been on night shift and are exhausted. The initial phase is quiet. Then just after 11 a.m. there is an announcement over the intercom system to the effect that all cars have to be removed from the hospital car park by 1 p.m. This announcement causes an outbreak of chaos, for this means that the doctors will have to leave the ward at some time during their work.

We can therefore divide the discourse unit very roughly into three phases: the initial quiet phase from 9.30 a.m. to 11.10 a.m., a problematic middle phase from 11.10 a.m. to 11.30 a.m., and the stressful final phase from 11.30 a.m. to 12.20 p.m.

Case-Study 1: The Experienced Patient

The first patient is 60 years old and has a stomach complaint. The problem that the doctor has to solve is one of non-compliance with official procedures: the patient has apparently taken medication on his own initiative. On the basis of Text 1, it is possible to analyse strategies of problem-solving, as well as specific power registers used with an experienced patient. Through an analysis of macro-strategies, we can identify specific patterns used to cover up a fault in the procedures on the ward which the patient discovers and exploits. At the same time, this discourse is representative of the quiet initial period, in which the patients have much more space than at the end of the morning session. It is important for an understanding of the closer context to know that an experienced patient is the subject of this case.

Text 1[1]

11   D M8. as pills from today
      P.   Yes – and yesterday you

12   D M8.   Well, I hope it works – hmm well
      P.   didn't give me any at all.

31   D M8. well – but that's always – that's only the Lasix 80 –

32   D M8. you can get that from us – yes AS WELL

33   D M8. – you know. Or   have you also taken
      P.   I haven't taken any — no

34   D M8. an 80 today as well?
      P.   I haven't had any – till now

35   D M8. It says here Lasix 80 milligrammes — as a daily

36   D M8. dose — on the chart. Lasix 80 — yes there was

37   D M8. a short break. Oh well
      P.   I didn't get it till – I didn't get it till –

38   D M8.   mhm
      P.   Dr X [doctor's name] was er down here

39   D M8. yes
      P.   with me you know. Gave me that er – Novarin. Apart from

40   D M8.   Yes I see – you didn't
      P.   that I haven't had any tablets – (. . . . . .)

[1] German originals of Texts 1 and 2, together with a key to the transcription symbols, can be found in the Appendix to this chapter.

41   DM8. get them again till today? — Yesterday was
     P.   Got them

42   DM8.   very good
     P.   today and YESTERDAY I took one of my own because

43   P.   if you remember – I asked you

44   DM8. yes   that's fine – GOOD –
     P.   (. . . . . .) that I stopped taking those –

45   DM8. fine – you can start taking them again from today –
     P.   the tablets you know.

46   DM8. because you don't need any more infusions.

111 DM8.   — yes
     P.   What's the next step – well –

112 DM8.   Well if you
     P.   at least here in hospital I mean

113 DM8. if you – can be discharged – as far as the ulcer's

114 DM8. concerned – from the surgery wing, which will clearly

115 DM8. be soon   That we could perhaps add
     P.   Yes that's what I meant

116 DM8. on – a short stay HERE – until

117 DM8. the pulse rate's settled down again a bit.
     P.   Is it possible for me –

118 DM8.   In the surgery wing?
     P.   to stay there?   Yes — is that OK?

119 DM8. Well, I mean theoretically that's OK – but but
     P.   that's OK

120 DM8. whether Dr X /name/ will allow it, you know –
     P.   whether he'll allow it

121 DM8. if you're not a surgical case any longer. –

122 DM8. We'll have a word about it – yes. At any rate

123 DM8. there's no question of you being discharged yet. – But

124 DM8. the liver results have got better – and the

125 DM8. pulse has gone down a bit — yeah?

141 P.   It's er like this – well — what's it called Mulox or something

142   D M8. Malox yes
        P.   I was prescribed the Malox – but I never got

143   P.   any. And that's only one example (. . .) first of all

144   D M8.   mhm
        P.   that. (. . .) But now – now I know

145   D M8.   That's right
        P.   if you don't ask (. . .).

146   D M8. So when did you have breathing problems – would you

147   D M8. tell me about it?

183   D M8.   Yes – it'll be sent up.
        P.   Will it be sent up?

184   D M8.   Yes?   X /surname/
        P.   Doctor – may I ask your name

185   D M8.   Not at all.   Goodbye.
        P.   Thank you   Goodbye.

The doctor begins with an indirect accusation: he accuses the patient of having taken a Lasix 80 tablet on his own initiative, even though he receives this drug from the ward anyway (lines 31–4). The accusing voice can be detected in the intonation, especially from the emphasis on the words AS WELL. In line 33, the patient tries to justify himself, but the doctor does not allow him to take his 'turn'. It is not until line 34 that the doctor actually pauses, enabling the patient to take his turn, and the patient then begins his story. In his account, the patient reminds the doctor that the doctor had personally given him the instruction to take his own tablets (lines 42–3). The doctor tries as early as in line 42 to interrupt by offering praise and positive comment, but the patient continues. In line 44, the doctor finally manages to interrupt the patient, again with praise, and then closes down the discussion ('fine, good'). The doctor's discursive strategy has therefore served two purposes: to cover up the institution's error and to terminate a potentially embarrassing discussion on an apparently positive note. In the final part of the episode, the doctor introduces a revised and largely self-serving description of his discussion with the patient, and in the end prescribes exactly the same medicine as before but seeks to legitimize this with ostensibly relevant new information (lines 45–6). In this way, he resumes his active role in keeping with his position of power.

This episode leads to the doctor being relatively open towards this (experienced) patient, which encourages the patient to take another initiative by asking: 'What's the next step? Well, at least here in hospital, I mean' (lines 111–12). The doctor replies at first in a rather noncommittal way (lines 113–17; here incidentally there is an interesting mixture of colloquial features such as 'a bißl' and the formal technical register, as in 'entlassungsfähig'), but then goes on to respond in more detail to questions concerning the patient's stay in hospital, a strategy that serves to block any non-medical questions or questions which relate to his life 'outside' (lines 118–25). So openness and personal relationships are possible, but only up to a certain point: that is, as long as they remain within the frame determined by the doctor.

However, when the patient takes his third initiative and complains again about the ward: 'I was prescribed the Malox, but I never got any. And that's only one example: first of all that . . . but now I know, if you don't ask . . .', he is praised ('that's right') but then the doctor abruptly changes the topic and asks: 'So when did you have breathing problems — would you tell me about it?' (lines 141–7). The same pattern as in the previous episode is repeated, but here it is much stronger as the experience of the patient becomes a little too threatening. However, the doctor maintains his formal manner and takes his leave politely (lines 183–5). A relatively close relationship has been established, there has been no clash between different 'worlds of knowledge' (see Section 2.2 above), and so the doctor has to work hard to maintain his authority by using several strategies (interruption, rationalization, and topic shift). One particularly subtle aspect was the discursive strategy of combining praise and mitigation and/or a change of topic, a clever 'packaging' of power.

Case-Study 2: The Inexperienced Patient
The second patient enters the ward at 9.45 a.m. She is considered a difficult patient because of her age (87). She is not prepared to take off her hospital gown as there are men present. Here there is a real clash between values and generations coupled with a total lack of experience on the part of the patient. Through the micro-analysis of this case study, it is possible to make explicit the systematic interaction of different aspects of discourse (speech acts, particles, forms of address, socio-phonological realizations) in the process of making the patient adapt to the institution.

Text 2 contains extracts from the beginning, middle, and end of the examination.

Text 2

| 1 | DF2. Right we'll have to take off the gown too |
| | P.   (. . . . . .) don't |

2 DF2. Why not? – We are in hospital you know. Right – now then

3 DF2. let's sit down here shall we?
   P. [*quietly*] (. . . . . .)

4 DF2: RIGHT please take off your gown.
   P.   Gown – but I've

5 DF2.   Take it off please – the gown. We've
   P. got nothing under the gown.

6 DF2. got to do an ECG. Right.   No one's
   P.   (. . . . . .) Gown

7 DF2. looking — well – it's only the doctor

8 DF2. – isn't it.   He's allowed
   P.   The doctor is allowed to look – but

9 DF2. to look isn't he – OK let's sit down here

10 DF2. shall we.   exactly
   P.   Sometimes he even has to look (. . .)

11 DF2. Right – tell me, which was the broken arm?

64 DF2. She keeps wobbling around — NOW JUST LIE STILL

65 DF2. DON'T KEEP WOBBLING AROUND – OR THE ECG WON'T WORK

66 DF2. QUITE STILL – JUST RELAX OK. Good — right:
   P.   All right – yes.

67 DF2. that's fi:ne.

221 DF2. But she's sore EVERYWHERE — she's sore

222 DF2. everywhere. DOES IT HURT THERE TOO?
   P. Ah: yes [*sighs*] (. . .)   no no

223 DF2. Ah, not there – only the back and there it does hurt, doesn't
   it
   P.   it's OK

224 DF2.   Yes – and there?
   P. It hurts there.   Well – I can feel it – but

225 DF2.   not too bad.
   P. it's bearable.

The doctor makes four attempts to get the patient to take off her gown. The first is in the form of an indirect speech act using a child-like form of address ('we'll have to . . .'), which I refer to as *pluralis hospitalis*. Then the doctor tries to persuade the patient to comply with her wishes by rationalizing the situation ('we *are* in hospital . . .'), but again without success. After a structuring signal ('right'), the doctor makes a fresh attempt, this time in the form of a polite but firm request ('please take off your gown'), using a more direct but still distant or polite form of address and a socio-phonological switch into standard German articulation. Finally, when the patient still refuses to oblige, the doctor repeats her request ('take it off please, the gown'), the imperative form indicating the more peremptory nature of this attempt. She reinforces this with a technical explanation, which the patient does not understand. As the patient still fails to remove her gown, the doctor tries to reassure her that no one else is looking other than her (male) colleague (lines 6–7); the particle 'right' is intended both as a concluding signal and a reassurance. The patient begins to give in, although she still does not seem to be totally convinced (line 8). The doctor picks up this more positive sign, echoing the patient's remark that 'the doctor is allowed to look' and switching to dialect, which helps to bring the two closer together. She follows this up with a further request, softened by the reassuring particle 'OK' (*gell*). The patient again repeats 'sometimes he even has to look', as if to convince herself, and the doctor confirms this before changing the topic with the structuring signal 'right' (*so*). Then the actual examination begins.

Later in the examination, the doctor is unable to control her irritation with the patient and begins to shout at her (lines 64–6). This is followed by a direct request, which is realized in standard German. Four more demands follow before she resumes a calmer, reassuring tone (line 67). The final section of the discourse (from line 221) is polite and gentle, and the patient is more subdued. The patient is the subject of the discourse but is referred to in the third person (line 221): this is a common pattern, but it causes a lot of uncertainty among patients since it is not clear who is actually meant. Only the use of a direct question (line 222) makes it clear that at that point it is the patient herself who is being referred to and that a response is required. Finally, the patient concedes that 'it's bearable': a sign that she has, at last, adapted to the institutional frame.

The doctor's manner with the patient ranges from a gentle approach through stiff formality and impatience to harsh authority. These changes are indicated by the character of the various requests (direct and indirect), the different forms of address, and the switches in socio-phonological style. The patient is provided with very little in the way of discursive orientation other than the particles 'OK', 'right', and so on, which are used to express reassurance and as structuring signals (beginning or concluding topics). There appears to be a clear correlation between these indicators, and certain patterns seem to emerge:

for example, indirectness of request is associated with *pluralis hospitalis*, the use of dialect, and certain particles.

The doctor's behaviour is a form of exercising power, but it also reflects a conflict that is caused by a number of factors. It is her first professional conversation with a patient which is being recorded. The patient is considered 'difficult', as her 'uncooperative' behaviour slows the examination down, and the doctor has great difficulty in dispelling the patient's fears. The case is therefore seen as a disruption of the normal routine: it conflicts with institutional expectations. As a result, no close personal relationship is established between doctor and patient. Helpful explanations and information are not forthcoming precisely when they are most needed, and even when the doctor does attempt a more personal approach, it is characterized by childlike language and *pluralis hospitalis*. This only serves to reinforce the difference in power and the patient's assumed mental inferiority. The frame conflict and the language barriers separating the two participants render co-operative face-to-face communication virtually impossible, and in the end the patient falls silent.

Chaos

Just after 11 a.m. the following announcement is made over the intercom: 'Your attention please. The army out-patient unit in Götzendorf is going to help clear the snow from the car park from 1 p.m. today. All staff are requested to move their cars to make the work easier.' At first, the announcement is greeted with laughter: 'Oh God, that just can't be true. At 1 o'clock today. We won't get out of here . . . what a joke.' But when the doctor realizes what this announcement actually means, it becomes a threat: 'Oh no, I've got my course at 1 o'clock, I don't think I've been to the course for three weeks. . . . Where am I supposed to put the car? I'll just leave it there. I've had enough', and his irritation is transferred to the waiting patients: 'And they've all got to be looked at yet. How many more are there, for God's sake?'

By this stage, the staff are no longer able to cope with the pressures of the situation calmly. It is particularly clear here how important it is to define the discourse unit and the context precisely. Taken out of context and seen as an isolated sequence of exchanges, the speech behaviour of the staff could only be interpreted as irrational and indicating a lack of self-control, even a reluctance to be concerned with the patients. Within the context of the whole morning session, however, this discourse appears in a different light: it illustrates the inability to cope with conflict and the lack of problem-solving strategies.

The final stage of this morning session becomes hectic and full of tension, the examinations are significantly shorter, and follow-up discussions are abandoned altogether. Approaches from patients are ignored and a general sense of distance between patients and staff prevails. In the penultimate examination an ECG is refused, even though the patient really ought to have one, as the machine does not work straight away. In the end, the doctor only appears to be concerned with his hunger: 'I'm going to die if I don't get something to eat soon . . .'

Life on the out-patients' ward is stressful, the more so when institutional routines are disrupted. As this study shows, it is 'difficult' patients who are typically held responsible for these disruptions: the scapegoat is sought and found outside the ward and not within the structure of the hospital. The clients become the external enemy; the institution protects itself and disguises its contradictions. Only an exact analysis of the context, as understanding of everyday life in the institution, and the sequential analysis of the discourses permit a full interpretation of events and the discovery of contradictions and of the ways in which power is exercised. Since what should not be cannot be, myths and rationalizations are kept alive, and this means that any form of reflection and the possibility of change are excluded. However, the particular study discussed here has shown how applied sociolinguistic research can be beneficial in precisely these circumstances, as it provided the basis for a successful seminar programme on improving communication.

4 CONCLUSIONS

Let us now return to the questions and assumptions we raised earlier (see Section 3.1 above) and to the concepts introduced by Elliot Mishler (1984) and Thomas Bliesener (1982): if we can extrapolate from the findings of our study of hospital discourses, what tentative conclusions might be drawn about institutional communication in general? It is not the quantity alone, in terms of time devoted to the client, which is important but the quality of the way in which the discourse is controlled, and this applies not only between the institution and its clients but also within the institution itself. The patients are certainly audible (experienced patients formulate initiatives, inexperienced patients express uncertainty and fear), but they are not always heard.

A whole range of strategies for exercising power can be observed:

from subtle topic- and style-shifts to shouting and ignoring, from indirect to direct speech acts. This power register, the language of the élite, is dependent on many context variables: on static variables relating to the patients (such as age), but also on very specific elements of the setting (such as space and time). It is possible to establish relatively close relationships, but they are not allowed to transcend the frame prescribed by the experts in their active roles. Orientation aids, such as social turns, structuring signals, and explanation are few and far between, and in the end only experienced clients have much chance of receiving satisfactory treatment. The client ultimately has to adapt and adjust to the structure of the institution.

On the one hand then, explicit sociolinguistic discourse analysis can contribute to basic research in linguistics, since important discourse patterns and typical correlations between discourse units and extralinguistic variables can be identified. The most important thing in this respect is the definition of the discourse unit, and this necessarily depends on the analysis of the whole context (in this respect, see also Jäger, this volume). On the other hand, this approach also makes it possible to explain and illustrate more abstract sociological concepts relating to institutions (see the adaptation of Bourdieu's concepts in Menz 1991). The crux of this kind of sociolinguistic research is the notion of 'text in context', and the value of its analysis depends on taking both of these elements in their dialectical interaction into account. Critical sociolinguistic discourse analysis is therefore an important discipline, which attempts to make the meaning of socially relevant everyday events transparent and allows proposals for change towards more egalitarian forms of communication.

APPENDIX

Transcription symbols used in Texts 1 and 2

DM (AM)	male doctor
DF (AF)	female doctor
P	patient
(. . . . .)	inaudible passage
–	break in intonation
——	short pause
[*sighs*] (e.g.)	non-verbal feature
a: (e.g.)	long vowel
YOU (e.g.)	word spoken with emphasis
.	falling intonation
?	question intonation

Text 1

11   A M8. Ab heute schon in Tablettenform:
     P.   Jo – und gestern hams ma

12   A M8.   Na hoffentlich gehts guat – najo
     P.   überhaupt ka gem.

31   A M8. Najo – das is oba immer – des is nur das Lasix 80 –

32   A M8. das kriegn Sie an und für sich von uns – ja AUCH

33   A M8. – wissen Sie. Oda   habens heut noch zusätzlich
     P.   I hob überhaupt — nein

34   A M8. a Ochziger gnumman.
     P.   I hob überhaupt kans kriegt – bis jetzt

35   A M8. Da steht Lasix 80 Milligramm — als tägliche

36   A M8. Therapie — auf da Kurvn. Lasix 80 — jo da wars

37   A M8. kurz pausiert. Naja
     P.   I hobs erst – i hobs erst kriagt –

38   A M8.   mhm
     P.   wia – da Doktor X. [Name Arzt] ah untn bei mir

39   A M8. jaja
     P.   gwesn is net. Hot a ma des ah – Novarin. Sunst hob

40   A M8.   Ja versteh – hams erst
     P.   i übahaupt kane Pulva kriegt – (. . . . .)

41   A M8. ab heute wieda kriegt? — Gestern war
     P.   Hab i heute

42   A M8.   sehr gut
     P.   kriegt und GESTERN hab i a eigenes genommen weu i

43   P. wenn Sie Sich erinnern können – hab ich Sie gefragt

44   A M8. ja   alles bestens – GUT –
     P.   (. . . . .) daß ich die abgesetzt hab – des

45   A M8. olles klor – ob heute könnens es wieder nehmen –
     P.   Pulver net.

46   A M8. weil Sie jetzt keine Infusionen mehr brauchen.

111 A M8.   — jo
     P.   Und wie geht das weiter –   No –

112 A M8.   Jo wenn Sie
     P.   zumindest jetzt amal hier im Spital net

113   A M8. wenn Sie – von Seiten der – des Ulcus entlassungs-

114   A M8. fähig sind – von da Chirurgie wos ja sichtlich bald

115   A M8. erscheint   Daß ma vielleicht doch noch
        P.   Ja das meint ich jo

116   A M8. kurz – a bißl bei UNS an Aufenthalt anschließn – bis

117   A M8. die Frequenz wieder halbwegs beinand is.
        P.   Is des möglich daß ich –

118   A M8.   Auf der Chirurgie?
        P.   dort bleiben kann?   jo — geht des?

119   A M8. Najo – i mein theoretisch gehts schon – jo aber aber
        P.   geht des schon

120   A M8. die ob die da Primarius X. [Name] des zulaßt net –
        P.   ob er ob ers zulaßt

121   A M8. wenn Sie eben kein chirurgischer Fall mehr sind. –

122   A M8. Wir werden des no besprechen – ja. Jedenfalls an

123   A M8. Entlassung ist noch nicht zu denken. – Aber die

124   A M8. Leberwerte sind auch besser geworden – und die

125   A M8. Frequenz ist etwas abgesunken — jo?

141   P.   Es is ah so – des is — wie haßt des Mulox oda wia.

142   A M8. Malox ja
        P.   Die Malox – san zwar aufgschriebm – hab i aba nie

143   P.   kriegt. Des is nur a Beispiel (. . .) zuerst amol

144   A M8.   mhm
        P.   des. (. . .). Inzwischen – inzwischen weiß ich des

145   A M8.   Na richtig.
        P.   schon – wenn ma net frogt (. . .).

146   A M8. Also wann war die Atemnot – sagn'S ma das bitte?

183   A M8.   Jo – wird raufgschickt.
        P.   Wird des raufgschickt?

184   A M8.   ja bitte   X. [Nachname]
        P.   Herr Doktor – dürft ich um Ihren Namen

185   A M8.   Bitte   Wiederschaun.
        P.   Danke schön   Wiedersehn.

Text 2

1   A F2. So das Hemd müß ma auch ausziehn

P.   (. . . . .) net

2   AF2. Warum nicht? – Wir san ja im Spital. So – da da

3   AF2. draufsetzn gell.
      P.   [*leise*] (. . . . .)

4   AF2. SO tun Sie das Hemd bitte ausziehen.
      P.   Hemd – i hob jo

5   AF2.   Ziehns das aus bitte – das Hemd. Wir
      P.   nua des Hemd an

6   AF2. müssen ein EKG schreibn. So   Na schaut
      P.   (. . .) Hemd

7   AF2. schon niemand ha — so – is ja nur der Herr Doktor

8   AF2. da. – Gell   Der derf
      P.   Da Herr Doktor der kann schaun–aba

9   AF2. schon schaun göll – so tun ma uns do rauf setzn –

10   AF2. gell.   eben
      P.   Der muß sogar manchmal schaun (. . .)

11   AF2. So – sagens welcher Arm war denn da Gebrochene?

64   AF2. De wockelt imma — JETZT BLEIBENS ABER RUHIG LIEGN

65   AF2. TUNS NET IMMER WACKELN HM – SONST WIRD DAS EKG NIX.

66   AF2. GANZ RUHIG – SCHÖN LOCKER JA.   Gut —— so:
      P.   Gut is – ja

67   AF2. is gu:t.

221   AF2. Aber es tut ihr ALLES weh bitte — überall tuts ihr

222   AF2. weh.   DA TUTS AUCH WEH?
       P.   Ah: ja [*seufzt*] (. . . . .)   nein nein

223   AF2. Ah da nicht – nur da Rücken und da tuts weh gell
       P.   geht schon (. . . . .)

224   AF2.   Ja – und da?
       P.   Da tuts weh.   Naja – spürns tu ichs – aber

225   AF2.   geht
       P.   zum Ertragn is es.

Further Reading

Van Dijk (1984)
Freeman and Heller (1987)
Lalouschek *et al.* (1990)
Menz (1991)
West (1990)
Wodak (1987*a*)
Wodak (1989)

References

AMMON, U., DITTMAR, N., and MATTHEIER, K. (eds.) (1987), *Sociolinguistics: An International Handbook of the Science of Language and Society* (Berlin, New York: de Gruyter).

BARBOUR, S., and STEVENSON, P. (1990), *Variation in German* (Cambridge: Cambridge University Press).

BARTHES, R. (1974), *Mythen des Alltags* (Frankfurt: Suhrkamp).

BENNET, A. E. (ed.) (1985), *Communication between Doctors and Patients* (Oxford: Oxford University Press).

BERNSTEIN, B. (1960), 'Language and Social Class', *British Journal of Sociology*, 11: 271–6.

BLIESENER, TH. (1982), *Die Visite—Ein verhinderter Dialog* (Tübingen: Narr).

BOURDIEU, P. (1979), *Entwurf einer Theorie der Praxis* (Frankfurt: Suhrkamp).

CICOUREL, A. V. (1981), 'Language and Medicine', in Ferguson and Heath (1981), 403–30.

—— (1984), 'Doctor–Patient Discourse', in Van Dijk (1984). iv. 193–202.

—— (1987), 'Cognitive and Organizational Aspects of Medical Diagnostic Reasoning', *Discourse Processes*, 10: 347–67.

DANET, B. (1980), 'Language in the Legal Process', *Law and Society Review*, 14/3: 447–564.

VAN DIJK, T. (ed.) (1984), *Handbook of Discourse Analysis* (New York: Academic Press).

—— (forthcoming), *Discourse and Elite Racism*.

DITTMANN, J. (1982), 'Konversationsanalyse—eine sympathische Form des Selbstbetrugs?' in *LAUT Papier 75*, Serie B, Trier University.

DITTMAR, N. (1973), *Soziolinguistik. Eine Einführung und kommentierte Bibliographie* (Frankfurt: Athenäum)

—— (1987), 'Sprache und soziale Ungleichheit', manuscript.

DRESSLER, W., and WODAK, R. (eds.) (1989), *Fachsprache und Kommunikation* (Vienna: Österreichischer Bundesverlag).

EHLICH, K. (ed.) (1980), *Erzählen im Alltag* (Frankfurt: Suhrkamp).

—— and REHBEIN, J. (1976), 'Sprache im Unterricht', *Studium Linguistik*, 1: 47–69.

—— —— (1986), *Muster und Institution* (Tübingen: Narr).

—— KOERFER, A., REDDER, A., and WEINGARTEN, R. (eds.) (1989), *Medizinische und therapeutische Kommunikation* (Opladen: Westdeutscher Verlag).

FERGUSON, C., and HEATH, S. B. (eds.) (1981), *Language in the USA* (Cambridge: Cambridge University Press).

FIEHLER, R., and SUCHAROWSKI, W. (eds.) (1992), *Kommunikationsberatung und Kommunikationstraining* (Opladen: Westdeutscher Verlag).

FISHER, S., and GROCE, S. B. (1990), 'Accounting Practices in Medical Interviews', *Language in Society*, 19: 225–50.

FISHER, S., and TODD, A. D. (eds.) (1983), *The Social Organization of Doctor–Patient Communication* (Washington DC: Center for Applied Linguistics).

FRANKEL, R. (1983), 'The Laying on of Hands: Aspects of the Organization of Gaze, Touch, and Talk in a Medical Encounter', in Fisher and Todd (1983), 19–54.

FREEMAN, S. H. (1987), 'Organizational Constraints as Communicative Variables in Bureaucratic Medical Settings—A Case Study of Patient-Initiated Referral Talk in Independent Practice Association-Affiliated Practices', *Discourse Processes*, 10: 385–400.

—— and HELLER, M. S. (eds.) (1987), 'Medical Discourse', *Text*, 7, special issue.

GREENE, M. G., ADELMAN, R., CHARON, R., and HOFFMAN, S. (1986), 'Ageism in the Medical Encounter: An Exploratory Study of the Doctor-Elderly Patient Relationship', *Language and Communication*, 6: 113–24.

HABERMAS, J. (1981), *Theorie des kommunikativen Handelns* (Frankfurt: Suhrkamp).

HEIN, N. (1985), *Gespräche beim praktischen Arzt.* MA thesis, Vienna.

—— and WODAK, R. (1987), 'Medical Interviews in Internal Medicine', *Text*, 7: 37–66.

—— HOFFMANN-RICHTER, U., LALOUSCHEK, J., NOWAK, P., and WODAK, R. (1985), 'Kommunikation zwischen Arzt und Patient,' *Wiener Linguistische Gazette*, suppl. 4.

HELLER, M., and FREEMAN, S. (1987), '1st Encounters—The Role of Communication in the Medical Intake Process', *Discourse Processes*, 10: 369–84.

HOFFMANN, L. (1991), *Rechtsdiskurse. Untersuchungen zur Kommunikation in Gerichtsverfahren* (Tübingen: Narr).

KÖHLE, K., and RASPE, H. (eds.) (1982), *Das Gespräch während der ärztlichen Visite* (Munich: Urban und Schwarzenberg).

LALOUSCHEK, J., and NOWAK, P. (1989), 'Insider–Outsider: Die Kommunikationsbarrieren der medizinischen Fachsprache', in Dressler and Wodak (1989), 6–18.

—— MENZ, F., and WODAK, R. (1990), *Alltag in der Ambulanz* (Tübingen: Narr).

—— —— NOWAK, P., and WODAK, R. (1987), 'Konzept einer Gesprächsausbildung für Ärzte', Vienna: unpublished.

LEODOLTER, R. [= R. Wodak] (1975), *Das Sprachverhalten von Angeklagten bei Gericht* (Kronberg: Scriptor).

230   Ruth Wodak

Lutz, B., and Wodak, R. (1987), *Information für Informierte* (Vienna: Akademie der Wissenschaften).

Menz, F. (1991), *Der geheime Dialog* (Bern: Lang).

Mishler, E. G. (1984), *The Discourse of Medicine: Dialectics in Medical Interviews* (Norwood, NJ: Ablex).

Mumby, D. K. (1988), *Communication and Power in Organizations: Discourse, Ideology and Domination* (Norwood, NJ: Ablex).

Oevermann, U. (1972), *Sprache und soziale Herkunft* (Frankfurt: Suhrkamp).

Pfeiffer, O., Strouhal, E., and Wodak, R. (1987), *Recht auf Sprache* (Vienna: Orac).

Quasthoff, U. (1980), *Erzählen in Gesprächen* (Tübingen: Narr).

Redder, A., (1982), *Schulstunden 1* (Tübingen: Narr).

Spranz-Fogasy, Th. (1987), 'Alternativen der Gesprächseröffnung im ärztlichen Gespräch', *Zeitschrift für Dialektologie und Linguistik*, 15/3: 293–302.

Strong, P. M. (1979), *The Ceremonial Order of the Clinic: Parents, Doctors, and Medical Bureaucracies* (London: Routledge).

Strotzka, H., Pelikan, J., and Krajic, K. (eds.) (1984), *Welche Ärzte brauchen wir? Medizinstudium und Ärzteausbildung in Österreich* (Vienna: Facultasverlag).

Todd, A. (1983), 'A Diagnosis of Doctor–Patient Discourse in the Prescription of Contraception', in Fisher and Todd (1983), 159–88.

Wassermann, R., and Petersen, J., (eds.) (1983), *Recht und Sprache* (Heidelberg: Müller).

Weick, K., (1985), *Der Prozeß des Organisierens* (Frankfurt: Suhrkamp).

West, C., (1984), *Routine Complications: Troubles with Talk between Doctors and Patients* (Bloomington: Indiana University Press).

—— (1990), 'Not just "Doctors' Orders": Directive–Response Sequences in Patients' Visits to Women and Men Physicians', *Discourse and Society*, 1/1: 85–112.

Wodak, R., (1984), 'The Interaction between Judge and Defendant,' in Van Dijk (1984), iv. 181–92.

—— (1986), *Language Behavior in Therapy Groups* (Los Angeles: University of California Press) originally published as *Das Wort in der Gruppe* (Vienna: Akademie der Wissenschaften, 1981).

—— (1987*a*), 'Kommunikation in Institutionen,' in Ammon *et al.* (1987), 800–20.

—— (1987*b*), 'And where is the Lebanon?', *Text*, 7/4: 377–410.

—— (ed.) (1989), *Language, Power and Ideology* (Amsterdam: Benjamins).

—— Andraschko, E., Lalouschek, J., and Schrodt, H. (1992), 'Kommunikation in der Schule—Schulpartnerschaft' final project report; Vienna.

—— De Cillia, R., Blüml, K., and Andraschko, E. (1989), *Sprache und Macht—Sprache und Politik* (Vienna: Österreichischer Bundesverlag).

—— Nowak, P., Pelikan, J., Gruber, H., De Cillia, R., and Mitten, R. (1990), *'Wir sind alle unschuldige Täter'. Diskurshistorische Untersuchungen zum Nachkriegsantisemitismus* (Frankfurt: Suhrkamp).

Wunderlich, D. (1976), *Studien zur Sprechakttheorie* (Frankfurt: Suhrkamp).

10 Political Discourse: The Language of Right and Left in Germany

SIEGFRIED JÄGER

I INTRODUCTION

In this chapter I shall analyse the political discourse of the right and of the left in Germany. Elements of these forms of political language may also be found in 'independent' newspapers and in the rhetorics of members or sympathizers of several political parties of the centre. However, I shall deal in particular with contemporary texts and talk of the right and, with some reservations, of the left. I say 'with reservations' because there is no (strong) movement of the left in Germany today. Public discourse in contemporary Germany is predominantly coloured by conservative and right-wing ideologies and ideas, especially as regards everyday thinking and debate. Leftist ideas and ideologies survive only in the niches of little groups and circles, and relics of them can be found in a small number of journals and pamphlets.

Thus it is a reflection of the political landscape of contemporary Germany if the analysis of right-wing political discourse dominates the following discussion. However, there is also some justification for this in the fact that right-wing thinking and debate in the new larger Germany can be seen as a danger for the democratic development of society in this country, and I think that it is also necessary for linguists such as myself to see their own political responsibility, especially in a country in which large sections of the population and many politicians today tend to deny its past political history and to suppress its responsibility for the Holocaust and the 'Third Reich' and other crimes. Furthermore, after the 'fall of the Wall', feelings of nationalism and chauvinism are growing again, and new right-wing political groups are emerging, while the whole political map is drifting towards the right. It is important to realize, for example, that political slogans of a right-wing party like the 'Republicans' (such as 'the boat is full',

I want to thank Susan Houlton, who read a first English draft of this paper.

232   *Siegfried Jäger*

meaning that no more refugees should be accepted) and concepts like placing refugees seeking asylum in so-called *Sammellager* (that is, concentration camps (!)) have been adopted by most conservatives and even by growing numbers of representatives of the Social Democratic Party (SPD).

But first of all I would like to address the question of what we mean when we talk of political discourse. Political statements occur in all sorts of contexts. You find them in everyday conversation at home, in restaurants, or at work, as well as reading or hearing more or less elaborate fragments of political discourse in the printed press or other media, in school-books and lectures, at universities, in parliaments, and on many other political platforms. But what is the political element of political discourse? What turns discourse into political discourse?

In this chapter I shall call a text or discourse a political one, first and rather trivially, if it deals with matters of public interest such as general public affairs, for example housing or health services, activities of politicians, political parties or trade unions, foreign or home affairs, political aims and concepts, philosophies, ideologies, etc. Secondly, I shall call a discourse political if it deals with political issues but is not classifiable as an instance of the specialized discourse of political science.

Indeed, political discourse is not characterized by any special structural, rhetorical, or stylistic features such as special metaphors, special sentence structures, special text types, etc. It draws on the full repertoire of a given language, although it may do so in special ways, depending on the context, aims, and topics of the specific text. In other words, the political nature of discourse does not so much depend on the competence of individual speakers, it is more a question of their use of language, of their particular aims and intentions, and so on.

So political discourse, as understood in this chapter, is not just the discourse of professional politicians or of political science, even though elements of these special discourses always do penetrate into political discourse, as is the case with any special discourse: political discourse can be described as that part of general or everyday 'interdiscourse' that deals with political topics and themes.

Before proceeding any further, therefore, we should now define exactly what we mean when we use the term 'discourse', and how a linguistic analysis of discourse should be conducted. The approach I shall follow here draws on the ideas of the French social philosopher Michel Foucault, which have been adopted and to some extent modified by German scholars like Utz Maas, Jürgen Link, and myself (see

Maas 1984, Link 1982, and Jäger 1991, which combine theoretical (sociolinguistic) and practical political interests). In the framework of this new approach, discourse is not simply equivalent to the notion of 'text'. Discourse analysis does deal with texts (in the old sense) but if we look at how texts are produced in everyday life by everyday people, we see that the older concepts of text linguistics are no longer adequate.

In accordance with the traditional definitions of 'text', I consider 'texts' to mean all forms of written or spoken language. But if you look at the process of producing texts, you find that they are products of human labour just like other products (for instance knives, roads, cars, etc.). In producing texts, individuals use a set of 'tools', which admittedly are different from the tools they use when building a house or manufacturing a car. But in both cases people think, remember, anticipate, and so on; in addition, they are learning when they are producing 'things', be they texts or other products of labour. In producing texts individuals use words and sentence structures as opposed to a hammer or an axe. Producing texts or other 'things' means using knowledge and facilities acquired in learning processes from other people, which they in turn have learned and/or picked up from others, from their parents and friends, in their peer groups, from their teachers or their neighbours. So the production of texts (and other things) is an intrinsically social and historical fact. Since texts are determined socially and historically, they both 'contain' socio-historical material and have an effect on the social processes in a given society.

But there is another possible aspect of the notion of text we must consider. In many texts there is a utopian component, too. Through 'texts' or in 'texts' you may anticipate possible (or even impossible) future events. Here again, however, there is no difference between texts and other products. If you build a house, you are doing so in order to live in the house or for other people to live in it now or in the future. On the other hand you may produce things that will never be used, or that will turn out to be 'utopian', like Leonardo's aeroplane, which was planned and built and then forgotten for hundreds of years until new 'utopian' products came to be designed, manufactured, and used or, like atom bombs (I would hope), once used, damned and forgotten again.

The concept of text adopted here can be summarized in the following way: it is a product of human labour in a given society and is therefore determined by the efforts of our ancestors and past processes of learning and planning, which are handed down from generation to generation and which to some extent anticipate possible future events.

234   *Siegfried Jäger*

This differs from the traditional concepts of text linguistics, where
the text is considered as an isolated object, and the linguist's aim is
to look for internal regularities and structures within texts (see e.g.
Brinker 1985). I would argue that we ought to take a text as a socio-
historical product, which is intrinsically both individual and social.
The individual aspect, if you see things this way, is reduced to the
individual being interwoven or caught up in the process of 'discourse'.
One piece of text is therefore an element or a fragment of all those
events which occur in a society and which can occur only because
they continue the process or flow of society in history. This is why I
propose that we no longer use the term 'text', and why I prefer to
speak of 'discourse' as the 'flow' of text and talk (and other products
of labour). This means combining Michel Foucault's understanding
of discourse with A. N. Leontiev's theory of activity (*Tätigkeitstheorie*;
see Foucault 1977*a* and 1977*b*, Leontiev 1982). Looked at in this way,
a single piece of text is nothing but an element of discourse, a fragment
of social discourse. In other words, if you analyse a piece of text, you
have to locate it in its social and in its (past and future) historical
contexts. In doing this, you go beyond merely analysing texts and you
leave the realm of text linguistics in favour of a new concept of discourse analysis.[1]

2   LANGUAGE AND IDEOLOGY OF THE FAR
RIGHT IN GERMANY

2.1   *The Social, Historical, and Ideological Context of Extreme
Right-Wing Propaganda*

In accordance with my concept of discourse analysis, I shall first of
all sketch in the political landscape of contemporary Germany and its
development in recent years.

[1] It is often the case that the term discourse is simply used instead of 'text'. Therefore I
consider it necessary to give a somewhat complex definition of my understanding of discourse.
This is even more necessary because there are some other new approaches to analysing and
understanding discourse, emerging from very different disciplines such as anthropology, semiotics, literary studies, linguistics, sociology, psychology, and the various branches of communication. Each of these approaches has to recognize the fact that an interdisciplinary approach is
unavoidable. This is one of the reasons why the term discourse sometimes sounds rather mysterious. If you take into account the fact that there are a variety of approaches and 'schools' in each
single discipline, you may understand why it is not possible to speak about discourse theory or
discourse analysis in terms of a firmly established field of research.

In Germany there are a number of different concepts of discourse and discourse analysis,
some adopting ideas of the philosopher Jürgen Habermas or other scholars like Karl Otto
Apel, others following US-American traditions (e.g. Dell Hymes 1979). Teun A. Van Dijk of
Amsterdam University and editor of the new journal *Discourse & Society* describes the growth
of discourse analysis during the last two decades and calls it 'a new discipline of sociolinguistics'

In 1981–2 the coalition of Social Democrats (SPD) and Liberals (FDP) had come to an end, and the Conservatives (CDU/CSU), in coalition with the Liberals, took over government. At the same time this so-called *Wende* (political change) initiated a political development that made room for extreme right-wing movements, and that culminated in their electoral successes between 1987 and 1989 on various levels, from the European Parliament down to local elections. Although the change from Helmut Schmidt to Helmut Kohl as Chancellor was accompanied by some social problems, one cannot say that these problems were the basis for the renaissance of the far right. The ground for this political change had been prepared by a strong ideological campaign, which was intended to demolish the welfare state and weaken the 'social security net'. At least some sections of the Social Democrats and their Liberal partners supported this development, and in doing so they were assisting the political change.

One cannot describe this in terms of political *hara-kiri*, or simply of treachery. To understand this process we have to look at global economic developments and the growing competition between several parts of the world and national or continental situations of capital, especially that of the USA, Japan, and in Europe the countries of the European Union. Along with this development there emerged radical technological change, which in turn was accompanied by changes to the social and psychological condition of large sections of the population (for a detailed analysis of this interrelation see Hirsch 1990.) At the same time certain concepts of social darwinist and biologistic ideologies were disseminated, in order to legitimize the decline of social security, at least for certain parts of the population, and the growing support of industry by the State.

It was the right-wing parties, and populists like Franz Schönhuber, leader of the 'Republicans', and Dr Gerhard Frey, head of the Deutsche Volksunion (DVU; German People's Union) as well as other small right-wing groups, who profited from this development. Their ideological concepts became socially acceptable again and were supported, since conservatives (of all parties) had prepared the ideological ground for them. Indeed, many voters preferred to elect the original instead of the copy, as the leaders of the renewed right-wing movement proudly proclaimed after their first electoral successes.

This development came to a head in 1989, and it was interrupted very suddenly when the second German state (GDR) ceased to exist

(Van Dijk 1990: 5). Van Dijk's own concept of discourse analysis is based on cognitive psychology and artificial intelligence research. My concept is developed in Jäger (1991 and 1993); it has been used in empirical analyses of right-wing texts in Jäger (1988) and in the field of everyday racist discourse in Jäger (1992*a*) and in various other publications.

(and the Soviet Union began to decline). Chancellor Helmut Kohl was allowed to present himself as a great historical victor, and as the architect of Germany's (re)unification. The events of October 1989 seemed to prove that most of the national aims of the far right had become reality, especially the aim of reunification. Large sections of the German public fell into a state of national intoxication. As a result, the room for manœuvre that right-wing parties had gained dramatically disappeared. Today, however, several years after these events, it is becoming evident that the new 'golden age' has brought serious social problems and a loss of social security for many people on both sides of the former Wall. Indeed, the elections in some federal states and at local level show that extreme right-wing parties are becoming popular again.[2]

But it was not the larger right-wing parties alone which profited from this development. It was (and is again) the circles of the New Right who were strengthened by it. They propagate their concept of a right-wing 'cultural revolution', as had been envisaged in recent decades by French intellectuals like Alain de Benoist, the leading figure of the New Right in Europe. In Germany's popular right-wing parties the influence of New Right ideas is also becoming apparent although not yet to the extent that official co-operation exists, such as can be seen between Le Pen of the Front National and leading figures of the intellectual right in France like Bruno Megret. In Germany the influence of the New Right can be seen in the form of ideological modifications, new themes like ecology, intellectual support in the formulating of party programmes, legal help in avoiding unconstitutional passages in manifestos, and so on.

The influence of New Right ideas that can be seen quite clearly in most of the right-wing groups and parties in the states of the 'old' Federal Republic has not yet reached the so-called *neue Länder* of the former GDR. Right-wing groups and leaders of far right parties in the East mostly refer to ideological tenets of the National Socialists of the Third Reich. That is why they seem to be much more militant than their counterparts in the West. One manifestation of this is the sudden and rapid spread in the East of racist skinhead groups, who are still very marginal in the West, even though attacks on the homes of refugees and other criminal acts of aggression towards foreigners

---

[2] This became clear in the elections to the senate in Bremen and the municipal elections in Schleswig-Holstein in Oct. 1991, and the elections to the parliaments of Baden-Württemberg and Schleswig-Holstein in Apr. 1992 showed dramatically that the temporary decline of right-wing parties was over. The 'Republicans', who seemed to have reached a dead end after the fall of the Wall, gained 10.9 per cent of the votes; in Schleswig-Holstein the DVU gained 6.6 per cent.

are widespread in both parts of the country. The number and form of these assaults is comparable in western and eastern German states. I shall come back to these events later on, because they can only be explained if we understand the ideological motives and discourse conditions that are the basis for these attacks.

2.2 The Discourse of the Right in Germany

Having outlined some of the general conditions and contents of right-wing discourse in Germany as elements of this discourse as a whole, I shall now examine the general discourse of the right-wing press, its weight and importance, and some of its linguistic characteristics.

The influence and the importance of the discourse of the far right and right-wing propaganda in Germany, its contents and its goals, the weight and importance of its ideology, its rhetoric, and its effects can best be judged by looking at its newspapers and pamphlets, which reach at least 5 million readers. In addition we should consider the public performances of members of the far right on television, in more widely read publications such as daily newspapers and, not least, the penetration of (large parts of) this special political discourse into the steady flow of the 'inter-discourse' (see Section 1 above).

First of all I shall describe the scope of the right-wing press and some of its major exponents, identifying the key ideological concepts of the far right in the process.[3] Then I shall analyse specific text material like headlines and passages from a typical article from the monthly *Nation Europa*.

Publications of the Right

In the western part of the Federal Republic (the so-called *alte Länder*) there are about 130 right-wing papers. Some of them are weekly newspapers, the rest are journals, magazines, reviews, and other periodicals, some of them quarterlies, but most appearing on a monthly basis. Most of these journals are published by political parties, as is the case with *Der Republikaner*, edited by Franz Schönhuber, or the *Deutsche Nationalzeitung*, central organ of the Deutsche Volksunion. Others are published by relatively small groups, or even by individual political mavericks and right-wing or fundamentalist Christian sects. Between them, these journals cover the broad spectrum of right-wing ideologies.

---

[3] For a broader analysis of the scope of the press of the far right and its ideologies see Jäger (1988) and Jäger and Jäger (1992).

Some of the bigger and more influential journals include the following:

Nation Europa appears monthly with a circulation of about 10,000 copies and may be considered the central organ of all party functionaries of the far right. The main focus is on 'race' and racism.[4] For the makers of *Nation Europa* there are such things as human 'races', and racism itself is inborn and a simple matter of fact: you cannot fight it; at best you can only mitigate its effects. The core of their ideology is ethnopluralism and apartheid; all other ideological tenets of this militant racist organ, like nationalism and national identity or its bio-ecological concept of society, are derived from its fundamental racist formula. The paper's main slogans are 'Ausländer raus!' (Foreigners out!), 'Deutschland den Deutschen!' (Germany for the Germans!), 'Das Boot ist voll!' (The boat is full!), and so on.

Another monthly publication, reminiscent to some extent of *Reader's Digest*, is *Mut*, which has been appearing for more than twenty years. It has a circulation of about 40,000 copies. Its target audience are members of right-wing conservative middle-class families, who seek confirmation of their 'values' and who want to forget the crimes of the Third Reich. This paper is very clearly an extreme right-wing one, although it tries to package its extreme ideology in a 'refined' writing style and high-quality appearance. It also tries to acquire a 'serious' reputation by commissioning contributions from recognized conservative writers. This enables articles by extreme right-wing writers to gain a spurious respectability by being placed alongside the work of eminent but more 'moderate' writers. The main authors of this periodical are Gerd Klaus Kaltenbrunner and Bernhard C. Wintzek, a former member of the National Democratic Party (NPD).

One other interesting journal that is worth mentioning here is *Criticon*, which was founded by Caspar Schrenck-Notzing in 1970 and acts as a kind of bridge between the conservative right and the far right.

All these publications, and most of their authors, co-operate very closely. However, there is some competition between them, so that the interests of readers from different social and political backgrounds are catered for.

---

[4] My notion of racism cannot be outlined in detail here. I basically concur with the views associated e.g. with the names of Stuart Hall (1989), Étienne Balibar (1989), Georg Auernheimer (1990), and Kalpaka and Räthzel (1990). A fully fledged racist disposition is given if the following three elements co-occur: (1) biologically or culturally founded differences are manifested; (2) these differences are seen negatively (or sometimes positively, e.g. if somebody says that black people make especially gifted jazz musicians); and (3) if these positions are based on real power, so that the person concerned or his/her group is able to act against foreigners. This power is normally given, if the person concerned is a member of the majority or dominant group.

In the 'new' German states there are no high-circulation right-wing journals. The existing extreme right-wing groups produce small pamphlets with fewer than a thousand copies each. However, magazines from the West are beginning to penetrate into the eastern parts of Germany.

Racism in Extreme Right-Wing Texts

Headlines such as the following appear regularly in all right-wing papers:

Ausländer kosten Milliarden (*Code*, Jan. 1992)
(Foreigners cost billions)

Multikulturalismus ist Völkermord (*Nation Europa*, Dec. 1991)
(Multiculturalism is genocide)

Asylanten-Invasion: Wer betreibt den Völkermord am deutschen Volk? (*Remer Depesche*, Nov. 1991)
(Invasion of asylum seekers. Who is perpetrating the genocide of the German nation?)

Bonn schläft. 1991 erstmals über 240 000 Ausländer — aber nichts geschieht (*Deutsche Rundschau*, Jan. 1992)
(Bonn is asleep. In 1991 for the first time more than 240,000 foreigners — but nothing is done)

'Deutschland den Ausländern!' Überfremdung: Die Union läßt die Maske fallen (*Deutsche Rundschau*, Jan. 1992)
('Germany for the foreigners!' Foreign infiltration: The Union [CDU] drops its mask)

Asylanten immer krimineller (*Deutsche Rundschau*, Jan. 1992)
(Asylum-seekers increasingly criminal)

Asylantenskandal: Sind unsere Politiker unfähig, das Problem zu lösen? (*Deutsche Stimme*, Sept. 1991)
(The scandal of the asylum-seekers: are our politicians incapable of solving the problem?)

Zustrom der Auslander wird immer größer (*Deutsche Stimme*, Sept. 1991)
(The influx of foreigners is growing and growing)

Nachrichten von der Überfremdungsfront (*Nation Europa*, column in each edition)
(Foreign infiltration: news from the front)

Migrationsfolgen und deren Einfluß auf die innere Sicherheit (*Nation Europa*, June 1986)

(Consequences of migration and their influence on internal security)

Rasse und 'Rassismus' (*Nation Europa*, Sept. 1986)
(Race and 'racism')

In the texts themselves there are many suggestive formulations like 'full boats', 'invasions of ants', 'bursting houses', 'exploding' or 'colliding cars', or 'aeroplanes flying out of control', and they are full of dubious theories and all the theoretical rubbish of social darwinists like Konrad Lorenz, H. J. Eysenck, Irenäus Eibl-Eibesfeldt, and Arthur Jensen. A detailed analysis of some hundreds of these articles shows that there is a limited stock of about thirty prejudices which are brought up again and again to support the central aim of these right-wing authors, which is to push foreigners out of Germany, or 'better still', out of Europe (see Jäger 1988).

In the following passage, I shall consider an article by Christian Mattausch, which appeared in *Nation Europa* in 1986 and which is quite typical of the more intellectual branch of right-wing racism.[5] Under the headline 'Consequences of migration and their influence on internal security' Mattausch begins:

The presence of non-Caucasian ethnic groups in western Europe creates problems of historical significance for all states which are affected by it. It is not just the economic pressures that are giving cause for concern; it is first and foremost the efforts of these groups to take root here, and to set up cultural and/or political branches of their homelands.

The expression 'non-Caucasian ethnic groups' points the finger at a central problem of the far right as a whole, namely that they were faced with the planned opening of the internal market within the context of a new 'Europe without frontiers'. They expect a new 'flood' of African and Asian immigrants to penetrate the neighbouring countries, and from there 'flow' into Germany without hindrance. That incidentally is the main reason why they are demanding the erection of a 'fortress Europe', a sort of closed shop with high walls at all European frontiers. Furthermore, the word 'Caucasian' might just as well be replaced by 'Aryan'. But the writer criticizes the reduction of the consequences of this unification to economic issues. There is a warning about 'economic pressures', which is the standard argument against immigrants, but it is not the only argument. Mattausch is concerned with 'culture' as a whole, and this points to one of the central ideological tenets of the far right: the priority of (their understanding of) politics over economics.

[5] The original German can be found in the Appendix to this chapter.

Mattausch goes on:

If this fact is put in the context of its psychological correlate, there is no mistaking the significance of one attribute that determines every aspect of human behaviour: the different nature of non-Caucasian populations, the number and variants of which are constantly increasing with the flood of Afro-Asian economic refugees.

We see that Mattausch does not only focus on the 'non-Caucasian populations' (incidentally the word *Population* is normally used only in connection with animals) but there is another danger: the expected 'flood of Afro-Asian economic refugees'. Mattausch here uses two of the keywords of far right propaganda, which unfortunately have since become part of the whole public discourse in Germany: 'flood' and 'economic refugees'. Flood is a so-called collective symbol, calculated to evoke fears of being drowned or getting killed in a wild outburst of the elements. Today this collective symbol appears in 'mainstream' newspapers, and deep in the centre left of our political landscape. It is used in connection with storms, invasions, and other dangerous occurrences, and it is often used in combinations with other metaphorical fields, such as medicine or disease. For example, formulations like the following are quite common: 'The flood of economic asylum seekers will destroy the body of our Germany like a tumour'. These mixed metaphors ('catachreses') paradoxically serve to hold the argumentation together, by linking different and often contradictory parts of the text.

Mattausch here adopts a central component of the official policy on foreigners in Germany: to distinguish between foreigners who come from (what is now) the European Union and those from the rest of the world. He supports this distinction with racist arguments: he applies the label 'non-Caucasian' to the large numbers of Turks and Kurds living in Germany, while 'Afro-Asian economic refugees' is used to describe asylum-seekers from Sri Lanka and from other countries in the Far East and from Africa.

The author takes advantage of the fact that there is a public discourse of differentiation between different sorts of foreigners, and he tries to establish a connection between his racist construction and this discourse, which itself is used as a means of distinguishing between 'good' and 'bad' foreigners: the good ones are the so-called *Gastarbeiter* (guest workers), who work hard and pay taxes, and can therefore be integrated more easily into German society; and the bad ones are those who create social costs, who are not able or willing to work or to adopt German customs and values. Here we can see in a detailed way how a special political discourse tries to infiltrate everyday inter-discourse, and that this inter-discourse is closely bound up with the ruling and therefore dominant hegemonic interests.

If we compare extreme right-wing texts which appeared some years ago (such as the Mattausch text), with texts from the 'mainstream' press today, we can see that the attempts of the right to exert an influence on the general inter-discourse have become more and more successful. Formulas like 'floods of asylum-seekers', 'our boat is full', expressions like 'economic refugees' and other discriminatory phrases, which until a few years ago belonged exclusively to the domain of right-wing extremists like Franz Schönhuber, can be found in almost any section of the contemporary press. If we compare articles by right-wing propagandists with those of mainstream journalists today, and at the same time look back at the way in which the latter used to write, we can see that the whole political landscape in Germany has been drifting to the right. I think that by making statements such as this possible, the concept of socio-historical discourse analysis shows its intrinsic analytical value.

But let us return to Mattausch's article from *Nation Europa*. Another interesting feature of his writing is his use of scientific-sounding vocabulary to make his text appear serious and credible: consider, for example, words like *Population* (instead of *Bevölkerung*), *europid*, *verhaltensbestimmend* (determining human behaviour). He continues:

The indigenous populations perceived these new conditions as a menace to their collective identities, which had grown over hundreds or even thousands of years. In those European countries which have been affected by these problems, a deeply rooted feeling of anxiety and in some cases a sense of serious crisis has been growing. There can be no doubt that the process of alienation of the native environment was increased by the influx of groups that cannot be Europeanized and this was responsible for the loss of a sense of security in the minds of many native inhabitants.

In this passage Mattausch discloses his real message: these foreigners are not just different, they cannot be integrated, they cannot be 'Europeanized'. The influx (*Zuzug*) of these people will result in riots and general crisis.

The biologistic and racist position expressed in this passage dominates the thinking of the whole group of extreme right-wing intellectuals like Mattausch, who base their thinking on the work of 'major' scholars like Konrad Lorenz, Irenäus Eibl-Eibesfeldt, and others. In the rest of his article Mattausch draws attention to the 'fact' that 'racial' mixture will lead to civil war and riots. Conservative politicians are depicted as idiots, and passages from all kinds of newspapers are quoted to show that Germany is becoming more and more *überfremdet* (overrun by foreigners). Immigrants with a black skin are pictured as especially dangerous because they have a different *Naturell* (mentality), inborn aggression, and deeply rooted criminal inclinations.

Articles like this contain the essential propagandist keywords and collective symbols for the everyday agitation of right-wing extremists. For example, Mattausch uses terms like *afroasiatische Völkerwanderung* (that is, Afro-Asian migration on the scale of entire nations) or like *Suche der Einwanderer nach dem Sicherheitsstaat* (the immigrants' search for a security state). But he also adopts some items of conservative vocabulary in order to build bridges in that direction. The *soziale Hängematte* (literally, 'social hammock'), for instance, is a phrase which was coined by conservative politicians in order to discriminate against the unemployed (cf. the English expression 'feather-bedding'). Another common expression is the *Belange des deutschen Volkes* (the interests of the German people), which can be found in the law on foreigners. The semantics of the word *Belange* is highly vague; it means 'concerns' in a very broad sense, and therefore if you say that foreigners are not allowed to disturb the *Belange des deutschen Volkes*, this can be applied to almost any behaviour of foreigners that is different from German customs and values.

Mattausch tries to relate his formulations to anxieties and fears which affect many people (of all social classes) in Germany. There is, for instance, a widespread fear in Germany of losing a sense of national identity or the homogeneity of the German people. In doing this Mattausch pours oil on the fire which is smouldering under the cover of normal everyday life. He confirms his readers in their prejudices, and thereby helps to strengthen feelings of racism in the inter-discourse. Furthermore, he constructs a 'natural' relationship between a fear of foreigners and the urge or compulsion to drive them out, which enables people to define and consolidate 'their' cultural territory. In this way, he claims that xenophobia and racism are innate human characteristics.

The article ends with the sharp warning that 'the stage is set for a future civil war'. His final words are: 'The shadows of a dark future can no longer be ignored.'

The Reproduction of Racism and Extreme Right-Wing Ideological Tenets in Conservative Texts and in the Daily Press

Following this brief review of the extreme right-wing press and my attempt to analyse an article by a well-known far-right author, I should add at least a few remarks about some other newspapers, such as the mass-circulation tabloid newspaper *Bild*, with more than 15 million readers (for a sociologist's analysis of *Bild*, see Albrecht 1982).

Racism as a concept which opens the door to extreme right-wing thinking is one of the most important focal points in *Bild*. There have been, and still are, many elaborate campaigns against foreigners in this racist paper, especially in the run-up to elections; but even at 'normal' times *Bild* never tires of attacking and discriminating against immigrants, using suggestive symbols and images like that of the threatening 'flood of asylum-seekers', who will make 'our boat' sink. On the front page on 5 September 1991, for example, the following set of banner headlines appears:

> Blutschande / Endlich / Die Töchter wehren sich
>
> Asylanten / Endlich / Bonn wacht auf
>
> Miethaie / Endlich / Die ersten bestraft
>
> (Incest / At last / The daughters put up resistance
>
> Asylum-seekers / At last / Bonn is waking up
>
> Rent profiteers / At last / The first are punished)

This racist and suggestive combination of headlines is by no means harmless. The word *Asylant* (asylum-seeker) is a neologism with deeply discriminatory overtones, and in the context of this set of headlines it is associated with a cluster of negative images. This operates not only on the surface level, but also by implicitly suggesting parallels with the past through the use of keywords like *Blutschande* and *Miethaie*: *Bild* conjures in the minds of its readers the threat that asylum-seekers will seduce 'our' daughters and wives (like the Jews in the propaganda of the Nazis) and that they will buy up our homes and property and charge extortionate rents (*Miethaie* was an abusive term for Jewish landlords used in the Third Reich). To this extent we can say that *Bild* is working with a kind of 'subcutaneous' anti-Semitism too, although the paper's founder, the late Axel Springer, and his followers always laid great emphasis on being a big friend of Israel.

In an edition of *Bild* which appeared some days before the elections in Schleswig-Holstein and Baden-Württemberg in April 1992 the following headline appeared:

> Die Flut steigt—wann sinkt das Boot?
> Fast jede Minute ein neuer Asylant (*Bild*, 2 Apr. 1992)
>
> (The tide is rising—when will the boat sink?
> A new asylum-seeker almost every minute)

This headline is a very typical example of *Bild*'s use of collective symbols: they simply take suggestive metaphors, put them together, and couple them with nonsense-facts. Logical connections are clearly considered unimportant: why, I ask myself, should a boat sink if the tide is rising?! Nevertheless, *Bild* succeeds in producing its dramatic message of danger, loss of space, and overpopulation.

But *Bild* is just one exponent of this sort of journalism. Racism 'occurs' and is produced in almost every other paper in Germany, even in some of the more serious papers like *Der Spiegel*, *Die Zeit*, or the *Süddeutsche Zeitung*. The racist articles or racist opinions published in these papers, however, sound less blatant and overt than in *Bild*, which goes on to incite its readers against immigrants and refugees (for a detailed discussion see Quinkert and Jäger 1991 and Gerhard 1992).

Racism in Everyday Discourse and its Mediation by the Press
In a detailed study of everyday discourse (see Jäger 1992*a*), I came to the conclusion that most people in Germany are more or less imbued with racist thinking, even though not all of them can be called blatant racists. That is to say, they produce and reproduce racist opinions and prejudices. These prejudices appear in our 'normal' newspapers as well. That is why I assume that the influence of the press in producing and reproducing racism is so great. Furthermore, in everyday life people share their experiences socially through talk. This also includes racist interpretations of these experiences as well as value judgements, attitudes, and norms.

I shall illustrate this contention with some examples from statements made by our informants, but first I should say a few words about the project itself. Together with some of my students I carried out twenty-two non-standardized open interviews each lasting 45 minutes on average. We conducted these interviews in five large towns in western Germany. The interviewees were men and women of all ages and from various occupational backgrounds. Rather than putting set questions we tried get our informants to speak on various subjects such as their neighbourhood, shopping, travelling by bus or train, the reunification of Germany, and the Gulf War. Normally our interviewees began of their own accord to talk about foreigners. If they did not, we gave them a 'prompt'. This sample, which is reasonably representative of German populations in larger towns, was analysed on the basis of the discourse-theory approach outlined in Section 1 above.

In one interview a 52-year-old woman is asked right at the beginning: 'How do you like the town you live in?' She answers without hesitation: 'I don't like my neighbourhood very much, because there are a lot of foreigners living here, especially Turks.'[6] Asked why she feels disturbed by foreigners in general, she answers:

It bothers me because the mentality of the foreigners is totally different from that of the Germans, especially as regards cleanliness, as regards . . . I also don't like the discrimination against women, at least that's the way it seems to me; if you see, er, these people in couples, er, the woman has to carry the heavy bags, the men walk some metres behind the women; and I have the impression that women have very little freedom in these countries.

Asked why she feels disturbed by foreigners living in her neighbour-hood, she says:

I don't like it very much. First of all, er, as far as school is concerned; admittedly my own children are a little bit older and they don't go to the local school any more, but I think it is bad, because the education of the children is very, very bad and the children fall behind, and when they go on to further education or training their basic education will be worse than in schools where there are mostly German children.

At this point, the interviewer intervenes, saying: 'But that's not an answer to my question', and the women continues:

Yes, I have nothing directly against foreigners [*laughing*], but somehow, er, it's not so nice, er, to be surrounded by foreigners; there's really very little contact; the foreigners shut themselves off, the Germans too of course; most of them don't want to have anything to do with the foreigners, and the other way round I think, it is the same way with the foreigners, who don't like to have anything to do with Germans. They have their own religion, they have customs of their own, and, er, somehow I don't like that. I'd prefer to live in a place where there are only Germans.

When the interviewer asks if she would like to improve her acqaint-ance with Turks, she answers rather brusquely: 'I don't want to have any contact with Turks at all.'

In this passage we find some of the typical attitudes towards foreigners, especially Turks, who are in a sense considered as proto-types of the foreigner: 'We just don't like them, our children suffer from them, they are different as regards their customs and especially their religion, they are not clean', etc. The presence of Turks is a general cause of anger and feelings of frustration. Attitudes like these are very often accompanied by a clear desire to drive foreigners out.

The following argument can also be heard very often: 'We can't take in all the distressed and burdened people of the whole world,

---

[6] The translations inevitably neglect the fact that the speakers use dialects or some other forms of the non-standard speech of the Ruhr area. For the original text see Appendix.

can we? I think if they all come, the boat will be exhausted, er, as regards the density of the population, won't it?'

Here we find some of the collective symbols which were mentioned earlier, and which are drawn from the discourse of the media. A great many of these collective symbols, which have been shown to be the 'cement' of media discourse (Link 1991), can be found in everyday discourse as well. In addition, this discourse is interspersed with journalistic keywords (like agglomeration, assimilation, identity, discrimination, interaction, level, rationality, veto) that stand out from the more colloquial nature of the rest of such texts. This and some other features of these texts, such as in some cases verbatim quotations from passages from the newspapers the interviewees habitually read, confirm my supposition that the influence of the media on everday thinking and speaking is enormous.

But how does it work in detail? Do the newspapers just reflect what people think, as is often claimed when the press is accused of instigating racist or other prejudices? Having analysed media discourse and everyday discourse in parallel for some years, I would venture to say that the influence of the press is to be seen in its ability to promote or play down racist discourses at will. For instance, racist discourses typically increase in prominence in the press and in everyday discourse in the run up to any election. It is true, however, that politicians, who have easy access to the media, play their part in this game, too; but it is the press itself that is capable of managing and influencing the inter-discourse as it pleases (for this notion, see Van Dijk 1991.) In short, it can at least be argued that racism is socially reproduced by discourse.

Finally, one further point should be mentioned here: racist (and other) discourse may by no means be harmless, or just a cluster of bad attitudes and worse opinions. As recent events in Germany have shown, there is a high correlation between the rise of racist discourse in the press (see above) and thousands of attacks against immigrants, including arson and even murders. Discourse theory, then, argues that there is a continuum between verbal and non-verbal action. In certain circumstances, verbal action may well be transformed into physical actions like attacks and riots (see Link 1982, Jäger 1992*a*, where this problem is dealt with in detail). In the interviews we conducted we found that some of the interviewees showed considerable willingness to attack foreigners if they did not behave in accordance with German customs and norms.

It goes without saying that racism is one of the central ideological tenets of right-wing parties (and of right-wing conservative politicians, too). Therefore it seems reasonable to conclude that an increase of racist discourse in the German population may signal an increase in the prospects of right-wing extremism too.

3 THE LANGUAGE AND IDEOLOGY OF THE LEFT IN GERMANY

3.1 *The Press of the Left in Contemporary Germany*

During the last two decades the German left, and together with it the left-wing press, has declined dramatically. Many journals disappeared, or became commercialized, or lost large numbers of their readers. It is not possible to analyse the causes of this decline in detail here, and I shall confine myself to just a few points. First of all we must differentiate between the two mainstreams of the German left since 1968: the orthodox left, which adhered closely to the canonical ideology of Stalin's and (parts of) Lenin's works, and the New Left. The orthodox left disappeared from the political stage after the fall of the Berlin Wall and the end of the Soviet Union. Some marginal groups still exist, however, such as the DKP (German Communist Party), which has about a thousand members and publishes the newspaper *Unsere Zeit*.[7] The PDS (Party of Democratic Socialism), successor to the GDR's official SED (Socialist Unity Party), still has some 200,000 members in eastern Germany, but it has not been able to establish a foothold in the West, where it has gained only about 700 members. The attempt to get rid of the old orthodox thinking, undertaken by leading members of the PDS, for instance by Gregor Gysi, was not very successful, and there are many members of this party who still adhere to the old ideologies and politics.

Neues Deutschland, the former central organ of the SED, was saved by the PDS, and its former bureaucratic style and ideological contents have been changed to a great extent. It now reflects the debates of the progressive wing of the PDS (and some other socialists) on socialist reforms and anti-capitalist perspectives. These changes can be illustrated by comparing two editions of *Neues Deutschland*, one before and one after the *Wende*. In the edition of 27/8 June 1987, the following headlines appeared:

> 4. Tagung der Volkskammer bestätigte Haushaltsrechnung für 1986: Eine Bilanz der erfolgreichen Wirtschafts- und Sozialpolitik

---

[7] *Unsere Zeit* is still being published in a style and format comparable to *Neues Deutschland* before the *Wende* (see below), but it has become absolutely insignificant.

(4th session of the Volkskammer [Parliament] confirms
budget for 1986: a record of successful economic and social
policy)

DDR-Planwirtschaft funktioniert gut
(GDR's planned economy is working well)

Wir haben keine Schuldenwirtschaft
(We do not have a budget-deficit economy)

Steigende Leistung — Sinkende Kosten
(Rising output — falling costs)

Dem Ministerrat Entlastung erteilt
(Report of Council of Ministers approved)

Dank für großen Beitrag zum Wohle des Volkes
(Thanks for major contribution to the good of the people)

Grußadresse des ZK zum 'Tag des Bauarbeiters'
(Message of greeting from Central Committee on
'Construction Worker's Day')

Festumzug ein Höhepunkt des Jubiläums von Berlin
(Procession a highlight of Berlin's jubilee celebrations)

Schwarzer Donnerstag für Westberliner Mieter
(Black Thursday for West Berlin tenants)

This anthology shows that readers were confronted with long and
tedious official statements in the style of official bulletins. The consequence of this sort of state journalism was that, with the exception of
some functionaries who were compelled to do so, nobody actually
read *Neues Deutschland*. Passages like the following were typical:

Abgeordnete R. W., Vorsitzende des Ausschusses für Haushalt und Finanzen,
bezeichnete in der Stellungnahme des Ausschusses die Ergebnisse bei der
Senkung der Kosten als unbestechlichen Maßstab für die Fortschritte auf
dem Weg der umfassenden Intensivierung.

(In the report of the Budget and Finance Committee, the Chair, delegate
R.W., described the results of the cost-cutting exercise as an unerring
measure of progress on the road to comprehensive intensification.)

The bureaucratic and clumsy style reflects a complete lack of content,
which seems to have been replaced by a series of empty phrases.

By contrast, in an edition of 24/5 March 1990, there are lively
debates about the future of the GDR, in which passages are quoted
from the work of Theodor W. Adorno, the great scholar from the
'Frankfurt School', who until recently had been accused of being an
Abweichler (deviator) from the true ideology of socialism. However,
whilst in this edition journalists and scholars are still fighting for a

new socialist democratic state of the GDR, this has given way more recently to running battles and attempts to criticize the capitalist way of life and the annexation of the former GDR by the West.

I shall now turn my attention from the orthodox left to other left-wing groups. In doing this I shall ignore the many small sects, some of which continue to produce their small papers (for instance the VSP (Unified Socialist Party), a Trotskyist group). Around the beginning of the 1990s many members of Trotskyist or Maoist and other 'Marxist' groups (like MG (Marxistische Gruppe)) or even of the DKP moved in the direction of the Greens and changed or modified their political ideologies to a large extent, in some respects adopting blatant political pragmatism (the so-called *Realos*). The political debate of the members and sympathizers of this political group can be found in journals like *Kommune* or in the *tageszeitung*. This means that there have been some changes in the style and form of debate, and there cannot be any doubt that this new 'movement' has contributed a great deal to the emergence of new themes like ecology, which have since been adopted (and often diluted) by other parties. To some extent it could be said that they have developed a new language, nearer to everyday communication, and that their discourse has gained some influence on the general inter-discourse.

At this point I should mention some journals of the moderate New Left on the one hand and the more radical New Left on the other. There are papers like *Links*, *kultuRRevolution*, *Wechselwirkung*, or *Sozialismus*, which have no particular association with any political parties. Like journals such as *Frankfurter Hefte*, an independent journal which is politically aligned with the left wing of the Social Democrats, these publications try to maintain an intellectual debate, based on the ideas of Adorno, Bloch, Althusser, Benjamin, Foucault, and others. They all argue on a relatively high intellectual level, and so far they have had no influence on inter-discourse whatsoever. But some of their authors have access to the major papers too, for instance Elmar Altvater, a well-known Marxist professor of economics, Jürgen Link, professor of literature and discourse theory at Dortmund University, and Joachim Hirsch, who teaches politics at Frankfurt University. So it is fair to say that a German New Left still exists, but it is not organized and has no meaningful influence on the process of politics and the general inter-discourse as a whole.

Finally, I should at least draw attention to two papers of the radical left in Germany: *Arbeiterkampf* and *Konkret*. *Arbeiterkampf* was founded many years ago as the organ of the Kommunistischer Bund (Communist Union), a former Maoist group that is now defunct. The paper itself, a strictly anti-capitalist monthly, still exists, with a circulation of about 3,000 copies. It regularly reports on affairs of the so-called underdeveloped countries, and it deals with all subjects in which Marxists and Anarchists may possibly be interested. Their analyses are founded strictly on Marxist theory, especially the 'original Marx', but this is not done in the sophisticated and academic manner of most of the other papers of the New Left. *Konkret*, a monthly magazine with a circulation of about 20,000 copies, is also a radical socialist paper with a long tradition. It has always been independent, and it specializes in criticizing the left from an even more left position. Its style is readable and its themes are many and various.

The total number of readers of the press of the left in Germany may be estimated at about 1 million.

3.2   *The Gulf War in the Discourse of the Left*

The way at least large sections of the left debated the Gulf War may serve as a typical example of debates within the left in contemporary Germany. Since it is not possible to go into detail on this topic here, a few remarks must suffice (for a detailed discussion see Kellershohn 1992.) The German left is, put simply, a left-wing movement in Germany. This means that it is also caught up in the last sixty years of German history and its discourse. During the Gulf War the problem of anti-Semitism became central. The German left was divided between pacifists on the one hand, who defended peace at any price, and on the other hand those who defended Israel at any price and who accused the 'pacifists' of anti-Semitism. The debate has continued to divide whole groups and journalists on political newspapers and periodicals. The charge of anti-Semitism is probably the most serious accusation left-thinking people in Germany could be confronted with, and so it is not surprising that this debate was conducted very vehemently and in some cases even violently.

This is just one of many delicate topics that the German left seldom likes (or dares) to discuss. One of the main reasons for this is the fact of Germany's past, which has also not been dealt with adequately by the left. Further subjects of this kind include abortion, eugenics, and (at least for some parts of the left) Stalinism. The left normally leaves it to the right to discuss these topics, but there is no getting away from them. Since the left appears to be unable to deal with these topics on a serious theoretical level, it is not prepared to discuss them at all in an adequate way. This is one of the reasons why the left in Germany has remained marginal and politically weak. Some years ago there was a popular slogan: 'Der Geist steht links!' (literally: 'the mind is on the left!', that is, the intellectual climate is dominated by the left). This may have been true then, but not any more: since then the initiative has shifted firmly to the right!

4   CONCLUSIONS

The bulk of this chapter has been devoted to the language of the right in Germany, and this is because at least since the early 1980s the focal point of German political discourse has been drifting to the right. This trend represents a significant new development in Germany's post-war history. Furthermore, the signs are that in the new 'unified' Germany the right will continue to dominate political discourse and the left will continue to decline.

Nevertheless, it is both legitimate and important to consider in conclusion whether there is any possibility of developing counter-discourses, in order to confront at least the most powerful discourses of the right, such as racism and anti-Semitism. In this respect, I think there may be some grounds for optimism. As discourses are linked with power and as they are not directly determined by purely economic circumstances, there is at least a chance that enlightened concepts will eventually prevail; and as far as the topic of this chapter is concerned, it may just be that discourse analysis and improvements in discourse theory will make a positive contribution to this process.

APPENDIX

Migrationsfolgen und deren Einfluß auf die innere Sicherheit
Christian Mattausch

Die Anwesenheit nichteuropider Volksgruppen in Westeuropa stellt alle betroffenen Staaten vor Schwierigkeiten von historischer Bedeutung. Es sind nicht allein wirtschaftliche Engpässe, die Kopfzerbrechen bereiten — es sind vor allem die Bestrebungen dieser Gruppen, hier Wurzeln zu schlagen und zu kulturellen bzw. politischen Ablegern ihrer Herkunftsländer zu werden. Setzt man diesem Sachverhalt sein psychologisches Spiegelbild gegenüber, fällt unweigerlich der Stellenwert eines in jeder Beziehung verhaltensbestimmenden Merkmals auf: Andersartigkeit nicht-europider Populationen, deren Zahl und Varianten sich noch durch die Flut afro-asiatischer Wirtschaftsflüchtlinge ständig vergrößert.

Diese Bedingungen, von den einheimischen Bevölkerungen überall als eine Infragestellung alter, über Jahrhunderte bis Jahrtausende gewachsener Kollektividentitäten empfunden, haben in den betroffenen Ländern Europas zu tiefer Beunruhigung, stellenweise sogar zu akutem Krisenbewußtsein geführt.

Es unterliegt keinem Zweifel, daß die Entpersönlichung der heimatlichen Umgebung durch den Zuzug nicht-europäisierbarer Gruppen geradezu beschleunigt wurde und bei vielen Bodenständigen zu einem Verlust an Geborgenheitsgefühl geführt hat. . . .
Die Schatten einer düsteren Zukunft sind nicht mehr zu übersehen.

German Text of Interviews (pp. 245–6)

F = Frage   A = Antwort   x = pause

F. Wie gefällt Ihnen denn die Stadt, in der Sie hier wohnen?

A. Die Wohnlage gefällt mir ja nicht so besonders, weil um mich herum sehr viele Ausländer wohnen; vor allen Dingen Türken.

F. Warum stört Sie das?

A. Mich stört das insofern, weil die Mentalität der Ausländer eine total andere ist als die der Deutschen, vor allen Dingen geht's da um Sauberkeit, um, was mich daran auch stört, die Diskriminierung der Frau, habe ich zumindest das Gefühl; wenn man die Leute paarweise, eh, sieht, ist es halt so, daß die Frau die schweren Taschen tragen muß, die Männer einige Meter hinter den Frauen laufen und ich hab eben das Gefühl, daß die Frau sehr wenig Freiheit genießt in diesen Ländern.

F. Warum stört Sie das, wenn hier Ausländer in der Umgebung wohnen?

A. Ich find das nicht besonders gut; einmal, eh, auf dem Gebiet der Schule; meine Kinder sind allerdings schon älter und gehen halt nicht mehr in diese hiesige Schule, aber, eh, ich finde es schlecht, weil die Ausbildung der Kinder sehr, sehr schlecht ist und die Kinder dann halt ins Hintertreffen geraten, wenn sie zu einer fortbildenden Schule gehen oder auch später in der Berufsbildung ist die Grundlage eine schlechtere als in einer Schule, wo überwiegend deutsche, eh, Kinder zur Schule gehen.

F. Aber das ist ja keine Antwort auf die Frage; ich mein, dann brauchte Sie das ja nicht zu stören, wenn hier Ausländer in der Umgebung wohnen.

A. Ja, ich hab nichts direkt gegen Ausländer [*Lachen*], aber irgendwie, eh, ist es nicht so sehr schön, eh, von Ausländern umgeben, eh, zu wohnen; der Kontakt ist, eh, ganz minimal nur; die Ausländer grenzen sich irgendwie ab, natürlich auch die Deutschen; die meisten wollen halt mit den Ausländern nichts zu tun haben und umgedreht glaube ich, ist es auch so, daß die Ausländer mit Deutschen recht wenig zu tun haben wollen. Die haben ihre eigene Religion, die haben ihre eigene Lebensweise, und, eh, irgendwie stört mich das. Ich würde also lieber in einer Gegend wohnen, wo nur Deutsche wohnen.

F. Möchten Sie eigentlich auch keinen Kontakt zu den Türken?

A. Ich möchte keinen Kontakt zu den Türken haben.

A. Wir können nicht alle mühselig und beladenen der ganzen Welt hier aufzunehmen, ne. Irgendwo is ja auch, wenn die alle kommen, denk ich ma, dat Schiff hinterher, eh, vonner Besiedlungsdichte her, eh, erschöpft, ne, denk ich mir.

Further Reading

Butterwegge and Jäger (1992)
Ehlich (1989)
Heringer (1990)
Institut für Migrations- und Rassismusforschung (1992)
Jäger (1992*a* and *b*)
Januschek (1985)
Klein (1989)
Liedtke *et al.* (1991)

References

ALBRECHT, R. (1982), 'Bild-Wirkung — Annäherung an die Wirksamkeit einer Institution, *Neue politische Literatur*, 27/3: 351–74.
AUERNHEIMER, G. (1990), *Einführung in die interkulturelle Erziehung* (Darmstadt: Wissenschaftliche Buchgesellschaft).
BALIBAR, E. (1989), 'Gibt es einen "Neuen Rassismus"?', *Das Argument*, 175: 369–79.
BRINKER, K. (1985), *Linguistische Textanalyse. Eine Einführung in Grundbegriffe und Methoden* (Berlin: Erich Schmidt Verlag).
BUTTERWEGGE, C., and JÄGER, S. (eds.) (1992), *Rassismus in Europa* (Cologne: Bund).
VAN DIJK, T. A. (1990), 'Discourse & Society: A New Journal for a New Research Focus', *Discourse & Society*, 1/1 (July): 5–16.
—— (1991) *Racism and the Press* (London: Routledge).
EHLICH, K. (1989), *Sprache im Faschismus* (Frankfurt: Suhrkamp).
FOUCAULT, M. (1977*a*) *Überwachen und Strafen* (Frankfurt: Suhrkamp).
—— (1977*b*), *Der Wille zum Wissen, Sexualität und Wahrheit* (Frankfurt: Suhrkamp).
GERHARD, U. (1992), 'Wenn Flüchtlinge und Einwanderer zu "Asylantenfluten" werden — zum Anteil des Mediendiskurses an rassistischen Pogromen', in Jäger and Januschek (1992), 163–78.
HALL, S. (1989), 'Rassismus als ideologischer Diskurs', *Das Argument*, 178: 913–21.
HERINGER, H. J. (1990), *'Ich gebe Ihnen mein Ehrenwort.' Politik — Sprache — Moral* (Munich: Beck).
HIRSCH, J. (1990), *Kapitalismus ohne Alternative? Materialistische Gesellschaftstheorie und Möglichkeiten einer sozialistischen Politik heute* (Hamburg: VSA).
HYMES, D. (1979), *Soziolinguistik. Zur Ethnographie der Kommunikation* (Frankfurt: Suhrkamp)
Institut für Migrations- und Rassismusforschung (1992), *Rassismus und Migration in Europa* (Hamburg: Argument).
JÄGER, S. (ed.) (1988) Rechtsdruck. Die Presse der Neuen Rechten (Bonn: Dietz).

JÄGER, S. (1991), *Text- und Diskursanalyse. Eine Anleitung zur Analyse politischer Texte*, 3rd edn. (Duisburg: DISS).

—— (1992*a*) *BrandSätze. Rassismus im Alltag* (Duisburg: DISS).

—— (1992*b*) *Faschismus, Rechtsextremismus, Sprache. Eine kommentierte Bibliographie*, 3rd edn. (Duisburg: DISS).

—— (1993) *Kritische Diskursanalyse. Eine Einführung* (Duisburg: DISS).

—— and JÄGER, M. (1992), *Die Demokratiemaschine ächzt und kracht. Zu den Ursachen des Rechtsextremismus in der BRD*, 3rd edn. (Duisburg: DISS)

—— and JANUSCHEK, F. (eds.) (1992), *Der Diskurs des Rassismus. Ergebnisse des DISS-Colloquiums November 1991 [Osnabrücker Beiträge zur Sprachtheorie 46]* (Universität Osnabrück).

JANUSCHEK, F. (ed.) (1985), *Politische Sprachwissenschaft. Zur Analyse von Sprache als kultureller Praxis* (Opladen: Westdeutscher Verlag).

KALPAKA, A., and RÄTHZEL, N. (1990), *Die Schwierigkeit, nicht rassistisch zu sein* (Leer: Mundo).

KELLERSHOHN, H. (1992), *'Frieden oder "rettet Israel"?' Die linken Kritiker der Friedensbewegung und ihr Beitrag zur neuen deutschen Normalität. Ein kritischer Rückblick auf die Golfkriegsdebatte* (Duisburg: DISS).

KLEIN, J. (ed.) (1989), *Politische Semantik* (Opladen: Westdeutscher Verlag).

LEONTIEV, A. N. (1982), *Tätigkeit, Bewußtsein, Persönlichkeit* (Cologne: Pahl-Rugenstein).

LIEDTKE, F., WENGLER, M., and BÖKE, K. (eds.) (1991), *Begriffe besetzen* (Opladen: Westdeutscher Verlag).

LINK, J. (1982), 'Kollektivsymbole und Mediendiskurse', *kultuRRevolution*, 1: 6–21.

—— (1991), ' "Der irre Saddam setzt seinen Krummdolch an meine Gurgel!" Fanatiker, Fundamentalisten, Irre und Trafikanten. – Das neue Feindbild Süd', in Jäger (1991), 73–92.

MAAS, U. (1984), *'Als der Geist der Gemeinschaft eine Sprache fand . . .'. Sprache im Nationalsozialismus* (Opladen: Westdeutscher Verlag).

QUINKERT, A., and JÄGER, S. (1991), *Warum dieser Haß in Hoyerswerda? Die rassistische Hetze von Bild gegen Flüchtlinge im Herbst 1991* (Duisburg: DISS).

11 Evaluation of Language Use in Public Discourse: Language Attitudes in Austria

SYLVIA MOOSMÜLLER

I INTRODUCTION

1.1 *Some Preliminary Remarks on Dialect and Standard in Austria*

The study of how language use is evaluated is a large field and this chapter will concentrate on one specific aspect and on one German-speaking context: attitudes towards linguistic variation in public discourse in Austria. In order to do this, it is necessary to begin with a brief outline of the general language situation in Austria.

The situation of the German language in Austria differs markedly from that in Germany. It is often described as a dichotomy consisting of dialect and standard (see Dressler and Wodak 1983), as for historical reasons the dialects (belonging, except in the extreme west, to the Bavarian–Austrian dialect group) developed independently of the standard variety. This dichotomy has been described by Austrian linguists in terms of a so-called two-competence model; that is, it is assumed that speakers of Austrian German have competence in two varieties, a dialect one and a standard one, each of which may be partial. Indeed, there are very few people, if any, who are fully competent in both varieties.

This concept implies that some dialect forms differ completely from 'corresponding' standard forms, and that speakers switch from one form to the other. These forms are described in terms of 'input-switch rules', which show that the dialect form and the standard form coexist independently. Examples of such forms are standard [tuːn] vs. dialect [tʊɐn] for *tun* (do) or standard [vaɪs] vs. dialect [vaːs] in (Vienna) or [voɐs] (in the other regions of Austria) for *weiß* ('know'). For individual speakers, there are no intermediate steps between the two forms in each of these examples. Sociolinguistically, this means that switches between these forms are easily controllable in terms of both production and perception: speakers readily switch from one to the other, depending on the circumstances, and usually notice when other speakers switch in this way (see also Dittmar, this volume, especially for the significance of this approach to the general development of variation theory in the German-speaking context).

Nevertheless, the situation in Austria is not diglossic (see Ferguson 1959, Fasold 1984), since in many cases intermediate forms between the two extremes of dialect and standard do in fact occur. For example, for the verb *leben* (to live) there are a number of intermediate variants between the formal standard form (see Barbour and Stevenson 1990) [le:bɛn] and the dialect form [le:m]: [le:bɛn] > [le:bən] > [le:bᵊn] > [le:bm̩] > [le:ᵇm̩] > [le:m]. So the language situation in Austria could in fact be described as both a dichotomy and a continuum.

This applies not only to the technical linguistic analysis of the situation, but also to 'popular' perceptions of it. On the one hand, for example, many people talk of 'two systems', both of which have to be learned: 'meine erste Fremdsprache war die Hochsprache' (my first foreign language was standard German — male politician in Vienna). On the other hand, some people see dialects merely as 'careless' variants of the standard variety, that is, standard and dialect are perceived as elements of a single system, in which the standard forms represent the 'correct' pronunciation, and the dialect forms the 'careless' one: 'ich würd sagen, ich sprech schon eher Hochdeutsch, ich laß mich allerdings öfter gehen' (I'd say I do generally speak High German, although I do often let myself go — female teacher in Salzburg). This second view implies a hierarchy with the standard placed on the upper end of the scale and the dialect on the lower end, a metaphor that is in fact frequently used.

For these reasons, and because there is no specifically Austrian codified norm for the standard variety (see Section 2 below), it is clearly difficult to draw a sharp line between dialect and standard in Austria, but the complex interaction between the two is an important factor in language evaluation. For despite the vagueness of the terms, Austrian informants are able to give quite unanimous judgements about 'what is standard' and 'what is dialect'. It is possible to account linguistically for native speakers' ability to categorize different speech forms (see Moosmüller 1991 for a detailed analysis, especially of the importance of prosodic features). However, our concern here is with the evaluation of language behaviour: as prestige, power, and policy are necessary preconditions for the definition of a standard variety (cf. Bartsch 1985), it is more appropriate in this context to analyse the notions of standard and dialect sociologically rather than linguistically. The fact that it is not the language that is being evaluated, but the person who speaks that language, has been evident since the work done by Trudgill, Lambert, Giles, and others (see e.g. Trudgill 1975, Lambert *et al.* 1960, Giles and Powesland 1975). In other words, linguistic evaluation depends largely on sociological evaluation.

1.2 *Dialect and Standard from a Sociological Point of View*

There are many dialects in Austria, but there seems to be only one accepted standard spoken variety, and this differs greatly from what is heard, for example, in the electronic media, which is widely perceived as artificial (cf. Moosmüller 1991). In order to interpret this view, the theory of 'centre and periphery' (Kreckel 1983*b*) may be useful. According to this theory, any geographical region can be divided into a centre and a periphery; a periphery can itself be divided into a centre and a periphery, and so forth. Given the political, cultural, and economic status of Austria, the centre of Austria is obviously Vienna, followed by the other main cities of the various *Länder* (federal states), each of which in turn constitutes a centre and a periphery. Thus, for example, Innsbruck is the centre of its region, but with respect to the whole of Austria, it is a periphery of Vienna.

This theory implies that the inhabitants of the various peripheries are more familiar with the various centres than the other way round, and this is exactly what appears to be the case in Austria. For example, Viennese speakers of all social classes are quite capable of differentiating standard from dialect in Vienna, but for them any speaker from, say, Innsbruck is a dialect-speaker, irrespective of his or her social status. This is not the case in Innsbruck: speakers from Innsbruck are not only able to distinguish both socially and linguistically between their different local varieties (for example the language of the higher social classes is perceived as being closer to the standard, whereas the language of the lower social classes is classified as dialectal), but they are also able to differentiate between the different Viennese varieties.

But even from this perspective what informants think about their differentiation between dialect and standard differs from what they actually reveal in experimental situations. For speakers of Viennese German the question of what the standard variety in Austria is, is very easily answered: for them, Austrian standard German is located in Vienna, and it is spoken by the Viennese middle and upper middle classes. Everything else is either a dialect or a form of the standard with dialect features. Inhabitants of the other culturally and politically important cities of Austria have more difficulty in deciding where the standard variety is located and who speaks it. For complex historical and economic reasons, the capital city inspires very negative feelings amongst the non-Viennese, including rivalry, jealousy, and feelings of inferiority. Not surprisingly, therefore, inhabitants of other cities do not necessarily locate the standard variety solely in Vienna. The reason for this seems to be obvious: the middle and upper social classes of the other cities also wish to be considered standard-speakers.

However, in evaluation tests (see Moosmüller 1991, 1994), both Viennese informants and informants from the capital cities of the other federal states locate the standard variety in the middle and upper social classes of Vienna. So there is a considerable discrepancy between what people appear to think and the judgements they actually make in practice.

1.3   *Dialect Evaluation in Austria*

This inconsistency applies to attitudes towards dialects as well as towards the standard variety. People in Austria have a very ambivalent attitude towards their dialects: on the one hand they are stigmatized, on the other hand they are romanticized. And although virtually everyone uses dialect to some extent, this is often denied by the speakers themselves.

This ambivalence towards dialect can be illustrated by an example. The dialect of Tyrol (which belongs to the South Bavarian group of dialects) is the most popular of all Austrian dialects, and yet Tyroleans visiting Vienna are often treated as something of a joke. On the other hand, while Tyroleans themselves evaluate their dialect positively, they try to suppress typical Tyrolean features in their speech whenever they spend any length of time in other parts of Austria. Thus it is not unusual for Tyrolean informants to claim that they become 'even more Tyrolean' in their language behaviour when in other parts of Austria, especially in Vienna, while in their actual language behaviour in such circumstances they in fact use fewer typical Tyrolean features.

This suggests that there are two sorts of discrepancies concerning attitudes towards dialect in Austria: first with respect to the evaluation of others and secondly with respect to self-evaluation. As far as evaluation of others is concerned, Trudgill's (1975) division in terms of dialect evaluation is often made by Austrian informants too: regional dialects seem to be evaluated positively, urban dialects negatively. But a closer look at the available data reveals a different picture: the positive or negative evaluation of a dialect depends mainly on social, historical, or geographical factors. For example, the Burgenland dialect is evaluated extremely negatively in Vienna (see Figure 11.1). This is mainly due to social reasons: the Burgenland is considered the least economically developed federal state in Austria. Many *Burgenländer* (mostly construction workers) cycle to Vienna to work, with the result that they have a very low social status. On the other hand, precisely

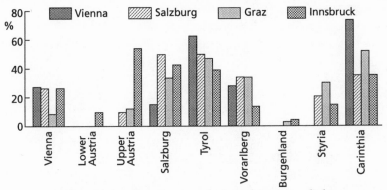

Fig. 11.1.   Relative popularity of Austrian dialects

because of its low economic status, the cost of living in the Burgenland is very low and it is therefore a popular holiday destination for the Viennese, especially for members of the lower social classes. So at least on holiday those Viennese who have low social status themselves in Vienna (their dialect is the most unpopular one throughout Austria) can feel superior.

Figures 11.1 and 11.2 give an overview of which Austrian dialects are popular (Fig. 11.1) and which are unpopular (Fig. 11.2) in the four cities of Vienna, Graz, Salzburg, and Innsbruck. As can be seen from the diagrams, the varieties of Tyrol and Carinthia are most popular in Vienna, whereas the varieties of Styria and the Burgenland as well as their own variety (Viennese dialect) are the most unpopular. Informants in Graz evaluate the variety of Tyrol positively, but they seem to be ambivalent about the Carinthia variety: roughly the same proportion views it positively as negatively. This may have something to do with the 'neighbourhood' factor, which seems to be significant in dialect evaluation: there seems to be a kind of rivalry between some neighbouring federal states. The reason for this mutual rejection is not quite clear and deserves further investigation. At all events, it does seem to be the case that, for example, people from Tyrol reject people from Vorarlberg and the other way round, and the same is true of people from Styria with respect to natives of Carinthia and the Burgenland. Not surprisingly, these social attitudes affect language evaluation too. For example, Graz informants reject the Viennese varieties (as Graz is the second largest city of Austria, there seems to be some sort of rivalry between the two cities). Similarly, informants from Innsbruck prefer the varieties of Salzburg and Upper Austria, but they reject the varieties of Vorarlberg and Vienna. Beside their own variety informants from Salzburg evaluate the varieties of Tyrol

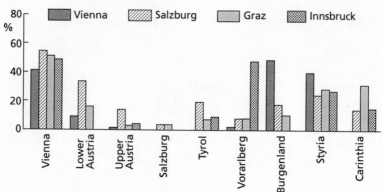

Fig. 11.2. Relative unpopularity of Austrian dialects

most positively, whereas they mainly reject the varieties of Vienna and Lower Austria.

From the data it can easily be seen that the popularity or unpopularity of a dialect or a variety does not depend on whether it is rural or urban. The fact that in Austria regional dialects are evaluated as negatively as urban dialects is to some extent explained by the views expressed in the following statement from a male teacher in Graz (his views are in fact fairly representative):

Well first of all the term dialect doesn't have any pejorative associations for me, of course, they are historically just as much legitimate forms of speech as what has become established as the standard German language as a result of a series of historical accidents. So, someone who talks a real and naturally acquired dialect [*Dialekt*], or perhaps I should say a *Mundart*, is in most cases a pleasure to listen to. To me, the intonation of someone from Carinthia has a slightly exotic and charming ring to it. I don't like listening to someone who is linguistically inept, but that's quite different, that has to do with natural intelligence and language education. So there are probably Viennese intonations that are decidedly painful and bad. For example, I think the working-class slang of Favoriten or of Floridsdorf or of Meidling is a degenerate form of language, but of course this exists in Graz as well. It disturbs me if people who become teachers or have some other occupation in the public domain don't learn to speak standard as well. It is depressing if someone can only talk in his local dialect (*Mundart*) and his sort of 'church' German, because he can't quite shake off a certain illiteracy in the area of language.[1]

First of all the informant uses the 'yes, but . . .' construction ('the term dialect doesn't have any pejorative associations for me, of course . . .'), a typical construction used in the expression of prejudices

---

[1] See Appendix to this chap. for the German original (text 1).

(cf. Quasthoff 1973, Van Dijk 1984). He starts with a positive evaluation of dialect, which in the course of his statement is converted into a totally negative one ('illiteracy').

Secondly, the dialect is denied the status of a dialect: urban dialects of the lower social classes are referred to as 'slang'. This terminological device is used to justify the negative evaluation of these varieties: it is not a dialect that is rejected, but something else. Furthermore, like many other informants, he uses different terms to refer to regional and urban dialects. For regional dialects, he uses the term *Mundart* (for an explanation of the terms 'dialect', *Mundart*, and 'standard' see Barbour and Stevenson 1990: 55–7), which is supposed to indicate that they developed from some kind of 'original form'. However, it never becomes clear what this original form really consists of and why regional dialects should be more 'original' than urban dialects. At any rate, dialects are divided into 'good dialects' and 'bad dialects' (note that the informant seems to know a great deal about the language situation in Vienna; Viennese informants would not be so knowledgeable about the situation in Graz). However, the informant does not even maintain his own differentiation, as in his last sentence he condemns the use of *Mundart* in the same way as he condemns the use of urban dialects ('illiteracy' vs. 'degenerate form of language' respectively).

Regional dialects are romanticized (they are 'exotic and charming'), that is, they do not have equal status with, for example, the standard, as might be assumed from the positive evaluation. Romanticizing such concepts typically has the function of stigmatizing minorities or lifestyles (Ehn 1989). For this reason, regional varieties are just as stigmatized as urban dialects. People who are 'only able' to talk in their local dialect (*Mundart*) are dismissed as illiterate.

In conclusion, then, we can summarize the evaluation of dialect and standard varieties as follows:

- dialects are generally evaluated negatively, regardless of whether they are rural or urban varieties, as they are associated with the language behaviour of the lower social classes;
- the standard variety is generally associated with Vienna, although not everyone would admit this or even consciously hold such a view, and more specifically it is identified with the middle and upper middle classes, despite the fact that the language use of these speakers often contains dialect features.

2   LANGUAGE EVALUATION IN OFFICIAL DISCOURSE

Austria has no codified phonological norm[2] of its own (except the concessions made for Austria by Siebs 1969, which do not, however, reflect the actual language situation in Austria: cf. Reiffenstein 1973, but see also the discussion in Barbour and Stevenson 1990). There is nevertheless a more or less clear conception of what Austrian standard German 'looks like'. As we have seen, the normal definition of the standard variety is more sociological than linguistic, in that speakers of the middle and upper classes of Vienna are considered its main exponents. Not surprisingly, therefore, it is this variety that is expected in official discourse. So people in official or semi-official positions (teachers in schools or universities, managers, politicians, etc.) typically try to conform to these expectations. But it is a well-known fact that there are discrepancies both between actual language behaviour and self-evaluation and between actual language behaviour and evaluation of others. Consequently people in official and semi-official positions are in a perpetual crossfire of criticism as far as their language behaviour is concerned.

2.1   *Evaluation of the Language Behaviour of Austrian Politicians*

Because of the dichotomous relationship between dialect and standard in Austria on the one hand and the negative evaluation of dialects on the other, the language behaviour of Austrian politicians is to a large extent evaluated negatively. This evaluation is not confined to the actual language behaviour of politicians: very often their language behaviour and the political content of what they say both contribute to the evaluation of their speech. Therefore, interpreting evaluation tests is difficult, since these two aspects are often not easy to separate.

Politicians themselves are of course aware of this close connection between language use and political evaluation, and try to take it into account in the way they operate. They have to be able to accommodate their speech to people's expectations, as their chances of being elected stand or fall with their public esteem. In other words, politicians try to accommodate (cf. Beebe and Giles 1984) to the group they want to address, although they are aware of the fact that they cannot be entirely successful, as nobody is competent in every variety. So whether their language behaviour is accepted depends on the tolerance, or rather, on the political attitude of the politicians' audience.

---

[2] Although in places this chap. refers to other linguistic levels too, it is primarily concerned with the phonological level.

Although politicians try to accommodate to different situations, they do not seem to be very successful, as there is no other social group whose language behaviour is subjected to such severe public criticism. They are generally criticized above all for using too much dialect, which as we have seen is associated with the language behaviour of the lower social classes. As a male university professor in Vienna put it, 'this working-class colouring, especially in the speech behaviour of many of the politicians from the eastern part of Austria, creates an ignorant rather than an educated impression.'[3] This criticism is directed principally at federal politicians who are speakers of Viennese dialect and who try to accommodate to the standard variety but tend to confuse the two systems.

But there are also regional politicians who always talk in their regional dialect and never try to accommodate to the standard variety in any situation. This is only a small group, who never even think of addressing anyone other than their regular audience (for example farmers). But not even the language behaviour of this small group enjoys an entirely positive evaluation, although in the end it is accepted. As a female university professor at Innsbruck said: 'But if they were to stick to their dialect: what I really liked was the former leader of our *Land*, Wallnöfer, who used the broadest Tyrolean dialect but was still perfectly well understood all over Austria.'[4]

Although this informant at least seems to accept the politician's dialect, she evaluates the dialect negatively ('broadest dialect'). Another university professor also accepts the dialect of the same politician, although he admits that he does not really like it:

The Tyrolean Wallnöfer was great of course; for me it was always amazing that something like this is still possible nowadays, that it is possible to appear in public with such a broad dialect. But of course he did this intentionally.

INTERVIEWER. And you liked that, that he talked dialect in public?

I thought it was original, but I didn't like it, I mean, it draws attention to the speaker and it's maybe amusing, but, well at any rate, I didn't really like it. Because I don't think it's appropriate to speak to the general public in dialect. I rather think that it's a trick, that it's a deliberate attempt to show the 'common touch' by choosing a particular form of language.[5]

This example shows particularly clearly how difficult it can be to distinguish between attitudes to language behaviour and the evaluation of a speaker's personality. One informant evaluates this politician (but not his language behaviour!) positively, precisely because he was able

---

[3] See Appendix, text 2, for original German.
[4] See Appendix, text 3, for original German.
[5] See Appendix, text 4, for original German.

to remain 'close to the people', while the other informant stresses the potential for manipulation in this behaviour. Either way, however, dialect is rejected, as it is not associated with an educated background, or as another informant, a female teacher in Vienna, put it in a very resigned way: 'Politicians are, as the law dictates, ordinary people, and ordinary people are not always capable of expressing opinions in public.'[6]

In fact, the great majority of Austrian politicians have a lower social class background and although many of them have a university education, this social background is reflected in their language behaviour. Thus although they try to accommodate to the standard variety, this standard language behaviour is often judged as sounding artificial. As dialect is often perceived as a slovenly form of articulation rather than as a system in its own right and as most informants, even teachers of German, have no idea of the syntactic characteristics of dialects, it is these characteristics in particular that are criticized as grammatical errors in the speech of dialect-speakers:

For me it's always awfully painful to listen to politicians in senior positions who have only an incomplete command of the German language. And there are loads of them. It starts with the finance minister, who keeps saying *Ziffern* when he means *Zahlen* (for 'figures'); this is not a dialect problem, but a lexical problem, and an educational problem, and it goes on to errors in grammatical cases, to members of parliament who really can't speak standard German, but can't get rid of their dialect.[7]

This informant (whose view is fairly representative) is a teacher of German in Graz and yet he does not seem to know that dative and accusative cases are neutralized in dialects. In dialect, a sentence like 'ich liebe ihm' (I love him) instead of standard 'ich liebe ihn' is not incorrect; nevertheless people who neutralize this case distinction are stigmatized as uneducated (and, needless to say, this has far-reaching consequences, especially in schools!). To use case neutralization in a standard utterance is of course a mixture of two systems, the dialect and the standard. The problem is that standard-speakers are allowed to mix the two systems without fear of ridicule (although, in fact, they do not use case neutralization), whereas dialect-speakers are criticized for doing so.

Furthermore, this example clearly shows the connection between judgements on language behaviour and on political competence: politicians' language, or rather their social background, is evaluated first, and only then is their political competence considered. So it seems

---

[6] See Appendix, text 5, for original German.
[7] See Appendix, text 6, for original German.

that the assessment of politicians' competence derives at least in part from the evaluation of their language behaviour.

For some informants, like this male teacher in Vienna, the use of 'incorrect German' also applies to politicians who want to avoid making concrete statements:

If every second sentence they start peters out, because they can't find the appropriate verb, if they keep using stereotypical phrases like 'you see' or 'that's right' or 'that is of course something that everyone should consider, if . . .' and those chains of subordinate clauses which don't lead anywhere and which are no more than empty phrases and camouflage, or strategies for playing for time or showing off, but are not in any sense 'language', that really is appalling.[8]

This language behaviour is regarded not only as a tactic to conceal true intentions, but also as insincere:

I think that the language used by politicians, whenever they try to display their knowledge of standard German, has the function of confusing people, that politicians are able to use a very complicated language when they want to talk their way out of a difficult situation or when they want to avoid making their position clear. They are able to throw a smokescreen over their uncertainty. And as far as their pronunciation is concerned, I don't mind if they speak the language of the people, because after all they are the representatives of the people.[9]

This informant, a teacher in Vienna, is one of the few who do not disapprove of the use of dialect ('language of the people'), but she emphatically objects to the use of an 'artificial standard' ('when they try to display their knowledge of standard German'), which she associates with insincerity.

There seems to be a multidimensional interrelationship between the evaluation of politicians' language behaviour, their membership of a certain political party, their political views, and informants' preference for a certain political party. The following informant, for example, rejects the dialect of politicians (dialect is often evaluated as vulgar, cf. Moosmüller 1988), but does not disapprove of the language of the Green Party, which he evaluates as 'straightforward' (it should be pointed out, though, that the language of some Green politicians is by no means free of dialect forms):

In the major political parties there are very few who are attractive, certainly none of the old ones, Kreisky perhaps was one of the better ones. Politicians use an incredible number of primitive expressions, unrefined, lacking any

[8] See Appendix, text 7, for original German.
[9] See Appendix, text 8, for original German.

form of expressiveness, they are simply very common and banal forms of expression without any rhetorical refinement. They're no orators, they haven't mastered the art of speaking and so they don't appeal to me, with the way they speak. I prefer politicians in the alternative parties who talk quite openly, and say what they mean, without any extravagant flourishes. And they do this in a way that isn't so clumsy, and is quite simply more honest. And if rhetoric is just a cover for lies, it is obviously always difficult to accept that in a politician.[10]

The informant, a doctor, seems to be criticizing the language behaviour of certain politicians ('they don't appeal to me, with the way they speak'), but in reality he is criticizing their political positions. Nevertheless, the persuasive power of politicians does not seem to depend solely on their language behaviour, as even politicians with crude speaking style are accepted (for example, by this male university professor at Graz) as long as their statements sound convincing:

I like it, for example, when a trenchant speaker like the member of parliament Rupert Gmoser makes a ranting speech and talks to us in powerful metaphors, and I dislike the superficiality, the phrase-mongering of those compulsive speakers who are very often incapable of reading out the texts drafted by their officials, in parliament or in a meeting.[11]

So the use of dialect is not always evaluated in the same way (cf. Moosmüller 1988), and this applies to both official and semi-official discourse. On the one hand, informants may conflate different factors in their evaluation of political discourse, but on the other hand they are able to differentiate the various possible functions of dialect in a very subtle way:

I think some politicians are good speakers, others are, consciously or unconsciously, rather folksy. For example, I accept the dialect of our *Land* leader Wallnöfer, because I think that, with him, the dialect is natural and he can hardly talk any other way and he doesn't want to, and after all he is a farmer from Mieming and that suits him. With a politician like the General Secretary of the ÖVP (Austrian People's Party), Graff, I dislike his excursions into dialect, because they are linked with a change in the level of argument and with punches below the belt.

INTERVIEWER. And is this the same with other politicians too? Have you observed that politicians who speak dialect tend to degenerate into insults?

In my opinion that depends on whether they normally use a higher level of language: if they do, and then switch into dialect, then what they're talking about normally moves on to a different level as well. But with politicians like Wallnöfer, but I don't want to mention him alone, well, Krainer in Styria,

---

[10] See Appendix, text 9, for original German.
[11] See Appendix, text 10, for original German.

he doesn't speak dialect as obviously as Wallnöfer, but enough for me, and that's appropriate for the level of language that he operates on.[12]

This is a very subtle observation by a male university professor at Innsbruck (note that the politicians he contrasts, Wallnöfer and Graff, belong to the same political party). Contrary to the above analysis, this informant rejects dialect when used by standard-speakers, because it symbolizes a change in the level of argumentation (see also below), whereas he accepts it when used by dialect-speakers. So this informant not only distinguishes between various functions of dialect use, he also seems to realize that it is not primarily the language behaviour but the speaker that is being evaluated.

2.2 *The Actual Language Behaviour of Politicians*

In the analysis of actual language behaviour, especially in official discourse (parliamentary debates, for example), a discrepancy between behaviour and evaluation becomes apparent. In particular, although politicians are heavily criticized for excessive use of dialect, their actual speech behaviour in official discourse does not seem to be characterized by extreme dialect use. In relatively informal interviews, of course, the use of dialect does tend to increase. This is true especially of politicians with a lower social background.

In a study of the language behaviour of a group of politicians in the federal parliament (Moosmüller 1987*b*), speech data were gathered from precisely these two contexts (debates and interviews). The data were then analysed to see if any correlation could be established between the politicians' social class membership and their realization of sixteen selected phonological variables (see Tables 11.1 and 11.2; only the eleven variables discussed here are included in the tables).[13]

---

[12] See Appendix, text 11, for original German.
[13] The following processes were investigated in the study:

1. /a/ to /ɔ/ e.g. [vasɛr] to [vɔsɛr] (water)
2. [ıç, dıç, mıç, sıç] to [i:, di:, mi:, si:] (I, yourself, myself, himself)
3. [nıçt] to [nɛːd̥] (not)
4. [aın] to [aː] (a)
5. [vir, mir] to [ma] (we, me)
6. [zınt] to [zan] (are)
7. /y,ø:/ to /i, e:/ e.g. [ʃøːn] to [ʃeːn] (beautiful)
8. [ıst] to [i:s] (is)
9. vocalization of /l/: e.g. [zolçɛ] to [zɔeçə] (such)
10. vocalization of /l/: e.g. [fiːl] to [fyː] (much)
11. reduction of the prefix 'ge-': e.g. [gəsaːg̥d̥] to [ksɔkt] (said)

WCB = politicians with working-class background
LMCB = politicians with lower middle-class background
MCB = politicians with middle-class background

Table 11.1. *Language use of politicians:*
realization of selected phonological variables in
interviews (%)

Variable	WCB	LMCB	MCB
1.	87.74	67.25	75.58
2.	69.67	10.78	16.92
3.	66.00	8.28	21.51
4.	33.15	14.10	10.46
5.	74.99	20.00	31.25
6.	31.11	7.14	0.00
7.	69.23	52.08	70.61
8.	78.52	44.16	50.26
9.	36.26	1.56	14.50
10.	20.38	1.56	1.45
11.	92.78	31.33	62.04

Notes: 100 = all dialect; 0 = all standard.
WCB = working-class background; LMCB = lower
middle-class background; MCB = middle-class
background.

Broadly speaking, all three groups use more dialect forms in the interviews than in the debates. Furthermore, dialect forms are used to a greater extent by politicians with a working-class background than by all other politicians in this context (Table 11.1). The only exceptions are the dialect input-switch rule /a/ to /ɔ/ (variable 1), which is also widespread in the middle and upper middle classes, and the dialect input-switch rule /ɪst/ to /iːs/ (variable 8), which in its output form is very similar to a natural phonological backgrounding process and therefore not salient either in terms of production or in terms of perception.

However, in the more formal context of parliamentary debates, there were fewer dialect realizations overall and there appears to be no correlation between social class background and language use (Table 11.2): in other words, politicians make greater efforts to keep as close as possible to the standard variety when their speech is subject to public scrutiny. On the other hand, this particular study was only concerned with the segmental level, and as other investigations have shown (cf. Moosmüller 1988, 1991), it is the prosodic level which has the greater influence on speaker evaluation. It is therefore not really surprising if informants in evaluation tests fail to make clear distinc-

Table 11.2. *Language use of politicians: realization of selected phonological variables in parliamentary debates* (%)

Variable	WCB	LMCB	MCB
1.	18.03	47.59	40.09
2.	1.21	4.54	12.77
3.	0.00	8.37	2.22
4.	0.00	8.89	7.69
5.	0.00	5.55	5.00
6.	0.00	0.00	0.00
7.	18.56	29.97	43.27
8.	13.63	19.08	31.47
9.	0.87	3.99	5.09
10.	0.00	2.94	0.00
11.	74.54	79.38	88.44

Notes: 100 = all dialect; 0 = all standard.
WCB = working-class background; LMCB = lower middle-class background; MCB = middle-class background.

tions, as politicians with a working-class background tend to have difficulty in realizing standard prosodic features accurately.

2.3 *The Function of Dialect in Parliamentary Debates*

Although the overall occurrence of dialect forms in parliamentary debates is relatively low, the distribution of the few forms that do occur is not random. Analysis of phonological variation on the text level reveals that dialect forms occur only in personal or 'private-level' utterances relating to other politicians (as opposed to 'subject-related' utterances: cf. Moosmüller 1989), especially in interjections and emotional outbursts.

Example: Politician, lower middle-class background:

[ʃɒːntsɪ ɪ 'leːs jɛtsd̥o ɪɐn æːgənən 'ʔand̥rɔːg d̥ən]
Schaun Sie, ich les' jetzt da Ihren eigenen Antrag, den

['kɛːnɐnts 'aː ned̥ ʃaɪmbɔɐ sɪ ʊnd̥aʃræm 'blaŋko]
kennen Sie auch nicht, scheinbar, Sie unterschreiben blanko,

[vɔs ɪɐrə segRe'd̥ɛɐrə 'aosɔɐbæːd̥n̥ ʊnd̥'vɪsn̩ɛt amɔe]
was Ihre Sekretäre ausarbeiten und wissen nicht einmal

[vɔs 'd̥rɪnʃd̥ed̥ ɔesosɪ:'vɪsn̩ aox ned̥ vɔsɪ 'sœ:bɐ]
was drinsteht, also Sie wissen auch nicht, was Sie selber
[ʊntɐ'ʃri:bm̩ ham̩]
unterschrieben haben.[14]

In this brief and emotionally very loaded passage, there is an extremely
high occurrence of stigmatized dialect forms, although this politician
avoids dialect throughout the rest of his statement. Besides the input-
switch rule /a/ to /ɔ/, he realizes forms such as [kɛ:nɐnts a:ned̥] ('you
don't even know') instead of [kœɐnəns aoxnɪçd̥], the latter being a
form of colloquial Viennese standard, which would be considered
appropriate in a situation such as this. The same holds for [nɛtamɔe]
('not even') and [ned̥ vɔsɪ sœbɐ] ('. . . not what you yourself . . .'),
contrasted with the possible realizations [nɪçt amal] and [nɪçt vasɪ
sɛlbɐ] respectively in colloquial Viennese standard. So in this passage
he uses a number of very salient dialect forms, which conflict with
the norm for this situation, but after this short digression he continues
his speech in the standard variety.

The same pattern can be observed in the speech of a female politician reacting to the accusation of a male opponent (the role of the
sex of the speakers is discussed in Moosmüller 1989):

> Politician, lower middle-class background:
> [d̥es hɔbɪ 'ned̥ 'ksogd̥ das əs ænə ɛndæ:gnʊŋ ɪs]
> Das habe ich nicht gesagt, daß es eine Enteignung ist.
> (I didn't say that — that it was an expropriation.)

The important point about this passage is that the speaker uses dialect
forms not only in unstressed, but also in stressed positions: [des]
instead of [das] ('that'), [ksogd̥] instead of [gəzakt] ('said'), [ned̥]
instead of [nɪçt] ('not'). These dialect forms will therefore be perceived particularly clearly.

As each of these utterances is a reaction to an accusation of a political opponent (and utterances of this sort can be found in almost
every speech by every politician), and as dialect is associated with
aggressiveness, low social status, and lack of education, it seems
reasonable to conclude that the use of dialect in these situations has
a clear function. As these particular politicians are not from a working-
class background, uncontrolled speech behaviour in emotional outbursts will not necessarily result in the use of more dialect, but more
likely in the increased use of phonological backgrounding processes.

---

[14] Look, I'm reading your own motion, you don't seem to know your own motion — you
sign whatever your officials work out for you and you don't even know what's in it — so you
don't even know what you've signed!

Therefore, given the negative connotations of dialect use, the use of dialect in such a formal situation suggests the implication that the opponent's objections are irrelevant, not even worthy of serious consideration, and that they are in fact being answered in a deliberate and controlled way. The speech situation is highly competitive, and the speakers' intention is to 'crush' the opponent (cf. the analysis of verbal conflict in Dittmar *et al.* 1986). The use of dialect in this context therefore has a very specific rhetorical function.

3 CONCLUSIONS

The language situation in Austria can be seen as both a dichotomy and a continuum. This complicated interaction between standard Austrian German and the Austrian dialects is very difficult to describe linguistically. Although informants seem to have quite concrete ideas about what constitutes standard and dialect, that is, informants believe they are able to decide whether a speaker is a genuine dialect- or a genuine standard-speaker, it is virtually impossible to draw a sharp line between the standard and the various dialectal varieties.

Consequently the typical characteristics of dialect and standard have to be determined sociologically rather than linguistically: the Austrian standard is spoken by the middle and upper middle social classes, whereas the various dialects are ascribed to members of the lower social classes. As a result, all dialects are evaluated negatively; the common division between positively evaluated rural dialects and negatively evaluated urban dialects does not apply here.

Furthermore, dialect is not only evaluated negatively, as it is associated with aggressiveness, brutality, low social status, and lack of education, it also has the function of actually expressing these negative characteristics: in other words, the use of dialect is one means of performing 'negative' speech acts. Given this function of dialect, it is not surprising that being spoken to in dialect in certain contexts is perceived as a sign of disrespect (cf. also Moosmüller 1991). For example, if people evaluate the speech behaviour of dialect-speaking politicians negatively, this is probably at least partly due to the fact that they feel personally offended by the speakers' language choice: it seems to be an indication that they are not being taken seriously. Clearly, a great deal of educational work remains to be done if these deep-seated prejudices against dialects are to be overcome.

274   *Sylvia Moosmüller*

Further Reading

Dressler and Wodak (1982, 1983)
Moosmüller (1987*a*, 1987*b*, 1989, 1992)
Wodak and Menz (1990)
Hoffmann (1982)
Holly (1990)
Ehn (1989)

References

BARBOUR, S., and STEVENSON, P. (1990), *Variation in German: A Critical Approach to German Sociolinguistics* (Cambridge: Cambridge University Press).
BARTSCH, R. (1985), *Sprachnormen: Theorie und Praxis* (Tübingen: Niemeyer).
BEEBE, L. M., and GILES, H. (1984), 'Speech Accommodation Theories: A Discussion in Terms of Second Language Acquisition', *International Journal of the Sociology of Languaqe*, 45: 5–33.
DARDANO, M. (ed.) (1983), *Parallela. Akten des zweiten österreichisch-italienischen Linguistentreffens* (Tübingen: Narr).
VAN DIJK, T. A. (1984), *Prejudice and Discourse. An Analysis of Ethnic Prejudice in Cognition and Conversation* (Amsterdam: Benjamins).
DITTMAR, N., SCHLOBINSKI, P., and WACHS, I. (1986), *Berlinisch. Studien zum Lexikon, zur Spracheinstellung und zum Stilrepertoire* (Berlin: Spitz).
DRESSLER, W. U., and WODAK, R. (1982), 'Sociophonological Methods in the Study of Sociolinguistic Variation in Viennese German', *Language in Society*, 11: 339–70.
—— (1983), 'Soziolinguistische Überlegungen zum Österreichischen Wörterbuch', in Dardano (1983), 247–63.
EHN, M. (1989), *Abweichende Lebensgeschichten. Angehörige sozialer Randgruppen erzählen* (Vienna: Österreichische Akademie der Wissenschaften).
FASOLD, R. (1984), *The Sociolinguistics of Society* (Oxford: Blackwell).
FERGUSON, C. A. (1959), 'Diglossia', *Word*, 15: 325–40.
GILES, H., and POWESLAND, P. (1975), *Speech Style and Social Evaluation* (London: Academic Press).
HASLINGER, A. (ed.) (1973), *Deutsch heute* (Munich: Hueber).
HOFFMANN, H.-R. (1982), *Politische Fernsehinterviews. Eine empirische Untersuchung sprachlichen Handelns* (Tübingen: Narr).
HOLLY, W. (1990), *Politikersprache* (Berlin: de Gruyter).
KRECKEL, R. (ed.) (1983*a*), *Soziale Ungleichheiten* (Göttingen: Schwartz).
—— (1983*b*), 'Theorien sozialer Ungleichheiten im Übergang', in Kreckel (1983*a*), 3–12.
LAMBERT, W., HODGSON, R., GARDNER, R., and FILLENBAUM, S. (1960), 'Evaluative Reactions to Spoken Language', *Journal of Abnormal and Social Psychology*, 60: 44–51.

LEODOLTER, R. (1975), *Das Sprachverhalten von Angeklagten bei Gericht* (Kronberg: Scriptor).

MOOSMÜLLER, S. (1987*a*), *Soziophonologische Variation im gegenwärtigen Wiener Deutsch* (Wiesbaden: Steiner).

—— (1987*b*), 'Soziophonologische Variation bei österreichischen Politikern', *Zeitschrift für Germanistik*, 4: 429–39.

—— (1988), 'Dialekt ist nicht gleich Dialekt. Spracheinschätzung in Wien', *Wiener Linguistische Gazette*, 40–1: 55–80.

—— (1989), 'Phonological Variation in Parliamentary Discussions', in Wodak (1989), 165–80.

—— (1991), *Hochsprache und Dialekt in Österreich. Soziophonologische Untersuchungen zu ihrer Abgrenzung in Wien, Graz, Salzburg und Innsbruck* (Vienna: Böhlau).

—— (1994), 'Assessment and Evaluation of Dialect and Standard in Austria', in Werlen (1994).

QUASTHOFF, U. M. (1973), *Soziales Vorurteil und Kommunikation — Eine sprachwissenschaftliche Analyse des Stereotyps* (Frankfurt: Fischer).

REIFFENSTEIN, I. (1973), 'Sprachebenen und Sprachwandel im österreichischen Deutsch der Gegenwart', in Haslinger (1973), 19–27.

SIEBS, T. (1969), *Deutsche Aussprache. Reine und gemäßigte Hochlautung mit Aussprachewörterbuch*, 19th edn. by H. de Boor, H. Moser, and C. Winkler (Berlin: de Gruyter).

TRUDGILL, P. (1975), *Accent, Dialect and the School* (London: Edward Arnold).

VOLLMANN, R., (1994), 'Soziophonologische Parameter der Wiener Monophthongierung', *Wiener Linguistische Gazette*, 48.

WERLEN, I. (ed.) (1994), *Verbale Kommunikation in der Stadt* (Tübingen: Narr).

WIESINGER, P. (ed.) (1988), *Das Österreichische Deutsch* (Vienna: Böhlau).

WODAK, R. (ed.) (1989), *Language, Power and Ideology* (Amsterdam: Benjamins).

—— and MENZ, F. (eds.) (1990), *Sprache in der Politik–Politik in der Sprache: Analysen zum öffentlichen Sprachgebrauch* (Klagenfurt: Drava).

APPENDIX

German Original Texts

1. Also erstens amal ist für mich mit Dialekt nix Abwertendes verbunden, selbstverständlich, sondern das sind ja historisch genauso legitime Sprechformen wie das, was sich dann durch eine Reihe historischer Zufälle als deutsche Gemeinsprache herausgestellt und durchgesetzt hat, also jemand, der einen echten und natürlich gewachsenen Dialekt spricht, eine Mundart spricht, vielleicht sagen wir lieber so, ist meistens auch angenehm zu hören. Im Kärntnerischen ist in meinen Ohren ein leicht exotischer und reizvoller Klang, sprachliche Unfähigkeit hör' ich nicht gern an, also das ist aber ganz was anderes und hat ja mit natürlicher Intelligenz und Sprachbildung zu tun, und deswegen gibt's wahrschein-

lich Wiener Tonfälle, die ausgesprochen schmerzhaft sind, schlecht und einfach, also ein Favoritner oder ein Floridsdorfer oder auch a Meidlinger Arbeiterjargon ist eine degenerierte Sprachform für mich, aber in Graz zum Beispiel gibt's des selbsverständlich genauso, und mich stört, wenn Leute, die zum Beispiel in den Lehrberuf gehen oder in sonst öffentlichkeitswirksame Bereiche, es nicht erlernen, die Hochsprache auch zu beherrschen. Wenn einer ganz afoch nur seine Lokalmundart und seine Art Kaplandeutsch wechselweise einsetzen kann, des find ich betrüblich, weil der einfach an gewissen Analphabetismus net ganz los wird im sprachlichen Bereich.

2. ... und gerade ein Großteil der aus Ostösterreich stammenden Politiker wirkt durch diese unterschichtige Färbung eher primitiv als gebildet.

3. Aber wenn sie ihren Dialekt beibehalten würden, da lobe ich mir unseren gewesenen Landeshauptmann Wallnöfer, der hat also den ärgsten Tiroler Dialekt gesprochen, der aber bitte quer durch Österreich sehr wohl verstanden wurde.

4. Klass war natürlich der Wallnöfer, der Tiroler, des war für mich immer einzigartig, daß so etwas überhaupt noch möglich ist, daß man mit an derartig intensiven Dialekt ah in der Öffentlichkeit auftreten kann. Und aber das hat er natürlich auch absichtlich gemacht.

INTERVIEWERIN. Das hat Ihnen gefallen, so in der Öffentlichkeit auch Dialekt zu sprechen?

Ich fand's originell, aber gefallen hat's mir nicht, na ich mein, es ist aufmerksamkeitserregend und belustigend vielleicht, aber jedenfalls gefallen eigentlich nicht. Weil's mir auch nicht adäquat erscheint, nicht, wenn ich mich an die große Öffentlichkeit wende. Ich glaub eher, daß da a Masche dahinter ist, daß da schon auch gezielt versucht wird, die Volksnähe zu demonstrieren durch die Wahl der Sprache.

5. Die Politiker sind also, wie es das Gesetz haben will, Leute aus dem Volk, und das Volk ist nicht immer befähigt, öffentlich Meinungen abzugeben.

6. Es ist für mich immer entsetzlich qualvoll, Politikern in gehobeneren Ebenen zuzuhören, die des Deutschen nur unvollkommen mächtig sind. Und solche gibt es reihenweise. Des beginnt beim Finanzminister, der dauernd von Ziffern spricht, wenn er Zahlen meint, des is eigentlich ka Mundartproblem, sondern ein lexisches Problem, auch ein Bildungsproblem, geht zu Fallfehlern, jetzt zu Nationalratsabgeordneten, die einfach tatsächlich nicht Hochdeutsch sprechen können, sondern aus ihrer Dialektform nicht herauskommen.

7. Wenn jeder zweite Satz, den man startet, irgendwo versandet, weil man das aussagende Verb nicht mehr findet, wenn man stereotype Phrasen immer hat, 'Schauen Sie' und 'das ist richtig' und 'das ist ein Ding, das sich natürlich jedermann überlegen muß, wenn man' und diese Nebensatzgliederketten (!), die nirgendwo hinführen, die nur Sprachkulissen und Sprachhülsen und Verkleidungen sind, auch Strategien zum Zeitgewinn, auch Imponiergehabe sind, aber absolut ka Sprache mehr darstellen, ja, dann graust's an.

8. Ich glaub, daß die Sprache, die die Politiker verwenden, wenn sie ihr Hochdeutsch versuchen anzubringen, dazu da ist, die Menschen zu verwirren, daß sie, wenn sie sich herausreden, oder wenn sie nicht ganz klare Positionen vertreten, eine sehr komplizierte Sprache aus ihrer Schublade ziehen können und letztlich sich damit über eine Unsicherheit hinwegschwindeln. Und von der Artikulation her stört's mich nicht, wenn sie die Sprache des Volkes sprechen, weil sie san ja schließlich unsere Volksvertreter.

9. Ja, also da gibt's sehr wenige, die attraktiv sind, von den Standardparteien, die Alten ganz sicher nicht, also auch der Kreisky war vielleicht eh noch bei den Besseren, es sind ganz einfach unheimlich viele primitive Ausdrucksformen unter den Politikern, nicht, nicht geschliffen, mit keiner Weise von, in irgendeiner Form starken Ausdruck gebend, sondern einfach, da was einfach ordinäre, ganz banale Ausdrucksformen, die also, die rhetorisch einfach unterm Hund sind, net, ganz banal gesagt, also, die keine Redner sind, die einfach die Kunst des Redens nicht beherrschen und einfach also mich nicht anziehen können in ihrer Sprachform, da gefallen mir dann schon eher Leute von den alternativen Lagern vielleicht, die sehr frank und frei reden, die keine großen Umschweife machen und einfach sagen, was gemeint ist und des in aner Weise tun, die net so plump is, ja und die ganz einfach auch ehrlicher, ja, des kommt dazu, ja, wenn Rhetorik Deckmantel für Lüge ist, ist es natürlich immer schwierig, das zu akzeptieren bei Politikern.

10. Es gefällt mir zum Beispiel, wenn ein pointiert Vortragender wie der Nationalratsabgeordnete Rupert Gmoser polternd auftritt und in kraftvollen Bildern zu uns redet, und es mißfällt mir, wenn man, wenn man das Oberflächliche, Floskelhafte dieser Zwangssprecher sieht, die oft nicht in der Lage sind, den von ihren Sekretären aufgesetzten Text vorzutragen am Pult oder in einer Sitzung.

11. Ja, bei den Politikern ist es so, daß ich eben meine, es gibt schon einige recht gute Sprecher, andere sind eben zum Teil bewußt, zum Teil unbewußt volkstümlich, also unserem Landeshauptmann Wallnöfer zum Beispiel nehm ich den Dialekt auch ab, weil ich das Gefühl hab, bei ihm ist er also echt und er kann kaum anders und er will nicht anders und er ist schließlich und endlich ein Bauer von Mieming und das paßt zu ihm. Bei einem Politiker wie dem Generalsekretär Graff von der ÖVP stören mich die Ausritte in den Dialekt bei ihm, weil sie gleichzeitig ah verbunden sind auch mit einem Wechsel in der Argumentationsebene und mit Schlägen unter der Gürtellinie.

INTERVIEWERIN. Und auch bei anderen Politikern ist das dasselbe, haben Sie das schon öfter beobachtet, daß Politiker, die Dialekt sprechen, jetzt, daß die eher ausarten in Beleidigungen?

Das hängt davon ab, meines Erachtens, ob sie also normalerweise sich auf einer höheren Sprachebene bewegen, wenn sie des tun und dann zum Dialekt greifen, dann ist es meistens also auch inhaltlich, rutscht es auf eine andere Ebene. Während bei solchen, wie eben Wallnöfer, aber i möcht net ihn allein nennen, ja, Krainer in der Steiermark, der spricht net so deutlich Dialekt wie Wallnöfer, aber doch auch für mein Empfinden, und da gehört's einfach zu der Sprachebene, auf der er sich bewegt.

12 Language and Gender

MARLIS HELLINGER

I INTRODUCTION

Almost ten years after language and gender studies had been established as an academic discipline in the United States, the topic was taken up in Europe. In 1978, the 8th World Congress of Sociology held at Uppsala University (Sweden) featured a section on language and sex, and in the same year the debate on male and female language began in Germany (Trömel-Plötz 1978). In 1979, the first German international conference on 'Sprache und Geschlecht' was held at the University of Osnabrück, with Cheris Kramarae, Muriel Schulz, and Dale Spender as influential participants (Andresen 1979). In 1980, the first German guidelines for the equal treatment of the sexes were published (Guentherodt *et al.* 1980), and with the publication of *Das Deutsche als Männersprache* (Pusch 1984) the topic was firmly established in German linguistics.

This chapter is not intended to provide a history of feminist linguistics in Germany: for an overview which includes a historical perspective, see Schoenthal (1985). Rather, I will concentrate on four major areas in the field: the question of whether women and men use the German language differently (Section 2), categories of gender in German (Section 3), the expression of sex of referent (Section 4), and language change under the influence of the women's movement (Section 5). I will take a contrastive approach to these issues, comparing German and English in terms of relevant structural properties (e.g. categories of gender), and strategies that are employed in the two languages to avoid sexist usage.

2   WOMEN AND MEN SPEAKING

In 1982, Trömel-Plötz published a book called *Frauensprache: Sprache der Veränderung* (Women's language: language of change), a collection of her own articles from newspapers, linguistics journals, radio interviews, and papers presented at various German and Swiss institutions. Pusch (1984) followed with her collection of articles and short commentaries on feminist linguistics entitled *Das Deutsche als Männersprache* (German as a male/man's language), a book which became extremely influential in both academic and public discourse (cf. also Pusch 1990). As with comparable publications on English (e.g. Key 1975), the titles of these books suggest that women and men use German differently to an extent which justifies the postulation of two separate codes, that is, a female and a male variety of German.

Sociolinguistic studies of English and German have indeed shown that there are differences in the way women and men use certain linguistic phenomena (for German see Ammon 1973, Mattheier 1980, Frank-Cyrus 1991). Women's speech is often described as orientated more towards standard varieties. For example, Labov (1966: 288) observes that women use fewer stigmatized forms of New York City English than men, for example deletion of postvocalic [r]. Labov maintains that women are more sensitive than men to sociolinguistic norms and prestige patterns, which contributes to a more 'correct' behaviour (for German see Werner 1983, Schmidt 1988).

Similarly, Trudgill (1974) found that Norwich women use more instances of [ɪŋ], as in *walking*, while men of the same social class display a higher frequency of non-standard [ɪn] (cf. also Horvath 1985). Trudgill explains this finding as a sign of women's social insecurity, which derives from women's subordinate status in society.

In her study of Belfast English, however, Milroy (1987) demonstrates that simply considering traditional sociolinguistic variables such as social class, sex, and age will not lead to acceptable generalizations about male and female language use. She insists that explanations for sex-preferential variation must be related to speakers' social networks, their social orientations and value systems.

For English as well as German, it is true that some differences in male and female language use have been found, but there is no empirical evidence for any sex-specific differentiation, that is, phenomena that are used exclusively by men or by women (*geschlechtsspezifisches Sprachverhalten*). Generally, differences can only be described in terms of frequencies or tendencies, that is, sex-preferential use (*geschlechtstypisches Sprachverhalten*).

Similar results have emerged from discourse analysis, which I take to include conversational analysis (for a terminological differentiation see Levinson 1983: 286 ff.; cf. also Jäger and Wodak, this volume). For English, Key (1975) and Lakoff (1975) have claimed that women use more polite forms and phenomena of 'uncertainty', such as tag-questions, question intonation, intensifiers, and hedges, forms which are supposed to characterize 'the female register' (cf. Crosby and Nyquist 1977, Macmillan *et al.* 1977). However, empirical evidence remains contradictory (see Coates and Cameron 1988, Hellinger 1990: 27 ff.). On the one hand, higher frequencies of uncertainty items have been found only in some types of discourse (typically experimental or more formal situations) and often as tendencies rather than with statistical significance. On the other hand, the occurrence of an item such as a tag-question may have different functions in actual discourse, so that an explanation as 'uncertainty' captures only one among several possibilities of interpretation (cf. Cameron *et al.* 1988). This is also true for categories of turn-taking, where a higher frequency of interruptions and overlaps, usually performed by male speakers, is widely interpreted as indicative of conversational dominance (see Zimmerman and West 1975; cf. also Hellinger 1990: 33 ff.).

In the next section I will discuss representative examples of German studies of four types of discourse: therapeutic discourse, TV discussions (talk-shows), academic discourse, and private conversation. While these studies are all concerned with standard varieties of Austrian, Swiss, or German German, they illustrate a variety of theoretical backgrounds and methodologies (for a recent collection which takes an intercultural perspective, see Günthner and Kotthoff 1991).

2.1 *Therapeutic Discourse*

Wodak's (1981) sociolinguistic analysis of group therapy sessions is of major importance in that it provides empirical support for assumptions about the interaction of sex, social class, and verbal behaviour. Wodak takes her data from the longitudinal field study of suicidal patients who participated in group therapy in a Vienna crisis intervention centre. She combines various elicitation procedures and measures of analysis which include sociological, psychoanalytic, and text-linguistic aspects.

Wodak sets out to answer the following questions: How does interaction in a therapy group differ from 'normal' interaction? Are sex-preferential mechanisms reproduced in group sessions? Does group therapy allow women and men to change their traditional sex-roles including their verbal behaviour?

The study produced some interesting and important results. For instance, contrary to common expectations, men do not take longer turns of speech than women. Also, while men interrupt and overlap women more frequently than vice versa, women do interrupt both male and female members of a lower social class. Women tend to use different strategies to achieve conversational interventions, for example, they use questions rather than more aggressive acts such as rejections, which are preferred by men. There are also differences in reactions to interventions: men tend towards a more competitive style, while women tend to avoid conversational conflicts. As far as topic selection is concerned, women will more often than men discuss such topics as relationships, children, illness, and medication; they will also express emotions more often and more intensely (cf. also Wodak-Leodolter 1977, 1979).

2.2   TV Talk-Shows

The popular collection by Trömel-Plötz (1984*a*), which bears the provocative title *Gewalt durch Sprache: Die Vergewaltigung von Frauen in Gesprächen* (Power/violence through language: the rape of women in conversations), contains three of her own studies of three Swiss TV discussions. In the first talk-show, eight men and one woman discuss riots at the Zurich opera house; in the second five men and two women discuss the situation after the riots. All participants are professionals from politics, the police, the media, and social organizations, with some differences in status (for example male politician vs. female TV editor) and social roles (male presenter vs. female panellist). In the third talk-show five participants, four female and one male, discuss feminist theology.

Trömel-Plötz analyses the discussions mainly in terms of turn-taking categories, for instance length of turn, interruption, and topic selection. She interprets her results as support for the following hypotheses about mixed-sex conversations (Trömel-Plötz 1984*b*: 58 ff.):

1. Men take turns more often and they talk more;
2. men interrupt women systematically, while women perform only few interruptions;
3. women must fight for their right of turn; and
4. men select topics, while women keep the conversation going.

Trömel-Plötz (1984*b*: 61) concludes that 'Männer kontrollieren den Gesprächsablauf, und Frauen leisten die Arbeit, um das Gespräch aufrechtzuerhalten' (men control the course of the conversation and women do the work to keep the conversation going).

Schoenthal (1985: 169 ff.) has pointed out that the study suffers from a number of methodological weaknesses, which makes any generalizations tentative. At least some of the results that Trömel-Plötz presents for such categories as speaking time, self-termination, and interruption cannot be interpreted as empirical evidence for sex-preferential variation.

However, there are differences in the way women and men attempt speaker change, and in the rate of success of such attempts. Women will have to make more attempts and even then will be less successful than men. The fact that women are interrupted more frequently than men is seen as 'massive Gewaltanwendung durch Beschneidung des Rederechts' (massive use of force by restricting women's right to speak; Trömel-Plötz 1984*b*: 59).

Interruptions have frequently been described as violations of (women's) conversational rights, but Schoenthal (1985: 170) proposes an additional, quite different interpretation: women play an active role in constructing their own subordinate position in conversations (cf. also Section 2.3 below). Finally, the study suggests that women and men react differently to interruptions: women tend to stop speaking, while men will resume their turn.

Trömel-Plötz (1984*c*) adds an interesting analysis of women's joining behaviour, that is, the technique of linking one's turn to that of the preceding speaker. Three types of joining are described: (1) conventional joining (*Ja-aber-Technik*/yes-but-technique); (2) formal joining (*Ja-und-Technik*/yes-and-technique); and (3) topical joining (*Ja-stimmt-Technik*/yes-that's right-technique). The first two types only superficially express agreement with the preceding turn, but actually introduce modifications or even rejection. The third type achieves joining by genuine agreement. According to Trömel-Plötz, joining is a strategy typically used by women in order to achieve (re-)entry into the conversation, a far-reaching hypothesis which should be tested in other types of discourse.

On the other hand, Trömel-Plötz (1984*d*) evaluates positively what she considers to be typical for female conversational behaviour, namely the relatively small number of interruptions, active involvement in other participants' contributions, explicit reference to other speakers, and the joint development of a topic. Many of these features she finds in the third talk-show on the topic of 'Zwischen Verehrung und Verachtung: Haben Frauen in der Kirche noch eine Zukunft?' (Between worship and contempt: do women still have a future in the church?). The situation is exceptional in that the four female participants all have a high social status (as does the one male participant) and all are feminists.

Lauper and Lotz (1984) analyse sex-preferential verbal behaviour in a German TV discussion on the issue of women in the armed forces, 'Von der Küche in die Kaserne' (out of the kitchen into the barracks). Even though both participants, a female feminist author and a male politician, have a high (although by no means equal) social status, the woman has more problems in establishing and maintaining her conversational rights. She makes an effort to avoid stereotyped female behaviour, but in the course of the discussion she loses much of her self-assurance and conversational authority, supporting the man's turns more actively and using more expressions of 'uncertainty'. Although she manages to interrupt the man, he reacts in such a way that the effect is practically lost: 'er hat es immer in der Hand zu entscheiden, ob er sie zu Wort kommen lassen will oder nicht' (it is always up to him to decide whether or not he will let her take a turn; Lauper and Lotz 1984: 253).

While Trömel-Plötz (1984*a*) remains valuable as a first attempt to describe the conversational behaviour of women and men in German-speaking contexts, Grässel (1991) is a genuinely empirical study, which is based on almost ten hours of analysed data from five TV talk-shows. Grässel uses 95 conversational variables (including some non-verbal ones), and subjects her results to rigorous statistical analysis (analysis of variance and Wilcoxon test). Her research leads her to reject many of the claims made by Trömel-Plötz and others, in particular the idea that sex of speaker determines conversational dominance: the relative status of speakers (for instance expert vs. non-expert) seems to be just as important.

For the large majority of the 95 variables no significant differences could be established between women and men: for example, according to her data men do not have more speaker time than women, nor do they interrupt or self-select more frequently. Where differences did emerge, however, these could often be interpreted in terms of male dominance and female subordination. For instance, men more frequently use turns that make no reference to a previous speaker's turn, and their turns are characterized by the absence of supportive elements. Women, on the other hand, are more active and cooperative listeners, and they generally feel more responsible for doing conversational work.

2.3   *Academic Discourse*

In this section I will briefly discuss two studies. The experimental study by Kotthoff (1984) was designed as an asymmetric situation, in which one student (female or male) made a request of a female or male lecturer in a role play. The student's task was to induce the lecturer to sign a letter composed by the students' body in protest against the installation of a video camera in the university library. The lecturers were instructed to reject the request so that a conflict was built into the situation.

The conflict was solved differently by female and male students. While the women used a more co-operative strategy, taking care not to risk a break-up of the conversation and making concessions to the point of accepting the lecturer's position, the men pursued their original goal and participated in more competitive arguments.

Wenn ein Mann sich im Gespräch offensiv und unnachgiebig zeigt, wertet er seine Männlichkeit auf. Es gibt keinen Bruch zwischen einer aktuellen Verhaltensnorm und seiner Geschlechtsidentität. (Kotthoff 1984: 110)

(If in a conversation a man behaves in an offensive and uncompromising manner, he asserts/upgrades his masculinity. There is no break between a current behavioural pattern and his sexual identity.)

Kotthoff concludes that women contribute towards their own conversational subordination and suggests that the use of different strategies might lead to more successful encounters.

Schmidt's (1988) analysis of academic discourse is also based on the assumption that in conversation women use a more co-operative style while men are more competitive. When these behaviours meet in mixed-sex conversation, they will be reinforced so that men will tend to dominate the interaction. The data for Schmidt's study consist of audiotaped material from seven same-sex and mixed-sex conversations between students discussing exam preparations. The categories of analysis are length of turn, speaker change, listener reaction, and use of supportive responses (*ja*, *stimmt*/yes, that's right; *genau*/exactly). Schmidt is particularly interested in the way in which these categories interact. For example, she does not simply count occurrences of interruptions but considers functional aspects and participants' reactions. While the results are statistically not significant, some tendencies emerge.

In mixed-sex groups sex-preferential variation was not found for number and length of turns. According to the study, men do not speak more than women, nor do they achieve speaker change more often by interruption: women also use this strategy. Schmidt also shows that an interruption cannot generally be interpreted as an attempt to establish dominance; it can also signal emotional involvement. However, women use more supportive responses than men, which confirms the co-operative hypothesis on this point. Surprisingly, men use more turn introducers (*ha ja*, *aber*/yes, but) than women, that is, they show a type of joining behaviour which Trömel-Plötz has described as typical for the female style. On the other hand, men respond less to a previous speaker's turn in terms of its content. Thus, while men superficially adhere to co-operative principles, they nevertheless direct the conversation according to their own interests.

Women also use more so-called softeners or downgraders (*irgendwie*, *eigentlich*/kind of, really). However, Schmidt offers an alternative interpretation for the use of such pragmatic particles: rather than signalling uncertainty or subordination, they may signal readiness to elaborate a point further.

The study illustrates the trend in current research to consider different functions of the same item, to include reactions to what speakers do, and to observe more social factors that constitute the particular situation.

2.4   *Private Conversation*

In order to show how power and dominance are actively negotiated in private discourse Thimm (1990) analyses two conversations, one with two participants (a married couple), the other with three participants, two male and one female member of a *Wohngemeinschaft* (group of people sharing a flat). In both cases, a conflict is the focus of the interaction. As a theoretical framework Thimm uses a model of strategic behaviour in which conversational strategies are described as the result of behavioural planning. *Strategisches Handeln* (strategic behaviour) can be identified by sequences of strategic steps. For example, in order to persuade a participant to accept an argument, the *Durchsetzungsstrategie* (assertive strategy) can be employed, with steps such as accusations or complaints.

Familiar categories from the turn-taking model are embedded within the larger framework of speech acts of three main types: acts by which a speaker claims dominance, acts which grant dominance to another speaker, and acts which are neutral with respect to the category of dominance. Thimm emphasizes the point that frequency of turn-taking or interruptions does not necessarily establish dominant behaviour. It is crucial to analyse reactions towards such behaviour so that conversational dominance is seen as the result of all participants' active involvement.

Participants have various options in attempting to reach their conversational goals. They may even change strategies; for example if the assertive strategy is insufficient for the maintenance of conversational dominance, the *Ausweichstrategie* (evasion strategy) may turn out to be more successful.

Given her small sample Thimm makes only tentative statements about sex-preferential use of the strategies. For instance, she suggests that women resort more frequently to the *Intimitätssicherungsstrategie* (strategy of securing intimacy) in order to avoid jeopardizing the relationship in a conflict situation, which supports tendencies observed in other types of discourse.

2.5 *Conclusions*

From this selective overview, no general conclusions can be drawn about how women and men behave verbally in German. Apart from sex, social class, age, and social networks, it is essential to consider the social and situational roles of speakers in particular interactions. Also, quantitative analyses must be supplemented by qualitative analyses which take into account the fact that the occurrence of the same linguistic phenomenon may have different functions in different contexts. Finally, the investigation of non-typical social roles will allow for a wider range of interpretations. Thus, in a paper presented at the 1992 conference of the Deutsche Gesellschaft für Sprachwissenschaft in Bremen, Wodak and Andraschko (1992) have shown that powerful women will not display the expected female style. On the other hand, underlying their conversational assertiveness may be a category of power that the authors describe as 'Macht durch Mütterlichkeit' (power through motherliness).

3   CATEGORIES OF GENDER

3.1   *Semantic Gender*

In non-technical language, the term 'gender' usually relates to the property of biological maleness or femaleness. In linguistics this category of gender has been called natural or semantic gender (*natürliches Geschlecht*); it is relevant in contexts where 'sex of referent' of a lexical item is specified. For example, in a conversation about British politics, the noun phrase *the Prime Minister* will have a female referent in 1982, but a male referent in 1992 (cf. Lyons 1977: chap. 7). In a model of semantics which differentiates between sense and reference, sex of referent is thus part of the semantic description of a particular referring expression with an animate or human noun as head. Sense, on the other hand, specifies the semantic relationships between lexical items; thus the semantic properties [+ female] and [+ male] are useful for the description of *mother* as opposed to *father*. The noun *Prime Minister*, like the noun *child*, includes the semantic feature [+ human] but not [+ female] or [+ male]; such nouns are unspecified for natural gender, since potential referents may be either female or male.

In English, human nouns such as *brother*, *king*, or *boy* will be pronominalized by *he*, while *sister*, *queen*, or *girl* require the feminine pronoun *she*. These nouns belong to a lexical category which is inherently specified for sex of referent; among these are kinship terms and terms of address (*Mr, Ms, Madam, Sir*). On the other hand, most English human nouns are unspecified for natural gender and can be pronominalized by either *he* or *she*: *individual, neighbour, criminal, engineer, secretary* (cf. Quirk *et al.* 1985: 315 ff.).

I will use the terms *male* and *female* in two ways: (1) as labels for semantic features in the case of personal (human) nouns that are inherently specified for natural gender, for example *aunt, grandson, stepfather*, but not *secretary* or *boss*; and (2) as labels for sex of referent in actual discourse; thus, sex of referent of the noun phrase *our chief engineer* may be female or male. These two functions make up the category of semantic gender.

3.2   *Social Gender*

The term gender is also used as a social or cultural category 'to refer to the socially imposed dichotomy of masculine and feminine roles and character traits' (Kramarae and Treichler 1985: 173). An illustration of social gender in English is the fact that many higher-status occupational terms, such as *lawyer*, *physician*, or *scientist*, will frequently be pronominalized by the masculine pronoun *he* in contexts where sex of referent is either not known or irrelevant. On the other hand, low-status occupational titles such as *secretary*, *nurse*, or *schoolteacher*, will often be followed by anaphoric *she*. But even for general human nouns such as *pedestrian*, *voter*, or *driver*, there is a traditional rule which prescribes the choice of *he* in 'neutral' contexts. This prescription, which is called 'generic *he*', has long been the centre of debates about linguistic sexism in English.

Social gender has to do with stereotypical assumptions about what are appropriate social roles for women and men, including expectations about who will be a typical member of the class of, say, *surgeon* or *nurse*. Deviations from such assumptions will often require formal markings, for example *female surgeon, male nurse*.

Underlying prescriptive *he* in English (and equivalent usages in German, which will be discussed below) is the ideology of MAN (male as norm), which considers the male/masculine as the higher, more prestigious category and the female/feminine as secondary and subordinate (cf. Baron 1986: chap. 6).

3.3 *Grammatical Gender*

Finally, the category of grammatical gender must be considered. Grammatical gender is in principle independent of semantic and social factors. It is a category of minor significance in English in comparison with German, where it is central in the construction of the nominal system.

In contrast to number and case, grammatical gender is an inherent and invariant property of the noun. Languages differ in the number of gender classes they possess: German, Russian, and Latin have three genders (masculine, feminine, and neuter), while French, Italian, and Dutch have only two (masculine/feminine; common/neuter). Typically, elements within the noun phrase (determiners, adjectives, pronouns), but also outside the noun phrase (predicative adjectives, anaphoric pronouns, some inflected verb forms) 'agree', showing morphological variation according to the noun's grammatical gender: consider the following German examples:

(1) eine klassisch*e* Sinfonie . . . *sie* (f.)
    (a classical symphony . . . it)

    ein hölzern*er* Taktstock . . . *er* (m.)
    (a wooden baton . . . it)

    ein japanisch*es* Klavier . . . *es* (n.)
    (a Japanese piano . . . it)

Since the selection of the appropriate inflection is syntactically motivated, the semantic specification of the noun, e.g. [− animate], [+ human], or [+ collective], is irrelevant:

(2) eine berühmt*e* Dirigentin . . . *sie* (f.)
    (a famous conductor . . . she)

    ein hervorragend*er* Solist . . . *er* (m.)
    (an outstanding soloist . . . he)

    ein großartig*es* Orchester . . . *es* (n.)
    (a brilliant orchestra . . . it)

As Huddleston (1984: 289) remarks, '[it] is clear that English does not have any gender agreement of this kind'.

In principle, the assignment of a noun to one of the three gender-classes in German is arbitrary (see Eisenberg 1986: 159 ff.); exceptions are (1) human nouns, especially kinship terms, where grammatical and semantic gender frequently correlate, as in *die Mutter* (f., female), *der Vater* (m., male); (2) some lexical subsets, for instance days and months, are grammatically masculine: *der Montag*, *der Januar*; and (3) some morphologically marked nominal classes, for example de-adjectival nouns ending in *-heit* or *-keit*, are feminine (*die Trockenheit*, dryness; *die Feuchtigkeit*, dampness), while de-verbal nouns in *-er* are masculine (*der Schraubenzieher*, screwdriver; *der Schornsteinfeger*, chimney-sweep), and diminutive nouns in *-chen* are neuter (*das Männchen*, little man; *das Mädchen*, girl).

There are very few exceptions to the rule that a noun's gender is invariant: only nominalized adjectives and participles may be assigned one of the three genders by the choice of dependent categories, for example the article:

(3) krank (sick)   *die* Kranke/*der* Kranke
     studieren (study)   *die* Studierende/
                            der Studierende
     versichern (insure)   *die/der/das* Versicherte

In German, gender conflicts are by no means rare. Examples such as *das Individuum* (n.), *die Person* (f.), and *der Mensch* (m.) are grammatically neuter, feminine, and masculine respectively, but sex of referent is either female or male. Sometimes, especially in contexts related to a particular individual, semantic gender may override grammatical gender so that *das Mädchen. . . sie* (the girl . . . she) or *das Individuum. . . er* (the individual . . . he) may be used.

3.4   *Generic Forms*

The prescription of the generic masculine (*der Wähler*/voter, *der Steuerzahler*/taxpayer) in neutral contexts which are intended to include women has increasingly been interpreted as sexist language use. An example is:

(4) Jeder Wähler sollte von seinem Stimmrecht Gebrauch machen
    (Every voter should exercise his right to vote)

The two major arguments against the generic use of masculine nouns are: (1) as opposed to nouns such as *Kind* or *Individuum*, words like *Wähler* or *Steuerzahler* generally constitute members of lexical pairs whose second members (in this case *Wählerin* and *Steuerzahlerin*) are grammatically feminine and semantically sex-specific, that is, female. Therefore, not using these feminine nouns is considered as contributing towards the invisibility of the female; (2) one cannot be sure whether a masculine human noun does or does not include women. This is illustrated by numerous examples of various types (cf. Hellinger 1990: chap. 6):

(5) 45 Millionen Bürger sind zur Bundestagswahl aufgerufen
(45 million citizens are called upon to vote for the Bundestag)

The grammatical masculine *Bürger* can only be interpreted as including women if the reader knows that the sentence relates to the former Federal Republic, which at the time had some 60 million citizens, of whom 45 million were of voting age; the majority of these were women. In fact the full quotation is this:

(6) 45 Millionen Bürger sind zur Bundestagswahl aufgerufen. Etwa 24 Millionen Frauen und 21 Millionen Männer sind wahlberechtigt, wenn . . .
(45 million citizens are called upon to vote for the Bundestag. Approximately 24 million women and 21 million men are entitled to vote, when . . .)

Additional information is often provided to ensure a generic interpretation of the masculine, for example in the following sentence:

(7) Apropos Navratilova: Sie hat als bisher *einziger Tennisspieler* mehr als 10 Millionen Dollar an Preisgeldern eingenommen — *egal ob weiblich oder männlich.*
(Apropos Navratilova: she is the only tennis-player — male or female — who has won more than 10 million dollars in prize money)

On the other hand, in contexts where women are not invited to feel *mitgemeint* (included), this may be marked by a sex-specific adjective, as in the following job advertisement:

(8) Wir suchen: *Männliche* Hausdetektive
(We are looking for *male* store detectives)

The recent tendency in German towards a correlation between semantic gender (sex of referent) and grammatical gender is supported by the usage of pairs of human nouns where the supposedly generic masculine noun, such as *Arzt* (doctor), is typically interpreted as having a male referent since it denotes a higher-status occupation, while the feminine noun, for example *Krankenschwester* (nurse), refers to a female in a subordinate social role (*Ärzte und Krankenschwestern*; *der Chef und seine Sekretärin*); cf. also the following examples:

(9) 'Funktioniert denn die Technik nicht?' fragen *Konferenzteilnehmer* verdrossen, wenn wieder einmal *eine Simultandolmetscherin* erschöpft ausfällt
('Why aren't the facilities working?' conference participants (m.) will ask angrily, when once again an interpreter (f.) drops out exhausted)

(10) Bei uns spielen Hausfrauen gegen *Juristen* und die Oma gegen *ihren studentischen Enkel*
(Here housewives play against lawyers (m.) and the granny against her student grandson)

Examples of this type are also frequent in English:

(11) It's a great secret of doctors, known only to their wives . . . that most things get better by themselves.
(Miller and Swift 1981: 52)

3.5 *Summary*

The major differences between German and English concerning gender can be summarized in the following way: in contrast to English, German has the category of grammatical gender, which is marked morphologically (this includes inflectional as well as derivational marking). Eisenberg (1986: 159) observes that 'Das Genus oder grammatische Geschlecht ist die durchgängigste und einheitlichste Kategorisierung der deutschen Substantivparadigmen' (*Genus* or grammatical gender is the most productive and consistent category of the German nominal paradigm). In the area of human nouns this system facilitates explicit specification of sex of referent rather than neutralization, and the relationship between grammatical and semantic gender is changing towards more agreement. It can no longer be assumed that masculine human nouns generally 'include' women.

In English, which lost the category of grammatical gender in the fourteenth century (cf. Strang 1970: 265), most human nouns can be used to refer to female or male individuals. These systematic structural differences do not make either language more sexist. In both languages women and men can be treated as equals, without demeaning either sex or making women invisible. However, the risk of linguistic discrimination is higher in a language such as German, where the well-established morphosyntactic markers of grammatical gender naturally lead to sex-specification. This means that the development of non-sexist alternatives and the achievement of equal linguistic treatment of the sexes require considerably more effort in German than in English (see Section 5 below). In the next section I will discuss the possibilities each language has for sex-specification and sex-neutralization as well as abstraction from sex of referent.

4 SEX OF REFERENT IN GERMAN

In the preceding sections, human nouns (*Personenbezeichnungen*) have emerged as the central issue in debates about language and gender in German. Human nouns are needed to communicate about the self and others; they are used to identify people as individuals or as group members, and they can transmit positive or negative attitudes. In the following example (cf. Hellinger 1990: 67), an individual is first identified by a proper noun: *Lise Meitner*; then various common nouns provide information about the person's regional origin, ethnic membership, professional status, etc.: *die Wienerin Lise Meitner; Jüdin; Atomphysikerin; Abteilungsleiterin am Kaiser-Wilhelm-Institut für Chemie; Mitarbeiterin Otto Hahns; Tante von Otto Robert Frisch*. Meitner's identification as a woman follows from the feminine first name (*Lise*), feminine derivations in *-in* (*Wienerin*, etc.), and the kinship term *Tante* (aunt), which is inherently specified for the semantic property [+ female].

The use of human nouns may have serious psychological consequences (cf. Pusch 1984: 24 ff.). While an appropriate use may contribute towards the maintenance of an individual's identity, inappropriate use, for example identifying someone repeatedly (either by mistake or by intention) by a false name, by using derogatory or discriminating language, or by not addressing someone at all, may not only cause irritation or anger but may affect the individual in a more serious way. And since an individual's identity includes an awareness of being female or male, and of the social norms connected with sex-membership, it is of importance to develop an understanding of the ways in which sex of referent can be specified in a language. This must be based on the description of the relevant structural and functional properties of the language. As a central category gender has been discussed above in Section 3; this section deals with the grammatical and lexical means of sex-specification and sex-neutralization in German.

4.1   *Specification of Sex of Referent*

Grammatical Means

In contrast to English, German uses nominal gender to specify sex of referent (*Geschlechtsspezifikation*). This is the case for singular human nouns which are derived from adjectives (e.g. *krank*/sick) or verbs (participles: e.g. *vorsitzen*/preside → *vorsitzend*; *abordnen*/delegate → *abgeordnet*). Sex-specification as female or male occurs simultaneously with the attribution of feminine or masculine gender: *die Kranke* (f., female), *der Kranke* (m., male); *die Vorsitzende/der Vorsitzende*; *eine Abgeordnete/ein Abgeordneter*. Grammatically neuter and semantically non-animate examples are *das Neue*, *das Beste*, *das Gelesene* (derived from the past participle of *lesen*, to read). This type of variable gender-membership has been called *Differentialgenus* (see Wienold 1967: 147 ff.).

Since articles and other determiners do not vary morphologically for grammatical gender in the plural, sex-specification in the plural must be achieved by other means, for example by use of the adjectives *weiblich/männlich* (female/male): *die weiblichen Abgeordneten*, *die männlichen Abgeordneten*. Of course, *die Abgeordneten* can have female or male referents although there may be a male bias in this term, whose potential referents are expected to be male rather than female; a similar example is *die Gefallenen* (soldiers killed in battle) which can also be said to have prototypically male referents. On the other hand, a plural such as *Büroangestellte* (office workers) will be associated more with female than male referents, while *die Reisenden* (travellers) or *die Behinderten* (disabled persons) may be interpreted as genuinely neutral. This kind of bias was called social gender in Section 3.2. above.

Indefinite pronouns (*jed-*/each, every, *kein-*/no, *jemand*/someone, *niemand*/no one) can also be described as having differential gender. In the case of *jede* (f.)/ *jeder* (m.)/ *jedes* (n.) and *keine/keiner/keines*, gender is marked morphologically and can thus be used to specify sex of referent:

(12) Das weiß doch jede
      (Everyone (f.) knows that)

(13) Das glaubt uns keiner
      (No one (m.) will believe us)

In the case of *jemand* and *niemand*, which are morphologically invariable for grammatical gender, sex of referent can be specified by various pronouns (e.g. relative, possessive):

(14) Da gab es niemand, *der* nicht zupacken wollte
     (There wasn't anyone, who (m.) didn't want to help)

(15) Kann mir jemand mal *ihr* Fahrrad leihen?
     (Can anyone lend me her bicycle, please?)

The essential difference between German and English is, of course, that English cannot specify sex of referent by means of grammatical gender. Both languages are similar in that they adhere to the principle of 'male as norm', when they prescribe the use of grammatical masculines in German, or semantically male expressions in English, in neutral contexts. According to this principle, both (16) and (17) suggest a generic interpretation:

(16) Das weiß doch jeder, der Steuern zahlt
     (Anyone (m.) who (m.) pays taxes knows that)

(17) Someone has lost his key in the corridor.

Lexical Means

In examples such as *weibliche Beschäftigte/männliche Beschäftigte* (female employees/male employees), sex-specification is achieved by adjectival modification of the neutral noun. The adjective is derived from a human noun which is in turn inherently specified for sex of referent: *Weib* (n./female) → *weiblich*, *Mann* (m./male) → *männlich*. While these adjectives are equivalent in that both have the same semantic function in the examples given, their lexical root forms *Mann/Weib* are characterized by asymmetries: *Weib* carries strong negative connotations, which make it unsuitable for use in neutral contexts, where *Frau* is required: *Männer und Frauen; Frau Müller und Herr Meier. Weiblich/männlich* can be combined with any human noun which is not sex-specific, regardless of the noun's grammatical gender, e.g. *eine männliche/weibliche Person* (f.); *ein weiblicher/männlicher Rockstar* (m.); *ein weibliches/männliches Genie* (genius) (n.).

A second type of sex-specification by lexical means is found in compounding. German — as well as English — has many occupational terms and other human nouns which are morphological compounds containing *-mann/*-man (in a few cases *-herr*) or *-frau/*-woman (with very few instances of *-herrin*) as a second element: *Kaufmann/Kauffrau* (businessman/businesswoman); *Feuerwehrmann/Feuerwehrfrau* (firefighter); *Amtmann/Amtfrau* (senior civil servant); *Ratsherr/Ratsfrau* (member of city council); *Bauherr/Bauherrin* (client for whom house is being built). Again, sex of referent is determined by the lexical items *-mann/*-frau, which are lexically specified for sex of referent.

In cases where the masculine/male terms were compounded first (this is true for practically all *-mann* compounds indicating a higher social status, as in *Amtmann, Staatsmann*, and for typically male occupations, as in *Feuerwehrmann*), German tends towards the formation of corresponding *-frau* compounds, that is, it follows the principle of female visibility. In contrast, English tends towards the principle of neutralization. While compounds such as *congresswoman, businesswoman*, or *saleswoman* are used in contexts which relate to a particular female individual, in neutral contexts terms that include members of either sex are preferred: *member of Congress, business executive, sales representative, insurance agent, fire-fighter, police officer*. Again, such examples leave open the question of social gender, that is, whether a typical referent will be expected to be male rather than female.

In German, there are examples that further illustrate the tendency towards a correlation of grammatical gender with sex of referent. Although under the rule 'male as norm' a human noun such as *die Chirurgen* (surgeons) is prescribed in neutral contexts, modification by the adjectives *männlich/weiblich* is often used to ensure that in fact women are included: *männliche und weibliche Chirurgen*. On the other hand, women can be excluded by the same means (cf. example (8) above). Such usages reflect the variable and indeterminate status of masculine human nouns in current German.

Morphological Means
German has numerous suffixes for the derivation of human nouns from nominal, verbal, or adjectival stems (cf. Wilmanns 1967; Fleischer 1982). These suffixes have three functions: syntactically they determine word-class membership: *Sport* (noun) →*Sport-ler*/athlete (noun); *grob*/rough (adjective) →*Grob-ian*/brute (noun); *singen*/sing (verb) →*Säng-er-in* (noun); they determine grammatical gender: *Läufer* (m.)/ *Läuferin* (f.); and semantically they specify sex of referent: *Läufer* (m., male)/ *Läuferin* (f., female). This specification is independent of the fact that grammatical masculines may be used generically to include male and female referents.

Major suffixes that derive masculine human nouns in German are: *-er* (*Maler*/painter), *-ler* (*Sportler*/athlete), *-ner* (*Rentner*/pensioner); of minor importance are *-ling* (*Lehrling*/trainee); *-ant* (*Intendant*/director), *-ent* (*Dirigent*/conductor), *-eur* (*Friseur*/hairdresser), *-ist* (*Marxist*/Marxist), and *-or* (*Inspektor*/inspector).

Feminine human nouns are almost exclusively derived by the suffix *-in*, which is extremely productive in contemporary German; examples are: *Malerin, Rentnerin, Sportlerin, Intendantin, Dirigentin, Friseurin*. In most cases these nouns are derived from existing masculine terms.

There are only a few other feminine (human) suffixes in German; in fact, most of these — or rather the derived words containing them — are borrowings from French, for example *Chansonnette* (singer-songwriter), *Souffleuse* (prompter), *Garderobiere* (cloakroom attendant), *Direktrice* (senior female employee in a fashion store); *Stewardess* was borrowed from English, the suffix itself being, of course, of French origin. None of these suffixes is equivalent to *-in*, either in terms of productivity or in terms of semantic specification. While *-in* derives feminine nouns which form largely equivalent pairs with the corresponding masculine nouns (*Rentner/Rentnerin*), derivations in *-ette* or *-euse* generally carry negative connotations; in pairs of words, the masculine (where this exists or is used in German) usually denotes an occupational activity of higher social status, as in *Direktor/Direktrice*. *Chansonsängerin* is 'more serious' than *Chansonette*; the term *Garderobenfrau* has become more frequent than *Garderobiere*; and the official feminine counterparts of *Friseur* and *Masseur* are *Friseurin/Masseurin*, and not *Friseuse/Masseuse*.

Thus the suffix *-in* is well established in German word-formation: it is an indispensable means of achieving female visibility. In contrast, suffixes for the derivation of female human nouns in English are no longer productive and the remaining words in *-ess* (*poetess*), *-ine* (*heroine*), *-ette* (*majorette*), or *-trix* (*aviatrix*) must be interpreted as lexicalized items which in most cases are marked for derogatory or trivializing connotations. Therefore, derivation is of negligible importance as a means of specifying female sex of referent in current English.

4.2 *Neutralization of Sex of Referent*

In German there are two main ways of neutralizing sex of referent, providing equal chances for men and women to feel 'included'. I shall refer to them as 'covert' and 'overt neutralization' respectively.

Covert or inherent neutralization is achieved by lexical means, that is, the choice of lexical items that are not specified for sex of referent, regardless of which grammatical gender class they belong to: *Person*/person (f.), *Fachkraft*/expert (f.); *Mensch*/human being (m.), *Gast*/guest (m.); *Kind*/child (n.), *Individuum*/individual (n.), *(Partei)Mitglied*/(party)member (n.), *(Unfall)Opfer*/(accident)victim (n.).

Another type of covert neutralization is illustrated by the use of plural forms of nominalized adjectives and participles (see Section 4.1 above): *die Alten* (the elderly), *die Studierenden* (students), *die Angestellten* (employees). These nouns can be used appropriately in neutral contexts where sex of referent is irrelevant or, rather, where both sexes may equally figure as potential referents.

Overt or marked neutralization is achieved by double sex-specification, which can take various syntactic forms of splitting, that is, the explicit co-ordination of a semantically female (usually grammatically feminine) and a male/masculine lexical element within one phrase. The nouns or pronouns may be co-ordinated by *und* (and), *oder* (or), or *bzw.* (respectively), resulting in expressions which have been called *langes Splitting* (long splitting) or *Paarformeln* (pair formulas):

(18) alle Lehrerinnen und Lehrer; Männer und Frauen; jeder Wähler oder jede Wählerin; jemand, die bzw. der
(all teachers; men and women; each voter; someone who)

In some cases, abbreviated splitting, sometimes called '*Sparformeln*' (economy formulas), may be used; this type is marked by orthographical symbols:

(19) Bürger/innen; LeserInnen; keineR
(citizens; readers; no one)

Currently, *Sparformeln* are not widely used, especially since the formation of acceptable forms is restricted. In most singular forms, feminine and masculine nouns as well as determiners (articles etc.) show morphological variation according to grammatical case. Thus only nominative singular expressions such as *jede/r Abgeordnete* (every delegate) or *kein/e WählerIn* (no voter) can be found occasionally, while oblique cases cannot be derived at all. These must be rendered by unabbreviated splitting:

(20) Dies dürfte jedem Mieter und jeder Mieterin bekannt sein (dative)
(This should be known by each tenant)

(21) Zu den Pflichten einer Staatsbürgerin bzw. eines Staatsbürgers gehört es . . . (genitive)
(It is one of the duties of each citizen . . .)

A third type of splitting is adjectival splitting by means of double adjectival modification of a neutral noun:

(22) männliche und weibliche Abgeordnete/Fachkräfte/Senatsmitglieder
(male and female delegates/experts/members of the senate)

While the double sex-specification in (22) is in principle unnecessary, since the nouns are semantically unmarked for sex of referent, adjectival splitting is sometimes used in order to emphasize that both sexes are included, especially in cases where the noun is socially marked for a male bias, as in the case of *Senatsmitglied* or *Führungspersönlichkeit* (executive). Double adjectival modification also occurs with masculine nouns, whose generic interpretation — despite prescriptive traditions — can no longer be assumed to be generally available. This motivates expressions such as *weibliche und männliche Chirurgen/ Piloten/Politiker*.

4.3 *Abstraction from Sex of Referent*

Finally, abstraction from sex of referent must be mentioned as a special type of neutralization. In order to ensure that women as well as men can be actual or potential referents, the use of *Personenbezeichnungen* (masculine 'generics' in particular) can be avoided altogether. Thus, instead of *der Minister, der Präsident, der Geschäftsführende Leiter* (executive director); *Redakteure* (editors), *Arbeiter* (workers), *Beamte* (civil servants), *Verkäufer* (sales people), the following collective nouns are possible alternatives:

(23) das Ministerium, das Präsidium, die Geschäftsführende Leitung; die Redaktion, die Belegschaft, die Beamtenschaft, das Verkaufspersonal
(the Ministry, the Presidency, the management; editorial staff, labour force, civil service, sales personnel)

Of course, these alternatives are not semantically equivalent to the original masculine human nouns:

(24) Der Minister führt die Verhandlung
(The Minister conducts the negotiations)

(25) Das Ministerium führt die Verhandlung
(The Ministry conducts the negotiations)

While (25) avoids any male bias, it also obscures individual responsibilities:

Die sächliche Behördenbezeichnung verhindert den gedanklichen Rückschluß, die leitende Person sei ein Mann. Sie verwischt jedoch die personelle Verantwortlichkeit. (BRD Report 1991: 27)

(The neutral term for public authority prevents the reader from interpreting the executive as being a male. At the same time, however, it obscures personal responsibilities.)

Similar examples are the following:

(26) Den Teilnehmern wird ein Tagesgeld ausbezahlt
(Participants will be paid a daily allowance)

(27) Bei Teilnahme wird ein Tagesgeld ausbezahlt
(Literally: In the event of participation a daily allowance will be paid)

The choice of an abstract noun in (27) may contribute towards semantic inaccuracies: neither agent nor beneficiary of the action are named, which creates more distance and less personal involvement. In many guidelines for the equal treatment of the sexes, which will be discussed in the next section, it is therefore suggested that sex-specification (female visibility) should take priority over neutralization in German.

5 LANGUAGE CHANGE UNDER THE INFLUENCE OF THE WOMEN'S MOVEMENT

5.1 *Variability in the Use of Human Nouns in German*

The Hanover guidelines *Empfehlungen zur Vermeidung von sexistischem Sprachgebrauch in öffentlicher Sprache* provide the following definition of sexist language:

Sprache ist sexistisch, wenn sie Frauen und ihre Leistungen ignoriert; sie ist ebenfalls sexistisch, wenn sie Frauen in Abhängigkeit von oder Unterordnung zu Männern beschreibt, wenn sie Frauen nur in stereotypen Rollen zeigt und/oder sie anspricht und ihnen so über das Stereotyp hinausgehende Interessen und Fähigkeiten nicht zugestanden werden; sie ist sexistisch, wenn sie Frauen immer wieder durch herablassende Ausdrücke demütigt und lächerlich macht. (Hellinger, Kremer, and Schräpel 1989: 1)

(Language is sexist when it ignores women and their achievements; it is also sexist when it describes women as dependent on or subordinate to men, when it depicts and/or addresses women only in stereotypical roles, thus denying them interests and capabilities beyond the stereotype; language is sexist when it humiliates and ridicules women by derogatory expressions.)

Examples (28–30) typify sexist usage illustrating the category of female invisibility, while (31) and (32) illustrate types of asymmetry:

(28) 1974 wurden in West-Berlin 102 000 ausländische Arbeiter beschäftigt, vor allem Türken, Jugoslawen, Griechen und Italiener
(In 1974 102,000 foreign workers (m.) were employed in West Berlin, in particular Turks, Yugoslavs, Greeks, and Italians)

(29) Jeder sollte auf seine Kosten kommen
(Everyone (m.) should enjoy himself)

(30) Staatssekretär Anke Fuchs; Herausgeber: Marion Gräfin Dönhoff
(Parliamentary Secretary (m.) Anke Fuchs; editor (m.) Marion Gräfin Dönhoff)

(31) Präsident Reagan und Frau Gandhi

(32) Sartre und Simone de Beauvoir

Examples (33) and (34) illustrate stereotyping, where the feminine noun typically denotes an occupation of a lower social status than the masculine noun in the same sentence:

(33) Ärzte und Krankenschwestern

(34) der Chef und seine Sekretärin

In German, the influence of the feminist movement is particularly salient in the area of terms for human referents, which is marked by increasing variability, above all in cases of intended mixed-sex reference; consider, for example, the variety of ways of representing 'teachers': *Lehrer, Lehrerinnen und Lehrer, Lehrer/innen, LehrerInnen, Lehrende, Lehrpersonen, Lehrkräfte.* This variability reflects serious changes regarding the generic interpretation of masculine human nouns, which has become increasingly unacceptable over the past two decades. The referential range of masculine human nouns is becoming narrower and there is a growing tendency towards agreement between grammatical gender and sex of referent.

Thus a masculine occupational term such as *der Rechtsanwalt* (lawyer) is increasingly interpreted as referring to males only. This type of agreement traditionally marks most feminine human nouns: while in neutral contexts *der Rechtsanwalt* may still be used to include women, *die Rechtsanwältin* cannot be used generically. However, in current German deviations from this prescription can be observed, usually in contexts which are marked for pro-feminist behaviour and attitudes. For example, in *EMMA*, a German feminist magazine, feminine nouns ('feminine generics') such as *Leserinnen* (readers) or *Demonstrantinnen* (demonstrators) are frequently used to include potential male referents.

Variability is manifested in the use of linguistic alternatives with the same intended meaning, for example *jeder Ausländer* (every foreigner) or *jeder Ausländer und jede Ausländerin*, where both expressions are intended to include potential male and female referents. Such variability may reflect a stable differentiation within the speech community (for instance with regard to socio-economic or situational parameters), but it may also signal the take-off phase of linguistic change which typically includes linguistic alternatives (cf. Aitchison 1981).

Linguistic change manifests itself on the quantitative and/or qualitative level of norms of usage. Thus new terms which may have been used only infrequently will be used more often by more members of the speech community (diffusion phase), while usage of conventional patterns, for example generic masculines, will decrease further. Examples of such new expressions are the indefinite pronouns *frau* and, less frequently used, *mann* and even *mensch* (cf. Pusch 1984: 6 ff.), new derivations in *-in* and new compounds with *-frau* as a second element; cf. the following examples:

(35) Über die neue Abtreibungsregelung muß frau sich genau
    informieren
    (Every woman (everyone) should gather precise information
    about the new abortion regulations)

(36) In Uniform fühlt mann sich ganz anders
    (Wearing a uniform a man (one) feels quite different)

(37) Was macht mensch damit?
    (What does one do with this?)

(38) Flugzeugbauerin; Bundeskanzlerin; Soldatin; Bankerin; Bischö-
    fin;
    (aircraft constructor, Federal Chancellor, soldier, banker, bishop
    ([all nouns are feminine])

(39) Kneipenfrau; Notruffrau; Landsfrau; Fachfrau; Ratsfrau.
    (woman working in a pub; woman on duty for emergency calls;
    countrywoman; female expert; female member of council)

5.2   *Feminist Language Planning*

In this section I shall discuss guidelines for the equal treatment of the sexes in German. First, some general remarks about guidelines will be made, then the major principles of German and English guidelines will be illustrated, and finally I will compare three guidelines relating to German.

My description of guidelines for non-sexist language use is based on the following understanding of language functions. Language as a tool of social interaction mainly serves referential functions, for example in the exchange of information; as a mirror language reflects social hierarchies and mechanisms of identification (social functions); and as a weapon it can be used to discriminate against people, for instance by derogation or stereotyping. Guidelines are directly connected to the social functions of language. They identify areas of conventional language use as sexist and offer alternatives following the principles of equal linguistic treatment of women and men. They reinforce tendencies of linguistic change by means of explicit directions. Thus guidelines are an instrument of language planning which offer solutions for a particular linguistic problem (cf. Frank 1989: 197).

Guidelines for non-sexist language are a reaction to changes in the relationships between the sexes, which have caused overt conflicts on the level of language comprehension and production. Guidelines offer solutions by suggesting non-sexist alternatives to traditional patriarchal language. In most cases guidelines explicitly articulate their political foundation by pointing out that equal treatment of women and men must also be realized on the level of communication.

As an instrument of language planning guidelines symbolize the dissonance between traditional prescriptions, for example the so-called generic *he* in English, and new regulations. MacKay (1980) suggests that a new prescription should be weighed in terms of benefits and costs. His analysis of alternatives for generic *he* (especially singular *they*) is based on the following principle:

a usage should be prescriptively recommended if and only if the benefits of the usage outweigh the costs, where benefits facilitate communication (i.e. the comprehension, learning, and production of the language) and costs make communication more difficult (relative to all other means of expressing the same concept). (MacKay 1980: 352)

However, the restriction to the criterion of facilitating communication does not do justice to the present conflict. Agreement between what is intended and what is actually said cannot be established by simply avoiding referential ambiguities. A modified model of the cost–benefit analysis would have to incorporate the criterion of female visibility, which has always been an essential issue in feminist linguistics.

Finally, guidelines are based on the assumption that a change in behaviour, that is, using more instances of non-sexist language, will be attended by a change in attitude, so that positive attitudes towards non-sexist alternatives will develop (cf. Smith 1973: 97). Conversely, positive attitudes will motivate speakers to use more non-sexist language (cf. Frank 1989: 112).

5.3   *Principles of English and German Guidelines for Non-Sexist Language*

In English guidelines—representative examples are McGraw-Hill (1972) and Unesco (1991)—the principle of neutralization has highest priority. On the one hand, neutralization means the avoidance of false generics, especially generic usages of *man*, as in *primitive man*, *to man a project*, or *congressman*. On the other hand, neutralization can be achieved by avoiding sex-specific terms for female referents, especially derivations ending in *-ess* or *-ette* (*stewardess, usherette*).

Second in the hierarchy of English guidelines is avoidance of stereotyping and third is the principle of symmetry. The reasons for the high ranking of neutralization in English have been discussed repeatedly (see e.g. Hellinger 1989). First, they concern the fact that the English system of human nouns is primarily structured by semantic categories (semantic and social gender) and not—in contrast to German—by grammatical gender. Secondly, they concern the derivation of female human nouns, which is no longer a productive process in English word formation. In addition, most derived female human nouns deviate from the corresponding male terms not only in connotation, but also in denotation; familiar examples are *governor/governess*, *major/majorette*, and *mister/mistress*.

As a subordinate strategy, neutralization includes visibility of (potential) female referents, cf. the example of pronominal splitting as in *a lawyer . . . she or he*. This example also illustrates symmetrical usage, which should be observed whenever specification of sex of referent is required, as in *male and female lawyers; camerawoman/cameraman*. An alternative which takes the generic meaning of *lawyer* for granted is the use of singular *they*: *a lawyer . . . they*.

In German guidelines the principle of visibility has highest priority, followed by the principles of symmetry and avoidance of stereotyping. As was shown above in Section 3, the existence of grammatical gender in German reinforces agreement between grammatical and semantic gender, which is typical for German kinship terms. Also, in contrast to English, processes of the derivation of feminine human nouns are deeply embedded in German word formation, and derivations in *-in* (e.g. *Wählerin*, voter) no longer carry negative connotations. Again, the principle of symmetry must be observed in cases where sex-specification is required and, due to the structural properties of the German nominal system, this is the norm rather than the exception in German.

Avoidance of stereotyping in German may include neutralization, for example by using the plural: instead of the stereotypically masculine generic such as *der Auszubildende* (trainee), or *der Versicherte* (policy holder), gender-neutral plural forms may be used, as in *die Auszubildenden; die Versicherten.* In the singular, of course, the principle of visibility must be followed for nominalized participles; this is achieved by means of *Differentialgenus,* marked, for example, by variation of the article: *der/die Auszubildende; die/der Versicherte.*

5.4   *A Comparison of Three Guidelines for German*

In this section I will first compare the Hanover guidelines (Hellinger, Kremer, and Schräpel 1989) with the report of the committee on legal language which was set up by the German Federal Government (BRD Report 1991); this will be followed by a comparison of the German report with a similar report by the Swiss Federal Government (Swiss Report 1991).

The Hanover Guidelines vs. the Report by the German Bundestag
The guidelines developed by the *Arbeitsgruppe Rechtssprache* (working party on legal language) differ considerably from the Hanover guidelines. Whereas the latter are conceived for public language in general, the Bundestag report deals only with legal language, that is, a variety of German for specific purposes.

In the opinion of the German Federal Government the extensive report can serve 'als Richtschnur für die künftige Rechtssetzung' (as a guideline for future legislation; BRD-Report 1991: 3). The first section of the report summarizes developments in the German federal states as well as in Austria and Switzerland; then the relevant 'Eigenheiten der deutschen Sprache' (peculiarities of the German language — above all the generic interpretation of masculine human nouns) are discussed, as well as the proposals for non-sexist language which were presented to the Bundestag by the parliamentary parties (CDU/CSU, FDP, SPD, and Greens).

The report rejects the view proposed in the Hanover guidelines, which was also shared by the Greens and the SPD, that the demand for equal treatment of the sexes can be derived from the German constitution (Art. 3 Para. 2 of the Basic Law). The report insists that constitutional rights naturally relate to both sexes; therefore even alluding to the constitutional interpretation of legal language is considered quite unnecessary. Thus the generic interpretation of masculine human nouns is in principle maintained, and although the working party concedes that linguistic asymmetries exist, it insists that these are constitutionally irrelevant. Therefore the notion of *sprachliche Un/*

Gleichbehandlung (un/equal linguistic treatment) is rejected, the preferred terminology being *sprachliche Asymmetrien*.

According to the report, private law (Art. 2 of the Basic Law) is the only area of legal language where the use of feminine human nouns is justified as an alternative to generic masculines. Private law applies to specific situations and individuals, and the report concedes that conventional usage does not reflect the changes which have occurred in the use and interpretation of German human nouns.

Of fundamental importance is the working party's differentiation of *Rechtssprache* (legal language) into two categories, the so-called *Amtssprache* (official language) and the *Vorschriftensprache* (legislative language). Administrative communication, judicial decisions, forms of all sorts, etc. are written in the *Amtssprache*, while laws and decrees are formulated in the *Vorschriftensprache*. As far as official language is concerned, the working party supports the principle of visibility of the female as suggested by the Hanover guidelines and elsewhere. Forms, personal documents, educational programmes, examination regulations, etc. are to be revised to include feminine occupational titles and terms of address.

As an alternative to the principle of linguistic visibility of the female (realized above all by the use of feminine nouns ending in -*in*), the report recommends the use of non-sex-specific wording or the avoidance of human nouns altogether; for example *der Minister* can be replaced by *das Ministerium*. In several places, it is argued that the use of *Paarformeln* (nominal splitting), for instance in a headline, need not exclude the use of generic masculine terms in the following text. This is in disagreement with the proposals of the Hanover guidelines.

In contrast to the Hanover guidelines, which apply to all types of formal German, the Bundestag working party excludes the second category of legal language (*Vorschriftensprache*, legislative language) from any changes. Visibility of the female in legal texts is rejected on several grounds. The working party believes that the occurrence of (sex-specific) feminine *and* masculine human nouns in revised or newly formulated texts will cause inconsistencies with the traditional usage (generic use of the masculine) of older texts. Of course, in support of the changes one could use the working party's own argument presented earlier in defence of the generic masculine. It was argued that since the law must conform to the constitution, it will necessarily relate to both sexes (no matter which wording was used). Therefore variation should pose no threat to consistency across legal texts.

The working party's second argument against the principle of visibility is a practical one: it would take too much time, money, and effort to revise the entire legal code. Again, the practical problems would appear much less threatening if one followed the Swiss Report's suggestion (see below): changes would have to be made gradually, that is, only new laws and revisions of existing texts would have to adhere to the principle of visibility.

Thirdly, the working party warns that legal texts with 'wenig schönen Formulierungen' (rather ugly formulations) would be created, for instance by too much nominal or pronominal splitting. In an analysis of arguments against non-sexist language, Hellinger and Schräpel (1983) have pointed out that the criterion of stylistic elegance and economy must be weighed against the socially and psychologically more salient criterion of visibility (cf. also Blaubergs 1980).

In summary, the report defends the status quo. For official language (*Amtssprache*) it simply acknowledges tendencies of language change which are considered normal usage in more and more domains of public communication. For legislative language (*Vorschriftensprache*) the report defends the prescription of the generic masculine and the underlying ideology of 'male as norm'.

The BRD Report vs. the Swiss Guidelines

In contrast to the German working party the Swiss *Arbeitsgruppe* explicitly refers to the Council of Europe guidelines of 1990, in which the member states are urged 'in Rechtstexten und im Unterricht eine nicht-sexistische Sprache zu verwenden' (to use non-sexist language in legal texts and in teaching; Swiss Report 1991: 64).

Both reports show a number of similarities. The Swiss Report was prepared by an interdepartmental working group, which was set up by the Swiss Federal Government in 1988 and which also deals with legal language. As in the German report, general linguistic problems of the German language are discussed first (gender, possibilities of sex-specification, neutralization of sex of referent), then the situation in the Swiss cantons and developments abroad are described, and suggestions for the realization of equal linguistic treatment are made. However, the Swiss Report is also concerned with French and Italian, the other two official languages of Switzerland.

An essential difference between the two reports concerns attitudes towards the notion of equal linguistic treatment. In contrast to the German working party the Swiss group explicitly adopts the principle of *sprachliche Gleichbehandlung* as its goal and — in agreement with the Hanover guidelines — visibility and symmetry receive highest priority. Thus a fundamental argument of feminist language criticism is endorsed, according to which the enforcement of neutralization in

German (generally the strategy preferred by the BRD Report) conflicts with the principle of women's visibility.

Also in contrast to the Bundestag recommendations, the Swiss guidelines explicitly refer to Art. 4 Para. 2 of the Swiss *Bundesverfassung* (Constitution), arguing that the avoidance of sexist usage in legal language is part of the overall societal goal of equal treatment of the sexes.

Another important difference concerns the scope of the recommendations. Although the Swiss Report also differentiates between two types of legal language, namely *Verwaltungssprache* (administrative language) and *Gesetzessprache* (legislative language), the guidelines are intended to apply to both varieties of Swiss standard German. The Swiss working party claims that the implementation of non-sexist language is 'eine sprachpolitische Maßnahme ersten Ranges, die weit bedeutender ist als beispielsweise die seit vielen Jahren umstrittene Orthographiereform' (a major priority in the formulation of language policy, far more important than, say, the issue of spelling reform, which has been debated for many years; 1991: 11).

More weight is attributed to changes in the everyday usage of Swiss standard German, 'in dem Maskulina in generischer Bedeutung zusehends durch Paarformeln ersetzt und somit immer geschlechtsspezifischer verstanden werden' (where more and more generic masculines are being replaced by splitting and are thus increasingly being interpreted as sex-specific; ibid. 40). These changes provide an important orientation not only for official and administrative usage but also for legislative language.

Finally, apart from the criterion of precise reference (which receives highest priority in the German report), the Swiss Report emphasizes the 'Notwendigkeit der symbolischen Identifikation von Rolle und Geschlecht von Frauen in allen Bereichen des öffentlichen Lebens' (necessity of a symbolic identification of women's social role and sex in all domains of public life; ibid. 19).

The coexistence of expressions containing generic masculines in existing decrees and sex-specific masculines in newly formulated ones is not considered to pose a threat to legal consistency. Nor do aesthetic arguments influence the evaluation of non-sexist alternatives. The Swiss Report recommends a 'creative solution' (*kreative Lösung*), which includes various morphological, stylistic, and orthographic possibilities for the realization of the principles of equal treatment. Abbreviated splitting (*Sparformeln*) such as *Antragsteller/innen* (applicants) as well as the capital I ('das große I') as in *AntragstellerInnen* are quite acceptable, although *Sparformeln* should not be used in legislative language.

Even the implementation of the guidelines is not considered as an insurmountable problem. Changes in the official and administrative language were scheduled to be made by 1994, while for legislative language gradual changes were recommended, beginning with the appropriate formulation of new laws. Of course, differences in the degree of awareness of linguistic sexism in Swiss French and Italian must be taken into account. In these languages the main concern is still with the derivation and diffusion of suitable feminine human nouns, especially occupational terms.

Summary
The present status of guidelines for German must be characterized as heterogeneous and variable. Acceptance of the recommendations varies between two extremes (see Kloss 1968), the level of discouraged usage where non-sexist alternatives are rejected (as is the case with the German Bundestag report's position on legislative language) and the level of sole official usage where non-sexist usage has become the norm. So far the latter development can only be observed in very few cases; for example, the German term of address for a female adult is *Frau* while *Fräulein* is no longer acceptable in official usage. The majority of recommendations currently have the intermediate status of tolerated or encouraged usage. Increasingly, the linguistic visibility of women and usage which conforms to the principles of symmetry and avoidance of stereotyping are not only tolerated but receive active support. However, this leaves open the question of what impact these changes in official (and predominantly written) language have on more informal domains of German.

6 CONCLUSIONS

The following concluding remarks can be made. In the section on female and male verbal behaviour in different types of discourse, it was shown that categories of the turn-taking system, which in studies of English discourse had been linked to the establishment of conversational dominance, are also relevant in the analysis of German discourse. At the same time it was argued that sex and social class membership are not decisive in determining a speaker's conversational status; the individual's social role in the interaction also contributes to conversational dominance or subordination. Furthermore, assumptions about conversational dominance cannot be based on frequencies alone, since the same pattern of behaviour may have different functions in different contexts. Thus, while an interruption will often signal conversational dominance, it may also express personal involve-ment. Finally, it was suggested that women play an active role in constructing their own subordinate position in many conversations.

Section 3 focused on categories of gender, which account for the most obvious differences in the area of human nouns in German and English. In contrast to English, grammatical gender is well established in German. In fact, recent changes in speaking about women and men in German, that is, using more grammatical feminines for female reference and fewer grammatical masculines for neutral reference, are contributing towards more agreement between grammatical gender and semantic gender or sex of referent. In the light of this trend, we can speculate that German masculine human nouns will lose more of their generic or non-sex-specific force in the future.

Similarly, in the area of word formation, there are substantial differ-ences between German and English. While the specification of sex of referent by morphological means plays only a marginal role in current English, German has the feminine derivational suffix -*in*, which is perfectly suited to express female reference: it is very productive, and, in contrast to English -*ess* or -*ette*, does not carry negative conno-tations.

In the last section, the tendencies towards change in German were described from the perspective of language policy, that is, in terms of female visibility and symmetry as the major strategies in German to avoid sexist language. There is still resistance to some of the changes as recommended in many guidelines, in particular to replacements of the so-called generic masculine, and the changes are only beginning to be made in the new federal states. However, the use of feminine terms for female reference is no longer an emotional issue, and it is by no means certain that prescriptions such as those made by the German Bundestag for legal language will actually save the generic masculine.

Further Reading

Braun (1991)
Günthner and Kotthoff (1991)
Häberlin, Schmid, and Wyss (1991)
Hellinger (1990)
Pusch (1990)
Trömel-Plötz (1984*a*)

References

AITCHISON, J. (1981), *Language Change: Progress or Decay?* (Bungay, Fontana).

AMMON, U. (1973), *Dialekt und Einheitssprache in ihrer sozialen Verflechtung* (Weinheim: Beltz).

ANDRESEN, H. (ed.) (1979), *Sprache und Geschlecht* [= *Osnabrücker Beiträge zur Sprachtheorie*] (Universität Osnabrück).

BARON, D. (1986), *Grammar and Gender* (New Haven, Conn., London: Yale University Press).

BLAUBERGS, M. (1980), 'An Analysis of Classic Arguments against Changing Sexist Language', in Kramarae (1980), 135–47.

BRAUN, F. (1991), *Mehr Frauen in die Sprache. Leitfaden zur geschlechtergerechten Formulierung* (Kiel: Die Frauenministerin des Landes Schleswig-Holstein).

BRD REPORT (1991), *Maskuline und feminine Personenbezeichnungen in der Rechtssprache. Bericht der Arbeitsgruppe Rechtssprache vom 17. Januar 1990* (Deutscher Bundestag. Drucksache 12/1041. 1991).

CAMERON, D., MCALINDEN, F., and O'LEARY, K. (1988), 'Lakoff in Context: The Social and Linguistic Functions of Tag Questions', in Coates and Cameron (1988), 74–93.

COATES, J., and CAMERON, D. (eds.) (1988), *Women in their Speech Communities: New Perspectives on Language and Sex* (London: Longman).

CROSBY, F., and NYQUIST, L. (1977), 'The Female Register. An Empirical Study of Lakoff's Hypothesis', *Language in Society*, 6: 313–22.

EISENBERG, P. (1986), *Grundriß der deutschen Grammatik* (Stuttgart: Metzler).

FISHMAN, J. A., FERGUSON, C. A., and DAS GUPTA, J. (1968) (eds.), *Language Problems of Developing Nations* (New York: Wiley).

FLADER, D., and WODAK-LEODOLTER, R. (eds.), (1979), *Therapeutische Kommunikation. Ansätze zur Erforschung der Sprache im psychoanalytischen Prozeß* (Königstein: Scriptor).

FLEISCHER, W. (1982), *Wortbildung der deutschen Gegenwartssprache* (Tübingen: Niemeyer).

FRANK, F. W. (1989), 'Language Planning, Language Reform, and Language Change: A Review of Guidelines for Nonsexist Usage', in Frank and Treichler (1989), 105–33.

—— and TREICHLER, P. A. (eds.) (1989), *Language, Gender, and Professional Writing: Theoretical Approaches and Guidelines for Nonsexist Usage* (New York: MLA).

FRANK-CYRUS, K. M. (1991), *Subjektive Varietätenwahl in pfälzischen Dorfgemeinschaften* (Frankfurt: Haag & Herchen).

GRÄSSEL, U. (1991), *Sprachverhalten und Geschlecht. Eine empirische Studie zu geschlechtsspezifischem Sprachverhalten in Fernsehdiskussionen* (Pfaffenweiler: Centaurus).

GUENTHERODT, I., HELLINGER, M., PUSCH, L. F., and TRÖMEL-PLÖTZ, S. (1980), 'Richtlinien zur Vermeidung sexistischen Sprachgebrauchs', *Linguistische Berichte*, 69: 15–21.

GÜNTHNER, S., and KOTTHOFF, H. (eds.) (1991), *Von fremden Stimmen. Weibliches und männliches Sprechen im Kulturvergleich* (Frankfurt: Suhrkamp).

HÄBERLIN, S., SCHMID, R., and WYSS, E. L. (eds.) (1991), *Übung macht die Meisterin. Richtlinien für einen nicht-sexistischen Sprachgebrauch* (Zurich: Netzwerk schreibender Frauen).

HELLINGER, M. (1989), 'Revising the Patriarchal Paradigm: Language Change and Feminist Politics', in Wodak (1989), 273–88.

—— (1990), *Kontrastive feministische Linguistik. Mechanismen sprachlicher Diskriminierung im Englischen und Deutschen* (Ismaning: Hueber).

—— and SCHRÄPEL, B. (1983), 'Über die sprachliche Gleichbehandlung von Frauen und Männern', *Jahrbuch für Internationale Germanistik*, 15: 40–69.

—— KREMER, M., and SCHRÄPEL, B. (1985), *Empfehlungen zur Vermeidung von sexistischem Sprachgebrauch in öffentlicher Sprache* (Universität Hannover).

HORVATH, B. (1985), *Variation in Australian English: The Sociolects of Sydney* (Cambridge: Cambridge University Press).

HUDDLESTON, R. (1984), *Introduction to the Grammar of English* (Cambridge: Cambridge University Press).

KEY, M. (1975), *Male/Female Language* (Metuchen, NJ: Scarecrow Press).

KLOSS, H. (1968), 'Notes Concerning a Language–Nation typology' in Fishman *et al.* (1968), 69–85.

KOTTHOFF, H. (1984), 'Gewinnen oder verlieren? Beobachtungen zum Sprachverhalten von Frauen und Männern in argumentativen Dialogen an der Universität', in Trömel-Plötz (1984a), 90–113.

KRAMARAE, C. (ed.) (1980), *The Voices and Words of Women and Men* (Oxford: Pergamon).

—— and TREICHLER, P. A. (1985), *A Feminist Dictionary* (Boston: Pandora Press).

LABOV, W. (1966), *The Social Stratification of English in New York City* (Washington, DC: Center for Applied Linguistics).

LAKOFF, R. (1975), *Language and Woman's Place* (New York: Harper Colophon Books).

LAUPER, H., and LOTZ, C. (1984), 'Also, wir müssen jetzt aufpassen, liebe Frau Struck. Untersuchungen einer Fernsehdiskussion zwischen Karin Struck und Hans Apel', in Trömel-Plötz (1984a), 246–57.

LEVINSON, S. (1983), *Pragmatics* (Cambridge: Cambridge University Press).

LYONS, J. (1977), *Semantics*, ii (Cambridge: Cambridge University Press).

McGRAW-HILL (1972), *Guidelines for the Equal Treatment of the Sexes in McGraw-Hill Book Company Publications* (Hightstown, NJ: McGraw-Hill).

MACKAY, D. G. (1980), 'On the Goals, Principles, and Procedures for Prescriptive Grammar: Singular They', *Language in Society*, 9: 349–67.

MACMILLAN, J., *et al.* (1977), 'Women's Language: Uncertainty or Interpersonal Sensitivity and Emotionality', *Sex Roles*, 3: 545–59.

MATTHEIER, K. J. (1980), *Pragmatik und Soziologie der Dialekte* (Heidelberg: Quelle & Meier).

MILLER, C., and SWIFT, K. (1981), *The Handbook of Non-Sexist Writing for Writers, Editors and Speakers* (London: Women's Press).

MILROY, L. (1987), *Language and Social Networks*, 2nd edn. (Oxford: Blackwell).

PUSCH, L. F. (1984), *Das Deutsche als Männersprache. Aufsätze und Glossen zur feministischen Linguistik* (Frankfurt: Suhrkamp).

—— (1990), *Alle Menschen werden Schwestern. Feministische Sprachkritik* (Frankfurt: Suhrkamp).

QUIRK, R., GREENBAUM, S., LEECH, G., and SVARTVIK, J. (1985), *A Comprehensive Grammar of the English Language* (London: Longman).

SCHMIDT, C. (1988), *'Typisch weiblich—typisch männlich'. Geschlechtstypisches Kommunikationsverhalten in studentischen Kleingruppen* (Tübingen: Niemeyer).

SCHOENTHAL, G. (1985), 'Sprache und Geschlecht', *Deutsche Sprache*, 2: 143–85.

SHUY, R. W., and FASOLD, R. W. (eds.) (1973), *Language Attitudes: Current Trends and Prospects* (Washington, DC: Georgetown University Press).

SMITH, D. M. (1973), 'Language, Speech and Ideology: A Conceptual Framework', in Shuy and Fasold (1973), 97–112.

STRANG, B. M. H. (1970), *A History of English* (London: Methuen).

SWISS REPORT (1991), *Sprachliche Gleichbehandlung von Frau und Mann in der Gesetzes- und Verwaltungssprache. Bericht einer interdepartementalen Arbeitsgruppe der Bundesverwaltung* (Schweizerische Bundeskanzlei).

THIMM, C. (1990), *Dominanz und Sprache. Strategisches Handeln im Alltag* (Wiesbaden: Deutscher Universitätsverlag).

THORNE, B., and HENLEY, N. (1975) (eds.), *Language and Sex: Difference and Dominance* (Rowley, Mass.: Newbury House).

TRÖMEL-PLÖTZ, S. (1978), 'Linguistik und Frauensprache', *Linguistische Berichte*, 57: 49–68.

—— (1982), *Frauensprache. Sprache der Veränderung* (Frankfurt: Fischer).

—— (1984a), *Gewalt durch Sprache. Die Vergewaltigung von Frauen in Gesprächen* (Frankfurt: Fischer).

—— (1984b), 'Gewalt durch Sprache', in Trömel-Plötz (1984a), 50–67.

—— (1984c), 'Die Konstruktion konversationeller Unterschiede in der Sprache von Frauen und Männern', in Trömel-Plötz (1984a), 288–319.

—— (1984d), 'Weiblicher Stil—männlicher Stil', in Trömel-Plötz (1984a), 354–94.

TRUDGILL, P. (1974), *The Social Differentiation of English in Norwich* (Cambridge: Cambridge University Press).

UNESCO (1991), *Guidelines on Non-Sexist Language*.

WERNER, F. (1983), *Gesprächsverhalten von Frauen und Männern* (Frankfurt: Lang).

WIENOLD, G. (1967), *Genus und Semantik* (Meisenheim: Anton Hain).

WILMANNS, W. (1967), *Deutsche Grammatik*, 2nd edn., pt. 2: Wortbildung. (Berlin: de Gruyter).

WODAK, R. (1981), *Das Wort in der Gruppe. Linguistische Studien zur therapeutischen Kommunikation* (Vienna: Akademie der Wissenschaften).

WODAK, R. (ed.) (1989), *Language, Power and Ideology* (Amsterdam: Benjamins).

WODAK, R., and ANDRASCHKO, E. (1992), 'Macht durch Mütterlichkeit: Eine qualitative soziolinguistische Untersuchung', paper presented at the 14th DGfS conference, Bremen.

WODAK-LEODOLTER, R. (1977), 'Interaktion in einer therapeutischen Gruppe', *Wiener Linguistische Gazette*, 24: 51–60.

—— (1979), 'Probleme der Unterschichttherapie: Aspekte einer empirischen Untersuchung therapeutischer Gruppen', in Flader and Wodak-Leodolter (1979), 186–207.

ZIMMERMAN, D. H., and WEST, C. (1975), 'Sex Roles, Interruptions and Silences in Conversations', in Thorne and Henley (1975), 105–29.

13 *Jugendsprachen*: Speech Styles of Youth Subcultures

PETER SCHLOBINSKI

I   INTRODUCTION

If one were to believe the statements of linguists thirty years ago, then youth speech is no more than the 'Jargon einer bestimmten Sondergruppe, der den größeren und wertvolleren Teil der Jugend erniedrigt und beleidigt' (Küpper 1961: 188) (slang of a particular subgroup, which is degrading and insulting to the larger and more reputable section of the young generation). Hartmut Engelmann was no less flattering in a radio programme broadcast by the Hessischer Rundfunk about 'young people and their private code': 'Das Klotzige, Protzige und Brutale ist wohl eines der wesentlichen Charakteristika des Jargons' (Engelmann 1964: 6) (It is overblown, showy, crude: that is one of the main characteristics of this slang). Those who take this point of view maintain that young people's speech is directly responsible for the 'decline of the German language', and it has been argued that for this reason they should be required to use standard German in school.

I wonder what these scholars would say now, if they were to read the following version of Kriemhild's dream (from the legend of the *Nibelungen*), which differs quite considerably from the still very widely distributed version:[1]

Eines Nachts kann Hildchen nicht ordentlich pennen, weil Vollmond ist. Vielleicht hat sie sich aber auch nur zu viel Wildschweinbraten reingepfiffen. Jedenfalls hat sie einen Alptraum, der glatt bei Freudi geklaut sein könnte: sie hat einen Falken hochgepäppelt, auf den sie total spitz ist, läßt ihn eines Tages in Richtung Himmel starten und sieht, wie zwei Adler auf ihn losdüsen und ihn allemachen. (Claus and Kutschera 1986: 23)

The orthodox version reads as follows:

---

[1] By its very nature, *Jugendsprache* is generally untranslatable, and a version in 'standard English' would serve little purpose here. The *Jugendsprache* examples will therefore not be translated, but their significance should be clear from the context.

Einmal trat Mutter Ute früh am Morgen in ihr Gemach und fand sie verstört und traurig. Sie forschte nach der Ursache ihrer Betrübnis. Da erzählte ihr die Maid, es habe ihr geträumt, daß sie einen edlen Falken aufgezogen, der ihr gar lieb geworden sei, als er aber aufgeflogen, hätten ihn zwei tückische Adler, aus einer Felsenkluft hervorbrechend, vor ihren Augen erwürgt. (From the *Sagenkreis der Nibelungen* in Wägner 1878)

Not only legends and fairy-tales have appeared in *Jugendsprache* versions, even the Bible has been translated. The following extract is a 'modern' version of Genesis 1: 3: 'Die Düsternis mißfällt dem GROSSEN BOSS. 'Man sieht ja nicht die Hand vor Augen!' räsoniert er. 'Licht! Aber ein bißchen dalli!' Prompt wird es hell' (Denger 1984: 17). Texts such as this are typical of what is commonly referred to as *Jugendsprache*, and they are widely read not only by young people but by adults too. The popularity of such texts is demonstrated by the numerous publications on the topic in recent years, the most famous example being Peter Müller-Thurau's *Laß uns mal ne Schnecke angraben* (1983), which rapidly shot up the best-seller lists and even merited an article in the Spiegel magazine. Müller-Thurau classifies words and phrases by topic areas, such as 'relationships' or 'expressions of consternation' without, however, having any basis whatsoever for these classifications, with the result 'daß er einen entsprechenden Wesenszug von Jugendsprache eher sucht als findet' (Januschek 1989: 137) (that he seeks rather than finds a particular *Jugendsprache* feature). The material from which books like Müller-Thurau's are constructed derives from overheard examples (but where, when, and under what conditions are they overheard?), and from fictitious texts, like the following piece, which appeared in the *Frankfurter Allgemeine Zeitung* of 4 January 1979, and which is often used by teachers to illustrate *Jugendsprache* (see e.g. Grimm and Sontheimer 1980):

Disco-Deutsch

Als ich neulich mit Peter in die City drückte, macht der mich unheimlich an aufs Tilbury. Na, schon bohren wir dahin, obwohl ich eigentlich aufs Lollipop stand. Ich Chaot hatte keine Matte mit, weil ich meinen Kaftan vergessen hatte, und sagte zu Peter, er solle mal ausklinken. In dem Schuppen zogen ein paar People schon eine heiße Show ab. Wir machten eine kurze Fleischbeschauung, und Peter machte sich sofort daran, eine riesige Tussi anzugraben. Die war echt einsam, aber ich hatte einfach keinen Schlag bei ihr. Peter schafft sich da also mächtig rein und wollte wahrscheinlich 'nen kleinen Wuschermann machen, blickt aber nicht durch, daß die Tussy einen Typ hat. Der hing zu dem Zeitpunkt allerdings schon völlig durch. Vielleicht zog er auch, jedenfalls konnte die Tussi darauf nicht. Aber als Peter so

---

[2] See Müller-Thurau (1983: 7) for a discussion of where writers on *Jugendsprache* find their source material.

ordentlich aufs Blech haut und mächtig mit seinem Busch wedelt, spannt der beknackte Gent seinen Glimmer, was läuft, und sagt Peter einen Satz heiße Ohren an. 'Ich glaub', mich streift ein Bus', tönt Peter daraufhin, 'paß lieber auf, daß du hier keine Taucherbrille erbst.' Na, ich hatte keinerlei Bock auf Terror, vor allem, weil der halbe Laden inzwischen zu war, weil jeder schon ein paar Wutschis und Lämmis drin hatte, und ich sagte zu Peter: 'Laß uns die Fliege machen.' Das konnte Peter nicht recht ab, logo, die Schnecke hat ihn voll angeturnt. Also hob ich leicht angesäuert allein ab und rief Heimat ab, denn draußen war's mächtig schattig, obwohl der Planet den ganzen Tag gestochen hatte wie irr.

Anyone who actually has anything to do with young people can see that this is not real *Jugendsprache* but a stylized form of it. None the less, parents and even some teachers are haunted by the succinct image (quoted at the beginning of this chapter) formulated by Heinz Küpper, the famous linguist and author of the *Wörterbuch der deutschen Umgangssprache* (Dictionary of Colloquial German). As one teacher put it to me: 'Jugendsprachlicher Sprachgebrauch ist OK, aber nur solange er sauber bleibt, was ich an der Jugendsprache hasse, das ist, wenn sie abgleitet in die Fäkaliensprache' (Young people's use of slang is OK, but only as long as it stays clean: what I detest about *Jugendsprache* is when it degenerates into filthy language). These negative images are reinforced when dictionaries like Schönfeld (1986) draw a parallel between *Jugendsprache* and prison slang 'die damit zu tun [hat], daß beide Gruppen am Rande der "normalen" Gesellschaft [stehen] . . . und ihre Randstellung auch in Worten, in einem anderen Sprachgebrauch, zum Ausdruck [bringen]' (due to the fact that both groups are on the margin of 'normal' society and that they express their marginal position in words and by a different use of language) (Schönfeld 1986: 5).

Distinguishing 'young people' from the rest of society is a popular way of demonstrating the linguistic mannerisms that characterize *Jugendsprache*. Consider, for example, the following extract from a sketch by the popular journalist and broadcaster Elke Heydenreich in the programme 'Espresso', broadcast by NDR II on 12 December 1983:

Oma bekam dieser Tage wieder Post von Harry: 'Liebe Oma', schrieb er, 'ich bin schottermäßig nicht gut drauf dies Jahr, da will ich gar nicht lange rumsülzen, und wenn jetzt Christmas in die Gänge kommt, Oma, is von mir aus gesehen, rein geschenkmäßig null Erwartung angezeigt. Hoffe, das säuert dich nicht an, Weihnachten is für mich sowieso out — totalo fatalo, Himbeerpudding für die Bürgerfraktion. Wünsche dir trotzdem frohes, ähm, dingns, Fest: Liebern Freak als Krieg, Freiheit für Nicaragua, so long, Harry!'
Nach einigem Grübeln schrieb Oma zurück: 'Mein lieber Harald! Vielen

Dank für Deine schöne Weihnachtskarte mit dem Gewehr drauf. Da ich weiß, wie beschäftigt Deine Eltern immer sind, möchte ich Dich einladen, die Feiertage doch bei mir in Wuppertal zu verbringen. Kannst auf dem Sofa im Wohnzimmer schlafen. Es gibt Gänsebraten mit Rotkohl, ich backe natürlich einen Stollen, und Onkel Otto bringt mir einen Baum, den Tante Grete dann schön schmücken will. Du brauchst kein Geschenk mitzubringen, lieber Harald, nur wenn Du vielleicht die grünen Haare ... aber ich will Dir nicht reinreden, lieber Junge, Deine Oma freut sich immer, wenn Du kommst, auch mit grünen Haaren. Bitte antworte mir bald, immer Deine Omi.'

2   THE EMPIRICAL STUDY OF *JUGENDSPRACHEN*

In contrast to the earlier, rather hostile and prejudiced views on *Jugendsprachen*, most linguists now see them in a more neutral light as varieties of the German language with particular linguistic characteristics. Helmut Henne, who conducted the most detailed investigation to date into *Jugendsprache* (see Henne 1986), defines the phenomenon as follows:

Jugendsprache bezeichnet spezifische Sprech- und Schreibweisen, mit denen Jugendliche u.a. ihre Sprachprofilierung und damit ein Stück Identitätsfindung betreiben. Eine der Möglichkeiten dieser Profilierung besteht nun darin, einen eigenen Jugendton in der jugendlichen Gruppe zu pflegen. (Henne 1981: 373)

(The term *Jugendsprache* applies to specific manners of speaking and writing, through which, amongst other things, young people define themselves linguistically and thereby construct part of their own identity. One way of doing this is to cultivate a specific tone within the group.)

It is significant that Henne distinguishes between speaking and writing, given that *Jugendsprache* is primarily found in the spoken language. Its function, according to Henne, lies in constructing identity and group dynamics; the structural characteristics are covered by the term *Jugendton* (youth tone). According to Henne (1986), this youth tone displays the following features, which together with particles, English words, and archaic expressions are also mentioned by other authors (e.g. Beneke 1982, Heinemann 1989) as prominent youth-specific features:

1. greetings, forms of address, and names;
2. catchy nicknames and aphorisms;
3. smart expressions and stereotypical catch-phrases;
4. metaphorical, mostly exaggerated manners of speaking;

5. responses with words of delight and condemnation;
6. prosodic language games, shortened sounds, and weakened sounds, as well as grapho-stylistic techniques;
7. word formation: new words, new meanings, new formations;
8. formation of prefixes and suffixes.

Before giving some examples of these individual features, we should consider how Henne's study was actually conducted, as it is one of the few to date with a well-founded empirical basis, and it is the best known and most comprehensive study on this topic in the German-speaking area.

Henne's investigation into *Jugendsprache* is based mainly on a questionnaire which he distributed in 1982 in various schools in the Braunschweig, Kassel, and Mannheim areas of Germany. The questionnaire was compiled in such a way that it could be completed within a school lesson in the presence of the teacher and interviewer; the questions were mainly open-ended. As an introduction to the questionnaire, the transcript of a fictitious conversation at a disco (cf. above) was attached, in order to make the students aware of particular characteristics of *Jugendsprache*. As evaluation criteria, Henne classified the students according to their age and sex, the class or grade they were in at school, and the region they came from. The investigation was conducted at various types of secondary school (*Gymnasien, Real-schulen, Hauptschulen*, and *Berufsschulen*). The questions and tasks were arranged in the following way:

1. questions about favourite reading matter, films, and plays;
2. questions on specific linguistic behaviour (with the emphasis on sound-words and their usage, sayings, forms of address, and greetings);
3. questions on expressions peculiar to schoolchildren (for instance synonyms for 'teacher' — *Pauker*, 'headteacher' — *Boß*);
4. the students were asked to explain or define given *Jugendsprache* expressions;
5. the students were asked to give an assessment of their own language.

Henne collected 534 questionnaires in all. The language data that he gathered in this way primarily relates to the written language; the so-called spoken language, which would have been considerably more relevant for the investigation, could only be collected indirectly (if at all) in this way. By asking a question like: 'What do you say if you're cross?', it is not possible to reproduce a real situation, in which a speaker reacts spontaneously. Unconscious formulation in a very specific situation cannot be compared with the conscious process of writing down words or sentences. In Henne's results there are no particles or adjectival intensifiers (such as *ey, total, voll, echt*): when they consciously write down their answers, the students omit these features, although they have an important communicative function. By investigating only the written language, therefore, Henne can only illustrate a subsection of *Jugendsprache*. Finally, Henne looked at patterns of language knowledge. The interpretation of data acquired in this way is also problematic, as it is taken out of context. *Jugendsprache* as a spoken language is dependent on the situation and the context, and should therefore be analysed pragmatically. In other words, the material under investigation can only be evaluated meaningfully if the context is known (cf. Januschek 1989: 130 ff.; for a further critical review of Henne's study see Brandmeier and Wüller 1989).

3   PROBLEMS IN DEFINING *JUGENDSPRACHE*

So, is what is classified as *Jugendsprache* really *Jugendsprache*? Are the characteristics of *Jugendsprache*, even if there is little or no empirical evidence for them, none the less specific to 'young people and their language', as suggested by the title of Henne's investigation? Let us check this against two features that are considered typical of *Jugendsprache*: English words and 'sound words'.

3.1   *English Words*

It is well known that English loan forms are very common in German. By and large, they can be integrated without much difficulty into German grammar (see Glück and Sauer, this volume). For example, verbs are formed by simply adding the infinitive suffix to English verbs or even nouns; and these verbs can easily be expanded through the use of affixes:

> ab +   power +   -n
> *Das powert voll ab!* (of a rock band): The music's great!
> rein +   move +   -n (to go in, enter)
> los +   cut(t) +   -en (to cut off, remove)

Adjectives ending in *-ig* can be derived from the corresponding English adjectives ending in *-y* or from nouns: for example *speedig* (fast, zappy) or *popig* (expresses approval of something new and fashionable, such as item of clothing). When used attributively, the Germanized adjectives are inflected in the normal way (i.e. in agreement with the noun): *eine coole Sache*. The formation of participles is equally straightforward: *'n total abgefuckter Typ* (a really messed-up bloke/guy).

As far as plural formation is concerned, the German paradigm is usually chosen (e.g. *Punk-er*), although the English plural *-s* is also used in some cases (e.g. *Millionenseller-s*, but also *Punk-s*). The choice of gender is highly variable and depends on grammatical and semantic factors. For example, the distinction between *die Power* and *der Power* is based on the fact that in the feminine form *Power* is used in the sense of strength and power (*der hat echt die Power*), whereas the masculine form conveys the meaning 'sound' (*er brachte den vollen Power*).

Young people readily adopt English words and phrases because of the strong influence that the mainly anglophone music business has on the German market, which can be seen in the names of musical instruments: for instance, *Drums*, *Hi-Hat*, *Percussion*, and *Woodwinds*, all of which are, of course, the prerequisites for *good vibrations*. A rapid survey of the English words in teenage magazines like *Bravo* or *Mädchen* reveals that most of them come from the fields of entertainment, music, fashion, leisure, and sport: for example, *Touch*, *Leggins*, *Show*, *Fashion*, *Styling*, *Sound*, *Freak*, *Feeling*, *Fun*, *Body*, and *Power*. These words also form the basis of compounds, such as *Shirt-Styling*, *Seventies-Look*, or (as a combination of an English defining word and a French base) *Styling-Accessoire*.

English words are also very important in advertising. Take the following advertisement for a Swiss watch, for example (taken from *Bravo* 51, 13): 'Just Imagine. . . Swatch. . . there are no limits.' The key elements here are that English words and young people are fashionable. A futuristic scene is depicted: in the foreground there is a hammock, in which a man wearing swimming trunks and a sun hat is stretched out comfortably; in the background, a lunar landscape and the blue planet Earth. A wrist-watch covered with colourful galactic graffiti is draped across the foreground of the picture, and above this is the headline: 'Just Imagine'. This is immediately understood by the *Bravo* reader:[3] who doesn't know John Lennon? On the back page of the same edition there is an advertisement for *West* fashions with the headline: 'My papa was a rolling stone.' These English set-phrases are symbols of a culture which is offered to the young *Bravo* reader as the norm. The question is, do these readers absorb individual symbols into their language, and if so, are they then speaking *Jugend-*

[3] With 1.7 million copies per week, *Bravo* has the highest circulation of all German youth magazines; its target readership is 14- to 18-year-olds.

sprache, or are they reproducing the language of teenage magazines or advertising?

Rock music lyrics (not only those of British or American bands) also play a particularly important part in the growing use of English words. For example, English words are firmly established in the German lyrics of the popular rock-singer and 'professional teenager' Udo Lindenberg (cf. Meik 1984): for example *Zurückhaltefeeling* (from the song 'Fliesenlied'), *Liebe ist wie'n Kartenhaus, wenn's together bricht, knockt mich das aus* (from 'I love me selber').

3.2 'Sound Words'

There are two types of sound words, sometimes referred to in German as *Päng-Sprache* (Künnemann 1976): (1) sound-imitating (onomatopoeic) words and (2) 'root-words' (Schlobinski and Blank 1990). Onomatopoeic words have various functions: they are used, for example, to simulate an action or to express an experience or emotion:

grrr   expressive function: aggression
psch   imperative function: 'Be quiet!'
puh   affirmative function: 'Wow, that was difficult!'

Root-words are composed of the stem of certain verbs:

verb	**minus infinitive suffix**
hechel (pant, gasp)	
stöhn (groan)	
würg (choke)	*-(e)n*
ächz (moan, groan)	
seufz (sigh)	

The function of root-words is to express specific actions and to comment on them:

ächz   (1) effort; (2) rejection; and (3) relief
würg   (1) dislike, revulsion; and (2) feeling ill
hechel   effort and exhaustion

Like modal particles, sound words often have the function of strengthening the force of utterances, and they too have found their way from comics into everyday speech:

Ähnlich wie im Falle der Redensarten und metaphorischen Wendungen handelt es sich dabei oft um Sprachgags von ursprünglich singulärem Charakter, die allmählich zum selbstverständlichen Bestandteil konventionalisierter Comic-Sprache wurden und teilweise über den Jargon Jugendlicher in die Umgangssprache Eingang fanden. (Dolle-Weinkauf 1990: 70)

(As with fixed expressions and metaphorical images, sound words often started as one-off linguistic jokes, which gradually became essential components of conventional comic language and then found their way via the slang of young people into everyday speech.)

It is interesting to note here that these root-words are an invention of writers working on the comics *Mickey Mouse* and *Mad*, who in the 1950s were faced with the problem of translating English sound words into German. For example, Erika Fuchs created a special 'Donald Duck style' (cf. Dolle-Weinkauf 1990), and Herbert Feuerstein, former editor-in-chief of *Mad*, claimed recently in a talk-show that he had considerably enriched the German language by introducing these root-words. Sound words are by no means to be found in every comic; however, they do appear constantly in the comic *Clever & Smart* (with an average of about 13 per page), and also the very popular *Werner* comic (see Fig. 13.1).

In the spoken language of young people, however, sound words constitute only a very small part of the vocabulary. In all the investigations of actual language use that have so far been carried out (see below), sound words occur very rarely. Interestingly, the only word that does occur frequently in these studies is *boah*, an expression which came into fashion through *Manta-Manni*,[4] and is now widely used.

So, as is the case with English words, sound words are expressions that already exist and have been adopted, rather than coined, by young people. The questions remain: who actually does this linguistic borrowing? When, where, and under what conditions? While specific English words are important for certain groups of young people, and other words for other groups, sound words are entirely marginal in spoken forms of *Jugendsprache*, and are in general confined to the language of certain individual comics and perhaps baby-talk.

4 NEW APPROACHES TO THE STUDY OF *JUGENDSPRACHE*

It should now be clear, then, that *Jugendsprache* as such does not exist any more than there is a single, universally valid definition of 'young people' or 'youth'. With a few exceptions, research that has been conducted to date is of doubtful value on both empirical and methodological grounds, and it has failed to go beyond a kind of lexical

---

[4] *Manta-Manni* is a caricature from so-called *Manta-Witze*, jokes about drivers of the Opel Manta car, who are characterized as showy, stupid, and big-mouthed. *Manni* is the abbreviated form of the name Manfred.

FIG. 13.1.   Cartoon from *Werner*: 'Sektenquatsch und Eiermatsch'.
© Brösel-Achterbahn

voyeurism. Recently, however, the programmatic article by Neuland (1987) and the volume edited by Januschek and Schlobinski (1989) have introduced a new approach to the topic. A number of new proposals have been made:

- Young people and their language can be seen as sociologically based categories, incorporating a complex set of different phenomena.
- Even if one can claim to know what a youth group is, then there are as many *Jugendsprachen* as there are groups.

- If these phenomena which are called *Jugendsprache* are group-specific, then they should be analysed as such.
- It is not feasible to describe and explain group-specific language phenomena on the basis of questionnaires and individual personal interviews. This methodology should be replaced by participant observation and methods developed in the ethnography of speaking.
- The main focus of group-specific speech analysis should be on 'manners of speaking', which have specific functions for specific groups, depending on the context. I would like to refer to such group-specific and context-bound manners of speaking as 'speech styles'. Most of what comes under the label *Jugendsprache* can be analysed on this level: the basis for the concept of *Jugendsprache* in this sense lies in the congruence or merging of different speech styles.

That a paradigm shift has taken place (at least in the minds of linguists), from the 'classical' approach to *Jugendsprache* research to a pragmatically based study of group styles, became evident at the colloquium on *Jugendsprache* research in March 1992 in Leipzig. The participants at this conference unanimously agreed that the way forward was to work empirically and to use discourse analysis procedures. But if this is a fresh start, what form should the new approach take in order to account for the observable variation in *Jugendsprachen*?

As I have already suggested, concrete speech styles should be the basis for analysis. Speech styles are 'forms of differentiating behaviour' (Hymes 1979: 176, Bourdieu 1982: 62), and this implies that characteristics of language use (specific sound characteristics, lexical elements, morphological markers, conversational patterns, etc.), which are closely connected to social groups, differ in quality and number. Furthermore, group members have correspondingly different expectations in terms of the structure and use of utterances. A speech style is therefore a particular configuration of structural attributes and their communicative functions, which is based on shared sets of norms and values and therefore on shared expectations, but which may vary according to situational and interactive factors such as intention, choice of topic, audience, and context.

The analysis of speech styles as a complex of distinct linguistic preferences needs to be embedded in the sociological analysis of social groups and their activities. The pioneering work of the Centre for Contemporary Cultural Studies at the University of Birmingham (Willis 1981, Hall and Jefferson 1976) can be used here as an all-embracing model. This work analysed empirically the subcultural group styles of rockers, hippies, and skinheads as expressive lifestyles. The concept of cultural style and Lévi-Strauss's notion of *bricolage* (Lévi-Strauss 1966, 1969; cf. also Neuland 1987: 68 ff.) are crucial for an analysis of speech style:

The generation of subcultural styles . . . involves differential selection from within the matrix of the existent. What happens is not the creation of objects and meanings from nothing, but rather the *transformation* and *rearrangement* of what is given (and 'borrowed') into a pattern which carries a new meaning, its *translation* to a new context, and its *adaptation*. (Clarke 1976: 178)

This transformation and rearrangement is achieved through manipulation, through 'the re-ordering and re-contextualisation of objects to communicate fresh meanings, within a total system of significances, which already includes prior and sedimented meanings attached to the objects used' (Clarke 1976: 177). Through the process of de- and recontextualization of an object 'a new discourse is constituted, a different message conveyed' (ibid). As Willis and Clarke show, the style of the subculture groups is a criterion of group identity, i.e. a lifestyle, which is an ensemble of various individual styles: appearance, music, dress, accessories, graffiti, catch-phrases, and sentence structures combine to form a relatively uniform group style. Linguistic (and other) markers imply simultaneously recognition of group membership (solidarity), and difference from other groups (distinction):

les usages sociaux de la langue doivent leur valeur proprement sociale au fait qu'ils tendent à s'organiser en systèmes de différences (entre les variantes prosodiques et articulatoires ou lexicologiques et syntaxiques) reproduisant le système des différences sociales dans l'ordre symbolique. (Bourdieu 1984: 7)

(the social uses of language owe their specifically social value to the fact that they tend to be organized in systems of differences (between prosodic and articulatory or lexical and syntactic variants) reproducing the system of social differences in the symbolic order.)

Speech styles provide information about group behaviour and commonly shared values and norms, in short: about the lifestyle of social groups and their members.

The sociological basis of this research is interesting in two respects for a pragmatically orientated analysis of speech styles:

- An important aspect of the study of stylistic differences between speech styles is linguistic markers, whose symbolic value is constructed from a specific matrix of group interests, which in turn relate to specific interests (for example music). This then raises the following questions: to what extent are speech styles group-specific, and under what conditions do such styles emerge?

- How and on which linguistic levels does the principle of *bricolage* function in relation to the constitution of speech styles, and which resources are used to carry out the transformation process?

5   THE PRAGMATIC ANALYSIS OF 'SPEECH STYLES': SOME RECENT STUDIES

An illustrative analysis of group-specific and situational speech styles is to be found in Schlobinski (1989), more extensive investigations in Last (1990) and Wachau (1990). In the Schlobinski study (1989), the language use of a group of young punks in various situations was recorded. One particular situation (an evening of convivial drinking) produced many examples of the kind of feature that *Jugendsprache* researchers' dreams are made of. During this two-hour recording, which was made by the group itself, the participants were placed in an interactive situation, which following Goffman (1975) we could describe as 'collusive communication'. This means that the participants negotiate an agreement about commonly shared values and norms by referring to commonly shared pools of knowledge. A closer look at this particular conversation reveals that this agreement is not reached by exchanging information, but rather by acting out a 'scene', in which the actors are their own audience. The specific form of 'reality modulation', in which real action is transformed into something playful (Goffman 1975: 514), is based on two transformation processes. On the one hand, the participants make use of cultural resources which constitute their own values and norms by 'quoting' and thereby reproducing them ('mimetic quotation'). On the other hand, patterns are quoted and 'defamiliarized' in the form of language games borrowed from the (rejected) dominant culture ('defamiliarized quotation').

A central technique of both mimetic and defamiliarizing quotation is 'formal quotation'. This involves quoting short sequences from films, comics, or music lyrics, which as group-specific formulas have two communicative functions. First, they are linked to direct speech acts and structure the discourse; and secondly they serve to introduce particular patterns or routines, which can then be manipulated interactively in unstructured improvisations. In other words, the quotations are key sequences introducing a 'theme', which the participants then use as the basis for an improvised 'performance'.

Even without a language game or ironic discourse developing, quotations based on the group's common store of cultural resources can be incorporated into the dialogue, as in the following example, where a characteristic sequence is quoted from the *Werner* comic ('*Ich fahr nich!*', in lines 4–5).[5]

1   c. wie spät is das überhaupt (.) wenn wir alle stöppern müssen

2   ey dann müssen wa ja mal ( )

3   r. Enno fährt doch morgens

4   e. ich fahr nich

5   r. ich fahr nich (.) und ich sag dir noch (.) ich fahr NICH

6   j. [*lacht*]

7   e. Werner fahr nich. hab ich gesacht

8   c. ( ) wer darf wohl nich=

9   s. =Ja.

10 c. aber Werner hört ja nich

11 s. nee wohl nich

The specific communicative function of quoting the *Werner* expression is to turn down a request. A sequence like this is funny for the participants, but because the 'raw material' is drawn from a source that reflects the group's own values, it cannot then be developed in an ironic or defamiliarized way. This is only possible if the cultural resources which are exploited do not correspond to the group norm. A good example in the context of our punk group is the television quiz programme 'Der große Preis' (also known as *Der große Scheiß*). The following sequence is an obscene pastiche of this programme.

1   c. ficken einhundert

2   e. ficken einhundert (.)

3   x. risiko

4   q. nee

5   j. glücksspiel

6   c. was denn was war denn daran risiko (.) Rita Süßmuth oder

7   was/

8   e. ficken einhundert

---

[5] For transcription symbols, see Appendix to Chapter 9 above.

9   c. Rita Süßmuth

10   x. risiko

11   [*Lachen*]

12   c. frau Meyer hat aids (..) herr herr Tropfmann hat herpes

13   (..) was möchten SIE einsetzen (. . .) öhöh (. . .) syphilis.

14   [*Lachen*]

15   c. also hier die frage (. . .) also hier die frage

16   e. welche frage

17   [*Lachen*]

18   s. sein=

19   r. =das ist hier die frage=

20   s. =sein oder nicht sein

21   r. schwein oder nicht schwein

22   [*Lachen*]

23   c. schwein (..) oder nicht schwein

24   q. dein/

25   j. sein

26   s. kein

27   r. kein rabe (.) genau das is es

The complex pattern of the quiz programme with its series of short, ritualized utterances is adhered to but defamiliarized. The first effect of defamiliarization is seen in the opening statement: through the phrase *ficken einhundert* (fucking: one hundred, line 1) the game is initiated, a sexual theme proposed, and the framework for the ensuing improvisation established. Key phrases, repeated in the course of a conversation, have a kind of background function, a feature which has been described in the study of stylistics: it provides 'a background against which the statements preceding the repeated unit are made to stand out more conspicuously' (Galperin 1977: 212).

The key phrase in this context, *ficken einhundert*, refers to the fact that the contestants in this quiz are free to choose from a variety of topic areas with bets ranging from DM20 to DM100. These bets are shown on a screen and the contestants then make their choice; for example *Film einhundert* means that from the topic 'films' a question equivalent to a bet of DM100 has been chosen. Clearly, the higher the bet, the more difficult the corresponding question will be; if the contestants answer correctly, they are credited with the amount, if they answer incorrectly, the amount is debited from their 'account'. The contestant with the highest winnings at the end of the game is declared the winner. So *ficken einhundert* refers to a DM100 bet in the topic area: *ficken*. While the familiar speech-act function is transferred from the quiz to the improvised game, the basic pattern is modified by the choice of an unfamiliar topic area. Right from the start of this 'episode', therefore, 'Der große Preis' is transformed in an ironic way by the very selection of the topic area.

Behind many questions in the quiz programme there is also a 'risk question', which may arise by chance when particular questions are chosen. This gives the candidates the possibility of winning DM500, if they answer the risk question correctly. Here, a risk question is established in line 3 and paraphrased in line 5, then developed further in line 6. The risk is then accepted directly by C and 'Rita Süßmuth' (the name of the speaker of the Bundestag, the lower house of the German parliament) is chosen as the new topic; the choice of topic was probably determined by alliterative association (*Risiko. Rita Süß-muth*) and the desire to maintain and develop the sexual leitmotif. The key phrases and the main points are copied like an echo in lines 8–10,[6] in sequence as a triplet, and the initiation pattern is thereby closed.

After the pattern of the risk question has been established, there follows in lines 12–13 a simulation of the quizmaster's method of questioning. Here again, the speech pattern of the original model is adopted, while the leitmotif established in the opening sequence is developed further. As in the original quiz, this pattern has a three-pronged structure: 'Frau X bet *a*, Herr Y bet *b*, now what would you like to bet ?' The topics are chosen from the general area *ficken*, and the contestants are given the names of neighbours living in J's area. There is a double parody at work here: on the one hand, the neighbours are the butt of the joke, while on the other hand venereal diseases/AIDS are classified as topics in the game. By referring to the neighbours the joke pokes fun at social targets: Frau Meyer and Herr Tropfmann are taken as representatives of a petit bourgeois milieu and become the object of the parody.

This brief illustration shows that the ritualized speech patterns that form an important component of *Jugendsprache* can be analysed within

---

[6] As Deborah Tannen puts it, 'humor is a common function of repetition with slight variation' (Tannen 1987: 590).

a structural framework. There are repeated structures, organized in a three-part sequence, which are varied through the interaction of the group members. The variation occurs on two levels: the paradigmatic level, where the same number of variables are used but with different variants, and the syntagmatic level, where the syntactic structures are reduced. The crucial thing in this example is that the 'serious' quiz 'Der große Preis' is juxtaposed with a playful version of it, in which sexuality becomes the object of the game, something which would have been considered vulgar and unacceptable by the audience of the television programme. The defamiliarization of the quiz, through the adoption and variation of its speech-act patterns, is intended to ridicule the mentality of the consumers of this type of entertainment. This is the key to this ironic language game, which remains a kind of conventional party game. While pretending to identify with the values and norms of the dominant culture, the group is able at the same time to ridicule these very norms and values by manipulating conventional patterns. The method of using defamiliarizing quotation as a means of ironic distancing operates in two ways: first, what in terms of the dominant culture would be seen as vulgar (to take this specific example) is integrated into the 'quoted' pattern; and secondly, the effect is achieved interactively, as a group process, using structural and systematic repetition and variation.

This analysis of a particular speech style presupposed that speech styles can be analysed as group- and context-specific processes. I have tried to show that the specific speech style under investigation here could be described in terms of mechanisms which operate by transforming, decontextualizing, regrouping, and changing the function of cultural resources. The principle of *bricolage* as an important structural mechanism in the conversation of groups of young people is also demonstrated by the studies by Wachau (1990) and Last (1990), which took a sociology of language approach.

Wachau's investigation was based on a long-term study over a period of two years of a group of teenage girls. She collected data both on the use of language within the group and on their written style (biographies, letters, essays in school), as well as on language attitudes. The analysis of their speech style shows that besides using expressions from advertising and music, the group also has its own special language games and rituals. One interesting observation here is that as in the previous example variable patterns of discourse are initiated by the use of key phrases which trigger off certain associations. In the following excerpt from a conversation, J tries to characterize the husband of M's teacher by comparing him with James Bond. This in turn is used as a means of making fun of the teacher.

1   A. ich hab den gar nich gesehen auf'm elternabend.

2   M. der sieht schrecklich aus.

3   A. ziemlich gammelig kann das sein?

4   M. ja (.) der sieht aus wie james bond.
       [*Lachen*]

5   I. wie welcher?

6   J. sag ich doch ziemlich gammelig.

7   M. conney (.) so total schleimig weiß ich auch nich so

8   total öh

9   J. wie heißt die lehrerin?

10   M. frau hinz.

11   J. mein name ist hinz [*lacht*] hans hinz [*lacht*].

12   A. hinz und kunz.

13   M. und ich hab die lizenz zum töten [*lacht*].

14   A. mit schlechten zensuren! [*Lachen*]

15   J. die lizenz zum töten teufel noch mal (.) ohgr!
       [*lacht*] du sollst nicht fluchen! (Wachau 1990: 54)

The sequence 'Mein name ist hinz (.) hans hinz' is spoken with the same intonation as the corresponding sequence in the James Bond films: 'My name is Bond, James Bond'.

The study of the written texts shows that in their biographies the schoolchildren used features characteristic of the spoken language, such as intensifiers like *echt turbo* and English phrases (e.g. *stoned*). One girl even used a comic word, *gähn* (yawn). These texts were surprisingly lively, so much so that the teacher involved could not believe that his students had written them.

In her ethnographically based investigation of two youth groups, Last (1990) reaches comparable results to Wachau. One group was accompanied over a number of months in a youth club and the other came from a village near Osnabrück. Interestingly, informants in both investigations very often used the communicative particle *ey*. This particle also appeared quite often in the speech of the speakers investigated in the research project 'Zum Sprachgebrauch von Jugendlichen in Osnabrück', and it was also frequently observed in an investigation into Berlin speech (cf. Hädrich 1988). It appears that this feature, which actually only occurs in the spoken language and which has not been elicited in any of the studies based on questionnaires, can be considered as a marker of *Jugendsprache*, because adults only use other particles, such as *wa*, *ne*, and so on. Admittedly it remains unclear by whom, where, and when these particles are used. The pragmatic functions associated with *ey* vary and depend on the position of the particle in the sentence:

preferred position in sentence	main function	example
initial	attention getter	*ey, wann kommst'n?*
initial	address signal	*ey, schulze*
final	structuring signal	*wenn wir malochen, ey, kommen wir da hin*
final	intensifier	*echt geil ey!*
final	evaluation	*scheiße, ey!*

(For details, see Schlobinski *et al.* (1993), chap. 3.)

In the final position, *ey* is also used as a signal of confirmation or to terminate a sequence. As it is such a common feature, it is interesting to note that while young people themselves mention this particle as a typical characteristic of *Jugendsprache*, their attitude towards it is ambivalent: although they know that they use it, they nevertheless reject it. Presumably, this is partly because of the criticisms of their parents but it also has something to do with the proletarian image of the Manta-jokes (see footnote 4), as Jan (14) declares: 'weil ich finde (.) das sind (.) eben so (.) daß das jetz sone mode damit (.) *manta da ey* und *boah ey*' (I think it's just a fashion to use phrases like *manta da ey* and *boah ey*; Schlobinski *et al*, 1993). Like Wachau (1989), we found in our investigation that young people describe their speech as more direct, more spontaneous, and characterized by 'cool' expressions. However, there are variations here too. For example, the use of cool expressions by 13-year-old boys to gain the favour of girls is by their own admission a cover for their own emotional insecurity, while older boys put more emphasis on individually differentiated speech styles within their own groups. The young people we studied evaluate their own language by contrasting it with the negative image they have of 'adult language', which while not actually rejected is characterized as being dry and serious. At the same time, they are aware of the fact that their parents and teachers do not speak a uniform language any more than they do themselves.

6 CONCLUSIONS

Recent studies on *Jugendsprache* show that it is not a monolithic concept, but can be divided up into various group- and situation-specific speech styles. The most important factors distinguishing different styles are which cultural resources are exploited and how individuals speak in relation to particular topics and interaction partners. However, one further factor, which has not been mentioned so far, is that there are important sex-specific differences. In the long-term project currently being carried out in Osnabrück, which is investigating the language use of two youth groups (see Schlobinski *et al* 1993), it is evident that there is a particular 'chatting-up language' used almost exclusively by boys, whereas girls have their own repertoire of verbal techniques to fight off these advances. It is interesting too that these rejection techniques are also used by Turkish and Yugoslav girls, who are well able to defend themselves in verbal duels. In the following sequence, for example, a group of pupils are working on a story board (the script for a commercial). Dieter initiates the sexual discourse:

MUHRAT. jetz mal ich ilona!
ILONA. was soll das denn sein?
DIETER. ja DU [*lacht*] ja da isse (.) brauchst gar nich zu denken das sind nich deine titten (.) das is dein bauch=
DETLEF. = is das'n arsch?
ILONA. nä (.) das sieht sich was anderes ähnlich
MIOSCH. das sieht aus ey ey (.) gib ma her
ILONA. na und mich stört das nich muhrat
MIOSCH. das sieht aus (.) wenn man den so nimmt (.) [*lacht*] ooh [*er dreht das Blatt um und zeigt ein sexuelles Bild*]
ILONA. oh was (.) ey mein gott was soll nen das?

Ilona knows how to defend herself against verbal sexual attacks by counter-attacks:

ILONA. nä (.) du sei leise (.) du weißt wie du aussiehst? (.) wie nen schwuler hengst
DIETER. isser auch
DETLEF. und du siehst aus wie ne lesbische sau
MUHRAT. [*ahmt Ilona nach*] schwuler hengst

It goes without saying that there are speech differences between punks and yuppies, between girls and boys, and between 12- and 50-year-olds. The difficult question, however, is to determine where these individual differences lie. The paucity of valid investigations makes it difficult to give a generalized answer. What should by now be clear is that the decisive factors are who speaks when, where, and under what conditions. The particle *ey* is in my opinion the only individual characteristic of *Jugendsprache* which is in widespread use. Other linguistic structures are strongly speaker- and situation-dependent, or are used in everyday colloquial speech by many groups: for example, *geil* ('brilliant, cool'; literally 'randy'), which is supposedly a typical *Jugendsprache* word, but has long been part of the bedroom language of the German petit bourgeois.

Therefore as adults we should be honest for once: the little quips during card games, the sexual innuendos contained in pub-jokes, the language of the brothel: surely these and other forms of 'adult' speech provide as many examples of language that conflicts with 'public norms' or the teacher's 'ideal model' as *Jugendsprache* does? In other words, before we criticize *Jugendsprache* and hold it responsible for the 'decline of the German language', we should remember this: dealing with the language of young people also involves reflecting on our own use of language and on ourselves as well.

Further Reading

Ermert (1985)
Heinemann (1989)
Henne (1986)
Januschek and Schlobinski (1989)
Neuland (1987)
Pörksen and Weber (1984)
Schlobinski *et al.* (1993)

References

BENEKE, J. (1982), *Untersuchung zu ausgewählten Aspekten der sprachlich-kommunikativen Tätigkeit Jugendlicher*, Dissertation, Akademie der Wissenschaften der DDR (Berlin, GDR).
BOURDIEU, P. (1982), *Die feinen Unterschiede* (Frankfurt: Suhrkamp).
—— (1984), 'Capital et marché linguistiques', *Linguistische Berichte*, 90: 3–24.
BRANDMEIER, K., and WÜLLER, K. (1989), 'Anmerkungen zu Helmut Henne: Jugend und ihre Sprache', *Osnabrücker Beiträge zur Sprachtheorie*, 41: 147–55.
CLARKE, J. (1976), 'Style', in Hall and Jefferson (1976), 175–91.
CLAUS, U., and KUTSCHERA, R. (1986), *Total krasse Helden. Die bockstarke Story von den Nibelungen* (Frankfurt: Eichborn).
COULMAS, F. (ed.) (1979), *Soziolinguistik* (Frankfurt: Suhrkamp).

DENGER, F. (1984), *Der grosse Boss: Das Alte Testament unverschämt fromm neu erzählt* (Frankfurt: Eichborn).

DOLLE-WEINKAUF, B. (1990), *Comics. Geschichte einer populären Literaturform in Deutschland seit 1945* (Weinheim, Basle: Beltz).

ENGELMANN, H. (1964), 'Das Rotwelsch der Jugend', manuscript. Hessischer Rundfunk: Jugendfunk, broadcast on 16 Nov.

ERMERT, K. (ed.) (1985), *Sprüche—Sprachen—Sprachlosigkeit. Ursachen und Folgen subkultureller Formen der Kommunikation am Beispiel der Jugendsprache* (Rehburg-Loccum: Evangelische Akademie Loccum).

GALPERIN, I. R. (1977), *Stylistics* (Moscow: Higher School).

GOFFMAN, E. (1975), *Frame Analysis* (Harmondsworth: Penguin).

GRIMM, H., and SONTHEIMER, I. (1980), 'Pennälersprache— "Schülerdeutsch" ', *Praxis Deutsch*, 40: 45–9.

HÄDRICH, D. (1988), 'Berlinisch unter der Lupe. Elemente des Berliner Sprachstils in ihrer sprachlichen und sozialen Bedeutung', Wissenschaftliche Hausarbeit im Rahmen der ersten Staatsprüfung für das Amt des Lehrers mit fachwissenschaftlicher Ausbildung in zwei Fächern (Berlin).

HALL, S., and JEFFERSON, T., (eds.) (1976), *Resistance through Rituals. Youth Subcultures in Post-War Britain* (London: Hutchinson).

HEINEMANN, M. (1989), *Kleines Wörterbuch der Jugendsprache* (Leipzig: VEB Bibliographisches Institut).

—— (1990), *Jugendsprache. Ein Beitrag zur Varietätenproblematik* (Leipzig: Karl-Marx-Universität).

HENNE, H. (1981), 'Jugendsprache und Jugendgespräche', in Schröder and Steger (1981), 370–84.

—— (1986), *Jugend und ihre Sprache. Darstellung, Materialien, Kritik* (Berlin: de Gruyter).

HYMES, D. (1979), 'Über Sprechweisen', in Coulmas (1979), 166–92.

JAKOB, K. (1988), 'Jugendkultur und Jugendsprache', *Deutsche Sprache*, 29: 320–50.

JANUSCHEK, F. (1989), 'Die Erfindung der Jugendsprache', in Januschek and Schlobinski (1989), 125–46.

—— and SCHLOBINSKI, P. (eds.) (1989), 'Thema 'Jugendsprache' [*Osnabrücker Beitrage zur Sprachtheorie*, 41] (Universität Osnabrück).

KÜNNEMANN, H. (1976), 'Comics in der Bundesrepublik. Eine Einführung, Übersicht und Anregung zum Kennenlernen', *Medien und Erziehung*, 1: 4–15.

KÜPPER, H. (1961), 'Zur Sprache der Jugend', in *Sprachwart*, 10: 188.

—— (1990), *Wörterbuch der deutschen Umgangssprache* (Stuttgart: Klett).

LAPP, E. (1989), ' "Jugendsprache"; Sprechart und Sprachgeschichte seit 1945. Ein Literaturbericht', *Sprache und Literatur in Wissenschaft und Unterricht*, 63: 53–75.

LAST, A. 1990, 'Empirische Untersuchung zum Sprachverhalten einer Gruppe Jugendlicher', MA thesis, University of Osnabrück.

LÉVI-STRAUSS, C. (1966), *The Savage Mind* (London: Weidenfeld & Nicolson).

Lévi-Strauss, C. (1969), *Totemism* (Harmondsworth: Penguin).

Meik, R. (1984), 'Udo Lindenbergs Rocktexte und der Jugendjargon', in *Medien und Erziehung*, 5: 281–7.

Müller-Thurau, C. P. (1983), *Lass uns mal 'ne Schnecke angraben. Sprache und Sprüche der Jugendszene* (Düsseldorf: Econ Verlag).

Neuland, E. (1987), 'Spiegelungen und Gegenspiegelungen. Anregungen für eine zukünftige Jugendsprachforschung', *Zeitschrift für germanistische Linguistik*, 15/1: 58–82.

Nowottnik, M. (1989), *Jugend, Sprache und Medien. Untersuchungen von Rundfunksendungen für Jugendliche* (Berlin, New York: de Gruyter).

Pörksen, U., and Weber, H. (1984), *Spricht die Jugend eine andere Sprache?* (Heidelberg: Schneider).

Schlobinski, P. (1989), 'Frau Meier hat Aids, Herr Tropfmann hat Herpes, was wollen Sie einsetzen? Exemplarische Analyse eines Sprechstils', in Januschek and Schlobinski (1989), 1–34.

—— and Blank, U. (1990), *Jugendsprache* (Berlin: Pädagogisches Zentrum).

—— Kohl, G., and Ludewigt, I. (1993), *Jugendsprache. Fiktion und Wirklichkeit* (Opladen: Westdeutscher Verlag).

Schönfeld, E. (1986), *Abgefahren—eingefahren. Ein Wörterbuch der Jugend- und Knastsprache* (Straelen: Straelener Manuskripte Verlag).

Schröder, P., and Steger, H. (eds.) (1981), *Dialogforschung. Jahrbuch 1980 des Instituts für deutsche Sprache* (Düsseldorf: Schwann).

Tannen, D. (1987), 'Repetition in Conversation: Towards a Poetics of Talk', *Language*, 63/3, 574–605.

Wachau, S. (1989), ' "... nicht so verschlüsselt und verschleimt!" Über Einstellungen gegenüber Jugendsprache', in Januschek and Schlobinski (1989), 69–96.

—— (1990), 'Empirische Untersuchung zur Sprache Jugendlicher: Über Sprechstile, Schreibstile und Sprachbewußtheit', MA thesis, University of Osnabrück.

Wägner, W. (1878/n.d.) *Deutsche Heldensagen* (Erlangen: Karl Müller Verlag).

Willis, P. (1981), *Profane Culture. Rocker, Hippies: Subversive Stile der Jugendkultur* (Frankfurt: Syndikat).

14 Language and Television

WERNER HOLLY

The connection between language and television is at least a twofold one. We may ask: (1) how does language appear in television? Or the other way round: (2) what is the impact of television on language? For both questions it is first crucial to consider the institutional structure of German broadcasting, which has been called a 'dual system' of public and private organization (Section 2 below).

To take the first question first: 'language in German television' is not a subsystem of the German language and it is not homogeneous. It cannot be described in terms of morphology, syntax, or lexis, nor as a special variety like a dialect or *Fachsprache* (technical register) (Burger 1984: 3); different varieties are adopted according to context (Brandt 1985: 1672–3). An adequate description must include a pragmatic dimension incorporating certain forms of 'language use' which are governed by special media conditions on the one hand (Section 3 below), and by different *Textsorten* (genres) on the other (Section 4).

If we consider media conditions which shape the general character of television language use, we may see the modern TV system as a form of communication which seems to 're-'establish (though in a secondary, mediated manner and operating fundamentally in a one-way direction) some qualities of face-to-face situations: the interaction of verbal and non-verbal/visual signs, orality, intimacy, actuality, 'mixed styles', and the social constitution of meanings in which the recipient (viewer) plays an active role.

By and large, the genres of German television are not originally German or, if they are, they are influenced by other international productions. Nevertheless, there are some cultural peculiarities, for instance the *Fernsehspiel* (television play), the still rather formal news bulletins, features peculiar to particular regions, style and placement of advertisements, etc.

Despite a veritable flood of television research, linguistic aspects in general are still neglected (see Straßner 1980: 335). Schmitz (1987: 820–1) observes that there appear to be two distinct perspectives: either research is focused on the texts without considering the conditions of production and reception and their social embedding and functions, or the topic of language is not dealt with at all.

However, since the 1970s, linguists (or philologists) have begun to take an interest in television genres, a field about which I can only offer some impressions. A short survey on language in the mass media is given, besides the handbook articles I have mentioned already (Straßner 1980, Schmitz 1987), in a monograph by Burger (1984). Television genres are discussed in Kreuzer and Prümm (1979) and more briefly in Kreuzer (1982).

Linguistic research started with genres which seemed to allow the visual component to be disregarded: for example 'Das Wort zum Sonntag' (a TV sermon) and, of course, the whole complex of television news (Straßner 1982, Huth 1977 and 1985, Kübler 1975, Ebner 1986, Schmitz 1990). In this area, the interrelation of language and images has been discussed, prompted by Wember's (1976) thesis of the '*Text-Bild-Schere*' (text–image division), resulting for instance in Bentele and Hess-Lüttich's (1985) anthology and in Muckenhaupt's (1986) study.

A second important complex of topics included conversational forms which were surveyed by means of ethnomethodological and pragmatic conversation analysis: interviews (Schwitalla 1979, Hoffmann 1982, Thomas 1988), discussions (Linke 1985, Holly, Kühn, and Püschel 1986 and 1989, Sucharowski 1985), and talk-shows (Mühlen 1985). Other genres were political magazines (Keppler 1985), political advertisements (Wachtel 1988), sports reports (Neugebauer 1986), and quiz and game-shows (Woisin 1989, Hallenberger and Foltin 1990; Hallenberger and Kaps 1991).

There are some pessimistic standard hypotheses about the second of the questions raised here, the impact of television on language use. These arguments dominate public discussion although they have by no means been proved. They concern the pernicious cultural influence of television on language in general, particularly on literacy, and on the conditions of public and private discourse. But despite many long years of expensive research on media reception, the only result which stands up to a more detailed examination is the rather trivial insight that the effects the media have depend on how they are used by different groups.

Even the more linguistically orientated statements on this issue (Section 5 below) have not been based on well-founded surveys; a summary is given in the article by Brandt (1985). But at least they are not short of evidence: it is plausible to argue that broadcasting played an important part in the diffusion of standard German, in the relaxing of norms in the standard, in the growing knowledge of a large portion of the public about different linguistic subsystems, and finally in the rapid propagation and reinforcement of new trends and developments, that is, in accelerating linguistic change.

But all this calls for further empirical linguistic research which is directed towards the processes of comprehension, interpretation, and the linguistic use of television products by active recipients in everyday communication situations. Previous methods of mass-communication research have come to an impasse: to overcome this, we need more work based on qualitative linguistic approaches.

2   THE HISTORY AND INSTITUTIONAL FORM OF WEST GERMAN BROADCASTING

When television was young, it first had to find its own place (as each new medium does); it had to compete with the radio and the cinema, which were well established in the 1930s, when the history of television began in Germany. After initial experimental transmissions in 1928, a regular programme was started in Berlin in 1935, but without any notable public response. By the Second World War, there were only about 500 receiving sets (compared to 20,000 in Great Britain). One reason for this 'false start' was the fact that in those days TV was normally watched in public *Fernsehstuben* (viewing rooms) rather than in the home. The poor technical quality was therefore not compensated for by the privacy and comfort of individual reception, so that TV inevitably suffered in competition with the cinema.

The second start was launched after the war in 1952, this time based on home viewing, and this was more successful, as shown by the statistics on households with TV sets in the Federal Republic (see Brandt 1985: 1670):

1953	2,000
1955	100,000
1958	> 2,000,000
1968	15,000,000
1983	22,100,000

This development culminated in virtual saturation of the market. As early as 1980, 97 per cent of households had at least one set. Together with this expansion, the range of channels and programmes increased continuously. Until 1963, there was a monopoly for the ARD (Arbeitsgemeinschaft der Rundfunkanstalten Deutschlands), whose constitution is a federalist variation of the BBC model. After the totalitarian use of broadcasting for the Nazi regime's propaganda, this structure was intended to prevent abuse by any political party. Nevertheless, West Germany's first chancellor Adenauer tried (unsuccessfully) to establish a government TV station under private law (Deutschland Fernsehen GmbH). A judicial quarrel finally led to the foundation of the ZDF (Zweites Deutsches Fernsehen) in 1963 by the West German Federal states; thus the non-governmental, non-private status of the broadcasting system was kept intact.

Nevertheless, the political parties could not be completely prevented from taking a hold on television. Over the years, they gained more and more influence with official (and informal) control groups, and on employment policy, and even particular programmes, so that *Ausgewogenheit* (balance) and *Schere im Kopf* (literally 'scissors in the head': a symbol for pre-emptive self-censorship) became famous catchwords for West German TV journalism.

Throughout the 1970s, this situation was used by conservative politicians in the controversy over private broadcasting; in private stations they hoped to have a forum to counter the alleged leftist tendencies of the public systems, championing the business world's interests, especially those of publishers. The development was accelerated by the advent of new technologies (cable, satellite), which created additional opportunities for private stations. Finally in the 1980s, following a pilot cable project in four cities and the granting of licences by individual states, and with a new treaty between all states (1987) and extensive support in the form of the controversial installation of a cable system by the Ministry of Postal Affairs, a mixed 'dual' system was established and legalized with a guarantee for the public stations and development opportunities for the private ones.

This decisive development in broadcasting policy increased the expansion of programmes and channels already started by the public stations. Although the market is still growing and the formation of private stations is continuing, the picture has changed almost completely in the last decade. With cable and satellite, more than 20 channels can be received: besides the public channels of the ARD (the 'first channel' and the 'third channel', which is broadcast in several regional variants) and the ZDF, which together also offer the satellite channel '3sat', there are two major private stations ('SAT 1 and

'RTL'), four minor ones ('RTL2', 'Vox', 'Pro 7', and 'Der Kabel-kanal'), and several international sport or music channels, as well as the French–German culture channel 'Arte' and a first pay-TV channel ('Premiere').

However, despite the growing flood of channels and programmes, the actual consumption of television seems to have stagnated after an initial rise. There is a general tendency towards selective strategies amongst the viewing public, although certain groups tend to adopt more extensive viewing patterns. As a broad generalization, it is fair to say that the popularity of television is in an almost inverse relation-ship to its availability and the range of what it has to offer: it is seen as a regular companion with a firm position in the household, but it becomes less attractive the more familiar and the more abundant it becomes.

3 SOME GENERAL CHARACTERISTICS OF LANGUAGE USE IN GERMAN TELEVISION

As I have already argued, there is no such thing as 'television language' in the strict sense of a homogeneous linguistic subsystem. Concrete features of language use can only be described within the parameters of individual television genres. But it is possible to identify some very general characteristics resulting from the technical, institutional, and cultural conditions of a particular broadcasting system, and in this section, by way of illustration, I shall discuss some of the salient features of language use in German television.

3.1 *The Interplay of Language, Sound, and Pictures*

Like face-to-face interaction and unlike the telephone, radio, or silent movies, television provides multi-channel communication. In the opinion of many professionals, practitioners and critics, it is above all a visual medium, so that most of the contents ought to be conveyed by pictures. In a tradition deriving from the art of silent movies, language is at most a secondary, if not disturbing factor. Conversely, others have emphasized that, in contrast to cinema, television is more orientated towards language: the pictures on the small TV screen in a living-room do not enjoy the importance and intensity of (say) cinemascope in a large, dark, and stage-centred cinema, and most TV genres (and many of those working in TV) come initially from radio, which has contributed to the fact that pictures often seem merely to be illustrations for spoken texts. Furthermore, many informational genres, but also soap operas, largely depend on language. Finally, pictures may have a stronger emotional effect, but (as has been shown in several experiments) attention and complex information call for the acoustic channel.

In any case, language, when combined with sounds and pictures, loses its hegemony. This may be illustrated with regard to two areas which also show the institutional and cultural impact of visual–acoustic interplay. Speakers 'on screen' reveal by gestures and facial expressions much more of the cultural and institutional background than radio or 'off-screen' speakers. So the verbal style of a Scandinavian or British public-broadcast TV speaker may appear less relaxed or vivacious than a Latin or North American one from a private station. In German television we can see a stylistic change taking place from a more official, formal language and body language behaviour to a more mundane and informal style, a tendency which is strengthened by commercial channels with attempts on the part of presenters to be more personal, emotional, and direct.

Further examples of the importance of the language–picture-interplay are the different types of news film. What is provided by international agencies or networks is mostly short sequences of pictures, which are given a verbal text by the editorial staff, whose job is to comment on the visual material. Reports by the station's own correspondents are often made the other way round: a commentary-like text is illustrated by pictures which are more or less fitting. In German television, only the public stations have a widespread network of their own correspondents at their disposal, while private stations depend upon film material coming from anonymous sources. Straßner (1982) found that news films in German television have less complex syntax than reports by correspondents, because the editors take into consideration the problems of understanding both verbal and visual material and try to reduce the complexity of their secondary text.

3.2   Secondary Orality

A visual medium facilitates graphic realizations of language, but television also strongly emphasizes the phonic ones, despite the use of headlines in news broadcasts, programme notes, or even videotext. So television (together with radio) is introducing a new age of orality, as Ong (1982) states, adding that after centuries of literacy this orality is a 'secondary one', at the same time both very similar and dissimilar to primary orality.

Looking at this more closely, we can see that this applies on two levels. First, the performance of texts may be oral (phonic), but not so the production and preparation of performance, which is mostly written or based on writing. In the production process there is a continuum from the written and well-reflected to the spoken and spontaneous: in performing the speaker may read in an overt and undisguised manner, recite, read from a concealed script or tele-prompter, improvise on the basis of written preparation or with key-words on a cue-card, or speak genuinely spontaneously. The oral character of a text may vary depending on whether it is a monologue or a dialogue, on whether it is recorded (with or without editing) or transmitted live, on the subject-matter it deals with (fictional or authentic), or on the formality of the situation it presents.

These pragmatic aspects of production and performance have already shown that orality is not only a question of phonic realization, but also of stylistic conception. This may be shown by certain verbal strategies, which are considered typical of a spoken or a written style. Thus features of oral language use might include: shorter, less complex sentences; paratactic clause complexes; elliptical sentences; anakolutha; elaborations; modal particles; speaker signals; first-person references and the expression of attitudes and opinions; limited vocabulary; fuzziness; direct quotations; hesitation phenomena; self-corrections; and less coherent but more personalized texts. In contrast, written language is characterized by more variability and complexity; hypotactic and compact structures; more embeddings in noun phrases; and abstract formulations using passives and nominalizations. Chafe (1982) sums up both styles with the labels 'fragmentation' and 'involvement' for oral, and 'integration' and 'detachment' for written uses; Ludwig (1986) employs the terms 'aggregation' and 'integration'. Halliday (1985: 97) sees speaking and writing as different ways of knowing and different perspectives on reality: in written style a synoptic view presents things that exist, while in spoken language a dynamic view presents phenomena that happen.

In German television an overall style shift appears to be taking place, going from a written style towards a more spoken one; the most striking example of this is the presentation of the news, which used to be very formal (as the evening programme 'Tagesschau' still is), but has now changed in the direction of a more conversational mode with a presenter ('heute-journal' from ZDF, or 'Tagesthemen' from ARD) or even (in private channels) several presenters, who interact with each other.

3.3   *Secondary Intimacy*

Electronic media jump over distances and bring us close to what is far away. They seem to re-establish the conditions of close communication that we had to give up under the dominance of written texts. Meyrowitz (1985) thus argues that the main impact of electronic media is that they have completely changed the social meaning of space: owing to the media, as the title of his book says, there is 'no sense of place'.

But the nearness and intimacy that electronic media create are also 'secondary' (Habermas 1962: 207), that is, different from intimacy in face-to-face communication. Television, however, gives us the illusion of being immediately present, 'live' on the spot, where the action is. We often fail to notice that our visual presence is 'mediated'; we can only see and hear what the professionals with cameras and microphones select for us; intimate, close presence is a staged occurrence, and language has to deal with that pseudo-presence and has to be more explicit than in real face-to-face communication.

Conversely the media producers have no idea of the recipient's situation; they have only a vague notion of the average person, which can be made a little more precise by research. Consequently, the producers communicate for the most part within a phantom framework and have to cater for different kinds of reception, so that anything that diverges too far from the tastes of the 'average' viewer has to be avoided, even when smaller target groups are addressed.

Although electronic media communication differs from normal social interaction in face-to-face situations, television (more than radio) establishes a 'para-social interaction' (Horton and Wohl 1956); in other words, it conjures up a personal relationship with the recipient in which the latter feels directly addressed. This personalized style of pseudo-intimacy has been intensified systematically by media professionals and politicians. So it is not surprising that language use has also changed from a rather literate, detached mode to a more oral, involved mode.

Ideally, we need an analysis of examples of this shift covering the entire time span over which electronic media have operated. In the absence of this, we can still find useful material in what Habermas has called the 'simultaneity of the non-simultaneous'. A nice illustration of this is provided by interviews with two German politicians from about the same time, which show remarkable differences: one with the German Chancellor Helmut Kohl, the other with Egon Krenz, who for a short but critical period in 1989 was General Secretary of the SED, the governing party of the GDR (Holly 1992*b* and 1992*c*). Whereas both use a rather 'integrating' syntax (that is, characteristic of written language), Kohl uses linguistic means of 'involvement' to contrive a spontaneous and personal effect, although the subject-matter and the politician's training would have permitted a more literate style. The most striking device deployed in this staging of spontaneity is the continual use of parentheses, which is characteristic of Kohl's utterances (7.8 per thousand words), to such an extent that one is tempted to see it as a spontaneity-marker. Parenthetical remarks enable speakers to fill intervals in their utterances with supplementary information, comments, attention-claiming devices, and the like, without their having to abandon the dynamic and loosely structured style which is appropriate in contexts where the recipients will only have one opportunity to hear the speakers' statements. Kohl uses parentheses (as well as elaborations) extensively, whereas Krenz hardly uses them at all (1.4 per thousand words). This fits the overall impression that Kohl stages an oral, made-for-TV style, which has not been mastered by the less media-experienced Krenz, who remains more in the wooden, written mode of official announcements. Two typical passages from both interviews may illustrate the differences between the two speakers' styles (parentheses are marked by '[ ]', elaborations by: '+ +'):

und das ist eine der fragen [das hat jetzt gar nichts mit mir zu tun] an unser system ob das richtig ist daß wir alle paar wochen wahlen haben daß die handlungsfähigkeit der bundesregierung +der jeweiligen+ [das hat mir wiederum nichts zu tun] immer daran gemessen wird wie die jeweilige kommunal- oder landtagswahl ausgeht (Kohl)

(and that is one of the questions [that has just nothing to do with me] for our system whether it is right that we have elections every few weeks that the ability of the federal government to act +the current one+ [that once again has nothing to do with me] is always judged by the way the respective local or state elections turn out)

aber zutreffend ist genauso daß sein satz man darf den zeitpunkt nicht verpassen wenn man mit dem leben schritt halten will daß dieser satz mich persönlich sehr stark motiviert hat auch nach dem vierzigsten jahrestag probleme für das politbüro aufzuwerfen und im politibüro zu diskutieren die letztendlich zu den beschlüssen der neunten tagung des zentralkommitees geführt haben (Krenz)

(but it is also correct to say that his sentence you must not miss the moment if you want to keep in step with life that this sentence has motivated me personally very strongly even after the fortieth anniversary to raise problems for the Politburo and to discuss them in the Politburo which finally led to the resolutions of the ninth conference of the central committee.)

3.4   *Entertainment und Segmentation*

Besides the visual aspect, orality, and intimacy, the emphasis on enter-tainment is often mentioned as a key characteristic of television. On the more general level of 'culture', this is a further aspect of the orality discussed above. Just like speakers in symmetrical everyday conversations, the electronic media have to ensure that the listeners are able and willing to follow what is said. That implies the constant need to be interesting, gripping, amusing. Written texts may be dry and functional, only presenting necessary information to be referred to when it is needed. Orality has to deal with 'the human factor', that is, human-interest stories. Economic constraints strengthen the media's concentration on entertainment, making it an 'ideology' that has often been criticized (e.g. Postman 1985).

There are numerous examples of this overall principle of entertain-ment in German television: since the private channels increase their viewing public by offering game-shows, TV series, commercial films with sex, horror, and crime, and since they even invade the infor-mation sector with relaxed forms of news-shows, offbeat political magazines and talk-shows, discussions with abrasive confrontations, 'reality TV' with shocking events from everyday life, and the like, a process of self-commercializing has started in the public channels as well. Serious topics are moved to the edges of prime time or have to be packaged in the new forms of 'infotainment' and 'confrontainment' (Holly 1992*b*). The linguistic arrangement of such texts is dominated by glamour and gags, one-liners and anecdotes, gossip and gentle provocations, the curious, the spectacular, the sensational.

This orientation towards entertainment goes hand in hand with a fragmented time structure. Programmes cut up into shorter and shorter pieces become a kind of kaleidoscope, a mosaic, and thus resemble the discontinuity of other oral forms of communication more than written ones, in which the author can presuppose an engaged and patient reader and therefore offer longer consistent texts. What was obvious on the syntactic level applies on the text level of media language as well: fragmentation is an oral feature, integration a feature of writing. The fact that many recipients today typically switch from one channel to another at random intensifies the impression of seg-mentation. In short, the oral culture of the electronic media is composed of quick changes.

3.5 Mixing Styles

There is a further reason why television texts are fragmented. The attempt to be interesting for an audience which is as big as possible requires a text which is as mixed as possible: the aim is to offer something for everybody. So more and more of the traditionally separated areas of social and private life (sports, religion, arts, sciences, technology, politics, everyday life, etc.) are merged (Meyrowitz 1985: 73–92), as are the categories of the TV programme. Being open to easy access, the television text may be watched and listened to not only by its specific target group, but by anybody: a fact that is reflected by producers, who are conscious of the expectations of an average recipient. This leads to a mix not only of programme categories, but also of genres, styles, and varieties of language.

In the course of its expansion, German television developed or imitated a range of programme types using this characteristic mixed format under the generic label 'magazine'. But there is also mixing within genres: consider, for example, the mosaic shape of a news-show like 'Tagesthemen' or conversational forms like talk-shows which combine several, sometimes conflicting, functions of talk. Even linguistic features are mixed; we often find code-switches and code-shifts to dialects, sociolects, all kinds of technical registers, and various modalities (such as seriousness, pathos, or joking).

To give an example relating to syntax, I would like to return to the two interviews with Kohl and Krenz. In the utterances of these politicians, we find a mixed style, which seems to be typical for people who are used to speaking in a rather sophisticated manner, if a complex subject requires it, but at the same time to making efforts to be both comprehensible and lively. There are passages with very dry syntax:

wir haben die gewaltige herausforderung des vollzugs der inneren einheit deutschlands mit riesigen wirtschaftlichen problemen (Kohl)

(we have the enormous challenge of achieving the inner unity of Germany with huge economic problems)

... haben wir sicherlich das problem der harten diskussion bei der ablösung von Bernd Vogel ... aus dem parteivorsitz und dann aus dem ... amt des ministerpräsidenten (Kohl)

(we certainly have the problem of the tough discussions in the process of removing Bernd Vogel ... from his position as party chairman and then ... from the post of prime minister)

zum beispiel habe ich im Saarbrücker rundfunk anläßlich meines besuches bei Oskar Lafontaine auf entsprechende fragen ihrer kollegen gesagt daß ich es nicht gut finde wie die berichterstattung in der Bundesrepublik über die ereignisse in der Volksrepublik China läuft (Krenz)

(for example I said on Saarbrücken Television on the occasion of my visit to Oskar Lafontaine in answer to the questions of your colleagues on this matter that I don't approve of how events in the People's Republic of China are being reported in the Federal Republic)

Beside these examples of a compact nominal style, there are rather short 'idea units' of about six words, which give the impression of resolution and clarity:

wir sind in einer schwierigen situation (Kohl)

(we are in a difficult situation)

mit amtsmüdigkeit hat das nichts zu tun (Kohl)

(that has nothing to do with being tired of being in office)

das ist ein wirklich prima kabinett (Kohl)

(that is a really first-rate cabinet)

ich habe nie blutige taten gerechtfertigt (Krenz)

(I have never justified violent acts)

wir lassen uns überhaupt nicht treiben (Krenz)

(we are by no means adrift)

das war immer ein problem der west-presse (Krenz)

(that has always been a problem of the Western press)

The stylistic mixture iconicizes both the ability to handle complex matters and the power to be clear, decisive, and credible. This may therefore be seen as a strategy for dealing with the contradictory requirements which characterize any mass communication.

3.6 *Openness*

Openness as a characteristic of the electronic media is closely connected with the notion of easy access mentioned above. It also concerns the quality of media texts, their being open to different 'readings'. Fiske (1987: 84–107) takes up what Eco (1979) and Barthes (1975) had developed: the idea of active readers who use gaps, polysemy, and contradictions for their own purposes, thus creating their own text. Fiske argues that this process of active reading may apply not only to high literature, but also to the simple, trivial texts of popular television programmes. So we deal with electronic media (far more than with written texts) as pieces of our immediate everyday experience; we talk about them, discuss them, we change their meanings, remodel them according to our own subcultural notions, and thus integrate them into our everyday knowledge. Electronic media have become a part of popular, that is, oral culture.

Openness is a strategy for reaching a bigger audience; it is also the basis for active forms of reception. So it is a twofold notion, which may be expressed in terms of text features or of recipient's actions:

1. Openness through the plurality, and variety, of texts creates 'semiotic excess' (Fiske 1987: 90–3), which may lead the recipient to 'selection'.
2. Openness through vagueness, polysemy, plurality of meanings may be reduced by the recipient through 'interpretation'.
3. Openness in the form of gaps (empty passages in the text) so that the recipient may 'fill in the blanks'.
4. Openness despite relatively 'closed' texts is created by recipients who 'reinterpret' or 'change over' to another channel ('zapping') and so open up 'their' text.

German television has produced a range of genres which show features of openness. Whereas soap operas, for example, as typical open forms are adapted from American or British models, the open form of German TV discussions, which go out of their way to give a range of opinions and thus avoid giving the final word on any topic, has to do with the special institutional situation of German public networks, especially the strict requirement of proportional representation of the various political parties.

4   SOME GERMAN TELEVISION GENRES

'Genre' is a traditional concept in literature. Most of the time, genres have been seen as abstract ideas or forms of textual codification which may be characterized according to features of the texts. Newer approaches in media studies see genres as a 'cultural practice' (Fiske 1987: 109), as 'systems of orientations, expectations and conventions that circulate between industry, text and subject' (Neale 1981: 6), as 'cognitive concepts' (Schmidt 1987: 371) which are used by producers as well as in the domain of mediation and reception (Rusch 1987: 431).

There is sufficient empirical evidence of the existence of genre concepts, for instance in the production domain, where broadcasting corporations are organized in departments corresponding to genre categories (Rusch 1987: 458–65), or in the domain of mediation where Hauptmeier and Rusch found more than a thousand items of genre terms in papers and magazines providing TV programme information (Hauptmeier 1987: 403; Rusch 1987: 466).

In this section, I shall look selectively at a range of German TV genres (news broadcasts, soap operas, quiz- and game-shows, advertising), and attempt to identify their most salient features. These genres are variable, socio-historical patterns or schemata which structure all the actions involved in the process of media communication; in fact, they are 'families', in the sense that (as Wittgenstein (1969: 324) put it) they are bound together by *Familienähnlichkeiten* (family similarities).

4.1 *News Broadcasts*

As one of the main objects of research into TV language, news broadcasts have been examined from several perspectives (Püschel 1992: 1– 2): comprehensibility, the interplay of language and pictures, objectivity, ritualization, and text/genre structures. I shall only deal with the latter here.

Most of the time, German television news is presented in the form of *Mischtextsorten* (mixed text types) or in combinations of text types. Except for some kinds of short news presentations, news broadcasts are constructed in a 'magazine' format, which has two variations (Burger 1984: 153). The first is the classical form (for example 'Tagesschau', ARD), where a speaker reads the news, accompanied and interrupted by different sorts of visual contributions, partly with other 'on' or 'off' speakers. Schmitz (1990: 42), in his careful quantitative analysis of 'Tagesschau', counts seven so-called *Textsorten*: (1) first-speaker-texts (mostly 'on', only 'off' when accompanied by stills); (2) second-speaker-texts ('off' during film reports); (3) correspondent 'on' texts; (4) correspondent 'off' texts; (5) interviewer texts; (6) interviewee texts; (7) speech extracts or statements. Buchwald (1990: 243) differentiates nine 'elements' of news magazines: speaker report, news film, correspondent's report, special report, live report, commentary, interview, statement, visual elements (typography, photography, graphics, trick film).

A second, more modern form of news magazine (for instance 'heutejournal', ZDF; 'Tagesthemen', ARD) is presented by a newscaster or 'anchorperson', which leads to a rather different style, though the elements are more or less the same. This newer version of news broadcasts following international models was introduced after heavy criticism in the 1970s of the stiff, barely comprehensible, quasi-official ritual of conventional news presentation in Germany. The whole field of news presentation is a good example of the stylistic 'principle of inertia' (Straßner 1980: 332): text types which originated in print media or radio need a certain time to develop their own TV-specific character.

Püschel (1992) argues that news broadcasts in Germany still show 'the simultaneity of the non-simultaneous' (see Section 3.3 above): comparing the main news issues on two public channels (ARD: 'Tagesschau', ZDF: 'heute') and two private ones ('RTL Aktuell – Die Bilder des Tages', 'SAT 1 Blick') he finds that there is a development in the type of text structure from the pyramid-shaped 'lead' principle, commonly used in newspapers, to a 'cluster' principle which fits the requirements of TV presentation. It is surprising that not only the two public channels have kept the traditional report structure (starting with the most important point, then continuing with further facts and details), but the news on the private SAT 1 station has done this too, indeed to an even greater extent. The RTL news-show, however, abandons this pyramid structure and presents a loosely connected cluster of texts starting off with the presenter, who merely introduces the topic, creating suspense and attention; the basic information is given by a correspondent, and further elements might then be added in the form of an interview with the correspondent, a film report, or a studio interview with an expert. All of these components are held together by organizing speech acts performed by the presenter.

Whereas the pyramid principle sets up a one-dimensional perspective fixed in the 'lead', the cluster principle is open to changing perspectives. However, there is often no clear-cut distinction between the two forms: the newsreaders in 'Tagesschau' and 'heute', though not in the role of 'anchorpersons', also use short explicit patterns of organizing utterances. Finally, it should be noted that other news broadcasts on the public channels ('Tagesthemen', 'heute-journal') changed over to the anchorperson-presentation some years ago.

The verbal strategies used in formulating the news are interesting because of the problem that the texts are written, but are supposed to be understandable simply by listening. Throughout the history of radio and television news research, there has been repeated criticism of the written orientation in news texts (e.g. Straßner 1975, 1982). This criticism may have contributed to the journalists' efforts to use shorter and less complex sentences. Straßner (1982: 190) points out, for example, that syntax in film reports is less complex because of the strain imposed on the viewer by the 'double channel' reception (visual and aural). Schmitz (1990: 42–4) finds that the longest sentences in 'Tagesschau' occur in statements by interviewees or public speakers, while the edited sentences in the reports are decidedly shorter.

It is true that many news texts are written with an eye to the spoken performance. Nevertheless, commentators and correspondents in their commentary-like reports employ rather long sentences in keeping with a written style (Straßner 1982; Schmitz 1990). The traditional news report also contains many features of a written news style, such as a considerable number of complex nominalizations and attributes in noun phrases, extensive lexical variation, and adherence to the 'lead' principle.

Sentence length in television news texts is shorter than in many other written text types, but is clearly different from that in spontaneous oral text types. Straßner (1982: 188) counts 17.5 words per sentence in 'Tagesschau' (ARD) and 13.5 in 'heute' (ZDF). Schmitz (1990: 41) calculates an average of 16.1 for 'Tagesschau', which corresponds to Leska's finding that most written sentences vary from 11 to 18 words (1965: 458), but has to be compared with Leska's statistics for oral utterances (6 to 8 words per sentence) and to the average sentence length of 8 words for dialogues in television plays (Schneider 1974; cf. Brandt 1985: 1676).

The tendency to produce news-shows with anchorpersons strengthens the oral element by conversation-like presentations. Püschel (1992: 19) illustrates this by contrasting two extracts from his data. 'Tagesschau' formulates complete sentences with technical terms (*Kurseinbrüche, Tagesverluste*):

Die internationalen Aktienbörsen haben heute schwere Kurseinbrüche hinnehmen müssen. In Frankfurt gab es sogar die größten Tagesverluste seit Bestehen der Bundesrepublik.

(The international stock markets had to suffer a heavy fall in prices today. In fact in Frankfurt there were the biggest losses in a single day in the history of the Federal Republic.)

The RTL news-show prefers a more elliptical formulation (like a headline), with a metaphorical idiom (*der schwärzeste Tag*), and less formal, more colloquial vocabulary (*verschleuderten; ohne Rücksicht auf Verluste*):

Der schwärzeste Tag seit Jahrzehnten an der Frankfurter Börse. Die Besitzer deutscher Aktien verschleuderten heute ihre Papiere ohne Rücksicht auf Verluste.

(The blackest day for decades on the Frankfurt stock market. The owners of German shares dumped their securities today without considering losses.)

Püschel argues that there is not only a stylistic shift here from a formal, written orientation to a more colloquial, oral one, but also a change in the way in which reality is constructed: namely a concrete presentation with the 'human touch', instead of an abstract, unemotional treatment of the subject-matter.

A further step in the process of 'oralizing' news presentation on

German television is the introduction of *Frühstücksfernsehen* (breakfast TV), first on the two private stations mentioned above, now on ARD and ZDF as well (following pilot programmes in 1990–1). The basic pattern of this new generation of news programmes has been transferred from radio programmes which serve as a kind of background accompaniment (*Begleitprogramme*). Burger (1984: 194–212) describes the main characteristics of this style as: presenter texts that do not necessarily relate to news reports but have an 'independent' function; the integration of 'service' texts giving information, for example, about the weather or traffic conditions; the subjectivity of the presenter; the blending of information and trivial, light-hearted items ('infotainment'). All these features, which are familiar from radio programmes, are used in these kinds of TV news programmes as well, which shows that television is increasingly becoming a background medium.

Finally, it is worth mentioning that for a few years now there has been a special programme which presents news for children and young people in the afternoon ('logo', ZDF). The aim of this programme is not only to provide facts as a means of introducing this target group to the world of politics, using a mixture of hard and soft news and presenting special topics concerning children; it is also to offer explanations of relevant concepts and problems in a clear and comprehensible manner (Aufenanger 1990).

4.2 *Soap Operas*

The TV soap opera has been characterized as the most TV-specific genre on account of its intimacy and continuity (Newcomb 1982*a*). Its origins lie in various literary forms (domestic novel, melodrama, romance) and American commercial radio, where its immediate precursor appeared, sponsored by various companies eager to make their product a household word. Over the years and with the immense growth of television, it has developed into a complex family of genres, which has spread internationally and taken on a multitude of variations.

The German soap opera is still a rather young and somewhat underdeveloped genre, although its precursors, the so-called *Familienserien*, paved the way for its success. Since the 1950s, the *Familienserien* have had a central place in the programming schedule, even in prime time (Schäfer 1973, Wichterich 1979, Rogge 1986), but they do not fit into a strict definition of soap operas (Geraghty 1981, Cathcart 1986, Mikos 1987): they were not 'serials' with a continuing open narrative structure, but limited 'series' of isolated episodes (albeit in one case more than 100).

It was only with the enormous world-wide success of American prime-time variations of the traditional soap opera ('Dallas,' 'Dynasty') during the 1980s that a change in German (and French) production and reception of serials began (Frey-Vor 1990). However, neither of the best-known German 'soaps' followed the American model. 'Die Schwarzwaldklinik' (ZDF), seventy partly isolated, partly connected episodes of a hospital series, was more in the tradition of the German 'family, *Heimat*, and doctor' genre. 'Lindenstraße' (ARD) follows the British model of socio-realistic soaps like 'Coronation Street', but without the humour, and in this respect it resembles the American daytime soaps. Nevertheless, it is the first German soap opera in the strict sense of the term; the manner of production, the dramatic structure, the interplay with print media, and the forms of reception are comparable to the Anglo-American situation. 'Lindenstraße' has been running since 1985 in episodes of thirty minutes, though only once a week. With its socially and politically orientated topics and its actuality, it is seen as kind of a mirror of (West) German reality.

'Lindenstraße' is produced by a private company led by the well-known film director and author Helmut Geißendörfer in co-operation with the public station Westdeutscher Rundfunk. The production process is organized along industrial lines (Paetow 1989). The producer and the authors form a team to provide new ideas. Story-lines are then submitted to the editor for discussion and must receive the approval of a special advisory council (Cieslik 1991); they are fixed two years in advance. Each director is responsible for a limited number of episodes. It is shot using video, which in comparison to film is an inexpensive, quick, and simple medium, but which lacks opulent visuals, concentrating instead on the actors and their dialogues.

The dramatic structure contains all the generic elements of the soap opera form (Geraghty 1981, Mikos 1987, Fiske 1987): the endless, open narrative structure, segmented into episodes which are broken off at a climactic point (the so-called cliff-hanger, usually psychological in nature) so that the recipient is encouraged to anticipate the possible outcome; a set of different characters connected spatially and socially, the focus being the street Lindenstraße; the fiction of real time which is partly parallel to the recipient's life; a plurality of plots partly interwoven with each other; short scenes which are interrupted in a moment of suspense ('mini-cliffs') so that they are mostly combined using abrupt segmentation; the emphasis on dialogue, problem-solving, changing emotional perspectives, and intimate conversation.

Taken together, the characters and the topics create the 'realistic' format of 'Lindenstraße'; the characters are mostly middle-class, but some of them represent so-called minorities (foreigners, gays, radicals, punks). This provides opportunities for dealing with recurring day-to-day topics (love, marriage, puberty, age, illness, housing, work, etc.), but also for the treatment of acute social and political problems, such as drugs, ecology, neo-nazism, racism, AIDS, or German unification. This personal and topical profile is clearly different from those American prime-time serials which are located in a milieu of luxury and based on rivalry between family clans, a concept which was imitated in the ZDF serial 'Das Erbe der Guldenburgs'. Geißendörfer, the creator of 'Lindenstraße', wants to be realistic and thus claims that the show has an informing, educative, and politically motivating effect on the audience (Cieslik 1991).

The mediation of events is mainly linguistic and is effected by dialogues, which are the most important element of dramatic structure. The conversational mode allows for emotional and personal perspectives and a simultaneous, non-evaluative plurality of meanings which opens up the whole story to different readings (Fiske 1987: 194–5). On the other hand, it is difficult to integrate problems in the plot if they cannot be connected with the biography of a character in a convincing manner (Cieslik 1991).

The success of 'Lindenstraße' (it has up to 10 million viewers) confirms that this TV genre is well established for the German audience and has introduced special forms of reception. The emotional access, the plurality of characters, topics, and readings, the framework of day-to-day events, the gossip-like style which is connected to the everyday experiences of its audience: all these features promote identification with the microcosm of the soap-opera world and thus give some moral orientation without hard and fast rules (Lang 1991). Ang (1985) analyses letters from 'Dallas' viewers and gives examples of the ways in which these opportunities of identification are actually used, and the following extract from an authentic cafeteria conversation among students about 'Lindenstraße' may illustrate how the gossip mode of the soap opera is continued in real life and helps to integrate this genre into the world of the recipients:

A.   Hab gestern abend die Lindenstraße verpaßt, konnt ich nich gucken, da war ich auf der Autobahn. . . . Was war denn gestern abend? Lebt Robert Engel noch?

(Missed 'Lindenstraße' yesterday evening, I couldn't watch it, I was on the motorway at the time. . . . What happened yesterday evening? Is Robert Engel still alive?)

B.   Robert Engel lebt. Ist gar nicht getroffen worden. Und Onkel Franz meint, er wärs gewesen. Er hätt geschossen, weil er an dem Abend ziemlich angeheitert war, wie er sagte. . . . Trug aber gestern mal wieder nen auffälligen Schal.

(Robert Engel is alive. Wasn't hit. And uncle Franz thinks he did it. He shot because he was rather tipsy that evening, as he put it. . . . Was wearing a striking scarf again yesterday though.)
c.   Wer? (Who?)
b.   Frau von der Marwitz. (Mrs von der Marwitz.)
d.   Hat die Narben am Hals oder so?

(Has she got scars on her neck or something like that?)
c.   Nee, Knutschflecke. (No, love bites.)

In 1992 RTL started a daily soap opera, 'Gute Zeiten, schlechte Zeiten', which is shown Monday to Friday, with episodes composed of even shorter scenes than usual. Although it was a new production, it was not an original concept but an adaptation of an Australian serial, rewritten by a team led by the very successful author Felix Huby. The initial reaction of the critics was scathing, but the launching of this new soap seems to be a sign that the genre is gaining a firmer foothold in German TV.

4.3   *Quiz- and Game-Shows*

Like soap operas, quiz- and game-shows are 'popular' genres. Critics disparage them for their triviality, but these programmes constantly achieve high audience ratings, particularly the Saturday night 'gala'-type shows (Holly 1992*a*). Like soap operas, they satisfy the main function of television: to entertain. Like soap operas, they are easily integrated into the everyday life of the recipients. In contrast to soap operas, however, they have held a central place in German television right from the start, though here, too, many shows have been 'imported', being adapted versions of foreign, mostly Anglo-American programmes, such as 'What's My Line?', 'Twenty Questions', 'To Tell the Truth', 'Blockbusters', 'Child's Play', 'The $100,000 Pyramid', 'Game for a Laugh' (Hallenberger and Foltin 1990: 74–5).

So quiz- and game-shows have become a typical genre family on German television, but their roots go back to other media and genres, to everyday forms of entertainment, and to serious real-life situations. For example, there were quiz broadcasts on radio (though not during the Nazi period) and various show elements (music, dance, and the like) come from operetta, music-hall, revue, vaudeville, or circus traditions. Hickethier (1979: 53) refers to the parallel between fictional dramatic structures and the problem- or conflict-solving situation of contestants in quiz- and game-shows. Fiske (1987: 266–80) points out the structural and ideological similarities to the educational system and to competition in professional and social life, where knowledge counts as what Bourdieu (1979) calls 'cultural capital' and luck makes success available to all; Fiske also argues that these genres are well established in the oral culture of family and party games.

The fundamental structure of the genres is quite simple: the 'dramatic core' is the test situation of the contestant, consisting of the quiz question or a task to be accomplished by using a skill (sometimes performing a 'crazy act' in a *Guinness Book of Records* manner). It is embedded in a 'narrative' frame where the single parts are connected and the audience is addressed (Woisin 1989: 62). As Fiske (1987: 265) points out, this is at the same time an interplay of 'games' and 'rituals' in Lévi-Strauss's sense: whereas games differentiate equal partners into winners and losers and thus have a separating effect, rituals equalize the different participants, in the beginning to equal contestants and in the end to equal members of their groups; rituals therefore have an integrating function. In order to make this fundamental structure a bit more complex and diverting, interludes are sometimes inserted into television game-shows, which, apart from bringing variety, serve as a delaying device helping to increase the suspense (Hickethier 1979: 53).

There are several types of show which share this common structural base but which may be distinguished in terms of subject-matter and sequence structures. The following typology is a modified version of the one proposed by Hallenberger and Foltin (1990: 120–8), with examples drawn from German broadcasts (for an alternative typology, see Fiske 1987: 269):

1 quiz broadcasts, concerning:
1.1   general knowledge (e.g. 'Einer wird gewinnen')
1.2   special knowledge (e.g. 'Der große Preis', 'Alles oder nichts')

2 TV party games, concerning:
2.1   concepts (e.g. 'Dalli-Dalli', 'Die Pyramide', 'Dingsda')
2.2   personal characteristics (e.g. 'Was bin ich?', 'Herzblatt')
2.3   opinions (e.g. some games in 'Mensch Meier')

3 action games, concerning:
3.1   'crazy acts' (e.g. 'Wetten daß . . . ?')
3.2   sporting acts (e.g. 'Spiel ohne Grenzen')
3.3   social behaviour (e.g. 'Wünsch Dir was', '4 gegen Willi')

For the sequence structure, they propose five types (Hallenberger and Foltin 1990: 141–2):

1 'simple game': 1 game or round (I), one contestant or group (A)
     →formula (I/A)
2 'repetition': the same game (I) with different contestants (A, B . . .)
     →formula (I/A, B . . .)

3   'addition': several games (I, II . . .) with the same contestants (A)
     →formula (I, II . . . /A)
4   'competition': several games (I, II . . .) with the same contestants (A),
     some of whom, however, are eliminated, so that the group of contest-
     ants becomes smaller (a)
     →formula (I/A, II/a)
5   'string of pearls': several games (I, II . . .), each with new contestants
     (A, B . . .)
     →formula (I/A, II/B . . .)

In the history of German television, all of these types and a range of mixed types have appeared, though games based on guessing public opinion (type 2.3) are rare and then only form one part of a composite game-show. In 1986, the two public channels broadcast 47 different shows, 19 in prime time, 16 in the early evening, 11 in the afternoon, and one in the later evening (Hallenberger and Foltin 1990: 100).

Are there general lines of development that can be observed in the history of German television? The survey by Hallenberger and Kaps (1991) provides a rich source of information but this still requires interpretation. For our purposes, a comparison of two major Saturday night shows from the years 1987 and 1989 may illustrate what seems to be a general tendency (cf. Holly 1992*a*).

'Einer wird gewinnen', with its very popular host (Hans-Joachim Kulenkampff), was one of the most successful shows on German television, and only ended in 1987 after a run of more than twenty years. It was a traditional verbal quiz with a few skill games (types 1.1 and 3), starting with eight contestants from different European countries, who had to compete with each other for six rounds.

Another show, which is still running, is called 'Wetten daß?' (equivalent to 'You bet!'). It was started by Frank Elstner, who was then succeeded in 1987 by Thomas Gottschalk, whose popularity increased enormously as a result. He is now one of the best-paid German TV stars. 'Wetten daß . . . ?' represents the more modern style of action-orientation. It has a carefully constructed 'string of pearls' structure, consisting of 'crazy acts', with the guests betting on the likelihood of these acts being performed successfully. In this case, the guests are well-known personalities, who by appearing on the show hope to promote their own latest media products or simply to enhance their image. At the end, the viewers' favourite 'crazy act' performer is elected by means of a telephone poll, which provides a thrilling finish (!). An additional element provides underlying suspense throughout the show: at the start of the programme, a task is imposed by a member of the live audience, and at the very end it has to be performed by the host and his assistants, often with the aid of viewers at home.

A detailed analysis reveals how this newer version of a familiar TV genre takes on a form more suited to the medium, namely a more intimate and intensive visual presentation (Holly 1992*a*: 22–5), the staging of a spectacular event which can be experienced 'live'. Even the structure of the linguistic patterns the host has to use underscores the differences. In the traditional verbal quiz, Kulenkampff has to repeat several times the same limited sequence of patterns: introducing and warming up contestants (8 times), explaining and imposing tasks, and finally judging the solutions (22 times). The job of providing entertaining variation also falls on the shoulders of the host, who constantly has to make small talk with the contestants, a rather arduous business. In contrast, the action-show repeats the betting sequence only four times, on each occasion introducing a prominent person, who of course is more experienced in the art of entertaining talk, while the contestants simply have to perform their well-rehearsed stunt. The different set-up also means that the number and type of 'gags' is different: the verbal quiz needs more than twice as many verbal gags, mostly of the delicate type 'teasing the guests' and, for reasons of ritual balance, self-deprecation on the part of the host. The action-show, however, almost entirely dispenses with this kind of exchange: Gottschalk, for instance, favours frequent erotic allusions, presumably hoping to support his image of being a 'charmer'.

The advent of private television has substantially increased the number of quiz- and game-shows. Their low production costs and relatively high audience ratings make them the ideal type of broadcast to sandwich between commercials. In fact, in some cases the demarcation lines are blurred, as the shows are only pretexts for presenting commodities, for example 'Der Preis ist heiß' (RTL) or 'Glücksrad' (SAT 1). Through this new development in the private channels, German televison, once praised for its high standards, has now adapted more to international conventions. Many critics complain that this increasing orientation towards entertainment has led the public channels to a process of voluntary self-commercialization.

4.4 *Advertising*

Advertising is above all an economic factor in television production. Both systems in Germany, public and private, depend on advertising income, though to a different degree. The ZDF gets over 40 per cent of its income through advertising; for the ARD, the proportion sank in 1990 from 20 per cent to 15 per cent (Ridder 1991), but has now stabilized. The losses correspond to the growing importance of the private channels; these already occupy three-fifths of the TV advertising market, whereas legal requirements put a limit on the growth of advertising on the public channels (Storck 1992). Overall, the television advertising business is still booming, increasing from DM200 million in 1963, to DM1,300m. in 1979, 3,700m. in 1990, and 4,900m. in 1991 (Lützen 1982, Pretzsch 1991, Storck 1992).

But apart from its economic importance, advertising has become a specific television genre in its own right, in the minds of the producers and of the recipients. There is already a range of traditional types of commercial (Lützen 1979), such as: 'product as hero', where the product is central; 'slice of life', with everyday scenes; 'problem-solving', often with a riddle involving the recipient; 'presenter format', with someone recommending the product in an old-fashioned informing style; 'testimonial', where consumers praise the product; 'demonstration', presenting the product and how it functions; 'jingle' in the form of songs or musical elements; and 'news format', imitating the news genre to convey information about the product.

The development of the genre consists in a tension between convention and innovation. On the one hand, the commercial is the most expensive form of media communication, so it has to be extremely short: ten to forty seconds must be sufficient to deliver the message. This brevity necessitates fixed patterns and allusions to other well-established genres. Like the eighteenth-century novel, commercials absorb and integrate all kinds of elements from other genres (Landbeck 1989), indeed they depend on intertextuality.

On the other hand, the commercial has to get and keep the attention of a recipient who is bombarded by more and more advertising and tends to switch off, either mentally or physically. So it has to offer something new and surprising. Commercials should therefore always be creative, innovative, breaking norms and conventions, deviating from ingrained habits of reception. Alternatively, they should offer pleasure and entertainment, allowing the viewer to identify with the figures on the screen, which quite apart from all commercial considerations makes the process of reception an agreeable experience. This tendency to provide more and more 'fun' and titillation leads to a greater variety of styles and types, and to an increasing degree of aesthetic sophistication. Increasingly, too, the recipient is challenged to be active, to participate: the necessary brevity of the individual commercial, the build-up of suspense, and the goal of catching and keeping the recipient's attention all require an 'open' text which has gaps to be filled in.

German commercials have hitherto been considered boring and not very creative. An international comparison measuring the aesthetic status of commercials in six European countries between 1985 and 1987 (Landbeck 1989) reflects this impression. Whereas British commercials were quite long, but perfectly conceived in their connection of varying, often humorous forms and the subtle involvement of the recipient, and whereas French commercials are clearly shorter, more risqué, and more experimental, sometimes even asking too much of the recipient, the German commercials still followed pseudo-informative and other traditional patterns.

But the picture is changing a bit today. With the growing market and increased audience participation, new aesthetic concepts are coming to the fore, including even humorous scenes; pleasure and lifestyle, colours and atmosphere are all emphasized, and the commercials are becoming more ambitious and are trying to meet international standards. To illustrate this, I shall describe what seems to be a typical example of the current trend.

It is a twenty-second traditional jingle-type commercial for a brand of diet food called 'Du darfst' (you may). The accompanying music together with one singing and one speaking female voice coming from off-screen provide the rhythmic framework. There are twelve scenes: nine of them present a young, slim, and attractive woman in a red dress sauntering through a park and through streets, and looking at herself in a shop window; finally, she turns around and looks into the camera. Only three very short close-ups embedded into the stroll scenes show the 'Du darfst' products (indicated in the transcription by *). The sung and spoken text is a kind of inner dialogue playing with the brand name and its inherent appeal, inviting identification:

[*sung*]	ich will so bleiben wie ich bin (I want to stay the way I am)	[scene 1]
[*whispered*]	du darfst (you may)	[2]
[*sung*]	will so bleiben wie ich bin (want to stay the way I am)	[3]
	du darfst hat alles was mir schmeckt (you may has everything that tastes good to me)	[4*, 5, 6]
[*whispered*]	du darfst (you may)	[7*]
[*sung*]	ich hab du darfst für mich entdeckt (I've discovered you may)	[8, 9]
[*spoken*]	du darfst — meine art zu leben (you may — my way of living)	[10*, 11]

[*whispered*]   du darfst   [12]
                  (you may)

The song consists of two verses, each with two lines and each interrupted by a whispering voice, a kind of good conscience giving permission and the brand name, both at the same time. The final spoken slogan repeats the pseudo-dialogue once again, naming the product twice. The text shows that, even in a modern TV commercial, language can have an important function, if it is stylized. The extreme emphasis on references to the first person (seven times: *ich, mir, mich, meine*) and to the individual will, together with the contrast to the brand name, creates a powerful message. The delicate topic of weight is controlled not by a severe superego but a permissive one, and the weight problem is overcome by a proud, self-confident attitude allowing the consumer to discover the pleasures of supposed individuality and narcissism.

A text that argued explicitly would reveal the weaknesses of the inference; viewers might become suspicious about this positive view of the problem of weight: it would either be euphemistic or reinforce anxieties about not being accepted by anyone, oneself included; thus the text would weaken the viewers' self-confidence instead of confirming it. The actual text is more subtle: it does arouse negative feelings about weight, which might encourage viewers to buy the product, but not in an explicit way that could easily be seen as a trick. The poetic techniques of sound structure and rhyme, the embedding of the text in the music and the atmosphere of the pictures shift the discourse away from delicate matters and into the area of pleasure and lifestyle. Intertextual relationships with other song texts mobilize a familiar ideology: 'I did it my way', 'I am what I am'. The product itself remains in the background.

However, despite the perfect match of language, music, and pictures, this commercial is still rather conventional. Although it avoids the simple recommending style, and is not verbally explicit and trivial, it has no humour, no refinements of story-line or of visual arrangements, no elements of parody, and no surprising twist. It illustrates quite well the fact that the German genre of TV commercials has not yet reached the standards set elsewhere.

5   THE IMPACT OF TELEVISION ON THE GERMAN LANGUAGE

As I have already suggested, research into the effects of television on the German language has so far not been particularly satisfactory. The reason for this is obvious: if it is difficult to determine which linguistic phenomena are TV-specific, it is almost impossible to identify any effects. However, some hypotheses and tendencies have been proposed, for both whole varieties and sub-varieties, and in terms of individual levels of linguistic description (cf. Brandt 1985; Holly and Püschel 1993).

5.1 *Popularization of the Standard Language*

After the Second World War, the use of the spoken standard varieties of German spread rapidly, while the use of dialects continued to decrease; this process can be explained at least partly by the continuing urbanization of rural regions (Mattheier 1980: 162). Standard varieties have increasingly become the normal means of oral communication; this applies to dialect-speakers as well, in public and formal situations. There seems to be no doubt that first radio, and later on television, supported the growth of passive competence in the standard forms, and then began to reinforce and perhaps even to accelerate the transition to active use (Drosdowski and Henne 1980: 620).

It may also be argued that the shift from dialect to standard is at the same time a shift from an oral to a written style in the spoken language. This development has been strengthened by the fact that a large proportion of texts on radio and television are not only written and read, but are orientated towards the formal standard variety.

5.2 *Relaxation of Norms in the Spoken Standard Varieties*

In addition to the view that there is a continuing trend towards a formal and written orientation in spoken language, it could also be argued conversely that the formal standard is no longer the exclusive variety in public situations. The electronic media have also supported a relaxation of prevailing linguistic norms and the diffusion of colloquial standard varieties, in both formal and informal situations (for the distinction between formal and colloquial standard, see Barbour and Stevenson 1990: chap. 5).

The reason seems to be that orality in the electronic media is no longer understood as just the phonic realization of language, but as a stylistic feature which seeks to turn away from written texts and towards colloquial and conversational forms, which are better suited to the acoustic medium. So we find more elements of spoken language in written media texts and more conversational genres (formal and informal interviews, talk-shows, statements by politicians) (Burger 1984 and 1991, Holly forthcoming), both of which influence general language use as models of colloquial standard forms.

5.3   *Knowledge of Other Varieties*

The different genres of television deal with all kinds of topics: there are programmes about science, economics, politics, and culture, where the terminology of science and technology and other technical registers are conveyed. In this way, at least the passive competence of the recipients is increased (Polenz 1978: 137), and some expressions will then enter their active repertoire as well. So television supports the tendency towards the 'technologizing' and 'scientification' of everyday language (Drosdowski and Henne 1980: 620).

A similar function may apply to sociolects, such as the language of youth subcultures (Henne 1986: 185–200), so that one may ask whether the media reflect the language of these subcultures or vice versa (see also Schlobinski, this volume). Sociolects occur not only in advertising, but also in commentary texts or soap operas (for example in 'Lindenstraße'), with the aim of addressing the respective target groups and creating a particular atmosphere. At the same time, other groups become familiar with these 'alien' varieties. For the same reason, regional varieties are spread via television in the literary form of folk theatre. Dialectal and sociolectal features are used in some series, although mostly in a rather stylized form. In this way, television supports, if not real competence, at least a certain tolerance of these varieties.

It should also not be forgotten that for many years radio and television provided a direct opportunity for people in the GDR to become acquainted with neologisms and stylistic changes in Western Germany, especially Anglo-American influences (Polenz 1978: 175–6). It is interesting to note, though, that this flow of linguistic information went only in one direction, as developments in the GDR were hardly noticed in the West.

5.4   *Diffusion and Reinforcement of Linguistic Changes*

Taken at the level of genres and styles, television is clearly innovative (see Section 4 above), and some television genres are even transferred into other media (for instance talk-shows in theatres). However, this innovative function hardly applies at the levels of sound, word, or sentence. It may be that speaking without any emotion, as practised by newsreaders, has been imitated by politicians, officials, scientists, or academics, even in situations which would allow or require more variation (Polenz 1983: 45). But the main effect of television on these linguistic levels seems to be in the diffusion and reinforcement of linguistic developments which are already taking place. The following examples may illustrate some of these processes:

- Whereas radio speakers are typically models of conventionally 'correct' speech behaviour, this does not necessarily apply to television presenters; this may have an effect on general attitudes towards norms in pronunciation.
- Since 1945, there has been a tendency not to integrate Anglo-American loanwords into the German system of pronunciation (Drosdowski and Henne 1980: 622; see also Sauer and Glück, this volume). This is not only because more people now learn English; the frequent occurrence of the original pronunciations on radio and television seems to support this tendency.
- Television language tends to use more oral texts and colloquial forms; this may lead to the diffusion of morphological and syntactic phenomena characteristic of colloquial standard and non-standard varieties, such as genitives of proper names without -*s* (Brandt 1985: 1674).
- An enlarging of the vocabulary seems to be an obvious effect of television; this medium reaches far more people than any other. Our passive knowledge of words is much greater than that of earlier generations; it is possible, however, that the price for this is problems with semantics (Polenz 1978: 137). At all events, neologisms can spread very rapidly.

5.5 Television and Language Culture

The academic discussion about the impact of television on the German language system is rather tentative; not so the public discussion about its effect on language culture in a wider sense. There is a traditional pessimism concerning the pernicious influence of any new medium: critics like Neil Postman have become popular in Germany and have found their place in the rhetoric of media education. Some of the standard hypotheses are:

- Television as a primarily visual medium represses the importance of language altogether.
- Television as an oral medium represses literacy.
- Television as a public medium endangers the quality of public discourse.
- Television as a family medium endangers the private conditions of communication.
- Television as an increasingly international, Americanized medium endangers the national language culture.

Of course, none of these extremely general statements can be verified or falsified, and they have provoked counter-attacks (e.g. Maletzke 1988). There is evidence that this debate has raised interesting questions, but the pessimistic position is characterized by exaggerated claims. In the current state of research, it is not possible to reach any firm conclusions; for this, we would need a more detailed understanding of the linguistic processes of individual and group reception. Only the concrete linguistic behaviour of the recipients can reveal how language in television is interpreted, transferred, and adapted into a language culture.

6 CONCLUSIONS

The topic of language and television is still largely unexplored territory. Traditional research on mass communication has ignored language, just as linguistics has ignored the media. A major problem for any research in this area may be that the technical and institutional situation is subject to rapid and continual change. In the course of the last decade the structure of German television has changed almost completely and is still in a state of flux, first because of new technologies (cable, satellite TV) and the installation of private channels, and more recently of course because of the new political situation following unification. The television station of the former GDR has been shut down, and in its place there are now two new members of the ARD, Mitteldeutscher Rundfunk (MDR) and Ostdeutscher Rundfunk (ODR). As yet, it is too early to say exactly what the consequences of these far-reaching developments will be, but future research will clearly have to take these new technological and institutional conditions into account, and should not only deal with both well-established and new genres (in the form of classical 'product studies'), but also include aspects of production, mediation, and reception.

Further reading

Biere and Henne (1993)
Brandt (1985)
Burger (1984/1990)
Kreuzer (1982)
Kreuzer and Prümm (1979)
Schmitz (1987)
Straßner (1980)

References

ALTHAUS, H.-P., HENNE, H., and WIEGAND, H. E. (eds) (1980), *Lexikon der germanistischen Linguistik* (Tübingen: Niemeyer).

AMMON, U., DITTMAR, N., and MATTHEIER, K. J. (eds.) (1987), *Sociolinguistics. Soziolinguistik* (Berlin, New York: de Gruyter)

ANG, I. (1985), *Watching Dallas: Soap Opera and the Melodramatic Imagination* (Amsterdam: Uitgeverij SUA).

AUFENANGER, ST. (1990), 'Kindernachrichten "logo". Politische Bildung auf neuen Wegen', *Medien praktisch*, 3: 17–19.

BARBOUR, S., and STEVENSON, P. (1990), *Variation in German* (Cambridge: Cambridge University Press).

BARTHES, R. (1975), *S/Z* (London: Cape).

BENNETT, T., BOYD-BOWMAN, S., MERCER, C., and WOOLLACOTT, J. (eds.) (1981), *Popular Television and Film* (London: British Film Institute/Open University).

BENTELE, G., and HESS-LÜTTICH, E. W. B. (eds.) (1985), *Zeichengebrauch in Massenmedien. Zum Verhältnis von sprachlicher und nicht-sprachlicher Information in Hörfunk, Film und Fernsehen* (Tübingen: Niemeyer).

BESCH, W., REICHMANN, O., and SONDEREGGER, S. (eds.) (1985), *Sprachgeschichte. Ein Handbuch zur Geschichte der deutschen Sprache und ihrer Erforschung* (Berlin, New York: de Gruyter).

BIERE, B. U., and HENNE, H. (eds.) (1993), *Sprache in den Medien nach 1945* (Tübingen: Niemeyer).

BOURDIEU, P. (1979), *La Distinction. Critique sociale du jugement* (Paris: Éditions de Minuit).

BRANDT, W. (1985), 'Hörfunk und Fernsehen in ihrer Bedeutung für die jüngste Geschichte des Deutschen', in Besch *et al.* (1985), 1669–78.

BUCHWALD, M. (1990), 'Nachrichtensendungen', in Schult and Buchholz (1990), 243–6.

BURGER, H. (1984/1990), *Sprache der Massenmedien* (Berlin de Gruyter).

—— (1991), *Das Gespräch in den Massenmedien* (Berlin: de Gruyter).

CATHCART, R. (1986), 'Our Soap Opera Friends', in Gumpert and Cathcart (1986), 207–18.

CHAFE, W. L. (1982), 'Integration and Involvement in Speaking, Writing, and Oral Literature', in Tannen (1982), 35–53.

CIESLIK, N. (1991), 'Beste Wohnlage. Produktion, Anspruch und Probleme der "Lindenstraße"', *Medien praktisch*, 4: 19–22.

DROSDOWSKI, G., and HENNE, H. (1980), 'Tendenzen der deutschen Gegenwartssprache', in Althaus *et al.* (1980), 619–32.

DYER, R., GERAGHTY, C., JORDAN, M., LOVELL, T., PATERSON, R., and STEWART, J. (1981), *Coronation Street* (London: British Film Institute).

EBNER, W. (1986), *Kommunikative Probleme tagesaktueller Berichterstattung im Fernsehen. Dargestellt am Beispiel der 'Landesschau Baden-Württemberg'* (Frankfurt: Lang).

ECO, U. (1979), *The Role of the Reader: Explorations in the Semiotics of Texts* (Bloomington: Indiana University Press).

Fiske, J. (1987), *Television Culture* (London, New York: Methuen).

Frey-Vor, G. (1990), 'Charakteristika von Soap operas und Telenovelas im internationalen Vergleich', *Media Perspektiven*, 8: 488–96.

Friedrich, H. (ed.) (1977), *Kommunikationsprobleme bei Fernsehnachrichten* (Tutzing).

Geraghty, C. (1981), 'The Continuous Serial – A Definition', in Dyer *et al.* (1981), 9–26.

Grewenig, A. (ed.) (1992), *Inszenierte Information. Politik und strategische Kommunikation in den Medien* (Opladen: Westdeutscher Verlag).

Gumpert, G., and Cathcart, R. (eds.) (1986), *Inter/Media: Interpersonal Communication in a Media World*, 3rd edn. (New York, Oxford: Oxford University Press).

Habermas, J. (1962), *Strukturwandel der Öffentlichkeit* (Darmstadt, Neuwied: Luchterhand).

Hallenberger, G., and Foltin, H. F. (1990), *Unterhaltung durch Spiele. Quizsendungen und Game Shows des deutschen Fernsehens* (Berlin: Spiess).

—— and Kaps, J. (eds.) (1991), *Hätten Sie's gewußt? Die Quizsendungen und Game Shows des deutschen Fernsehens* (Marburg: Jonas).

Halliday, M. A. K. (1985), *Spoken and Written Language* (Oxford: Oxford University Press).

Hauptmeier, H. (1987), 'Sketches of Theories of Genre', *Poetics*, 16: 397–430.

Henne, H. (1986), *Jugend und ihre Sprache. Darstellung, Materialien, Kritik* (Berlin: de Gruyter).

Hess-Lüttich, E. W. B. (ed.) (1992), *Medienkultur – Kulturkonflikt. Massenmedien in der interkulturellen und internationalen Kommunikation* (Opladen: Westdeutscher Verlag).

Hickethier, K. (1979), 'Fernsehunterhaltung und Unterhaltungsformen anderer Medien', in Rüden (1979), 40–72.

Hoffmann, R.-R. (1982), *Politische Fernsehinterviews. Eine empirische Analyse sprachlichen Handelns* (Tübingen: Niemeyer).

Holly, W. (1992*a*), 'Die Samstagabend-Fernsehshow. Zu ihrer Medienspezifik und ihrer Sprache', *Muttersprache*, 102: 15–36.

—— (1992*b*), 'Was kann Kohl was Krenz nicht konnte? Deutsch-deutsche Unterschiede politischer Dialogrhetorik in zwei Fernsehinterviews', *Rhetorik*, 11: 33–50.

—— (1992*c*), 'Zur Inszenierung von Konfrontation in politischen Fernsehinterviews', in Grewenig (1992), 164–197.

—— (forthcoming), 'Secondary Orality in the Electronic Media', in Quasthoff (forthcoming).

—— and Püschel, U. (1993), 'Sprache und Fernsehen in der Bundesrepublik Deutschland', in Biere and Henne (1993), 128–157.

—— Kühn, P., and Püschel, U. (1986), *Politische Fernsehdiskussionen. Zur medienspezifischen Inszenierung von Propaganda als Diskussion* (Tübingen: Niemeyer).

———————— (eds.) (1989), *Redeshows. Fernsehdiskussionen in der Diskussion* (Tübingen: Niemeyer).

HORTON, D., and WOHL, R. R. (1956), 'Mass Communication and Para-Social Interaction: Observation on Intimacy at a Distance', *Psychiatry*, 19: 215–28.

HUTH, L. (1977), 'Ereignis, Objektivität und Präsentation in Fernsehnachrichten', in Friedrich (1977), 103–23.

—— (1985), 'Zur handlungstheoretischen Begründung der verbalen und visuellen Präsentation in Fernsehnachrichten', in Bentele and Hess-Lüttich (1985), 128–36.

KEPPLER, A. (1985), *Präsentation und Information. Zur politischen Berichterstattung im Fernsehen* (Tübingen: Narr).

KREUZER, H. (ed.) (1982), *Sachwörterbuch des Fernsehens* (Göttingen: Vandenhoeck and Ruprecht).

—— and PRÜMM, K. (eds.) (1979), *Fernsehsendungen und ihre Formen. Typologie, Geschichte und Kritik des Programms in der Bundesrepublik Deutschland* (Stuttgart: Reclam).

KÜBLER, H. -D. (1975), *Unterhaltung und Information im Fernsehen. Dargestellt am Beispiel der Abendschau in Baden-Württemberg* (Tübingen: Vereinigung für Volkskunde).

LANDBECK, H. (1989), 'Ästhetik in der Fernsehwerbung. Eine europäische Studie', *Media Perspektiven*, 3: 138–45.

LANG, P. CH. (1991), 'Wir wohnen alle in der Lindenstraße. Bemerkungen zum Fernsehen als moralische Anstalt', *Medien praktisch*, 4: 23–4.

LESKA, CH. (1965), 'Vergleichende Untersuchungen zur Syntax geschriebener und gesprochener Gegenwartssprache', *Beiträge zur Geschichte der deutschen Sprache und Literatur (Halle)*, 87: 427–61.

LINKE, A. 1985, *Gespräche im Fernsehen. Eine diskursanalytische Untersuchung* (Bern: Lang).

LUDWIG, R. (1986), 'Mündlichkeit und Schriftlichkeit. Felder der Forschung und Ansätze zu einer Merkmalsystematik im Französischen', *Romanistisches Jahrbuch*, 37: 15–45.

LÜTZEN, W. D. (1979), ' "Das Produkt als 'Held' " – und andere Typen der Fernsehwerbung', in Kreuzer and Prümm (1979), 230–48.

—— (1982), 'Werbefernsehen', in Kreuzer (1982), 203–6.

MALETZKE, G. (1988), *Kulturverfall durch Fernsehen* (Berlin: Spiess).

MATTHEIER, K. (1980), *Pragmatik und Soziologie der Dialekte* (Heidelberg: Quelle und Meyer).

MEYROWITZ, J. (1985), *No Sense of Place: The Impact of Electronic Media on Social Behavior* (New York: Oxford University Press).

MIKOS, L. (1987), 'Fernsehserien. Ihre Geschichte, Erzählweise und Themen', *Medien und Erziehung*, 31: 2–16.

MUCKENHAUPT, M. (1986), *Text und Bild. Grundfragen der Beschreibung von Text–Bild–Kommunikation aus sprachwissenschaftlicher Sicht* (Tübingen: Narr).

MÜHLEN, U. (1985), *Talk als Show. Eine linguistische Untersuchung der Gesprächsführung in den Talkshows des deutschen Fernsehens* (Frankfurt: Lang).

NEALE, S. (1981), 'Genre and Cinema', in Bennett *et al.* (1981), 6–25.

Neugebauer, E. (1986), *Mitspielen beim Zuschauen. Analyse zeitgleicher Sportberichterstattung des Fernsehens* (Frankfurt: Lang).

Newcomb, H. (1982*a*), 'Toward a Television Aesthetic', in Newcomb (1982*b*).

—— (ed.) (1982*b*), *Television: The Critical View* (New York: Oxford University Press).

Ong, W. J. (1982), *Orality and Literacy: The Technologizing of the Word* (London: Methuen).

Paetow, M. (ed.) (1989), *Lindenstraße — das Buch: Geschichten, Bilder, Hintergründe* (Düsseldorf: Zeitgeist).

Polenz, P. von (1978), *Geschichte der deutschen Sprache*, 9th edn. (Berlin: de Gruyter).

—— (1983), 'Deutsch in der Bundesrepublik Deutschland', in Reiffenstein *et al.* (1983), 41–59.

Postman, N. (1985), *Amusing Ourselves to Death. Public Discourse in the Age of Showbusiness* (New York: Delacorte Press).

Pretzsch, D. (1991), 'Werbefernsehboom hält an. Die Entwicklung in den klassischen Medien 1990', *Media Perspektiven*, 3: 147–60.

Püschel, U. (1992), 'Von der Pyramide zum Cluster. Textsorten und Textsortenmischung in Fernsehnachrichten', in Hess-Lüttich (1992).

Quasthoff, U. (ed.) (forthcoming), *Aspects of Oral Communication* (Berlin: de Gruyter).

Reiffenstein, I., Rupp, H., Polenz, P. von, and Korlén, G. (1983), *Tendenzen, Formen und Strukturen der deutschen Standardsprache nach 1945* (Marburg: Elwert).

Ridder, Ch.-M. (1991), 'Fernsehen und Werbemarkt in Deutschland. Prognosen, Fakten und Argumente zur Auseinandersetzung um die Werberegeln für den öffentlich-rechtlichen Rundfunk', *Media Perspektiven*, 8: 489–503.

Rogge, J.-U. (1986), 'Tagträume oder warum Familienserien so beliebt sind. Zur Geschichte, Machart und psycho-sozialen Funktion von Familienserien im deutschen Fernsehen', *Der Bürger im Staat*, 36: 201–6.

Rüden, P. von (ed.) (1979), *Unterhaltungsmedium Fernsehen* (Munich: Fink).

Rusch, G. (1987), 'Cognition, Media Use, Genres: Socio-Psychological Aspects of Media and Genres; TV and TV-Genres in the Federal Republic of Germany', *Poetics*, 16: 431–69.

Schäfer, H. (1973), *Struktur-Untersuchungen zur Situation der Familie vor und auf dem Bildschirm* (Marburg).

Schmidt, S. J. (1987), 'Towards a Constructivist Theory of Media Genre', *Poetics*, 16: 371–95.

Schmitz, U. (1987), 'Sprache und Massenkommunikation', in Ammon *et al.* (1987), 820–32.

—— (1990), *Postmoderne Concierge: Die 'Tagesschau'. Wortwelt und Weltbild der Fernsehnachrichten* (Opladen: Westdeutscher Verlag).

Schneider, P. (1974), *Die Sprache des Sports. Terminologie und Präsentation in Massenmedien. Eine statistisch vergleichende Analyse* (Düsseldorf: Schwann).

SCHULT, G., AND BUCHHOLZ, A. (eds.) (1990), *Fernsehjournalismus. Ein Handbuch für Ausbildung und Praxis*, 3rd edn. (Munich, Leipzig: List).

SCHWITALLA, J. (1979), *Dialogsteuerung in Interviews: Ansätze zu einer Theorie der Dialogsteuerung mit empirischen Untersuchungen von Politiker-, Experten- und Starinterviews in Rundfunk und Fernsehen* (Munich: Hueber).

STORCK, M. (1992), 'Werbefernsehboom — ein Geschäft für die Privatsender. Der Werbemarkt 1991', *Media Perspektiven*, 3: 158–71.

STRAßNER, E. (ed.) (1975), *Nachrichten. Entwicklungen — Analysen — Erfahrungen* (Munich: Fink).

—— (1980), 'Sprache in Massenmedien', in Althaus *et al.* (1980), 328–37.

—— (1982), *Fernsehnachrichten. Eine Produktions-, Produkt- und Rezeptionsanalyse* (Tübingen: Niemeyer).

SUCHAROWSKI, W. (ed.) (1985), *Gesprächsforschung im Vergleich. Analysen zur Bonner Runde nach der Hessenwahl 1982* (Tübingen: Niemeyer).

TANNEN, D. (ed.) (1982), *Spoken and Written Language: Exploring Orality and Literacy* (Norwood, NJ: Ablex).

THOMAS, J. (1988), *Denn sie leben ja voneinander. Analyse von Sport-Interviews im Zweiten Deutschen Fernsehen und im Fernsehen der DDR* (Frankfurt: Lang).

WACHTEL, M. (1988), *Die Darstellung von Vertrauenswürdigkeit in Wahlwerbespots. Eine argumentationsanalytische und semiotische Untersuchung zum Bundeswahlkampf 1987* (Tübingen: Niemeyer).

WEMBER, B. (1976), *Wie informiert das Fernsehen?* (Munich: List).

WICHTERICH, CH. (1979), *Unsere Nachbarn heute abend — Familienserien im Fernsehen* (Frankfurt: Campus).

WITTGENSTEIN, L. (1969), *Philosophische Untersuchungen* (Frankfurt: Suhrkamp).

WOISIN, M. (1989), *Das Fernsehen unterhält sich. Die Spielshow als Kommunikationsereignis* (Frankfurt: Lang).

Index

abbreviated splitting 298, 308
abbreviations 88, 98, 99, 102, 108
  bisyllabic 96; with specific sex
    indicators 97
  emotional connotations in 97
  monosyllabic 97
  often cause difficulties 125
  tendency towards 101
ABV (Abschnittsbevollmächtigter) 125
Abweichler 249
Abwicklung 123 n.
Académie Française 16
accusative case 104–5, 106, 107, 120, 266
activity theory 7–8, 161, 234
adaptation 326
Adelung, Johann Christoph 75, 76, 77
Adenauer, Konrad 342
adjectives 74, 75, 99–101, 102, 296
  comparison of 108–9
  derived: from English 109, 320; from
    human nouns 294, 295
  Germanized, inflected in the normal
    way 320–1
  governed by the genitive 103
  nominalized 290, 297
  post-noun position of 149
  predicative 104, 106, 289
  previously classified as indeclinable 108–9
  sex-specific 291
  splitting 298–9
Adorno, Theodor W. 249, 250
adverbial case 105
adverbs 102, 109, 149
  formed with *-mäßig* 100
advertising/advertisements 87, 100, 361–4
  campaigns 90, 128
  copywriters 88, 113–14
  English words in 321
  -ig suffix often used 101
  job, sex-specific 291
  political 340
  subordinate clause pattern in 110
affectionate forms 97
affixes 99, 320
Afghanistan 38

Africa 57, 174, 241
afroasiatische Völkerwanderung 243
aggression 236–7, 243
Aitchison, J. 302
Akademie der Wissenschaften 118
Aktantenwissen 209
Aktivist 123
Aktuelle Probleme 8
Albrecht, J. 172
Albrecht, R. 243
alliterative association 330
allusions 362
  erotic 361
alphabet 85
  Latin 71, 77, 84
Alphabetisierungskurse 176
Alsace 70
Altersheim 128
Altersübergangsgeld 124
Althusser, Louis 250
Altvater, Elmar 250
ambiguities 108, 303
Ammon, Ulrich 5, 26, 33, 37, 41, 46, 58, 280
Amselmännchengesang/Amselgesang 78
Amtssprache 306, 307
anacolutha 111, 345
anaphora 152, 288, 289
Anarchists 251
Andraschko, E. 287
Andresen, H. 279
Ang, I. 357
Anglo-American influences 358, 366, 367
anti-Semitism 244, 251, 252
AntragstellerInnen and Antragsteller/innen 308
Anwerbeländer 174
apartheid 238
Apel, Karl Otto 234 n.
aphorisms 318
apologizing 188
apostrophes 104
appreciation 193
Arabic 29, 47
Arbeiterkampf (radical left paper) 250–1
Arbeitsgruppe Rechtssprache, see BRD Report
Arbeitskreis für Rechtschreibregelung 84

ARD (Arbeitsgemeinschaft der
    Rundfunkanstalten Deutschlands) 342,
    355, 361, 368
Argentina 39
arguments 171, 185, 192–3
*Armleuchter (Arschloch/*arsehole) 159
'Arte' (French–German TV culture
    channel) 343
articles 72, 74, 75, 81, 294, 298
    deletion of 180
    masculine form in the wrong case 113
articulation 221, 266
Ärzte und Krankenschwestern 291
Asante, M. 171
Asia 174
assertive strategy 286, 287
assimilation and reduction 162
asylum 172, 232, 244
    economic 241
    political 169, 173, 174
attacks 236, 244, 247
    verbal sexual 334
attitudes 257–78, 345
    changing, towards linguistic
        variation 129–32
    negative 293
    positive 293, 303
Auer, Peter 6, 13, 14, 151, 152, 154–5, 157,
    170
Auernheimer, Georg 238 n.
Aufenanger, S. 355
Ausgewogenheit 342
Ausländerbehörde 183
Aussiedler 169, 173, 174
Austria 38, 44, 46, 47, 50, 63, 79
    dialects 257–63, 265, 266, 271–3
    evaluation of language behaviour of
        politicians 264–9
    German 26, 140, 141, 207; as a foreign
        language 34; official status 31;
        standard 153, 273
    language attitudes 257–78
    large waves of immigrants into 209
    legal language 305
    Lower 262
    monarchy, and the *Duden* 82
    Nazis and 62
    reform 86
    theories of linguistic variation 143–57
    see also *Burgenländer*; Graz; Innsbruck;
        Salzburg; Tyrol; Vienna
Ausweichstrategie 287
authority 154, 221
Auwärter, M. 145
Azubi (Auszubildende/r) 125

'baby-talk' 97, 323
Babys 90
Bach, Adolf 3
Backa, S. 187

Bade, K. 173
Baden-Württemberg 244
Bailey, Charles-James 13 n.
Baldauf, R. 42
Balibar, Étienne 238 n.
Banks, A. S. 33, 48
Barbour, S. 3, 4, 12, 65, 119, 173, 176, 177,
    208, 209, 258, 263, 264, 365
Barden, B. 172, 195
Barkowski, H. 184
Baron, D. 289
barriers:
    language 146, 212, 222
    social class 65
Barth, F. 170
Barthes, Roland 207, 350
Bartsch, R. 258
Bauer, H. L. 37
Bausinger, H. 194
Bavaria 12, 81, 257, 260
BBC (British Broadcasting Corporation) 342
Becker, A. 185
Beebe, L. M. 264
Begleitprogramme 355
behaviour:
    acquired habitual, style as a form of 156
    analytical/descriptive model of 186
    argumentative, linguistic differences
        in 192
    body language 344
    communicative 191, 206, 207
    conventional 188, 192
    conversational 284
    cultural 194
    description of 139
    differentiating 325
    doctors' 221
    female, stereotyped 284
    foreigners' 243
    human, determining 241, 242
    inappropriate 197
    institutional 187
    joining 283, 286
    listening 188–9
    non-linguistic 153
    non-verbal 189, 191
    patterns of 175, 209
    pro-feminist 301
    qualitative evaluation of 163
    social 170
    specific expectations in terms of, associated
        with various roles 182
    strategic 286
    'uncooperative' 222
    verbal 281, 309; sex-preferential 284
    see also linguistic behaviour; speech
        behaviour
Belange des deutschen Volkes 243
Belfast English 280
Belgium 34, 41, 44, 29

German-speaking community 31
Beneke, J. 318
Benjamin, Walter 250
Benoist, Alain de 236
Bentele, G. 340
Bericht 33, 48
Berlin 12, 61
  Brandenburg 118, 121
  language/speech 332; deferential manner
    towards strangers 131–2; dialect 98,
    132, 142; emphatic use of titles 131;
    recently coined terms 127; redundant
    terms 131; use of *wa* 177;
    vernacular 117, 119, 120, 121, 140, 150;
    see also Berlinish
  *Mietskasernenviertel, Bankenviertel, Villen-
    viertel*, and Jewish quarter 121
  Orthographical Conferences 79, 81–2, 83
  Potsdam 56
  social change and linguistic
    variation 117–32
  Soviet occupied zone 122
  television 88, 341
  Wall 236; fall of 123–9, 231, 248
  Western sectors 122
Berlin Academy 77
Berliner Zeitung 124, 128, 131
BerlinerInnen/Berlinerinnen/Berliner 88
Berlinish 117–22, 129, 130, 131, 132
  dictionaries of 118, 126, 127
  labels, terms, and words from 127
Berlitz language schools 34, 35
Bernstein, Basil 3, 4, 6, 144, 145, 208
Berufsverbot 1
Besch, W. 12
Besserwessis 97, 118, 195
Betonitis 98
Bewegungssprache 108
Bible 70, 316
bibliographies and databases 43, 44
Bickerton, Derek 13 n.
Bild 243, 244–5
Bildungskatastrophe 4
bioecological concept of society 238
Bismarck, Otto von 76, 79, 80, 81
black people 188–9, 243
Blank, U. 322
Blaubergs, M. 307
Bliesener, Thomas 212, 223
Bloch, Ernst 250
Blommaert, J. 172
Blut und Boden 63
Blutschande 244
boab 323, 333
Bodemann, M. 180
body language 152, 196, 344
Born, J. 39, 173
borrowed words 90, 113, 126, 297, 323
Bourdieu, Pierre 146, 147, 181, 207, 214,
  224, 325, 359

Braga, G. 28
Brandmeier, K. 320
Brandt, W. 339, 341, 354, 365, 367
Braun, Peter 103, 107
Braunschweig 319
Brazil 39
BRD Report (1991) 299, 305, 307–9
Bremen 174, 236 n., 287
Bremer, K. 178, 184
Brenner, Oskar 83–4
Breslau 84
bricolage 325, 326–7
Brimes, B. F. 29
Britain 37, 38, 44, 205
  newspapers' feelings on German
    language 50
  television: commercials 363; number of
    sets 341; verbal style of a
    public-broadcast speaker 344
broadcasting, *see* radio; television
'Broiler' 128-9, 196
broken German 142
Brown, P. 159
Brunt, R. 172
Büchle, K. 191
Bulgaria 36
Bundesanzeiger 82
Bundestag 310, 330
bureaucratic jargon/language 124, 210
Burgenländer 260-1
Burger, H. 339, 340, 352, 355, 365
Büroangestellte 294
Bürokratismus 98
Buscha, J. 105
Byrnes, H. 192

Cameron, D. 6, 281
Canada 34
capitalization 77, 80, 85, 91
  book titles 71
  categorizing all instances of 73
  initial 71; 'I' 112, 308
  innovators who sought to abolish 72
  nouns: abolishing 70; syntactically
    motivated 72
  proper names 75
  rules of 73, 74
  strict limitation of 84
  within a word 88
Carinthia 261
cases 113, 266
  see also accusative; dative; genitive
catachreses 241
catch-phrases 318, 326
Cathcart, R. 355
CC (*Congress Centrum*) 88
CDU/CSU (Conservatives) 2, 140, 235, 305
Celtic 61
Center/Centrum 79
Chafe, W. L. 345

chauvinism 231
children 143, 179, 246
  born in Germany in 1990 have parents
    who are not German nationals 173
  news for 355
  second and third generation 175
  Turkish and German, narratives
    of 189–90
  worlds of knowledge which determine
    interaction between teachers and 209
  written texts used to test linguistic
    behaviour of 208
Chile 38
China 350
  Chinese language 30, 47; speakers of 189
  deficits in communication between
    Germans and Chinese learners of
    German in 192
Chiquitos 137
Cicourel, Aaron 211, 213
Cieslik, N. 356, 357
Cigarette/Zigarette/Ziggi 87, 97
cinema 341
civil war 242, 243
Clahsen, H. 177
Clarke, J. 326
class, *see* social class
Classe/Klasse 79
Claus, U. 315
clauses:
  adverbs formed with -*mäßig* at the end 100
  conditional or concessive, not introduced
    by conjunctions 109–10
  dependent 109
  elliptical 162
  main 109, 110
  optative 109
  paratactic complexes 345
  relative 101
  subordinate 109, 110, 111, 267
  weil, 'incorrect' 110
clients 210, 214, 223
  middle-class 211
  Turkish 185
Clyne, M. G. 10, 179, 190–1, 194
Coates, J. 281
Code (right-wing paper) 239
codes, *see* linguistic codes
cognitive systems/concepts 151, 152, 209,
  351
Cold War 63–4, 86
colloquial forms 127, 132, 272, 367
  diffusion of standard varieties 365
  see also colloquial language; expressions;
    speech; texts; words
colloquial language 95, 103, 142, 186
  'educated' 157
  emergence of 159
  supra-regional 140
  vocabulary 159

comics 322–3, 324, 327, 328
common beliefs 170
Commonwealth of Independent States
  174
communication:
  across all social strata 57
  administrative 306
  basic conditions of 162
  business, German–Finnish 192
  close 346
  collusive 327
  concrete 155
  counselling sessions 186–7
  deficits in 147, 192
  doctor–patient 211–13
  effective, prerequisite for 76
  egalitarian forms of 224
  everyday 250, 341
  face-to-face 14, 222, 346
  facilitating 303
  favourable conditions of 138
  group-specific 327, 328
  inadequate foreign language competence
    hinders 197
  institutional 197, 205–27; most important
    features 210–11
  interactive process of understanding
    and 154
  intercultural 14, 169–98
  interference in 76
  international, use of German for 37–50
  linguistic-communicative activity 160–1
  media 350, 352, 362, 368;
    multi-channel 343
  nationalism, in terms of theory 58
  obligatory, in public agencies 183
  oral 95, 365; considerable impact on 164
  pragmatic styles 154–5
  private conditions of 367
  problems and other developments resulting
    from unification 127
  public 163
  quick and simple 128
  social 58, 163
  vertical and horizontal 65
  written, efficient 91
  see also communication communities;
    international communication;
    miscommunication
communication communities 146
  'diacommunicative' variation, specific
    to 142–3
  distinction between 'speech communities'
    and 161
  socio-political history within linguistic
    communities 143
  with different languages 144
comparative forms 109
  semantically nonsensical 97, 108
competence, *see* language competence

complaints 215, 286
  employment office 184
  shop 184–5
compliments 191
compounds 99–100
  capital letters for each of the
    components 88
  formation of 84
  linking two components 87
  morpheme boundaries in 88
  morphological 295–6
  new, with -*frau* 302
comprehensibility 210, 214
Comrie, B. 29
concentration-camps 88, 232
conflict 170, 221, 285, 287, 303
  conversational 282
  'frame' 212, 214, 222
  gender 290
  inability to cope with 222
  intercultural 198
  potential 176
  solving 358
  verbal 273
conflicts of interests 197
'confrontainment' 348
congratulations 192
conjugation 101
conjunctions 109–10, 111
connotations:
  derogatory or trivializing, lexicalized items
    marked for 297
  emotional, in abbreviations 97
  negative 98–9, 272, 295, 297, 304, 310
  ranging from chic to extremely positive 99
conquest and colonialism 30
Conrad, A. W. 51
conservatives 2, 4, 232, 235, 242, 305
  ideologies and ideas 231
  phrase coined by, to discriminate against
    the unemployed 243
  recognized writers 238
  right-wing 238
consonants 71, 77, 121
consumer goods 114
contact 130, 174, 179, 197, 211
  German–Spanish 193
  little, between different ethnic groups 173
context 114, 153, 273, 291
  constitution or construction of 13
  determination of, flexible and reflexive 151
  discourse and 151, 213–23
  holistic analysis of 205
  important elements of 152
  informal 'speech situations' 162
  institutional 213–14, 215
  interpretation of a particular utterance 152
  new, translation to 326
  official and informal 123
  political statements 232

related to a particular individual 290
  social 88, 147, 150
  text and 151, 205
  variables 223
  see also neutral contexts
contextualization 14, 151, 152–3, 170
contextualization cues 171, 182
  non-referential, non-lexical 151–2
contradictions 350
contrastive pragmatics 188–94
conversation 212, 284, 331–2
  activities 156; 'energy' transformed
    into 155
  analysis 281, 340
  argument in 171
  bank counters, counselling sessions, job
    interviews 182–3
  closing 193
  conflicts 282
  contrasts in styles 192
  discrete 214
  doctor-patient 212, 222
  dominance 281, 284, 287, 309
  individual types 211
  interventions 282
  intimate 356
  key phrases repeated in the course of 329
  media 349, 354, 357, 365
  mixed-sex 282, 285
  narration 189
  patterns 325
  private 281, 286–7
  rape of women in 282
  styles 192–3
  subordinate 283, 285, 310
  tendency to break off 181
  women's rights 284; violations of 283
co-occurrence restrictions 156
Cooper, R. L. 51
co-operative principles 286
copula deletion 180
'Coronation Street' 356
corpus 149, 150, 160
correspondence 191
Coseriu, Eugenio 18, 161
Coulmas, Florian 10, 30, 58, 59, 188
Council of Europe 48–9, 172
  guidelines (1990) 307
counselling 187, 210
crimes/criminals 231, 236–7, 238, 243
critical analysis 163–4
critical linguistics 205–27
criticism 191, 192
  feminist language 307–8
Criticon (right-wing journal) 238
Croatian 62
Crosby, F. 281
'cross-cultural'/'cross-linguistic'
    comparisons 144
Crowley, T. 17

culture 155, 162, 196, 240, 348
  assumptions 171
  background knowledge 187
  differences 182, 189
  group 175, 327–8
  identity 170
  knowledge 187, 358–9
  language and 198
  linguistic proximity 189
  popular, oral 351
  socialist monoculture 146
  speech 163; specific acts 137
  see also intercultural factors
customs 137, 241, 243, 247
Czechoslovakia 7, 36, 37, 62

'Dallas' 356, 357
Danet, B. 206
Das Deutsche als Männersprache (Pusch) 279, 280
'Das Erbe der Guldenburgs' 357
dative case 103–4, 105–7, 120, 266, 298
declension 106
decomposition 181
defamiliarization 327, 329, 331
'default assignments' 152
deficits 143, 147, 192
Dehnungsschreibungen 87
deixis 152
deletions 180, 280
Delors, Jacques 50
Denger, F. 316
Denmark 40, 70, 73
denotation 304
'Der Kabelkanal' (private TV station) 343
Der Republikaner (right-wing publication) 237
Der Spiegel 245, 316
Der Tagesspiegel (newspaper) 173
derivations/derivatives 107, 297, 302, 304
  adjectival 100; from English 109, 320; from human nouns 294, 295
derogation 38, 297, 302
descriptions 139–43
  hybrid 114
  representative 150
  structural, of utterances 110
determiners 294, 298
Deutsch, K. W. 58
Deutsch als Fremdsprache 172
Deutsche Rundschau (right-wing paper) 239
Deutsche Sprachlehre für Schulen (Adelung) 75
Deutsche Stimme (right-wing paper) 239
Deutsche Gesellschaft für Sprachwissenschaft conference (1992) 287
Deutsche Nationalzeitung (right-wing journal) 237
Deutscher Fernseh Funk 88
Deutsches Universalwörterbuch 95, 102
Deutschland Fernsehen GmbH 342

developing countries 30
'diachronic' dimension 142, 148
diagnosis 212
dialect 104, 221, 366
  Austrian 259–63; Burgenland 260; dative and accusative cases neutralized 266; deep-seated prejudices against 273; function in parliamentary debates 271–3; negative connotations/characteristics 272, 273; perceived as a slovenly form of articulation 266; rhetorical function 273; syntactic characteristics 266; Tyrolean 260, 265; urban 262, 263
  Berlin 98, 122, 142
  code-shifts to 349
  coexisting varieties 153
  decreasing use 157, 365
  functionally appropriate use of 164
  Kaiserslautern 150
  local 120, 157, 159, 160, 262
  Low German 120, 121
  Palatinate 148
  preservation and decline 157
  prestige variety 123
  regional 262, 263, 265; superordinate 159
  replaced by vernacular influenced by standard variety 162
  Rhine Franconian 71
  romanticized 260, 263
  Ruhr 246 n.
  rural, positively evaluated 273
  Saxon, amongst East Germans who have moved to the West 195
  South Bavarian 260
  stigmatized 122, 260, 263, 272
  strong personal associations with 118–19
  'territorial' variants 140
  Thuringian 157
dialectology 7, 150
  'pragmatization' of 11–12
dialogue:
  description of structures 153
  German–German, contrasting speech patterns 14
  group cultural resources incorporated into 327–8
  linguistics of 3
  television 345, 356, 357; inner 363–4; sentence length in plays 354
  women's 143
'dichotomistic fallacies' 145
Dickgießer, S. 39, 173
dictation 85
dictionaries 95, 101, 102
  Berlinish 118, 126, 127
  negative images reinforced in 317
  see also *Duden* spelling dictionaries

Die lautlichen und geschichtlichen Grundlagen unserer Rechtschreibung (Brenner) 83
'Die Schwarzwaldklinik' 356
Die Zeit 117, 132, 245
Dieckmann, W. 10, 135
Differentialgenus 294, 305
diglossia 11, 183, 258
digressions 191
di Luzio, A. 14
diminutive forms 74, 97
diplomacy and international organizations 46–50
discourse 154, 162, 177, 182, 184, 327
  academic 280, 281, 285–6
  argumentative 185
  context and 213–23
  contrastive 188
  courtroom 209
  different types of, female and male verbal behaviour in 309
  English 309
  expressive 156
  institutional 206–8; school 209
  ironic 327
  legal 185
  multidimensional relationship between context and text or 151
  of the right in Germany 237
  official 10
  political 15, 231–53
  pragmatic analysis of structures 171
  private 340
  racist 245–8, 247
  sexual 334
  social 234
  study of 205, 210
  therapeutic 212, 281–2
  transfer from Turkish to German 182
  variable patterns initiated by the use of key phrases 331
  with and about foreigners 164–5
  written 210
  see also discourse analysis; public discourse
discourse analysis 108, 161, 197, 212
  linking of sociolinguistics and 205
  procedures 325
Discourse & Society (journal) 234 n.
'Disco-Deutsch' 316-17
discrimination:
  against foreigners 207, 242
  against women 246, 292, 293, 302
  against working-class children 208
Dittmar, Norbert 3, 5, 8, 12, 13, 14, 117, 118, 119, 143, 144, 146, 147, 148, 150, 154, 155, 170, 173, 177, 178, 185, 208, 258, 273
DKP (German Communist Party) 248, 250
doctors 214, 292
  communication with patients 171, 211–13, 216–20; dealing with initiatives 215;

examinations 185; exercise of power 215, 221; language barriers 212; manner 221
'explicit' and 'implicit' functions 206
problem-solving procedures 215, 216, 222
Dokumentation über die VI. Konferenz der Europäischen Kulturminister (1990) 172
Dolle-Weinkauf, B. 322, 323
dominance 144, 154
  conservative 2
  conversational 281, 284, 286, 287, 309
  male 284
Donath, J. 8, 162
doublets 107
downgraders 286
Dresden 195
Dressler, W. 153, 154, 207, 257
Drewitz, I. 84, 85
drinks 125
Drosdowski, Günther 89, 365, 366, 367
du 193, 194
  instead of *Sie* for singular 'you' 180
Duden, Konrad 77, 79, 80–3 *passim*, 90, 91
Duden spelling dictionaries 88, 89, 90, 111
  adopted as definitive 82
  entrenched opposition of editorial staff 91
  establishment of 70, 78–9, 81, 83
  new words included 96, 97
  plural forms 107, 108
  public perception of 16
  see also Deutsches Universalwörterbuch
duplicate forms 102, 105, 107
Durchsetzungsstrategie 286
Dutch 62, 289
  business negotiators/managers 40, 41
DVU (*Deutsche Volksunion*) 235, 236 n., 237
'Dynasty' 356

East–West German question 172
eastern Europe 50, 174, 204
  former GDR orientation towards 196–7
  German minorities in 62
  people coming to Germany from 173
Ebner, W. 340
EC (European Community) 36, 173
  Commission 49, 50
  see also EU
Eco, Umberto 350
economic and social policies 196
economic pressures 240
economy formulas 298, 308
education 4, 211, 358
  bilingual, unsatisfactory 175, 179
  differing approaches to 189
  intercultural 172
  media, rhetoric of 367
Ehlich, K. 179, 182, 208, 210, 212
Ehn, M. 263
Eibl-Eibesfeldt, Irenäus 240, 242

'Eigenheiten der deutschen Sprache' 305
'Einer wird gewinnen' (TV show) 360
Einwanderungsland 175
Eisenberg, Peter 110, 290, 292
elaborations 345, 347
élites 59
    French language and 56, 58
    power 56, 57, 214
    typically consisting of white males, in
        dominant positions 207
ellipsis 111, 162, 345, 354
Elstner, Frank 360
emotions 159, 282, 322, 357
    connotations in abbreviations 97
    insecurity 333
    involvement 286
    outbursts 271, 272
    ways of talking about 191
empirical methods/studies 142–4 *passim*,
        157, 162, 171
    Jugendsprachen 318-20
    research 138–9, 341
employment 179, 207
endings, *see* suffixes
Engel, Eduard 62, 64, 111
Engelmann, Hartmut 315
English 16–17, 44, 45, 61, 73
    adjectives 108, 109, 320
    American 114, 189; New York City,
        stigmatized forms 280
    Asian 171
    Belfast 280
    campaigns against words and pop
        songs 125
    capitalization rules 71, 72
    descriptions of consumer goods 114
    expressing thanks and apologizing in 188
    gender issues 279; borrowed words 297,
        323; endings, *-ess* or *-ette* 310; essential
        difference between German and 295;
        guidelines 302, 304; human nouns 288,
        289, 292, 303, 304, 310; linguistic sexism
        in 288; studies in discourse linked to
        conversational dominance 309; women
        use more polite forms and phenomena of
        'uncertainty' 281
    German language infiltrated by words and
        expressions 89
    influence 107, 108, 112
    names on official forms 113
    native language 27
    neologisms and adjectives derived
        from 109
    official language of international
        organizations 47, 48, 49
    ranking of German behind 30, 31, 33, 34,
        35
    scientists having difficulty in reading
        45–6
    spelling 90, 102
    studied as a foreign language 35, 36, 37,
        40–1
    texts written by German-speaking
        academics 191
    trade 38, 39
    verbs 101, 102
    words adapted from 113
    worldwide spread 36
    youth language 318, 320–2, 332; sound
        words 323
Enlightenment 59, 60
Enninger, W. 172, 192
entertainment 358, 361
    and segmentation 348
equal treatment:
    linguistic 307
    opportunities in 175
    sexes 305, 308
Erben (grammar) 111
Erickson, F. 182, 188
errors:
    grammatical cases 266
    hyphenation 87
    writing 84; tolerance towards 91
ESL (Eastern sociolinguistics) 139
ethnic groups/minorities 152, 171, 181, 293
    little contact between 173
    non-Caucasian 240
ethnography 137, 151–3, 154, 172, 332, 325
ethnomethodology 154, 340
ethnopluralism 238
etymological elements 81
EU (European Union) 49, 50, 174, 235, 241
eugenics 251
euphemisms 15, 123 n., 159
European Parliament 235
European Science Foundation 178
evasion strategy 287
everyday language 127
    instructions to speak in 124
    reducing to 'baby-talk' 97
    technologizing and scientification 366
excursiveness 191
experience 211, 322
    hierarchy, knowledge, gender and 213
    social 145
experts 210, 214, 224
    in higher positions 211
    inexperienced 211
expressions 87, 114, 153, 154, 156
    appreciation 193
    archaic 318
    colloquial 104, 127
    'cool' 333
    East German 147
    English and American 89
    facial 186, 344
    fashionable 125
    fixed 106; direct translation of 192
    metaphorical 159

new 302
nominative singular 298
prepositional, using *von* 103
primitive 267–8
smart 318
stylistic means of, functional
    determination 157
urban 159
verbal 155, 178
extra-linguistic factors 143, 162
eye contact 152, 189
Eysenck, H. J. 240

facial expressions 186, 344
Fairclough, N. 17
fairy-tales 316
Familienserien 355
Far East 241
far right groups 234–48
fascism 9, 15
fashion 100, 125
Fasold, R. 258
Favoriten 262
FDP (Liberals) 140, 235, 305
Federal Post Office 92
Federal Railways 92
female visibility 291, 296, 300–10 *passim*
    indispensable means of achieving 297
feminine nouns 291, 298
    derivation of 97, 296, 297, 304, 309
    occupational titles and terms of
        address 306
    plural form 112
    sex-specific 306
    typically denotes an occupation of a lower
        social status than masculine 301
feminists 14, 17, 284, 301
    authors 113
    fundamental argument of language
        criticism 307–8
    language planning 302–3
    theology 282
Ferguson, C. A. 179, 258
Fernsehspiel 340
Fernsehstuben 341
Feuerstein, Herbert 323
Fichte, Johann Gottlieb 60–1, 63
ficken einhundert 328, 329, 330
Fiehler, Reinhard 210
Fienemann, J. 190
Finkenstaedt, T. 29
Finland 41
Finno-Ugric 37
First World War 44, 46–7, 62
first-person references 345, 364
Fisher, S. 211
Fishman, J. A. 51, 60, 61, 144
Fiske, J. 350, 351, 356, 358, 359
Fleischer, W. 97, 101, 296
Floridsdorf 262

fluency 196
Foltin, H. F. 340, 358, 359, 360
food 125, 128–9
foreign words 318, 320–2
    borrowed/adapted 90, 113
    frequently used, ⟨ph⟩ and ⟨f⟩ in 87
    German language infiltrated by 89
    growing use of English words 322
    pop songs with English lyrics, campaigns
        against 125
    replacement of ⟨c⟩ by ⟨k⟩ or ⟨z⟩ in 82, 87
    spelling, integrated into German from
        another language 90
    writing 84, 86
foreigners 171, 174, 238 n., 240
    actively discriminated against 207
    children 175, 189–90, 209
    'complementary knowledge' of
        students 184
    discourse with and about 164–5
    double burden for 183
    elaborate campaigns against 244
    fear of 243
    German classes for 184
    'good and bad' 241
    hostility towards 175
    law on 243
    legal security of many 175
    many live and work in Germany
        illegally 173
    murder of 176
    newspapers stating Germany overrun
        by 242–3
    'Ossis' and 'Wessis' still said to regard each
        other to a certain extent as 169
    percentage in Germany 173
    practical survival strategies for 184
    present policy towards 175
    prototypes of 246
    public discourse of differentiation between
        different sorts 241
    second language acquisition 169
    simplified speech style used by
        native-speakers when talking to 177
    tendency among indigenous population to
        blame 175
    see also FT
forenames 97
forms of address 193–4, 215, 219, 318
    child-like 220
    different 221
    distant or polite 221
    feminine 288, 306
    variations in 191
'fortress Europe' 240
'fossilization' 176, 178
Foucault, Michel 232, 234, 250
Fowler, R. 17
'fragmentation' and 'involvement' 345
'frame' conflict 212, 214, 222

France 34, 38, 79
  bibliographies and databases 43, 44
  right-wing figures 236
  see also French
Franconian 71, 140, 141
Franco-Prussian War (1870) 46, 62
Frank, F. W. 303
Frank, K. 14
Frank-Cyrus, K. M. 280
Frankel, Richard 211–12
Frankfurt am Main 11
'Frankfurt School' 249
Frankfurter Allgemeine Zeitung 86, 114, 316
Frankfurter Hefte (journal) 250
Frau/Frauen 113, 295, 302, 309
Frauensprache: Sprache der Veränderung
  (Trömel-Plötz) 280
Fräulein 309
Freeman, S. H. 211
French 42, 46, 141, 289, 307
  awareness of linguistic sexism in 309
  borrowings from 297
  cultural domination 58
  élite's affinity to 56, 58
  high prestige and obvious refinement of 61
  language of diplomacy 38
  narrative 190
  official language of international
    organizations 47, 48, 49
  ranking of German behind 33, 34, 35, 41
  reaction to supremacy 59
  regional international language 36
  scientific publications in 43, 191
  spelling 90
  studied as a foreign language 35, 36, 37
  superior to German in terms of logic and
    clarity of expression 135
  Swiss 26
  trade 39
  TV commercials 363
French Revolution 55, 56, 57, 58
Frey, Gerhard 235
Frey-Vor, G. 356
Front National 236
Frühstücksfernsehen 355
FT ('foreigner talk') 141, 142, 177
  German, typical features of 179–80
  register variation 180
  xenolects: native speakers' adaptation to
    non-native speakers'
    competence 179–81
fundamentalist Christian sects 237
Funk-Kolleg Sprache (Wunderlich) 6
fuzziness 345

Gabelentz, Georg von 137
Galperin, I. R. 329
Gastarbeiter 169, 176–7, 183, 241
Gastarbeiterdeutsch 14, 176–7
Gastarbeiterlinguistik 13

gate-keeping situations 182, 189
Gaumann, Ulrike 110
Gdaniec, C. 15, 17
GDR (German Democratic Republic) 1–2,
    6–9, 34, 47, 63
  expressions 147
  fight for new socialist democratic state 250
  former, orientation towards eastern
    Europe 196–7
  ideological concepts 235, 236
  language behaviour 147
  'language culture' 138, 165
  language in transition 10–11
  meaning of *du* 194
  newly created communicative norms 162
  'official linguistics' 138
  official status of German 32
  old terms replaced 123–6
  orthography 79, 86, 88
  radio 118
  social dialectology 157–60
  sociolinguistic research 160–1
  specific neologisms 162
  specific variants 141
  Sprachkultur 8, 9, 15, 91, 163–4
  state journalism 249
  teaching language in connection with
    business and technology 38
  television 366, 368
  Volk concept 64–5
  wooden, literate mode of official
    announcements 347
  see also Berlin; Krenz; Leipzig; '*Ossi*'; SED;
    Stasi; Ulbricht
Gefüge der Existenzformen 7
Gehnen, M. 49, 50
geil ('brilliant, cool'; literally 'randy') 335
Geißendorfer, Helmut 356, 357
Gellner, E. 58
gemäßigte Kleinschreibung 84
Gemeinschaft 60
Gemeinsprache 157
gender 108, 152, 209, 211, 321
  attribution of feminine or masculine 294
  differential 294
  hierarchy, knowledge, experience, and 213
  incorrect assignment 107
  language and 279–310
  marking 112; social and age factors 113
  neutral forms 88, 305
  semantic 287–8, 291, 292, 310
  social 288–9, 294, 304
  specific characteristics/strategies 89, 96,
    209
  studies 14
  see also grammatical gender
'genderlect' 144
generic forms 290–2
  der Grammatiker used for both female and
    male authors 112

feminine 301
interpretation 295, 301, 305
masculine 302, 306, 308, 310; alternative
   to 306; *he* 288, 303; replaced by
   splitting 308; stereotypically 304–5
genitive case 102–4, 298, 367
Genus 292
Geraghty, C. 355, 356
Gerhard, U. 245
Gerighausen, J. 172
German Empire 76, 79
German language, *see* German-speaking
   countries; High German; language; Low
   German
German Post Office 81
German Rail 97–8
German-speaking countries 45, 62, 89, 212
   economic strength 51
   institutional discourse 208
   origins and development of
      sociolinguistics 135–9
   see also Austria; Belgium; Liechtenstein;
      Luxembourg; Switzerland
Germanistik 4, 76
'Germanness' 55-66
Geschlechtsspezifikation 294
geschlechtsspezifisches/geschlechtstypisches
   Sprachverhalten 280
Gesellschaft 60
'Gesellschaftlichkeit der Sprache' 8
Gesetzessprache 308
Gestalt 154, 155
Gestapo 98
gestures 151, 152, 153, 344
   simple to repair misunderstanding by 186
Gewalt durch Sprache: Die Vergewaltigung von
   Frauen in Gesprächen 282
Giles, H. 259, 264
Glück, Helmut 16, 40, 89, 96, 107, 108, 112,
   118, 195, 320, 367
Gmoser, Rupert 268
Goethe, J. W. von 64, 75, 91
   Eckermann, secretary of 59
Goffman, E. 327
Good, C. 10, 196
'good' German 95
Gorbi (Gorbachov) 97
Gorter, T. R. 41
'Gothic' print 70
Götter in weiß 207
Gottschalk, Thomas 360, 361
Gottsched, Johann Christoph 72, 73, 74, 75,
   76
Gräßel, U. 14
Graff (Gen. Secretary, ÖVP) 268, 269
grammar 75, 81, 111, 112, 155, 177, 321
   accepted conventions, trends away from 96
   context-free 148
   errors in cases 266
s explicit 148, 153

flexible or rigid 145
limitations 156
low-level variants, empirical description
   of 142
means of specification 294
norms 114; codification of 78
reference 148–9
rules 70
structures 184
stylistic variation: input-conditioned
   variety switches 153–4
suitable type for description 150
trend for features to lose distinctive
   form 105
variety 143, 148–50, 154
grammatical gender 289–90, 291, 295, 304,
   310
   class 297
   correlation of 296
   inherent and invariant property of the
      noun 289
   morphosyntactic markers of 292
   pronouns morphologically invariable
      for 294
   tendency towards agreement between sex
      of referent and 301
graphemes:
   bound 87
   phonemes and 71, 84, 91
grapho-stylistic techniques 319
Grässel, U. 284
Graz 261, 262, 263, 266, 268
Greece 36
Greek characters 71
Green Party 97, 113, 250, 267, 305
Greene, M. G. 211
greetings 192, 318
   and departure 191
Grießhaber, W. 182, 183, 184
Grimes, B. F. 30, 31
Grimm, H. 316
Grimm, Jacob 62, 76, 77–8
Grimm, Wilhelm 62, 76
Groce, S. B. 211
Große, R. 6, 7, 8
Großkopf, B. 172, 195
'*grossdeutsch*' orthography 86
groups:
   cultural 175
   different, contact between members of
      197
   ethnic 152, 173, 240
   far right 234–48
   'foreigner talk' 142
   identity and integration 210
   in-group and out-group identification
      170
   left, marginal 248
   Marxist 6, 160, 250, 251
   membership of 170, 326

groups (*cont.*):
  patterns 110
  racist skinhead 236
  'social' 148, 194
  specific knowledge 142
  Trotskyist, Maoist 250
  varieties/registers 140
  see also youth groups
'*Grundlegung einer deutschen Sprachkunst*'
  (Gottsched) 72
Guentherodt, I. 279
guest workers, see *Gastarbeiter*
Gulf War (1991) 251-2, 245
Gumperz, John J. 14, 143, 151, 170, 171,
  172, 182, 187, 195
Günthner, H. 71
Günthner, S. 14, 16, 17, 111, 143, 144, 192,
  281
Gutenberg 71
Gysi, Gregor 248

Habermas, Jürgen 208, 213, 234 n., 346
habits 156
  older speech 131
  reception, deviating from 36
  verbalization of speech acts and text
    forms 188
Habsburg monarchy 56
Hädrich, D. 332
Haefs, H. 30, 31, 33
Halbmundart 120
Hallenberger, G. 340, 358, 359, 360
Halliday, M. A. K. 345
Hall, E. T. 194
Hall, M. R. 194
Hall, Stuart 238 n., 325
Hamburger, F. 173
Handlungsmuster 186
handwriting 70, 71
Hanover guidelines 300
  vs. Bundestag Report 305-7
Hartig, M. 3
Hartung, W. 2, 8, 9, 135, 160, 161, 162,
  163-4
Haselhuber, J. 49, 50
Hauptmeier, H. 351, 352
headlines 239-40, 248-9
  banner 244
hearer's expectations 145
hedges 281
Heidelberger Forschungsprojekt
  'Pidgin-Deutsch' 148, 150, 177, 178-9,
  180
heim ins Reich 62
Heimat 86, 356
Hein, Norbert 212
Heinemann, M. 318
Helbig, G. 105
Helbig-Buscha (grammar) 111
Heller, M. 211, 213

Hellinger, Marlis 14, 17, 88, 112, 281, 291,
  293, 300, 304, 305, 307
Hellmann, M. 1, 10, 11, 172, 194, 195
Henne, Helmut 318, 319, 320, 365, 366,
  367
Herberg, D. 1
Herder, J. G. von 59, 60, 61, 65
Heringer, H.-J. 15, 16, 17, 104
Hermann-Winter, R. 8, 162
hesitation phenomena 345
Hessischer Rundfunk 315
Hess-Lüttich, E. W. B. 340
'heute' 353-4
'heute-journal' 345, 352-3
Heydenreich, Elke 317-18
Hickethier, K. 358, 359
hierarchies 209, 210, 213
High German 76, 258
Hindi 62
Hinnenkamp, V. 11, 13, 14, 153, 154, 170,
  171, 179, 180, 181, 183, 187, 188
Hirsch, Joachim 235, 250
Hitler, Adolf 62-3, 64, 70
Hobsbawm, E. J. 55
Hochdeutsch 76, 258
Hochsprache, see standard German
Hoffmann, L. 209
Hoffmann, R.-R. 340
Holly, Werner 17, 340, 346, 348, 358, 360,
  361, 365
Holocaust 38, 231
Holy Roman Empire 46
homophones 80
Horton, D. 346
Horvath, B. 280
House, J. 188
housing shortages 175
Huby, Felix 358
Hufschmidt, J. 12
human nouns 112, 290, 293, 310
  adjectives derived from 295
  avoidance of, altogether 306
  can be used to refer to female or male
    individuals 292
  derived from verbs 294
  English 288, 304
  gender-neutral 88
  general 288
  pairs of 291
  pronominalized 288
  semantic categories English system is
    primarily structured by 304
  singular, derived from adjectives 294
  variability in use 300-2
  see also feminine nouns; masculine nouns
humanism 64
Humboldt, Wilhelm von 61, 65, 135-6, 137,
  139, 145
Hungary 36, 37
Huth, L. 340

Hymes, Dell 5, 234 n., 325
hypercorrections 108, 137
hyphenation 84, 86, 91
  incorrect 87
  rules 87; in compound words 88
hypotactic and compact structures 345
hypotheses 146, 195, 283, 365, 367
  current 144
  'decodability' of non-verbal behaviour 189
  mixed-sex conversations 282
  pessimistic standard 340
  'rule block' example 149
  'social significance' 156

identity 55, 132
  collective, menace to 242
  constructing 318
  cultural 170, 172
  group 210, 326
  individual 293
  language 61, 63, 65
  national 56, 60, 61, 62, 63, 238, 243
  sexual 285
  state 61, 63
ideologies 236, 237, 238
  biologistic 235
  canonical 248
  conservative 231
  leftist 231
  male as norm 289
  old 248
  right-wing 231, 237
  socialist 249
idiolect 137
idioms 103, 129, 147
  metaphorical 354
Ilkhan, I. 192
illiteracy 262, 263
'*Imagepflege*' 147
images 340
  suggestive 244
immigrants/immigration 171, 245
  African and Asian, 'flood of' 240
  Afro-Asian, on the scale of entire
    nations 243
  attacks against, including arson and
    murders 244, 247
  black 243
  children of 143; considered
    foreigners 175; language
    acquisition 209; second and third
    generations, born in Germany 179
  country open for 175
  discriminating against 244
  domains for possible encounters between
    indigenous population and 173–4
  experience of 183
  intensive courses for unemployed 176
  large wave of 209
  life in Germany 173

living conditions of population 175
  policy 175
  Polish 177
  'returning' 173
  search for a security state 243
  second language acquisition 176–9
  standard argument against 240
  workers 176; southern European 172, 179
imperatives 109, 221
imperialism 64
improvised 'performance' 327
'incorrect German' 267
Indonesia 37
industrialization 58, 65, 121, 157
  influence of 159
infinitives:
  nominalized form 102
  overgeneralization of 180
  verbs formed by adding suffix 320
inflection 113, 289, 320–1
  personal/temporal 101
  syntactically motivated 289
'infotainment' 348
initial letters 71
  small 74, 75, 77
'inner perception' 139
Innsbruck 259, 261, 265, 269
insider's perspective 213
institutions 11, 181, 206–8, 215
  intercultural communication in 182–8
  medical 185–7, 207
integration 164–5, 173, 175, 176
  linguistic, of outsiders 119
  political policies towards 169
intensifiers 100, 281, 332
  adjectival 320
interaction 285, 286, 287, 309
  classroom 208, 209
  encounters between '*Wessis*' and '*Ossis*'
    195
  face-to-face 343
  group members 331
  para-social 346
  partners 334
  process of communication and
    understanding 154
  sex, social class, and verbal behaviour,
    assumptions about 281
  social, language as a tool of 302
  specific norms 162
  verbal, openings and closings of 162
intercultural factors:
  communication 14
  conflict 198
  contrasts between East and West
    Germany 194–7
  education and learning 172
  general contrasts 191–2
interjections 271
Interkulturelle Germanistik 172

'interlanguages' 142, 178
international communication 25–6, 27–8,
    31, 51
  use of German for 37–50
interpretation 151, 152, 154, 194, 351
  generic 295, 301, 305
  see also misinterpretation
interruptions 281, 282, 283, 285, 286
intertextuality 362
intimacy 154, 348, 356
  pseudo- 346
  secondary 346–7
  strategy of securing 287
Intimitätssicherungsstrategie 287
intonation 182, 218, 262, 332
  question 281
  typical pattern 145
Ireland 38
irgendwie, eigentlich 286
Irish Gaelic 27
ironic distancing 331
Ising, E. 163
Ising, G. 6
isolation 173, 176
Israel 38–9, 244, 251
Italy/Italians 14, 31, 55, 185
  awareness of linguistic sexism in 309
  Italian language 49, 62, 289, 307
  restaurants 108

Ja-aber-/Ja-stimmt-/Ja-und-Technik 283
Jäger, M. 237 n.
Jäger, Siegfried 11, 17, 165, 224, 233, 235 n.,
    237 n., 240, 245, 247, 281
Jahresendflügelpuppe/Weihnachtsengel 123
Januschek, F. 316, 320, 323
Japan 37, 235
  Japanese language 29, 30, 37, 47;
    expressing thanks and apologizing
    in 188; forms of politeness 137
  learners of German 184
  paragraphing of texts by students 191
jargon 123, 124
Jefferson, T. 325
Jernudd, B. 42
Jespersen, O. 29
Jews 244
job interviews 182–3
Johnson, S. 117
'journalese' 123
journals 231, 234 n.
  disappeared 248
  militant racist 238
  New Left 250, 251
  published by political parties 237
Jugendsprachen 119, 127, 315–35
  sexual matters: discourse 334;
    innuendos 335; TV
    themes/leitmotifs 329, 330
Jung, W. 107

Kahane, H. 58
Kaiserslautern 150
Kallmeyer, W. 12
Kalpaka, A. 238 n.
Kaltenbrunner, Gerd Klaus 238
Kalverkämper, H. 41
Kaps, J. 340, 360
Kassel 319
Kaufhalle/Supermarkt 128
Keim, I. 12
Kekse 89
Kellershohn, 251
Kelman, H. 56, 57
Keppler, A. 340
Key, M. 280, 281
keywords 81, 241, 243, 244
  journalistic 247
kinship terms 288, 290, 293, 304
Kita (Kindertagesstatte)/Kindergarten 124, 128
Klann, G. 145
Klein, J. 16
Klein, W. 144, 148, 149, 150, 154, 176, 177
kleindeutsch solution 79
Kloss, H. 309
Knapp, K. 172
knowledge 164, 213, 233, 358–9, 366
  background 187
  common-sense 152
  commonly shared pools of 327
  'complementary', of foreign students 184
  culture-specific 187
  'default' 152
  everyday 351
  group-specific 142
  hierarchy, experience, gender and 213
  language, patterns of 320
  passive, of words 367
  pattern 209
  routine 209
  scientific 185
  social 151; recording in written form 164
  structuring of 212
  worlds of 209, 218
KOB (*Kontaktbereichsbeamter*) 125
Kohl, Helmut 2, 50, 235, 236, 346, 347, 349,
    350
  '*Kohlismus*' 99
Köhle, K. 212
Kolde, G. 15
Kollektiv/Team/Gruppe 131
Kommune (journal) 250
Kommunikation und Sprachvariation 8
Kommunistischer Bund (Communist
    Union) 250
Kongress-Zentrum 87-8
Korea 37, 137
Kornfleks 90
Kosog, O. 84-5
Kotthoff, H. 14, 16, 17, 143, 144, 191, 192,
    193, 281, 285

Krainer (Styrian politician) 268-9
Kramarae, Cheris 279, 288
Kraus, Karl 16, 17
kreative Lösung 308
Kreckel, R. 259
Kreisky, Bruno 267
Kremer, M. 300, 305
Krenz, Egon 346, 347, 349, 350
Kreuzer, H. 340
Kubler, H.-D. 340
Kuglin, J. 193
Kuhberg, H. 177
Kuhn, F. F. A. 144
Kühn, P. 340
Kukuckinnenei/Kukucksei 78
Kulenkampff, Hans-Joachim 360, 361
Kultur des Mißverständnisses 196
Kulturnation 57
kultuRRevolution (New Left journal) 250
Kunnemann, H. 322
Küpper, Heinz 315, 317
Kurds 241
Kurzarbeit 124
Kutschera, R. 315
KWV (*Kommunale Wohnungsverwaltung*) 125
KZ (*Konzentrationslager*) 88

labour:
  divisions of 207
  products of 233, 234
Labov, W. 5, 6, 12, 13, 137, 143, 144, 148,
  149, 154, 156, 280
Ladys/Ladies 90
Lafontaine, Oskar 350
Lakoff, R. 281
Lalouschek, J. 205, 208
Lambert, W. 259
Landbeck, H. 362
Länder 82, 130, 259, 360
  alte 237
  neue 104, 195, 236, 239
Lang, P. Ch. 357
langes Splitting 298
language:
  administrative 308
  adult 333
  asymmetric use of 26-7
  attitudes in Austria 257-78
  barriers 146, 212, 222
  bedroom, German petit bourgeois 335
  bilingual contact 151
  body 152, 196, 344
  bureaucratic, comprehensibility of 210
  central Indo-European groups 189
  change: directions of 95-114; under the
    influence of the women's
    movement 300-9
  'chatting-up', used almost exclusively by
    boys 334
  childlike 222

class-specific 143, 145, 146
'common' 157
conflicts with public norms 335
cultivation of 163
'culture' and 198
degenerate 262, 263
derogatory 293
dialectical relationship of society and 6
discriminating 292, 293
division of state and 63-5
economic strength 30-1, 47, 51
élite 214, 223
European standard 60
evaluation of use in public
  discourse 257-78
feminist 302-3
filthy 317
fixing the form of 69-92
formal 344
gender and 279-310
German as a Foreign Language 172
Germanic 37, 189
group-specific phenomena 325
gutter 159
identity of 61, 63
ideology and 234-52
images and 340
indirect uses of 147
infiltrated by English and American words
  and expressions 89
intercultural communication 169-98
international 25-51
legal/legislative 207, 209, 305, 306, 307,
  308, 309
'legitimate' 146, 147
literary 163
'living' and 'dead' 61
lower class 145, 259
maintaining the 'quality' of 163
mediation 179, 186-7
men's 137, 143
nation and 55-66
national 56-9, 367
'native' 27
'naturalness' of 155
nearer to everyday communication 250
new norms for use 131
non-sexist 302, 303, 304-5, 307, 308
numerical strength 28-30, 47, 49
opportunity or incentive for learning 173
oral 344-5
patterns of knowledge 320
political 231
political strength 31-3
'poverty' of 145
power/violence through 282
pragmatic aspects of use 18
preferences of use 156
'prestige' 58, 60
public, lexical contrasts in 194

language (*cont.*):
  race vs. 62-3
  real 1–18
  'refined popular' 157
  relativity of use 150
  'richness' of 145
  sexist 290, 300, 303, 308, 310
  social aspects of 137, 146, 147, 148, 163, 303
  specific patterns 145
  study strength 33–7
  television and 339–68
  thought and 135–6, 139, 146
  tool of social interaction 302
  women's 137, 143
  words integrated into German from another 90
  working 48, 49
  youth 142, 143, 164, 366
  see also under various headings, e.g. colloquial language; dialect; everyday language; Jugendsprachen; learners of German; official language; registers; sociology of language; spoken language; standard German; variants; variation; vernacular; written language; *also under following entries prefixed 'language' and 'linguistic'*
language acquisition:
  adults 176, 178
  age of learner at beginning of process 179
  interrupted 176
  measures to support 175
  natural, immigrant workers 176
  relative success of 178
  second 150, 169, 176–9; immigrant children 209; studies of 185
  social factors 178
  successful 17, 174
  untutored 174, 176
language classes/instruction 176, 184
  asylum seekers excluded from 174
  explicitly designed for *ausländische Arbeitnehmer* 174
  teaching 169, 172; 'compensatory' or 'emancipatory' programmes 208; new programmes 209; non-sexist language in 307
language communities 27, 49, 137, 139, 170
  interlingual 26
language competence 184, 198
  approximate native-speaker 178
  assessment of 178
  bilingual 183
  degrees of 187–8; achieved by non-native speakers 171
  inadequate 197
  low 171
  passive, in the standard forms 365
  socio-pragmatic 192
language culture 163–4

East German ethic 165
  socialist 138
  state-run 146
  tasks of 163
  television and 367–8
language games 327
  ironic 331
  prosodic 319
language varieties 159, 160, 161
  'age-specific' 141
  alternation of 162
  changes in the use of 130
  class-specific 145
  colloquial 132
  descriptions of 139–43
  'ethnic' 171
  extra-linguistic parameters and 148
  learners 141, 149, 150, 177
  native-speakers' 179
  non-standard 367
  regional 12, 157
  relaxation of norms in the spoken standard 365
Lasch, Agathe 117, 118
Last, A. 327, 331, 332
Latin 28, 37, 46, 61, 70, 289
  alphabet 71, 77, 84
  neuter nouns with the *-a* suffix of the Latin feminine singular 107
Lauper, H. 284
'Laut-Buchstaben-Beziehung' 71
law 209–10, 243
  Basic 305
  middle-class defendants 210
  private 306
lawyers 185
League of Nations 47
learners of German 141, 149, 150, 177
  Chinese, in China 192
  factors 178–9
  Japanese 184
  language competence 178
  learning process 233
  of the language as a foreign language 33–7
  US-American 193
  who speak Russian as their native language 192
left-wing groups 113, 231, 248–52
legal consultations 185
'legalese' 207
  see also language (legal)
legends 315, 316
'legitimacy' 62, 146
Lehrende/Lehrpersonen/Lehrkräfte 301
LehrerInnen 91, 301
Leipzig 1, 16, 325
  Bibliographisches Institut 16, 81
  trade fair 195
Lenin, V. I. 248
Leodolter, *see* Wodak-Leodolter

Leontiev, A. N. 161, 234
Le Pen, Jean-Marie 236
Lerchner, G. 161
Lernvermögen/Intelligenz 131
Leser/Leserinnen/LeserInnen 88
Leska, C. A. 354
Lessing, G. E. 64
letters:
  capital 88; names written in 89
  fixed combinations 87
Levinson, S. 159, 281
Lévi-Strauss, C. 325-6, 359
lexical factors:
  characteristics 142
  codification of norms 78
  contrasts in public language 194
  grammatically masculine subsets 290
  means of specification 295-6
  variation 142, 149, 354
Liebe Reséndiz 131
Liechtenstein 31, 47
Liedtke, F. 16
lifestyles 326
  expressive 325
  stigmatizing 263
Lindenberg, Udo 322
'Lindenstraße' 356, 366
lingua franca 26-7
  extensive 159
linguistic behaviour 147, 198, 258, 260,
    264-71
  characteristic 185
  children's, written texts used to test 208
  concrete 368
  contrasting patterns of cultural and 194
  instrument for describing 148
linguistic change 125, 162, 302
  diffusion and reinforcement 366-7
  preservation and decline 157
'linguistic cleansing' 15
linguistic codes 143, 144-8, 160
  dominant, in the classroom 4
  private, young people 315
  'restricted' and 'elaborated' 145, 147, 164
  separate 280
  shifts/switches 349
linguistic nationalism 58, 62, 63, 65, 66
  authenticity maxim of 64
  success of 59-60
  uncritical acceptance of dogmas of 65
linguistic norms 138
  conflicting cultural and 198
  double character of 161
  meaning of 119
  nationally recognized and codified
    standards 141
  prevailing 365
linguistic pluralism 60
linguistic pragmatics 3, 8, 11, 153, 172, 182
  contrastive 188-94

see also communication; conversation
    (analysis); dialectology; norms;
    sociolinguistics; speech behaviour;
    speech styles
'linguistic relativity' 135-6, 147, 165
  psychological reality of 154
Link, Jürgen 17, 232, 233, 247, 250
Linke, A. 340
Links (New Left journal) 250
listening 95, 188-9
literacy 176, 367
Literatursprache 7, 9
loanwords 89, 107, 113, 320
  Anglo-American 367
  writing 91
Löffler, Heinrich 3, 5
'logo' 355
Lorenz, Konrad 240, 242
Lotz, C. 284
Low German 118, 120, 121
lower classes 145, 146, 148, 261, 273, 282
  inferior academic capability 144
  linguistic poverty 143
  urban dialects 263
Ludwig, R. 345
Luitbert, archbishop of Mainz 71
Lutz, B. 214
Lutzen, W. D. 362
Luxembourg 31, 37
Lyons, J. 287
lyric poetry 73

Maas, Utz 232, 233
McDonalds 99
McGraw-Hill 304
'Macht durch Mütterlichkeit' 287
MacKay, D. G. 303
Macmillan, J. 281
macro-sociological categories 157
magazines 237, 239
  feminist (*EMMA*) 301
  Konkret 250, 251
  political 340
  providing TV programme
    information 352
  teenage (*Bravo* and *Mädchen*) 321
Mainz 71
  *Sprachverband Deutsch für ausländische
    Arbeitnehmer* 174
majuscule 71, 72
Maletzke, G. 368
man 113, 289
  generic usages 304
Mann/Männer 295, 302
  -mann compounds 296
mannerisms 317
Mannheim 12, 83, 89, 90, 319
  Bibliographisches Institut 16
  Institut für deutsche Sprache 1, 161
Manta-jokes 323, 333

markers 326
  morphological 325
  morphosyntactic 292
  'spatial' 162
  spontaneity 347
Maron, Monika 194
Marui, I. 191
Marx, Karl 64
Marxists 6, 160, 250, 251
masculine nouns 106, 292, 297, 298, 321
  alternatives not semantically equivalent
    to 299
  double adjectival modification with 299
  feminine noun typically denotes an
    occupation of a lower social status
    than 301
  generic 302, 308, 310; interpretation
    of 301, 305; replaced by splitting 308;
    use of 290–1
  referential range becoming narrower 301
  sex-specific 306
  variable and indeterminate status 296
Mattausch, Christian 240, 241, 242, 243
Mattel-Pegam, G. 185
Mattheier, Klaus J. 12, 280, 365
Mauthner, Fritz 111–12, 114
MDR (Mitteldeutscher Rundfunk) 368
meanings 153, 154, 194, 350
  denotative 99
  generic 304
  linguistic norms 119
  morpho-syntactic, verbal encoding of 156
  new 123 n., 319, 326
  plurality of 351
  prior and sedimented 326
  richness of 163
  social 153, 161, 346
media 127, 352, 367, 368
  electronic 346, 350, 351
  mass 340
  print 352–3
  see also press; radio; television
mediation 351, 368
Megret, Bruno 236
Meidling 262
Meik, R. 322
Meinecke, F. 57
Meisel, J. 177
memorization 190
mensch 302
mentality 243
Menz, F. 206, 207, 212, 224
metaphors 145, 159, 318, 354
  mixed 241
  power 268
  special 232
  suggestive 244
metre 110
Meyer, H. 118
Meyrowitz, J. 346, 349

Middle Ages 71
middle classes 144, 145, 210, 211, 264
  lower 137
  right-wing conservative families 238
  sociolect stigmatized by 118
  suggested treatment for patients 212–13
  upper 263, 270, 273
Mieming 268
Miethaie 244
migration, *see* immigrants
Mikos, L. 355, 356
Miller, C. 292
Milroy, J. 17
Milroy, L. 17, 280
Ministry of Postal Affairs 342
minorities:
  ethnic 171
  German, in/from eastern Europe 62, 173
  German-speaking, countries with 38–9
  stigmatizing 263
  'minorization' 195
Mischtextsorten 352
miscommunication 169, 194
  different ways of explaining 197
Mishler, Elliot 212, 223
misinterpretation 170, 189, 194, 197
mistakes 84, 85
misunderstanding 169, 209, 212
  consciously recognized 182
  'culture of' 196
  different ways of explaining 197
  difficult to resolve 188
  far-reaching 197
  major 189
  potential for 170
  simple to repair by gestures 186
  single word may give rise to 195
mitgemeint 291
mobility 196
  regional and social 130
modality 177–8, 349
Mongolia 38
monologue 345
moods 108
Moosmüller, Sylvia 16, 258, 259, 260, 267,
    268, 269, 270, 271, 272, 273
Morocco 174
morphemes 96, 97
  boundaries 87, 88
  free 99, 100
morphology 97, 155, 290, 294, 308, 367
  compounds 295–6
  inflectional 177
  markers 325
  means of specification 296–7, 310
  nouns 102–8
  rules 149
  variation 289, 298
morphosyntactic markers 292
Moser, H. 157

'mother tongue' 27, 34, 35, 86
Muckenhaupt, M. 340
Mühlen, P. 340
Müller, K. 71
Muller, S. H. 29
Müller-Thurau, Peter 316
multicultural society 172–6
Mumby, D. K. 207
Mundart, see dialect
music business 321
Muslims 175
Musterwissen 209
Mut (monthly publication) 238
Myers-Scotton, C. 57
myths 207, 213, 215

Nabrings, K. 142
names 73, 318
  company 98
  false 293
  God, famous people, countries and
    towns 72
  product 125
  proper 72, 75; diminutive forms 74, 97;
    practice of emphasizing 71; without
    -*s* 36
  written in capital letters, ß used for reasons
    of clarity 89
Namibia 30, 32, 38
narratives 189–90, 211, 356, 359
nation 26
  language and 55–66, 140
  national consciousness 76
  nationality 26, 62
  see also nationalism; native-speakers
Nation Europa (militant racist journal) 237,
  238, 239, 240, 242
National Socialism, *see* Nazi Germany
Nationalcharakter der Sprachen (Humboldt) 65
nationalism 55, 56, 58, 238
  far-reaching consequences 176
  feelings of 175, 231
  language not a 'natural' vehicle for 66
  militant 60
  see also linguistic nationalism
native speakers 29
  adaptation to non-native speakers'
    competence 179–81
  approximate competence 178
  language varieties 179
  simplified speech style used by when
    talking to foreigners 177
Naturell 243
natürliches Geschlecht 287–8
Nazi Germany 39, 45, 47, 56, 66, 97, 358
  National Socialism 62, 86, 98, 236
  Nazi-speak 64
  propaganda 244; totalitarian use of
    broadcasting for 342
  Third Reich 63, 64, 231, 238, 244

Volk the key term of ideology 64
Neale, S. 351
negation by *nix* 180
Neogrammarians 136
neologisms 99, 366, 367
  derived from English 109
  discriminatory 244
  expressive 101
  GDR-specific 162
Nerius, Dieter 9, 75, 79, 82, 83
Netherlands 37
Neubert, A. 6, 7, 8
Neues Deutschland 11, 248–9
Neugebauer, M. 340
Neuland, E. 323, 326
Neuner, G. 172
neuter form 107, 113
neutral contexts/neutralization 290, 295,
  296, 301
  choice of *he* in 288
  covert 297–8
  denominalization 101
  enforcement of 307–8
  inherent 297–8
  overt or marked 298
  sex 293
  special type of 299
  standard utterance 266
New Left 248, 250
New Right 236
New York City 280
New York Times 50
Newcomb, H. 355
newspapers 1, 88, 242, 243–4, 247, 248
  daily 237
  'independent' 231
  mainstream 241
  more serious 245
  National Socialist 64
  political 251
nicknames 318
Nietzsche, Friedrich 55
Nobel Prizes 45
nominal system 304
  see also adjectives; human nouns; infinitives;
    neutral contexts; NPs; participles;
    particles; splitting
non-Caucasian populations 240, 241
normative systems 85
Normen in der sprachlichen Kommunikation 8
norms 164, 245, 247, 272
  commonly shared 326, 327
  communicative 161–2
  cultural 175, 197, 198
  default 152
  grammatical 78, 114, 162
  group 328
  language 129, 139, 163
  male 289, 295, 307
  new 131, 147, 162

norms (*cont.*):
  old, ridiculed 147
  orthographical 69–92
  pretending to identify with, of the
    dominant culture 331
  public, language that conflicts with 335
  relaxing of 341, 365
  semantic 162
  social: connected with
    sex-membership 293; differences in 129
  sociolinguistic 163, 280
  standard, derived from the written
    language 95
  strict writing-based 103
  tolerance of 141
  violations of 111
  see also linguistic norms
North America 33, 37
  see also Canada; United States
'North German' 108
Norwegians 73
Norwich women 280
nouns 71, 73, 84, 96–9, 101, 320
  capitalization of: abolishing 70;
    syntactically motivated of 72
  abstract 300
  adjectival, ending in -*heit* or -*keit* 290
  arbitrary assignment to gender-classes 290
  attributive adverbs added to 149
  collective 299
  common 293
  compound, writing of 88
  'concrete' 74, 75
  derived from existing masculine terms 296
  de-verbal, in -*er* 290
  diminutive, in -*chen* 290
  followed by adjective 149
  grammatical gender an inherent and
    invariant property 289
  gender-neutral 88
  morphology of 102–8
  neuter 106; Latin 107
  neutral 295, 297–8
  proper 293
  regular patterns of plural formation 107
  repeated or replaced by pronouns 113
  simple 149
  with articles 149
  see also human nouns; NPs
NPD (National Democratic Party) 238
NPs (noun phrases) 149, 287, 288, 289
  complex nominalizations and attributes
    in 354
  embeddings in 345
  heads of 71
nurses 211
Nyquist, L. 281

objects 102, 109
  direct/accusative 104, 107

genitive 103, 106
indirect/dative 105, 106, 106–7
obligatory/optional 105, 106–7
prepositional 103, 106, 107
'obligations' 184
occupations/occupational terms 179, 295,
    309
  activity of higher social status 297
  feminine titles 306
  higher-status 288, 291
  low-status 288, 301
  typically male 296
ODR (Ostdeutscher Rundfunk) 368
Oevermann, U. 208
official contexts 124
  bulletins 249
  forms 113
  'officialese' 123
official language 47–9, 306, 307, 309
  countries with German as 31–3
'official linguistics' 138
Ohama, R. 184
Old English 71
Ong, W. J. 344
onomatopoeia 322
orality 348
  primary 344
  secondary 344–5
Orthographical Reform (1901) 82
orthography 69–92, 101, 298
  American 85
  English 102
  see also *Duden* spelling dictionaries
Orwell, George 17
Osnabrück:
  research project 'Zum Sprachgebrauch von
    Jugendlichen' 332
'Ossis' 96, 124, 146, 147
  'Wessis' and: interactive aspects of
    encounters between 195; still said to
    regard each other to a certain extent as
    foreigners 169
Ostow, R. 180
Ostrower, A. 46, 47
Otfrid 70–1, 84
ÖVP (Austrian People's Party) 268, 269

Paarformeln 298, 306, 308
pacifists 251
Pädagogik 4
Paetow, M. 356
pair formulas 298, 306, 308
Palatinate 148, 120
Paleit, D. 174
pamphlets 231, 237, 239
Päng-Sprache 322-3
Paraguay 39
paraphrases 180
parentheses 112, 347
participles 101, 294

clearly unmarked terms deriving from 113
  formation of 321
  nominalized, plural forms of 297
particles 125, 219, 318, 320, 333
  communicative, *ey* 332, 335
  'comparative', adverbs as 109
  modal 345
  nominalized 290, 305
  pragmatic 286
  so 100
  to modify or tone down utterances 147
patients 205, 206
  considered 'difficult' 222
  experienced 223
  fears 222
  inexperienced 218–19, 223
  initiatives 216, 218, 223; ways in which
    doctors deal with 215
  referred to in the third person 221
  suicidal 281
  working-class and middle-class, suggested
    treatment for 212–13
PDS (Party of Democratic Socialism) 11, 248
pedantry 78, 88
Peel, P. C. 172
Pei, M. 25
Perdue, C. 185
periodicals 237, 251
Personalchef/Kaderleiter 124
Personenbezeichnungen 293, 299
Petersen, J. 210
Pfeiffer, O. 209, 214
Philippines 174
'philological studies' 4
phonemes 71, 84, 91
phonetics 83, 142, 162, 180
phonics 344, 345
phonology 90, 155
  alternative realizations of rules 149
  backgrounding processes 272
  characteristics 120–1
  variables 150, 269–71
phrases 129
  classified by topic areas 316
  discriminatory 242, 243
  empty 249
  English 321, 332
  key 329, 330, 331
  stereotypical 267, 318
Picht, Georg 4
pidgin, *see* Heidelberger Forschungsprojekt
Pienemann, M. 177
pluralis hospitalis 220, 222
plurals 89, 107–8, 297
  feminine form 112
  formation 321
  Graeco-Roman 107
  intentionally ambiguous 108
  nominative 81
  regular patterns 107

-s forms 97, 108
sex-specification in 294
-ta forms 108
P-Moll Project 177–8
Poland 36
Polenz, P. von 10, 16, 366, 367
politeness 159, 162, 192, 220
  different societies 137
  Japanese, Korean, Polynesian, and
    Kri-specific forms 137
  rituals of 215
political parties 342
  centre 231
  centre left 241
  journals published by 237
  proportional representation of 351
  see also CDU/CSU; DKP; DVU; FDP;
    Green Party; NPD; ÖVP; PDS;
    Republican Party; SPD; VSP
politics 69, 99
  autonomy 63
  blatant pragmatism 250
  discourse: language of right and
    left 231–53
  ideologies 250
  language 231
  magazines 340
  newspapers and periodicals 251
  policies towards integration 169
  slogans 231, 238
  see also political parties; politicians
politicians:
  evaluation of language behaviour 264–9
  lower middle-class background 271–2
  working-class background 270, 271
PolitikerInnen 88
Polynesia 137
polysemy 350, 351
PONS test (Rosenthal *et al*) 189
Population/Bevölkerung 241, 242
populists 235
Pörksen, U. 17
Portugal/Portuguese 29, 32, 38
possessives 102, 103, 104, 106, 107, 294
'post-stratification' 146
'post-stress -*e*' 120
Postman, Neil 348, 367
postvocalic [r] 280
power 238 n., 258, 286
  doctors' exercise of 215, 218, 221, 222
  élite 56, 57, 214
  feminine form 321
  hierarchical patterns/relationships 209,
    210
  metaphors 268
  'packaging' of 218
  registers 214, 223
  relationships 195, 207
  through language 282
  through motherliness 287

Powesland, P. 259
pragmatics/pragmalinguistics, *see* linguistic
    pragmatics
Prague School of linguists 8
prefixes 101, 102, 121
    formation of 319
    pseudo- 99, 109
prejudice 197, 198, 240, 243, 245, 262
    deep-seated, against dialects 273
    racist or other, press instigating 247
    views on *Jugendsprachen* 318
'Premiere' (pay-TV channel) 343
prepositions 101, 102, 104, 105, 106
    expressions using *von* 103
prescriptive *he* 289
press:
    instigating racist or other prejudices
        247
    left, in contemporary Germany 248–51
    mainstream 242
    see also comics; journals; magazines;
        newspapers
prestige 58, 60, 132, 258, 280
Pretzsch, D. 362
'Pro 7' (private TV station) 343
pronouns:
    address 75
    anaphoric 289
    co-ordinated by *und, oder* or *bzw* 298
    deletion of 180
    indefinite 105, 294, 302; 'feminine
        alternative' to 113
    masculine 288
    nouns repeated or replaced by 113
    personal 194
    possessive 294
    relative 294
pronunciation 90
    'careless' 105
    correct 258
    German system 367
    original 367
propaganda:
    essential keywords 243
    right-wing 242; extreme 234–7, 241
    Nazi 244, 342
prosody 152, 155, 270, 326
    features 151, 182, 258; standard 271
    language games 319
Prumm, K. 340
Prussia 46, 78, 79, 81, 82
pseudo-English 114
psychological factors 139, 154, 293
public agencies 183–4, 187
    obligatory communication in 183
public discourse 280, 340
    evaluation of language use in 257–78
    quality of 367
publications 238
    right-wing 237

scientific 28; in German, French and
    English 43
see also comics; journals; magazines;
    newspapers; pamphlets
punctuation 84, 88, 89
punks 327, 328, 334
Pusch, L. F. 14, 17, 113, 279, 280, 293, 302
Püschel, U. 340, 352, 353, 354, 365
Putzi/Putzfrau 97
pyramid principle 353

Quasthoff, Uta 210, 263
Quasthoff-Hartmann, Uta 189
questions 211, 218, 282, 319
    complementary 109
    direct 221
    risk 330
    tag- 281
    teachers' 208
    yes/no, requiring confirmation 109
Quinkert, A. 245
Quirk, R. 288
quotations:
    'defamiliarized' 327, 329, 331
    direct 345
    formal 327
    mimetic 327

race 62–3, 238, 242
racism 238
    growing 172
    linguistic equivalent of 62
    press: daily, conservative texts and 243–5;
        instigating 247; mediation of everyday
        discourse by 245–8
    public, blatant 175
    skinhead groups 236
radio 129, 341, 344, 353, 365
    commercial, American 355
    East German 118
    speakers 367
Raspe, H. 212
Räthzel, N. 238 n.
Raumer, R. von 78
reading:
    active 350
    courses 176
    listening before 95
Realos 250
'receiver' factor 160
Rechtssprache 306
'recipient passive' 142
Redder, A. 171, 172, 182, 189, 208
reference grammar 148–9
referential ambiguities 303
reforms 69–92
reformulations 180, 181, 210
refugees 172–3, 232, 245
    attacks on homes 236
    economic 242; Afro-Asian 241

registers of language 145, 148, 207
  'age-specific' 141
  female 281
  formal technical 218–19
  power 214, 223
  rigid 146
  technical 108, 207, 349, 366
  variation 179
Rehbein, J. 171, 182, 184, 185, 186–7,
  189–90, 208
Reich, Jens 196
Reiffenstein, 264
Reinelt, R. 191
Reiners, Ludwig 103
rejection techniques 334
religion 246
Remer Depesche 239
Renan, Ernest 63
repair sequences 180–1
repetition 145, 180, 181, 330–1
Republican Party 97, 231, 235
'restricted' and 'elaborated' codes 145, 147,
  164
reunification 79, 245
  euphoria over 196
  strengthened feelings of nationalism 175
Reuter, E. 84, 85, 172, 192
Rezipientenpassiv 142
rhetorical tone 195–6
Rhine Franconian dialect 71
Ridder, Ch.-M. 361
right-wing groups 234–48
  initiative shifted firmly to 252
rights 185
  conversational 283, 284
  reciprocal 184
riots 242, 247, 282
rituals/ritualization 191, 210, 215, 330,
  331
Roche, J. M. 179, 180
rock music lyrics 322
Rogge, J.-U. 355
Röhl, E. 1, 194
roles 146, 156, 184
  dominant and authoritarian 144
  social 145, 152, 288; subordinate 291
  specific expectations in terms of behaviour
    associated with 182
Romaine, S. 6, 179
Romania 36
Romanticism 58, 59, 60, 66, 76
Römer, R. 100
root-words 322–3
Rosenberg, P. 173
Rosenkranz, H. 7, 157, 159
Rosenthal, R. 189
Rost, Martina 179, 180, 181
Rost-Roth, Martina 11, 142, 143, 209, 210
Rothenhäusler, R. 192
routine formulas 142

Routinewissen 209
RTL 343
  'Der Preis ist heiß' 361
  'Die Bilder des Tages' 353, 354
  'Gute Zeiten, schlechte Zeiten' 358
Rückkehrer 173
Ruhr 246 n.
rules:
  bidirectional/unidirectional options 153
  capitalization 73, 74
  customs and 137
  frequencies of occurrences 150
  grammatical 70
  hyphenation 87, 88
  incorrect 149
  input-switch 257, 272
  'legitimate' 145
  morphological 149
  phonological 149
  probabilistic weighting of 148
  syntactic 149
  variable 148–50, 154
  'write as you speak' (or as you hear) 85
'Rules of Spelling' (pamphlet, 1902) 82
Rusch, G. 351, 352
Russia/Russian language 29, 43, 44, 47, 289

Sachs, Hans 110
Sachtleber, S. 191
Salzburg 258, 261, 262
Sammellager 232
Sandig, B. 157
Sapir, E. 145
SAT 1 (private TV station) 342
  'Blick' 353
  'Glücksrad' 361
Sauer, Wolfgang W. 16, 81, 89, 96, 107,
  108, 112, 118, 195, 320, 367
Saxons 103, 118, 195
Scandinavian countries 37, 344
scapegoats 223
Schadenfreude 1
Schäfer, H. 355
Scharnhorst, I. 82, 83
Scharnhorst, J. 163
Scheibenhonig (*Scheiße*/shit) 159
Schere im Kopf 342
Schiller, F. 59
Schiller, J. C. F. von 64, 75
Schleswig-Holstein 236 n., 244
Schlieben-Lange, Brigitte 3, 4, 6, 11, 12, 14,
  17
Schlobinski, Peter 11, 111, 117, 118, 119,
  120, 126, 143, 150, 157, 322, 323, 327,
  333, 334, 366
Schlosser, H. D. 10, 11, 63, 64, 129, 195
Schmidt, C. 280, 285
Schmidt, Helmut 235
Schmidt, S. J. 351
Schmitz, U. 340, 352, 353, 354

Schneider, P. 354
Schnuller 97
Schoenthal, G. 279, 283
Schönfeld, E. 317
Schönfeld, H. 6, 7, 8, 11, 117, 119, 120, 126,
    127, 160, 162
Schönhuber, Franz 235, 237, 242
schools 75, 124, 175, 208–9, 246
    secondary 319
    standard German in 315
Schräpel, B. 300, 305, 307
Schrenck-Notzing, Caspar 238
Schröder, J. 104
Schröder, K. 29
Schröter, Chrysostomus Erdmann 73–4, 75,
    76
Schuchardt, Hugo 136–7
SchülerFerienTicket 88
SchülerInnen 91
Schultz, J. 182, 188
Schulz, Muriel 279
Schwitalla, J. 340
science 28, 38, 41–6, 51
    terminology of 366
'secessionists' 153, 155–7
Second World War 47, 86
secret state police, see *Gestapo*; *Stasi*
SED (Socialist Unity Party, governing party
    of the GDR) 346
    Neues Deutschland, former central
        organ 11, 248–9
Seilschaft 123 n.
self-corrections 345
self-termination 283
Selting, M. 13, 14, 153, 154
semantics 170–1, 177, 243, 290–1, 295, 321
    appropriateness 162
    gender 287–8, 291, 292, 304, 310;
        specification 289, 293, 297
    inaccuracies 300
    limitations 156
    nonsensical comparatives and
        superlatives 97, 108
    problems with 367
    references more likely to be obscured 96
    role of linguistic expressions 154
Semenyuk, N. N. 161
semi-dialect 120
'semilingualism' 143, 175
semiotics 208, 351
Senft, G. 148, 150
Seniorenheim 128
sentences 73, 326
    active/passive 104
    elliptical 345
    length for dialogues in television plays 354
    long 353, 354
    semicolon after certain complex
        structures 88
    shift of emphasis from 205

shorter, less complex 345
    special structures 232
    written 354
Serbian 62
sex of referent 290, 291, 293
    abstraction from 299–300
    neutralization of 297–9, 307
    preferential use 280, 281, 284, 285, 287
    tendency towards agreement between
        grammatical gender and 301
    see also sex-specification
sex-specification 280, 292, 293–7, 306, 308,
    310, 334
    adjectives 291
    plurals 294
    possibilities of 307
sexism 290, 300, 303, 307, 308, 310
    linguistic 288, 309
sexuality 331
'side sequences' 181
Sie 193, 194
    du instead of, for singular 'you' 180
Siebs, T. 264
silent movies 343
Simon, G. 5
simplification 177, 180
singulars 180, 294
    feminine 107
    genitive 81
    nominative 298
skinheads 236, 325
Skudlik, S. 41
slang 263, 323
    prison 317
    working-class 262
Slavic 37
slogans 231, 238
Slovak 62
Smith, D. M. 303
Smith, S. 181
social action 156
social change 117–32, 137
social class 4, 209, 211, 243, 281
    differences, East-West contrast 147
    the 'driving force' of social change 137
    see also barriers; lower classes; middle
        classes; upper classes
social darwinists 235, 240
social problems 236
social security 235, 236
social structure 146, 148
socialization 175
    class-based differences 4
    linguistic, inadequate 144
socio-economic and political conditions 162
sociolects 118, 160, 349, 366
sociolinguistics 3, 5–6, 7, 8, 9, 118
    American 144
    analysis 208, 281
    'dialectical' 11

discourse 205
'interactional' or 'interpretive' 14
linking of discourse analysis and 205
Marxist 160
micro- 11
new discipline of 234 n.
origins and development in
    German-speaking countries 135–9
pragmatic 11–15
role of research in the GDR 160–1
study of institutional
    communication 205–27
'subversive' and 'authoritarian' 138–9
theories of variation 135–65: East German
    perspectives 157–60; West German and
    Austrian perspectives 143–57
    Western 139
sociological variables 211, 208
sociology of language 11, 146, 161, 331
    German contributions to language 136–8
socio-phonological realizations 219, 221
socio-psychological processes 197, 210
    categories 154
softeners 286
Sontheimer, I. 316
sounds 320, 322–3
    initial 137
    shortened 319
    specific characteristics 325
    spelling relations 84, 86–7, 88
    weakened 319
South German 108
southern Europe 179, 209
Soviet Union 7, 37, 43, 173, 236, 248
    'appropriateness' in linguistics 9
    psychology 161
    see also Commonwealth of Independent
        States; Russia
'sozial-linguistiches Prinzip' 3
soziolinguistisches Differential 8, 9
Sozialismus (New Left journal) 250
'space' 148, 216, 224
    changed social meaning of 346
Spain 38
    Spanish language 33, 34, 37, 47, 49, 193;
        emphasis in North America 35–6
Spangenberg, K. 7, 157
Sparformeln 298, 308
spatial distance 191
SPD (Social Democratic Party) 174, 232,
    235, 250, 305
speakers:
    Bavarian 12
    body language revealing origin of 196
    characteristics 145
    Chinese 189
    East Berlin dialect 132
    intention of 153
    lower class 145
    middle class 145, 264

negative evaluation of 125
non-native, degrees of competence
    achieved by 171
'on' or 'off' 352
personality 265
possible to adjust to interlocutors 159
preferences of language use of 156
radio 367
relative status of 284
Russian, learners of German 192
signals 345
social categories and characteristics 154
Turkish and Italian, learning German 178
upper class 264
    see also German-speaking countries; native-
        speakers; speaking; *also under entries
        prefixed 'speech' and 'Sprach-'*
speaking 283
    assumption of a close relationship between
        social roles, thought, and ways of 145
    breaks while 145
    different ways of 171
    ethnography of 172, 325
    functional equivalence of culturally
        different ways of 143
    manners of 12, 325; metaphorical, mostly
        exaggerated 318; rather
        sophisticated 349
    styles of 191
    women: men and 280–7; more dialogical
        and more polite 143
speech 121, 128, 131, 148, 152
    adult 335
    affects linguistic usage 95
    Berlin 332
    colloquial 7, 100, 105, 107, 218, 335
    contrasting patterns in German-German
        dialogues 147
    cultivation/culture of 163
    everyday 10, 322–3, 335
    formal 145
    habits 130, 131
    non-standard 246 n., 280, 367
    reproduction of 110
    ritualized patterns 330
    situations 162, 273
    social 154
    social distribution/evaluation of
        varieties 119
    strategies 145
    turns of 282
    types of situations 160, 162
    women's 280
    written reproduction of 76
    youth subcultures 315–35
    see also group speech; *also under following
        entries prefixed 'speech'*
speech acts 196, 219, 330, 331, 353
    culture-specific 137
    direct 223, 327

speech acts (*cont.*)
  'habits' in the verbalization of 188
  indirect 220, 223
  theory 3
speech behaviour 130, 222
  conventionally correct 367
  individual 11
  pompous and vacuous 129
  politicians' 265, 269, 273
  pragmatic contrasts in 10
  uncontrolled 272
speech communities 110, 146, 159, 302
  distinction between 'communication
    communities' and 161
  'German', affiliation to 161
speech forms 127, 130
  ability to categorize 258
  functional adequacy in given situations 208
speech styles 12, 152, 326
  concrete 325
  constitution of 327
  differences in 129
  group-specific 327, 333, 334
  pragmatic analysis of 327–33
  simplified, when talking to foreigners 177
  situation-specific 327, 334
Speicher, J. K. 193
spelling 69–92
Spender, Dale 279
Spillner, B. 172
splitting:
  abbreviated 298, 308
  adjectival 298–9
  generic masculines replaced by 308
  long 298
  nominal 306, 307
  syntactic 298
  unabbreviated, oblique cases rendered
    by 298
spoken language 96, 103, 110, 345
  definition of 'texts' to mean all forms
    of 233
  features of 162, 332; common 100
  freedom to record 139
  information transmitted to a greater extent
    by 91
  Jugendsprache as 318, 319, 320, 323
  trend towards a formal and written
    orientation 365
  vernacular in Berlin 117
Sprachbarrieren 3, 6, 5, 13
Sprachgesellschaften 16
Sprachglossen 17
Sprachkonfusion 80
Sprachkritik 8, 15, 16, 17
Sprachkultur 8, 9, 15, 91, 163–4
Sprachkunst (Gottsched) 73
Sprachlenkung 8, 15
sprachliche *Asymmetrien* 306
sprachliche *Gleichbehandlung* 305-6, 307

Sprachmitteln 179
Sprachpflege 8, 15, 16
Sprachpurismus 15
Sprachreport 86
Sprachwissenschaft 15
Spranz-Fogaszy, Th. 212
Sprechweisen 12
Sri Lanka 241
Staatsnation 57
Stalin, J. 248
Stalinism 251
standard German 110, 140, 159
  Austrian 153, 257–60, 266, 267
  broadcasting important part in diffusion
    of 341
  dialect being replaced by a vernacular
    influenced by 162
  functionally appropriate use of 164
  popularization of 365
  school 315
  socio-phonological switch into 221
  superior position 164
  Swiss 26, 31, 34, 141, 305, 308
  variety with marked syntactic difference
    from 142
standardization of orthography 75, 78, 79,
    89
  absolute, attempts to achieve 91
  primary influence on 70
Stasi 97, 98
status 195, 284, 296
  occupational/professional 288, 291, 293,
    297, 301
Steger, H. 2
stereotypes 131, 197, 198, 214
  catch-phrases 318
  female 284, 288, 300–1, 302; avoidance
    of 304, 309
  individual cultural groups 194
  masculine generic 304–5
  verbal 145, 267
Stevenson, Patrick 3, 4, 10, 12, 65, 119, 135,
    173, 176, 177, 208, 209, 258, 263, 264,
    365
Stickel, G. 1
stigmatized forms 122, 132, 260, 263, 266,
    272
  New York English 280
Stolt, B. 191
Storck, M. 362
Strang, B. M. H. 292
Straßner, E. 340, 344, 352, 353, 354
Strategisches Handeln 286
Strecker, B. 195
Streeck, J. 170, 171
Strong, P. M. 208
Strotzka, H. 212
structuralism 4
structuring signals 221
students 113, 184, 191

Stutterheim, C. von 173, 177
'style' 154-5, 156, 157, 191, 192–3
　'Donald Duck' 323
　'functional' 160
　see also groups; speech styles
Styria 261, 268
'subcultural' factors 197, 325–6
subject 109
'subject genitive' 102-3
subordination 284, 285, 289, 291, 309
substantives 74
Sucharowski, Wolfgang 210, 340
Süddeutsche Zeitung 99, 114, 245
suffixes 79, 88, 98–9, 108, 319
　dative -*e*, decline of 106
　derivative 100, 296, 320
　de-verbal nouns 290
　diminutive nouns 290
　familiar 107
　feminine 107, 297, 310
　genitive singular and nominative plural 81
　infinitive 320, 322
　monosyllabic abbreviations with -*i* 97
　productive 101
superlative forms 97, 108
supportive responses 285, 286
Süßmuth, Rita 328, 330
Sütterlin-Schrift 70
swearing 137
Sweden 40, 41, 63, 279
　communicative behaviour, contrasting
　　German and 191
Swift, K. 292
Switzerland 37, 47, 63, 86
　Arbeitsgruppe 307
　awareness of linguistic sexism in Swiss 309
　Bundesverfassung 308
　German language 26, 34, 141; changes in
　　everyday usage of standard 308; Federal
　　Government Report (1991) 305; official
　　status 31
　institutions 280
　TV discussions 282
　see also Zurich
syllabification 84
syllables 96, 137
　contraction of 98
　repetition of 180
　stressed and unstressed 79
'symbolic capital'/'symbolic profitability' 181
'symbolic market' 207
symbols 321
　collective 247
　orthographical 298
　suggestive 244
symmetry 304, 309, 310
synonyms 15
syntagmatic level 331
syntax 109–12, 155, 177, 289
　characteristics 162, 266

difference 142, 148
diffusion of phenomena 367
film reports 353
forms of splitting 298
German system 101
integrating 347
motivated capitalization of nouns 72
rules 149
structures 331
very dry 349
'system linguists' 153

'Tagesschau' 345, 352, 353, 354
'Tagesthemen' 345, 349, 352
Tagezeitung 90, 250
tags 177
Tannen, Deborah 330 n.
Tätigkeitstheorie 7-8, 161, 234
teachers 208, 301, 317
　female leadership 209
　worlds of knowledge which determine
　　interaction between children and 209
technical terminology 113
Techtmeier, B. 9
teenage girls 331
Teichmann, C. 10
television 129, 237
　Americanized medium 367
　anchorpersons 354
　attention-claiming devices 347
　breakfast 355
　cable 342, 368
　commercial channels 344
　commercials 362–4
　genres 339, 343–4, 351–64
　households with sets 341–2
　impact on the German language 364–8
　international sport or music channels 343
　mixing styles 349–50
　news bulletins/broadcasts 340; children
　　and young people 355; magazine
　　format 349, 352; oralizing
　　presentation 354–5; texts written with
　　an eye to the spoken performance 353;
　　verbal strategies used in
　　formulating 353; written style 354
　obscene pastiche of *Der große Preis* (quiz
　　programme also known as *Der große
　　Scheiß*) 328-32
　official, formal language 344, 347
　openness 350–1
　private channels/stations 342–3, 345, 355,
　　368
　public channels 353, 360; voluntary
　　self-commercialization 361
　quiz- and game-shows 340, 358–61
　reception 351, 357, 362; 'double
　　channel' 353; individual and group
　　368
　satellite 342, 368

television (*cont*.):
  serials 355, 356, 358
  soap operas 344, 351, 355–8
  sports reports 340
  talk-shows 281, 282–4, 340, 349
  technical registers 366
  verbal gags 361
  'zapping' channels 351
television texts 349, 354
  correspondent 'off'/'on' 352
  empty passages in 351
  first-speaker/second-speaker 352
  interviewee/interviewer 352
  less coherent but more personalized
    345
  linguistic arrangement of 348
  loosely connected cluster of 353
  news 353
  'open' 362
  oral 354, 367
  performance of 344
  plurality and variety of 351
  presenter 355
  relatively closed 351
  secondary 344
  special types 232
  spoken, illustrations for 343
  trivial 350
  verbal 344
  written 365
Tendenzwende 2
Terborg, H. 178
text linguistics 234
textbooks in German 42–3
texts 162
  academic 103, 191
  administrative 103
  analysis 188
  anaphorical relations have to be discovered
    in order to understand 152
  colloquial nature of 247
  conservative, racism and extreme
    right-wing ideological tenets in 243–5
  definitions of, to mean all forms of written
    or spoken language 233
  English, written by German-speaking
    academics 191
  everyday 108
  extreme right-wing, racism in 239–40
  for public consumption 87
  formal 104
  German and French scientific, comparison
    of 191
  'habits' in the verbalization of 188
  in context 205
  internal regularities and structures
    within 234
  legal 103, 306; comprehensibility of 214;
    non-sexist language in 307;
    reformulation of 210

multidimensional relationship between
  context and 151
news 214, 353
production of 233
socio-historical product 234
written 95, 96, 190–1, 208, 354
see also television texts
Textsorten 339, 352
'*Text-Bild-Schere*' (Wember) 340
thanks 188
theories 138, 150, 161
  communication 58
  current 144
  high-quality research guided by 139
  'linguistic codes' 144–8
  postmodern framework 146
  socio-linguistic codes 143
theotisce 71
Thibault, P. 156
Thimm, C. 286–7
Third Reich, *see* Nazi Germany
Thogmartin, C. 42
Thomas, J. 340
thought 135–6, 139, 145, 146, 150
Thuringia 157, 159
Tietz, Hermann 98
'time' 148, 152, 155, 224
Todd, A. D. 211
tolerance 197
Tomuscheit, Peter 97
tone of voice 171
Torres, G. M. 192, 193
Townson, M. 10
Trabbi 124
Trachsel, C. F. 118
trade 37–41
traditionalists 155–7
Trägerdativ 106
Tram/Trambahn/Tramschiene/
  Straßenbahn 128
Tränenpavillon 127
transfer 182, 192, 193
translations 47, 246 n., 323
  bad 49
  direct, of fixed expressions 192
  literal 192
Treaty of Maastricht 49
Treichler, P. A. 288
Trömel-Plötz, S. 14, 279, 280, 282, 283,
  284, 286
Trudgill, Peter 3, 259, 260, 280
Tsunoda, M. 43
Tucholsky, Kurt 16, 17
Turks 14, 209, 241, 246
  conventional behaviour 192
  German-Turkish
    differences/contrasts 192, 193
  girls 334
  learning German 178
  narratives of schoolchildren 189–90

talking to, in broken German 142
women 176, 184
turn-taking 281, 282, 283, 284, 285, 309
  frequency of 286
  introducers 286
Tyrol 260, 261, 262
  broadest dialect 265
  South 31

Uesseler, M. 6, 8
Uhlisch, G. 192
Ulbricht, Walter 64, 127
Ulijn, J. 41
Umgangssprache 7
UN (United Nations) 48
  Economic and Social Council 47
  General Assembly 47
  Security Council 47
  Unesco 304; *Statistical Yearbook* 34
uncertainty 221, 223, 281, 284
understanding 159
  difficulty in 186
  expression of lack of 189
  interactive process of communication
    and 154
  medical terminology and scientific
    knowledge 185
  mutual 176, 181, 187, 196, 197, 198
  optimal 157
unemployment 175, 243
United Kingdom, *see* Britain
United States of America 45, 112, 150, 191,
    235
  American English 189, 280
  bibliographies and databases 43, 44
  commercial radio 355
  Constitution 55
  differences in listening response between
    black and white Americans 188–9
  language and gender studies 279
  learners of German 193
  linguistic differences in argumentative
    behaviour between Germans and
    Americans 192
  sociolinguistics 144, 205
  television medium 367
  textbooks in German used at
    universities 42–3
universalism 55
Unsere Zeit (Communist Party
    newspaper) 248
upper classes 146, 148, 260, 264
  sociolect stigmatized by 118
urban centres 159
Urdu 62
Urvolk 61, 65
Uske, Hans 2
USSR, *see* Soviet Union
utterances 12, 142, 149–50, 163, 196, 272,
    325

accentuation of certain segments 182
acquisition of tags at end of 177
actual, structural descriptions of 110
context for the interpretation of 152
difference between pragmatic and
    grammatical 112
interpretation of 151, 152
linguistic characteristics of 145
making mutually understandable 187
oral, television 354
parenthetical remarks and intervals in 347
particles to modify or tone down 147
personal or private-level 271
prosodic and gestural features to contexts
    of 151
quantitative distribution of types of 150
repetition of parts of 180
shift of emphasis from 205
short, ritualized 329
standard, neutralization in 266
strengthening the force of 322

vagueness 351
value judgements 245
value systems 175, 212
values 215, 219, 238, 241, 243
  commonly shared 326, 327
  communicative 147
  cultural 183; dominant, pretending to
    identify with 331
  pragmatic and interaction-specific 162
  social 129, 147
Van Dijk, T. A. 207, 234–5 nn., 247, 263
van Hoof-Haferkamp, R. 49
variants 142, 160
  Berlin 150
  'GDR-specific' 141
  'national' 140
  phonetic 162
  'territorial' (dialectal) 140
variation:
  changing perceptions and attitudes
    towards 129–32
  culture-specific 198
  'diachronic' 142
  'diacommunicative' 142-3
  'diatopic' or spatial 141, 142
  'diastratic' 142
  forms of address 191
  'free' 155
  German–Swedish/German–Spanish, in
    letter-writing 191
  lexical 142, 149; extensive 354
  morphological 289, 298
  phonetic 142
  register 179
  sex-preferential 280, 285
  social, 'internal counterparts' to 139
  structural and systematic 331
  study of 148–50

variation (*cont.*):
  stylistic, in grammar 153–4
  theory 135–65, 258
'variety space' 139-40, 143, 149
  'diasituative' dimension 142, 148
  three-dimensional 148
verbal barbarisms 114
verbal infix -*le*- 101
verbal representations 155
verbs 105, 258, 296, 322
  adding prefixes to English root forms 102
  appropriate 267
  English 101, 102
  finite 110
  formed by adding the infinitive suffix 320
  inflected forms 289
  intransitive 104
  nouns derived from 97, 294
  obsolescent 103, 106
  phrasal 103
  position of 110, 111
  transitive 100, 104
  weak 101
  which take an obligatory dative
    object 106–7
VerfasserInnen/Verfasserinnen/Verfasser 112
Verkehrssprache 159
vernacular 59, 60, 70–1, 130, 157, 159
  functionally appropriate use 164
  influenced by the standard variety, dialect
    replaced by 162
  North German 162
  Upper Saxon 120
  'urban' 144
  see also Berlin
Versailles 79
'*Verschnullerungskampagne*' 97
Verschueren, J. 172
'vertical' social stratification 159
Verwaltungssprache 308
variety grammar 143, 148–50, 154
Vienna 258, 263, 266, 267, 281
  Applied Language Studies research
    group 205, 212
  colloquial standard 272
  dialect 153, 257, 261, 262, 265
  hospital outpatient ward 205, 213–23.
  middle and upper class speakers 264
  negative feelings among
    non-Viennese 259–60
  schools 209
vocabulary 81, 95, 114, 127, 184
  broader 145
  colloquial language 159
  conservative 243
  differences in 147
  enlarging of 367
  highly localized 159
  limited 345
  scientific-sounding 242

Vogel, Bernd 349
Vogt, W. 78
Volk terms 59, 62, 63, 64–5
Volkssprache, see vernacular
Völkischer Beobachter (National Socialists'
    newspaper) 64
*Vollständiges Orthographisches Wörterbuch der
    deutschen Sprache* 81
Voltaire 56
Vorarlberg 261
Vorschriftensprache 306, 307
vowels 71, 79
  accumulation of 77
  lengthening 84, 87
  marking 86
  writing 87
VSP (Unified Socialist Party) 250
vulgarity 159
VW (*Volkswagen*) 65

Wachau, S. 327, 331–2, 333
Wachs, I. 117
Wachtel, M. 340
Wachtel, S. 10
Wagner, Richard (*Sagenkreis der
    Nibelungen*) 315-16
Wallnöfer (Austrian *Land* leader) 265, 268,
    269
Wassermann, R. 210
Weber, Max 56
Wechselwirkung (New Left journal) 250
*weibliche Beschäftigte/männliche
    Beschäftigte* 295
Weick, K. 207
Weinrich, H. 41
Weissenburg 70
Wells, C. J. 15
Weltanschauung 75
Wember, B. 340
Wende 1-2, 195, 235, 248
  '*pragmatische*' 2
  Sprache der 10
Werlen, I. 136, 145
Werner, F. 280
'*Wessi*' 96, 97, 124, 146, 147
  '*Ossi*' and: interactive aspects of
    encounters between 195; still said to
    regard each other to a certain extent as
    foreigners 169
  see also *Besserwessis*
West, C. 211, 281
Westdeutscher Rundfunk 356
Weydt, H. 11, 12, 173
white speakers 189
  male 207, 211
Whorf, B. 145
Wichterich, Ch. 355
Wienold, G. 294
Wierlacher, A. 172
Wiesbaden Recommendations (1958) 84

Wiese, J. 118, 126
Williams, G. 6
Williams, Raymond 17
Willis, P. 325, 326
Wilmanns, W. 296
Wilson, J. 17
Wimmer, R. 16, 17
Wintzek, Bernhard C. 238
WIP (*Wohnungsbaugesellschaft in Prenzlauer Berg*) 125
wissenschaftlich 15
Wittgenstein, L. 352
Wodak, Ruth 153, 154, 171, 182, 185, 207, 208, 209, 212, 213, 214, 257, 281, 287
Wodak-Leodolter, Ruth 209, 282
Wohl, R. R. 346
Wohngemeinschaft 286
Woisin, M. 340, 359
Wolff, J. 193
women 174
    conversations: active role in constructing subordinate position 283, 310; polite forms in 281; rape of, in 282; violations of rights 283
    dependent on or subordinate to men 300
    joining behaviour 283
    language 137
    more dialogical and more polite speaking 143
    more sensitive than men to sociolinguistic norms and prestige patterns 280
    speech 280
    working-class 210
    see also female visibility
women's movement 300–9
words:
    adapted 113
    analogy to other 108
    Anglo-American 367
    archaic sounding 100
    bisyllabic 97–8
    borrowed 90, 113, 126, 297, 323
    capitals within 88
    class membership 296
    classified by topic areas 316
    coined 123 n.
    colloquial 97–8
    compound, rules on hyphenation in 88
    concrete 75
    derived 297
    everyday Berlinish 127
    formation patterns 96–102, 113, 310, 319
    initial sounds of 137
    integrated 90, 101–2
    key 81, 241, 243, 244, 247
    loan 89, 91, 107, 113, 320
    magic 99
    monosyllabic 106
    new 123 n., 124, 128, 319
    onomatopoeic 322
    pairs of 297
    passive knowledge of 367
    polysyllabic 96, 106
    repetitions of 145, 180, 181
    responses of delight and condemnation 318
    root 100
    'sound' 320, 322–3
    specific, for certain terms 137
    see also foreign words
Wörterbuch der deutschen Umgangssprache 95, 317
'*Wossi*' ('*Wessi*' and '*Ossi*') 96-7
Wotjak, B. 1
Wrede, Ferdinand 3
written language/writing 103, 107, 160, 319, 347
    'aggregation, 'integration' and 'detachment' 345
    changes in 309
    characterized by more variability and complexity 345
    communicative function of 89
    compound nouns 88
    'correct' 91
    definition of 'texts' to mean all forms of 233
    errors 84
    expression of creative pleasure in manipulating 89
    foreign words 84, 86
    loanwords 91
    no longer the primary means of communicating over long distances 91
    norms derived from 95
    particular significance in the age of industrialization 157
    performance mostly based on 345
    specific manners of 318
    standard form based on 111
    standard practices 85, 89
    vowels 87
Wüller, K. 320
Wunderlich, Dieter 4, 5, 6, 206
Wurzel, Wolfgang Ullrich 81

xenolects 180
xenophobia 176, 243
    linguistic equivalent of 62

yes- techniques 283
Ylönen, S. 172, 195, 196
young people 175, 179
    language 142, 143, 164
    see also *Jugendsprachen*; youth groups
youth groups 324
    criterion of identity 326
    dynamics 318
    recognition of membership 326

youth groups (*cont.*):
  speech: pragmatic analysis 327–33;
    specific styles 325, 334
Yugoslav girls 334

ZDF (Zweites Deutsches Fernsehen) 342,
  361

Zeitgeist 61, 75
Zimmer, Dieter 17, 117, 132
Zimmerman, D. H. 281
Zimmermann, P. 172
Zurich 82, 282
'Zwangskommunikation auf der
  Behörde' 183

Index compiled by Frank Pert

✪

A Frontier Childhood

Jonathan Edwards was born in the East Windsor parsonage on October 5, 1703. This was precisely nine years after Timothy Edwards had come with his bride, Esther Stoddard, to the newly gathered congregation across the river. He was now thirty-four years old and his wife was thirty-one. Four daughters had already been born into their home and six more were to follow. Jonathan was their first and only son. Was he named Jonathan for what the name means, "Gift of Jehovah," for some English ancestor now lost to view, or for the Welsh theologian and controversialist, Jonathan Edwards of Jesus College, Oxford, whose *Preservative Against Socinianism* had been completed and published earlier in the year 1703? Any one of these reasons might have seemed the best reason to Timothy Edwards.

One cannot but remember that three months earlier, in a Lincolnshire parsonage on the Isle of Axholme, another son had been born to another minister and his godly, strong-minded wife. The two great religionists were never to meet, or even to know why such a meeting would have seemed significant to historians of another century. On two continents John Wesley and Jonathan Edwards were to go their separate and quite different ways, changing the meaning of religion for many thousands, and with it also the cultural pattern of their generation.

Reprinted with permission of the Macmillan Company from *Jonathan Edwards, 1703–1758: A Biography* by Ola Elizabeth Winslow. Copyright 1940 by The Macmillan Company; renewed 1968 by Ola Elizabeth Winslow.

In 1703 the East Windsor parish was still a young enterprise, full of promise. During the nine years he had been among them as their pastor, Timothy Edwards had definitely succeeded. The bitter controversy incident to the separation of his small flock from the parent congregation across the river had gradually slipped into the background and, in spite of occasional reminders that the new parish was made up of both factions in the dispute, withdrawal had abundantly justified itself in the growth and contentment of the new congregation. To go safely to meeting on their own side after years of perilous canoe crossings in all weathers was blessing untold. Month by month new families had come to reside on their fertile holdings across the river and the six-year-old meetinghouse was already too small. The new parish was now a separate township with full power to order its own affairs. This too was a great blessing. For the most part the pews liked the minister, and though no revival had yet come to bless his labors among them they believed God was merely testing their faith; and they waited confidently.

By 1703 life in this far-flung settlement had taken on a fairly settled character and was growing steadily safer year by year in spite of periodic alarms and very real dangers. When Jonathan Edwards was four months old the ever present Indian peril came close to the parsonage in the murder at Deerfield, Massachusetts, of Eunice Williams, half-sister of Esther Edwards. Two of Mrs. Williams' children were also killed, her husband and four more children carried into captivity. The news brought deep personal grief; it was also a grim reminder of the time when churchgoing Windsor had been fined for not carrying muskets to meeting, according to order. Indians were not very numerous in Connecticut by this time, and they were for the most part friendly; but there was still cause for fear. Not for another generation could a child grow up without the memory of a thousand cautions as to what was by no means a phantom danger. From all perils within and perils without, the village must be sufficient unto itself, for this part of the "Lord's Waste" was still a remote frontier. Except for Timothy Edwards and a few other great ones of the village, who occasionally took horse and rode away to Boston, the town limits were the very boundaries of life. One was born, had children, and died without ever going so far as Hartford—two centuries later, only twenty minutes away. As for the vast worries of the land of their grandfathers—Whigs and Tories battling over the nature of the monarchy at home, the War of the Spanish Succession raging abroad, and a stupid queen on the throne—these things were no longer the realities of life. Connecticut colony, East Windsor in particular, was all the world.

Agricultural pursuits made up the background of village life and, as in all country parishes of the day, the minister was perforce a farmer among farmers. He divided his time between his study and his acreage, directing the spring plowing or taking a hand at skinning a cow quite as naturally as he expounded the Scriptures or conducted a funeral. The isolation of East Windsor made the separation between parsonage and parish, sacred and secular, even less sharp than would have been true of Hartford or Northampton. In consequence, Timothy Edwards, for all his austere dignity, was not a man apart from his people. They cut and carted his wood as part of his "rate," made his children's shoes, brought him sugar and mutton and spice as they happened to have abundance, and advised him when to cut the hay. He gave them credit in his *Rate Books* for their services and donations, and in his turn taught their children for pay, bought their cider, distilled it into brandy and sold it back to them again, and engaged in many other sorts of barter convenient to both parties. He was their pastor whom they respected, to a degree feared, and sometimes opposed bitterly; but he was also their neighbor whom they knew in his second-best clothes. On Sundays and Thursdays he preached and assumed the full dignity of his priestly office; on other days he was one of themselves, taking part with them in the exchange of commodities and services by which this isolated little community maintained its independent life.

According to family tradition he was irked by these weekday details and inclined to delegate responsibility for them to his capable wife. Possibly, for he had been town-bred and as a boarding pupil in Mr. Glover's home had escaped chores at an early age; but as an East Windsor husbandman he could not have claimed

immunity from farm tasks. The Edwards acres were fairly extensive: there were fields to be fertilized, crops to be harvested, woods
to be cut down and put under cultivation, stock to be cared for,
hides to be tanned, extra acres to be rented for pasture, and
numerous routine chores to be performed daily. Some supervision
of all these multiple concerns fell to him as head of the house, no
matter how distasteful it may have been. Besides, there is plenty of
evidence that he knew the details first-hand and had some share in
the actual labors which came with the seasons.

By his son in his own country parish days, these tasks would be
assumed far more naturally. Born part villager, part farmer, he
would be able throughout life to accept the routine of field and
barnyard as a necessary, normal part of life, to be performed
without protest or apology. The difference between father and son
in this as in so many other directions was a difference of emphasis.
Jonathan Edwards hewed his life to the line of his main interest,
consciously subordinating those things which he considered lesser;
Timothy Edwards often became confused under tasks hostile to his
main interest, scattered his energies in a fretful and futile busyness,
and was at times defeated by the very details he hated. Both men
handled minutiae with a conscience; only the son chose to split
hairs in an argument, not to measure corn to the half pint.

Jonathan Edwards grew up in the house built as the gift of
Richard Edwards of Hartford at the time of Timothy's settlement
in East Windsor. It stood on the east side of the present highway,
about a quarter of a mile from the old burying ground. As described by Sereno E. Dwight, who saw it in 1803, and by John
Stoughton, who added memories of the oldest settlers in the mid-
century, the house conformed to the general plan of substantial
middle-cláss dwellings of the 1690's. It was a severely plain, two-
story structure of moderate size, built low to the ground and with
the second story projecting slightly beyond the first. A single
chimney separated the two first-floor rooms, one of which was the
kitchen-living-room—possibly also a bedroom as the family increased—the other, called by Timothy Edwards the "parlor," was
really the schoolroom. In this room, which was equipped on three
sides with benches fastened to the wall, Jonathan Edwards and his

ten sisters, together with the village boys who aspired to college and some who did not, received their elementary education. Like other Connecticut houses of the period the parsonage grew with the family, various lean-tos being added, and also an eight- or nine-foot projection at the middle front, spoken of as the "porch" but really a vestibule.

Tradition has built this house of somewhat better materials, more ample proportions, and more expensive appointments than the other houses in East Windsor. Possibly, although its alleged "elegant ornaments" would hardly seem consistent with the character of the donor, Richard Edwards. More probably like the Grant Mansion built in the same decade, it merely introduced architectural improvements hitherto unknown along the "Street." Any house built in the 1690's would naturally have been superior to the log houses of the first residents. Extant expense accounts show the parsonage to have been built of hewn lumber, probably brought by sledge from the nearest mill at Scantic, and of bricks carted from Podunk. The labor of building was the donation of the parishioners, who put a year of their spare time into the task. How well they did their work became a village legend to be repeated confidently generations afterward when the house was being torn down. During all its one hundred and eighteen years, said the great-grandchildren of the pioneer builders, this house had but one covering of shingles—those originally nailed in place by the brethren. Such statements are best left unchallenged, if only to perpetuate the picture of deacons in their old clothes, armed with hammer and saw to a godly end.

In this frontier parish and in this house, its recognized center, Jonathan Edwards lived for the first thirteen years of his life. In many ways he was fortunate, not only for what he missed but for what he gained by such isolation. No wonder the beauty and majesty of nature stamped themselves unforgettably on his early thought. In such a setting nature would have been the most important daily fact to a sensitive child. With a horizon in all four directions he could hardly have escaped impressions of a spacious world: a world of meadows, unending forests, the river; a world of ever changing beauty, not a world of man's making. Even today,

standing on the slight eminence which marks the site of the
Edwards parsonage, the virgin forests gone and the meadows
turned into tobacco fields, one still has a sense of spaciousness and
isolation amounting almost to loneliness. Before 1716 isolation
meant also helplessness, for danger lurked beyond the dark line of
the forest, and miles beyond there were still no habitations.

From the "Street" running in front of the house he could see to
the west, beyond the meadows and beyond the river, the turret of
the Windsor meetinghouse—larger than his father's—and the
more numerous dwellings of the parent settlement. Trips to Windsor in the homemade canoes, so much feared by the older folk,
would have been events in his boyhood. In the foreground, a little
to the right of the parsonage and just across the ravine from his
father's meetinghouse, stood the small fort or Palisado built a
generation earlier as a place of rendezvous in time of Indian
attacks, but in his boyhood used for more peaceful purposes. Even
so, to every boy in the village, acquainted with the tales of earlier
raids, a blast on the infrequent Palisado trumpet would have
sounded a hope of high adventure for his generation also.

Scattered along the "Street" beyond the meetinghouse and
beyond his own home were the houses of the other families of the
parish, fewer than one hundred in all. They stood scarcely closer
together than the farmhouses along the present highway, on which
life now goes so rapidly by; for East Windsor was not a huddled
village. Each house was built on its own acres; and the tracts,
small for farms, were large for town plots. The house nearest the
Edwards home was that of Captain Thomas Stoughton who, in the
year the parsonage was built, married Abigail, sister of Timothy
Edwards. In the Stoughton home there were also eleven children,
with ages corresponding almost exactly to those of the Edwards
eleven. Seven of these were boys—three older, three younger, and
one almost the exact age of Jonathan Edwards—so that the
companionship with boys which he missed in his own home he had
with his seven boy cousins next door. The assumption that, as the
only son in his father's house, he had to endure being petted by his
ten sisters and made to share their girl games is absurd. In addition
to the Stoughtons similar hosts could have been mustered from

almost every one of the hundred houses in the village, for in spite of the "throat distemper," upsetting canoes, and home remedies East Windsor, as well as the rest of colonial New England, was full of children.

At the rear of the house, toward the east, there was scarcely a suggestion of man and his concerns. The land slopes gently down to a brook on the Edwards side, then up a hill—at that time densely wooded. Somewhere along this brook Jonathan Edwards built the booth in which he and his boy companions used to meditate and pray. These were the fields in which "multitudes of times" he had "beheld with wonderment and pleasure" the spiders marching in the air from one tree to another, "their little shining webbs and Glistening Strings of a Great Length and at such a height as that one would think they were tack'd to the Sky by one end were it not that they were moving and floating." One may be sure he had also watched other living and growing things with the same philosophic eye. He may even have committed his observations to paper frequently, for the spider essay, so often cause for the marvel of posterity, can hardly have been his only excursion into a realm so minutely known and so confidently possessed. When he wrote of spiders, he wrote not of something which transiently caught his eye but of a world which belonged to him by right of long and deep intimacy.

Inevitably in his speculations about the universe he shared the belief of his contemporaries that the processes of nature went on by personal manipulation of the Almighty and therefore had a logical relation to the shortcomings of man; but having accepted this major tenet his mind went freely on to other queries. Although when he wrote of the rainbow he was probably still young enough to believe that the ends of it stood in basins of gold, his orthodoxy had been corrupted by no such pleasant fables. But to believe instead that it was the symbol of God's covenant with Noah did not paralyze his boyish inventiveness when it came to making a little rainbow of his own. There were several ways. He could take water in his mouth, stand between the sun and "something that looks a little Darkish," spurt the water into the air, and make a rainbow as complete and perfect as any ever seen in the heavens.

He could get the same result by dashing up drops of water from a puddle with a stick. Unfortunately (and unforgivably) he had been deprived of a visit to the sawmill at Scantic; but he had heard his "Countrymen that are Used to sawmills" say that rainbows could be seen in the violent concussion of the mill waters. It is pleasant to imagine the picture of this serious-faced and persistent small boy catechizing his sawmill countrymen for purposes of his own philosophic speculation. In the spider essay he accepted the current notion that spiders are the most despicable of the insect kind. They are the "corrupting nauseousness of the air," and yet this assumption, borrowed from his elders, did not vitiate his own clear-sighted observation as to the spider's ballooning habits, or his inspired guess (for a twelve-year-old) as to the liquid character of the unspun web.

In an eager desire to discover the child as father of the man, this unit of boyish composition, possibly written even earlier than his twelfth year, has been dignified more than once into a truly remarkable piece of scientific observation for its day and assumed to contain proofs that Jonathan Edwards had potentialities for a career in science as great as, if not greater than, in theology. Such enthusiasm is pardonable, and the conjecture is perhaps warranted. Argument spends itself vainly on such matters. The fact is that Jonathan Edwards' observation of flying spiders is accurate so far as it goes, even when tested by the findings of mature observers in a later day. As the findings of a boy who had no training in scientific observation, no microscope, no body of specialized knowledge by which to test his own observations or his conclusions from them, this juvenile effort is indeed arresting. It might do credit, in the observation alone, to an amateur twice his age.

The deductions leading from his observations are even more arresting: the basis for classification, the theory of equilibrium by which he explains the spider's navigation of the air, the character of the web, even his naïve justification of nature in providing creatures with just such equipment. That he took great pains with the essay is apparent, especially in the extant manuscript which was probably a first draft. The erasures and substitutions suggest that he had set himself to deserve a hearing from his learned correspondent, not realizing that the boyish letter accompanying his effort would easily have gained the hospitality of one not interested in spiders.

Forgive me, sir, [he wrote] that I Do not Conceal my name, and Communicate this to you by a mediator. If you think the Observations Childish, and besides the Rules of Decorum,—with Greatness and Goodness overlook it in a Child & Conceal Sir, Although these things appear very Certain to me, yet Sir, I submit it all to your better Judgment & Deeper insight. . . . Pardon if I thought it might at Least Give you Occasion to make better observations, on these wondrous animals, that should [be] worthy of Communicating to the Learned world, respecting these wondrous animals, from whose Glistening Webs so much of the wisdom of the Creatour shines. Pardon Sir

> your most Obedient humble servant,
> JONATHAN EDWARDS

As to spiders, how many kinds were there? Why did they always fly in a southeasterly direction? How was it possible for them to navigate the air? Determined to satisfy his curiosity as to the "manner of their Doing of it," he became, as he said, "very conversant with Spiders," spending in their interest days in the woods—exploring rotten logs, tracking them down, classifying them, and trying to understand how they stretched their webs from tree to tree. Like any other wide-awake boy he was sufficiently inventive to devise ways and means of finding out what he wanted to know; but unlike most boys his age he was unable to rest until he had finished what he had begun. After he had evolved a satisfactory technique of observation, he "Repeated the triall Over and Over again till I was fully satisfied of his way of working." When presently he saw the second string issuing from the tail of the spider he held on his stick, he concluded that he had "found out the Whole mystery." Stick in hand, he gave demonstrations to his companions of the spider's habit of "mounting into the air," discussed his theory with others and no doubt set his sisters and the Stoughton cousins to watching spiders and reporting their observations. One hopes also that he hoarded a collection of specimens on the parsonage windowsill.

Years later when he preached on the spider as one of the four things on earth which are exceeding small and yet exceeding wise, how did he remember this boyish attempt to solve the spider's mystery for itself, not as the prop to doctrine? Perhaps he did not remember it at all, for long before that time the door to this early world was shut, and he had lost the key.

Whatever its precise date, this precocious essay, as perhaps the earliest of his voluminous writings, is of unquestioned biographical importance. More than precocity is involved. The quality of mind revealed in these boyish observations and deductions would be equally significant whether he was eleven or thirteen when he wrote them down. The essay is a chapter in his mental development, a glimpse into the world he lived in, a world of speculative thought reached through objective fact. It is illuminating also as a personal document out of his East Windsor boyhood, testifying to long afternoons in the meadow when as a little boy he lay on his back, apparently idle, but his mind and eye intent on the life of the fields. There was no reason two and a half centuries ago for any East Windsor neighbor to set down a description of Jonathan Edwards as a child; but if his portrait were to be imagined in characteristic pose, the open fields should be the background, the figure that of a healthy boy dressed in sturdy homespun, sitting alone, doing nothing with his hands, but mentally as active as though bent over his books. Aged eleven or twelve he was no daydreamer, or even Boy of Winander, taking sensitive pleasure in bird calls and cloud movements or in listening to the rhythms of nature, heard and unheard; he was already a thinker, pushing his natural boy's curiosity about the universe as far as infinity. On such days his East Windsor boyhood was indeed "fair seed-time" for the soul of a philosopher.

In the light of his mature development one need scarcely wonder why he did not continue to devote his great powers of mind to scientific thought. The answer is that science would not have satisfied him. The physical universe was to him only the skeleton of reality, and scientific investigation was the means of stripping off only the outer layers of the mystery. From the utmost bounds of material science other speculative minds likewise have been teased along until they have leaped from the known and measurable to the intangible and infinite. To such minds only the ultimate questions as to the whence and whither of being seem worth the asking. Even as a boy, Jonathan Edwards was one of this company. Why a world at all? he was saying. "What need was there that any thing should be?" To Pascal, Newton, Swedenborg, and other giants in scientific reasoning his intellectual history would be an open book. These men also turned from physical science to religion; but they turned late in life after they had made contributions which changed the direction of scientific thought in their day. Jonathan Edwards turned away before he had made more than a bare beginning, but he obeyed the same impulse.

Had he been the son of Josiah Franklin he might have carried his boyish observations further; but as the son of Timothy Edwards he was not allowed to grow up in the meadow watching spiders, unsupervised. Like Aunt Mary Emerson's famous nephew he was "born to be educated," and indications are that the process began as early as speech. The setting was favorable. Whether Timothy Edwards had begun to prepare boys for Harvard College as early as Jonathan's infancy is not clear, but there were already four other Edwardses needing his services, and the "parlor" was in daily use. Under a discipline more rigorous than obtained in any "dame school" of the period, Jonathan Edwards laid substantial foundations for his ministerial career from the time he could first read. He began with the "Tongues."

Some few hints of the pedagogical process survive in several letters written by Timothy Edwards to his wife, when she was obliged during his absence on military duty in the fall of 1711 to take over his schoolroom duties. He admonished her not to let Jonathan, aged seven, lose the Latin he had already learned by heart, suggesting that she have him "say pretty often" to the girls from the Latin *Accidence* and both sides of "propria Quae moribus," and also that he help his younger sisters to read as far as he had learned. More than economy of effort for the teacher was back of this law of the Edwards schoolroom, by which the older child taught the younger. Timothy Edwards knew that, by the time the young Latinist had said "pretty often" to one group and heard "pretty often" the "sayings" of another group, he would have the Latin *Accidence* and both sides of "propria Quae moribus" for life; and to learn them less permanently was not to learn them at all.

His parent-teacher could hardly have been one whose teaching brought joy of the vision or made discipline seem more than an end in itself; but by his tireless persistence, which brooked no indolence and no half-knowledge, Timothy Edwards fortified his son for life against textual errors, major and minor, and made thoroughness one of the ten commandments. Unlike the tutor of Cotton Mather he did not encourage his pupils to compose poems of devotion in the tongues they were set to master. He preferred that they be letter-perfect in their verbs. Jonathan Edwards accepted his father's standard when he was too young to question it, and several years later, when the unlucky "Stiles" who was also a "parlor" product could not tell the "Preteritum of Requiesco" in a Yale examination, Jonathan shared his father's humiliation. The fact that Stiles committed no error in Tully's Orations, which "he had never Construed before he came to Newhaven, nor in any other Book, whether Latin, Greek or Hebrew," would seem to a modern college board examiner something of an extenuating circumstance, if indeed he could believe the sight of his eyes; but not to Timothy Edwards. No wonder Harvard and Yale were glad to accept his pupils. The lesson of strict accuracy was perhaps the most valuable lesson which Jonathan Edwards learned in the East Windsor schoolroom, along with his own unforgettable preterits.

Parental discipline was not limited to schoolroom exercises. The other minutiae of daily life were likewise under a supervision all but omniscient though never harsh, and filial obedience was the first law of the household. Timothy Edwards' elaborate catalogues of instruction, written on march and sent back to his partner in authority, re-create more fully than it has been elsewhere preserved the panorama of parsonage life as it was lived under the watchful eyes of the heads of the house. These letters are therefore an important part of Jonathan Edwards' childhood story. Written in homesick mood, they constitute a kind of last will and testament of affection to those Timothy Edwards had left behind and might not see again. In the light of his phrase "If I Live to come home," his exaggerated worries become understandable. As he called up the familiar round, his homesickness took the form of imagined disaster for each child of the flock. In his absence something might go wrong. Hence the pyramid of hypothetical woes and multiple cautions which, taken out of their emotional context, appear almost ludicrous.

The letter of August 7 is the richest in household detail. It is also a strange medley and a revealing glimpse into a man's mind.

Tuesday
Newhaven Aug/ 7th/ 1711

My Dear

This comes to express my Dearest Love to thee, and to Informe Thee yᵗ I am (Through the goodness of God) yet in Good health, & do expect to Go towards Albany in a few days; yᵉ Govn:ᵒʳ Intends yᵉ part at Least of yᵉ Regiments Shall March to morrow, & talks of Going himself on Friday next at furthest.

I desire thee to take care yᵗ Jonathan don't Loose wᵗ he hath Learned but yᵗ as he hath got yᵉ accidence, & above two sides of *propria Quae moribus* by heart so yᵗ he keep what he hath got, I would therefore have him Say pretty often to yᵉ Girls; I would also have yᵉ Girls keep what they have Learnt of yᵉ Grammar, & Get by heart as far as Jonathan hath Learnt: he can help them to Read as far as he hath Learnt: and would have both him and them keep their writing, and therefore write much oftener than they Did when I was at home. I have left Paper enough for them which they may use to yᵉ End, only I would have you reserve enough for your own use in writing Letters &c.

I hope thou wilt take Special care of Jonathan yᵗ he dont Learn to be rude & naught &c. of wᶜʰ thee and I have Lately Discoursed. I wouldnt have thee venture him to ride out into yᵉ woods with Tim.

I hope God will help thee to be very carefull yᵗ no harm happen to yᵉ little Children by Scalding wort, whey, water, or by Standing too nigh to Tim when he is cutting wood: and prithee take what care thou canst about Mary's neck, which was too much neglected when I was at home, & Let her also sometimes read over what She hath Learnt in the Grammar yᵗ she Maynt Loose it: and Let a new rope be speedily put upon yᵉ well pole, if it be not done already: And Let Esther & Betty Take their powders as Soon as the Dog Days are Over, & if they dont help Esther, talk further with yᵉ Doctʳ: about her for I wouldnt have her be neglected: Something also Should be done for Anne who as thou knowest is weakly: & Take Care of thy Self, and Dont Suckle little Jerusha too Long.

My horse Got a bad wound her in Broth^r: Mathers Pasture, I would have due Care taken y^t he May be well lookt to, and thoroughly cured, If he Should be Neglected, or Ridden much before he be pretty well, It may be of very Ill consequence.

I herewith Sent you a Bill of 40^{sh}, because I would not have thee want mony in My Absence; this & ye other I Left with thee thou knowest are Loose papers, & if they be not carefully Laid up they may Soon be Lost. ye Lord Jesus Christ be with thy Spirit my Dear, & Incourage thee to hope and trust in him, & discover his Love to thy Soul to whom I commit thee & all thine and mine, to whom Remember my Love, & also to Mercy Brooks & Tim: Demming & tell him y^t I shall much Rejoice If I Live to come home to know y^t he hath been a Good Boy, & tell my Children y^t I would have Them to pray dayly for their Father, and for their own Souls, and above all things to Remember their Creator and Seek after ye Lord Jesus Christ now in ye Days of their youth. God be with & bless you all.

I am my Dear, ever Thine in y^e
Dearest Love and affection

Tim^o: Edwards

If any of ye children should at any time Go over ye River to meeting I would have them be exceeding carefull, how y^y Sit or Stand in ye boat Least they should fall into ye River.[1]

I like thy Letter so well my dear both as to ye hand, and ye framing of it, y^t I Desire more of y^m one at Albany would be exceeding well-come to me towards which I am going tomorrow.

Let care be taken y^t ye cattle dont get into ye orchard & wrong ye trees.

& y^t ye barn ben't left open to ye Cattle, thy^r Dung be carried out & Laid in ye orchard where there is most need before winter, & y^t ye flax be not spoiled.

The fleet sailed Last Monday was Sev'n night (consisting of 100 Sail of all Sorts, & as is computed here of about 20 men of war. this I had from ye post Last Friday evry Sev'n night: Col. Whiting also sent it to ye Govn:^{r} in a Letter, as I have been told.

Let Mary write pretty often as well as the Rest of ye Girls &c.

---

[1] These last instructions are written in the margins.

If the legend of Esther Edwards' strong-mindedness be true, these marginal additions must have somewhat mitigated her joy in the pleasure her letter had given. For one of her instincts and her breeding to be reminded of what she could not possibly forget—her children's safety, and, on one later occasion, her manners—would seem to have been a severe strain on her Christian forbearance.

In these multiple admonitions Timothy Edwards sat for his own mental portrait. Like his son he had the kind of mind which visualizes its concepts, an excellent kind of mind to possess if one would be a preacher of the Last Judgment, but requiring sterner terrors than scalding whey, flying chips, and neglected medicine to summon its powers appropriately. The son's resources of imagination were, by contrast, reserved for the agonized suspense of the final day and the subsequent tortures of the damned, not unleashed to conjure up minor injuries to the children around the kitchen stove. As he lays bare his characteristic ways of thought in these intimate letters Timothy Edwards shows himself to be a man careful and troubled about many things, one who forgot nothing and yet assumed that everyone else forgot everything continually, one who busied himself unnecessarily with the obligations of others and half enjoyed the self-imposed burden of details innumerable. In all these counsels, which by long habit he usually delivered in the negative, there is not the slightest hint of a peevish or unpleasant spirit. He merely could not help thinking for everyone else and compiling ubiquitous lists of tasks to be done, with all conceivable hazards present to his mind at every turn.

Instead of quieting childish fears he raised them, as though parental guidance consisted in advance notice of potential disaster. A letter written to his daughter Mary when she was attending school in Hatfield, begins with cautions against wet feet and going "too thin to meeting," proceeds through warnings against losing her good name (especially since she is a woman), and ends with an injunction to remember she has an immortal soul lodged in a frail mortal body. This letter might well stand as a father's legacy to his daughter in the days when one was permitted to live in order to get ready to die.

Such counsels were by no means unique. Children of Jonathan
Edwards' generation, who were not sons and daughters of ministers, were made to live in the ever present consciousness of death.
Every Sunday might be the last. Every parting was for eternity.
Newspaper accounts of accident were invariably framed to suggest
that no one dare boast himself of tomorrow. It was as though life
were indeed lived in the formula of the Middle Ages: "What is this
our life but a march toward death?" Children might as well learn it
early as late. The chance legend

REMEMBER YOU WAS BORN TO DIE

surviving as a child's copy on the flyleaf of an old almanac, and
painstakingly scrawled nine times down the page, was no morbid
reflection. It was merely the inevitable truth brought home afresh
with each new onslaught of pestilence or other disaster, born of
isolation and man's impotence.

Jonathan Edwards like other children of his day grew up with
this as a settled conviction, although his own childhood was
singularly protected from loss of those near him. Almost phenom-
enally the Edwards family circle remained unbroken for thirty-five
years, and when Sister Jerusha died in 1729, aged twenty, Jonathan Edwards was a man grown and had been away from East
Windsor for thirteen years. This unusual record, be it said in all
fairness, may have owed something to Timothy Edwards' tiresome
vigilance, and that vigilance in turn may have owed something to
the supposed mythology of his own boyhood, reputed to have been
a succession of remarkable deliverances from drownings, freezings,
scaldings, killing of playmates, and swallowing of peach stones. If
these tales be not the sheerest invention memory doubtless aided
imagination whenever he saw his own children set foot in a rocking
boat or ride away on horseback.

The potential naughtiness of Jonathan, mentioned in the Albany
letter, may have been only another parental chimera, although
allusion to the late conference on the subject suggests that, thanks
to Tim the chore boy, Jonathan, aged seven, may have manifested
symptoms of taint. One hopes so, since his story includes all too
few hints of a childlike childhood. East Windsor would have had

its corrupting influences of course, like all towns small and large; but by comparison with less remote communities these would certainly have been less numerous. Samuel Hopkins, born in 1721, made the astonishing statement that up to his fifteenth year he had never heard a profane word from any of the children with whom he grew up in Waterbury, Connecticut. The answer is of course that he had not been listening for profanity. His ears were stopped against all sinful matter because his head was full of something else. Jonathan Edwards at no time in his life was given to such impressive personal statements, but he too had been protected in childhood by the strength of his impulses in the opposite direction. From his birth he had lived in an atmosphere of respect for all things holy and had deep concern for the exercises of piety after the earlier American pattern. Until he rationalized and justified these attitudes by his own thought he accepted them as unquestioningly as he accepted the sunrise and the seasons.

It is in the light of such boyhood training that his later guidance of the Northampton boys and girls must be judged. Playing leap-frog in the parsonage yard while they waited their turn to be reproved by the minister was a gigantic impropriety in comparison with his own boyhood standard. Had he or any one of the Stoughton boys felt inclinations toward such blasphemous behavior they would not have dared indulge them in the very shadow of the meetinghouse turret. Satan would have been too much pleased.

Of Jonathan Edwards' earliest religious experiences there is no contemporary record; only his own later allusion to his first "awakening" which, as he wrote, took place "some years before I went to college," and the well-known detail of the booth in the swamp, belonging to the same period. This may have been at any time from his eighth to his tenth year, for Timothy Edwards was having annual revivals during that period. Looked back upon, this first awakening did not seem to him a profound experience. It was rather a greatly quickened delight in the outward duties of religion which he had been performing all his life, but in which he now took intense new pleasure. The building of the booth in the swamp was a group response to the same quickening of religious interest and is not so strange as it has sometimes seemed to later generations. In part a boy enterprise, interesting in the doing, and in part imitation of adult action during a revival season, it probably surprised none of the parents whose sons were associated in the scheme. In a sense the boys who went to the booth to pray and to talk about their own salvation were playing at religion, as children of a later generation played at vast Tory and Continental hatreds, and re-enacted the drama of adult action. The significant detail in this episode for the understanding of Jonathan Edwards is that praying with his companions did not satisfy him. Even as a child he felt religion as too personal an experience to be shared so intimately; hence, unknown to his companions, he had his own place of secret prayer deeper in the woods. This was years before his mind acknowledged that religion must be an individual experience, else it was nothing; but even as a child he felt it so, and in this solitary quest was responding to the deepest instinct of his nature. Going back and forth to the meetinghouse, keeping the Sabbath as the son of Timothy Edwards was expected to keep it—these things were not enough. Religion was more than the mere observances of it. What it was, he could not have said, except that his mind was "deeply engaged in it" and no other delights were comparable.

There is not the slightest suggestion that either at this time or at any time later in his life he courted austerity for its own sake, or that in his solitary devotions he sought deliberately to mortify the flesh in order to develop the spirit. Always his mind was on the end, not the means, and the discipline itself was of so little importance that he was usually oblivious of it. Moreover, at this time, going to the woods to pray was something of a practical necessity in a household so numerous that privacy was all but impossible at any hour in the twenty-four. In addition to the Edwards flock guests were frequent, sometimes staying for weeks and paying board as was the custom. Some of Timothy Edwards' pupils from other towns also lived at the parsonage. One wonders how or where. Certainly there must have been times when, between parental supervision, sisterly criticism, and the presence of perhaps twenty persons under one sloping roof, those "little nervous strings" which, according to Jonathan Edwards' boyhood reasoning, proceed from the "soul in the brain" must at least have been "jarred" by these external things. For one to whom solitude was an unquestioned necessity, to be obliged—not only in his boyhood but throughout his life—to live in houses which were more like hostelries than private dwellings seems unkindness indeed.

Particularly in connection with these earliest religious experiences one would like to know more than the records tell of his relation to his mother Esther Edwards. No letters to her or from her are extant for any time in his life. She takes on individuality only in his father's numerous epistles filled with everyday details testifying to her resourcefulness in the minor crises of frontier life, and to her unsparing vigilance as she nursed one after another of them through serious illnesses. "We find your absence, (especially So Long) makes a great empty place in the house," he wrote on one occasion. One might think it would. She was the shadow of a great rock to them all. Did she, in addition to her practical gifts, her intellectual vigor and zeal in good works, have also an understanding of her son's deeply spiritual nature, his sensitive approach to religious experience? There is no recorded answer to these questions. An unauthenticated tradition that during one of her husband's revivals she made public profession of conversion would certainly suggest, if true, that she had not only the courage of her convictions in a difficult test but also a capacity for religious emotion which might have given her a sympathetic understanding of his young ecstasies; but, if so, the evidence does not appear. He spoke freely of his experiences to his father, but there is no record that he confided them also to his mother. She lives only in the filial idiom "Remember my Duty to my honored Mother" unfailingly included in his letters to his father.

The fact that Sister Jerusha, six years younger than Jonathan, was also given to solitary walks and prolonged devotions, with corresponding abhorrence of "froth and levity in conversation" and delight in weighty discourses, particularly books of divinity, may mean that she was consciously or unconsciously imitating him, or more probably that something in their joint heritage prompted these similar yearnings and in a sense unfitted them both to live in the world as they found it. In Jerusha there is no hint of mystical raptures. She was merely engaging in devotions, with more than a hint of childish asceticism in the manner of them. It was her custom on Saturday nights to stay up later than the rest of the family, in preparation for the Sabbath, and in the morning to walk alone to the house of God in solemn meditation. When she returned from the afternoon service, if the weather were not too severely cold, she diligently improved the remainder of the holy day in an unheated upper room, as the saying was, "filling in all the chinks of the Lord's day with useful thoughts." When she attended any merry meeting of young people she took no part in the merriment, but instead sat on "one side of yͤ Company with some person who would entertain her upon some sollid and profitable subject." Not that she was an "enemy to something of innocent Jesting," her sisters protested; she merely chose to use her wit as sauce, thinking it "very improper food, for yͤ soul."

Even after allowance is made for sisterly overstatement, this picture of Jerusha Edwards with her beautiful countenance, her blameless life, and "Quiet Virtue" has its ludicrous side, even for a minister's daughter in a godly age. Her extravagant pieties, however modestly she attempted to hide them, must have made her something of a village oddity and none too welcome at the merry meetings she rebuked by her soberness. She judged by a standard too high for weekday living, yet thought it her duty, for the good of others, to speak her criticism frankly. When on one occasion she so far overshot the mark as to attempt to improve the virtue of her sweetheart and "preserve him against yͤ infection of vice" by telling him what was wrong with his behavior, the sally cost her a budding romance. But she had done her duty as she saw it and in the sequel bore herself like a true Edwards, her calm unruffled by the ferment of gossip which ensued. So great was her personal triumph (so said her sisters) that she took no pains to contradict the story that he had jilted her but went serenely to meeting, all eyes upon her. Such was the Edwards code.

With all his flat-footed good sense Timothy Edwards applauded these unyouthful rigidities. He was, in fact, responsible for most of them. Jerusha was the eloquent embodiment of the Christian virtues he preached; only, being her father's child, she had taken them somewhat too literally. Had she lived, her wit might have helped her to attain a better balance, but she was denied the chance. Even before her death she seems to have been all but canonized in the Edwards household where she lived with her sisters "in love, not unlike to y^t which is in ye heavenly regions."

Had Jerusha been nearer to him in age Jonathan Edwards might have found much in common with her; but when he left home for college she was a child of six, whom he was to know later only in brief vacations. It was to his practical sister Mary, two years older than himself, that he turned for companionship through all his young life. When she went away to school in Hatfield he sent her the family news; and when later he went away to Yale College she did the same for him. These letters tell a story of affectionate comradeship and mutual dependence pleasant to read. It was to Mary that his first extant letter, written when he was twelve years old, was sent. In the news of the revival with which he begins, he talks more like a deacon than a twelve-year-old boy; but with his own awakening behind him he was already on the side of the pulpit and yearning toward the unconverted. By his twelfth year, also, he had learned the formalities of polite correspondence, and out of respect for a missive which must be carried by hand (sometimes by several hands) he did not fill his pages with light matter but appropriately subordinated the trivialities of chickenpox and toothache to lists of the newly converted and the newly dead. He wrote in a neat hand and made the customary epistolary flourishes. The letter reveals much as to his childhood background of thought and his standards of value.

Windsor May 10 1716

Dear Sister

Through the Wonderful Mercy and Good[ne]ss of God there hath in this Place Been a verry Remarkable stirring and pouring out of the Spirit of God, And Likewise now is But I think i have Reason to think it is in Some Mesure Diminished but I hope not much. About thirteen Have been joyned to the Church in an estate of full Comunion These are those which by Enquiry I Find you have not heard of that have joyn'd to the Church, viz; John Huntington, Sarah Loomas the Daughter of Thomas Loomas, and Esther Elsworth. And their are five that are Propounded which Are not added to the Church, namely, John Loomas, John Rockwell's wife, Serg.ᵗ Thomas Elsworth's wife, Isaac Bissels wife, and Mary Osband I think there Comes Commonly a Momdays above thirty Persons to Speak with Father about the Condition of their Souls.

It is a time of Generall Health here in this Place. There Has five Persons Died in this Place Since you have been gone, viz. Old Goodwife Rockwell, Old Goodwife Grant, and Benjamin Bancroft who was Drowned in a Boat many Rods from Shore wherein were four young women and many others of the other Sex, which were verry Remarkably Saved, and the two others which Died I suppose you have heard of, Margaret peck of the New Town who was once margaret Stiles hath Lost a Sucking Babe who died very Suddenly and was buried in this Place.

Abigail Hannah and Lucy have had the Chicken Pox and are recovered but jerusha has it now but is almost well I myself Sometimes am much Troubled with the tooth ack but these two or three Last Days I have not Been troubled with it but verry little so far as i know the whole famaly is well except Jerusha.

Sister i am glad to hear of your welfare So often as i do I should be glad to hear from you by a Letter and therein how it is with you as to your Crookedness.

                              Your Loving Brother Jonathan E.
Father and Mother Remember their Love
unto you. Likewise do all my Sisters and Mercy and tim

It is plain to see that already the meetinghouse had first place in all his boyhood plan of life. It was his one extramural interest, his larger world. He saw the whole drama of village life from the angle of the parsonage and the pulpit. East Windsor was a parish, a little corner of the Lord's vineyard, not a center of secular interests. What he knew of the world outside the town limits came chiefly from visiting clergymen who brought news of the Lord's work in other corners of the same vineyard. The ministerial language of the hour was as natural an idiom to him as the language he spoke in the schoolroom. Likewise the pulpit controversies of the hour: the Halfway Covenant and Grandfather Stoddard's bitterly opposed amendment thereto, the old and the new way of singing in the churches—all this was familiar territory in his thought.

When East Windsor had its own village quarrel over where to set the new meetinghouse, it is safe to imagine that he listened to the long and bitter arguments detailed in nightly sessions at the parsonage, had his own opinion on the subject, discussed it with his father and was respectfully heard, and that when the church finally voted to rebuild on the old site (the usual decision after the peace of a village had been sadly frayed) he was one of those present at the demolition of the old structure and thereafter watched week by week the new meetinghouse take shape under parish labor. It would have been an absorbing drama more intensely personal to him than to the other village boys, a major event from which to date his own smaller concerns. Before the still greater village excitement of "dignifying the seats" came to pass he was a student at Wethersfield and for the first time in his life met with new scenes, new thoughts, and new ways of thinking them.

Childhood ended for Jonathan Edwards just before his thirteenth birthday. He had recited his last lesson in the "parlor" and was now ready for college. These first thirteen years had determined many things: his sober view of life, his reflective bent, his refinement of self-discipline, his pursuit of religion as the unquestioned goal of life. To some extent his mind was already his servant; he could think for himself. He had learned the benediction of solitude amid the quiet beauty of woods and fields. His calling was a straight path before him. The foundations of a deep understanding sympathy had been laid with the man who was to mean more to him throughout his life than any other human being he was ever to know—Timothy Edwards, his own father. Poles apart in temperament, in natural endowment, and in ways of thought, father and son were to enjoy for life a rare fellowship.

Outwardly, there would not be much change in the look of life. College would mean no quadrangles, no spires and deep-toned bells. Matriculation day to Jonathan Edwards meant merely exchanging a schoolroom in one Connecticut farmhouse for a similar room in another, slightly more pretentious. He was not even to be taught by strangers. His own cousin Elisha Williams, nine years his senior, was to be his tutor. In such terms the distance between life as it had been and life as it was to be does not seem very great; but to Jonathan Edwards, as to any child standing on the threshold of independence, it was a chasm. In the fall of 1716, just before his thirteenth birthday, he took horse and rode away to Yale College, leaving his childhood behind him.

PERRY MILLER

✪

The Objective Good

In vulgar modern terms Newton was profoundly neurotic of a not unfamiliar type, but—I should say from the records—a most extreme example. His deepest instincts were occult, esoteric, semantic—with profound shrinking from the world, a paralyzing fear of exposing his thoughts, his beliefs, his discoveries in all nakedness to the inspection and criticism of the world. . . . His peculiar gift was the power of holding continuously in his mind a purely mental problem until he had seen straight through it. . . . Anyone who has ever attempted pure scientific or philosophical thought knows how one can hold a problem momentarily in one's mind and apply all one's powers of concentration to piercing through it, and how it will dissolve and escape and you find that what you are surveying is a blank. I believe Newton could hold a problem in his mind for hours and days and weeks until it surrendered to him its secret. . . . He looked on the whole universe and all that is in it as a riddle, as a secret which could be read by applying pure thought to certain evidence, certain mystic clues which God had laid about the world to allow a sort of philosopher's treasure hunt to the esoteric brotherhood. He believed that these clues were to be found partly in the evidence of the heavens and in the constitution of elements (and that is what gives the false

From Perry Miller, *Jonathan Edwards* (New York: William Sloan Associates, 1949), pp. 71–99. Reprinted by permission of William Morrow and Company, Inc. Copyright © 1949 by William Sloan Associates, Inc.

*suggestion of his being an experimental natural philosopher), but
also partly in certain papers and traditions handed down by the
brethren in an unbroken chain back to the original cryptic revelation in Babylonia. He regarded the universe as a cryptogram set by
the Almighty—just as he himself wrapt the discovery of the
calculus in a cryptogram when he communicated with Leibnitz. By
pure thought, by concentration of mind, the riddle, he believed,
would be revealed to the initiate.*

—JOHN MAYNARD KEYNES

A stray copy of Locke might have found its way to Wethersfield in
1717, but that Williams' little band owned a copy of *Principia* is
unlikely. Many entries in the "Notes" probably attained their
present form only after Edwards had moved back to New Haven
in 1719, and some passages may date from his tutorship in
1724–26. After the College forcibly reclaimed the Dummer collection from Saybrook, Edwards could hold in his hands the actual
Principia and *Opticks* Newton himself took down from his shelves
and gave to Dummer for a gift to the new college in the wilderness.
(In 1723 Harvard owned the *Opticks* but no *Principia*.) We know
that Samuel Johnson read this *Principia* and vainly tried to teach
himself enough mathematics to understand it. Edwards never
understood fluxions or other higher mathematics, but to the extent
that a man can read Newton without such proficiency, he read
him, and though like most admirers he accepted the "sublime
geometry" on Newton's say-so, he appreciated the more literary
"Scholia" with a profundity not to be rivaled in America until the
great John Winthrop took over the Hollis professorship as successor to Greenwood in 1738, or until a printer in Philadelphia
succeeded in keeping his shop until it kept him in sufficient leisure
to allow time for reading. Consequently, when we go behind
Edwards' early publications to find the hidden meanings, we discover in the "Notes" not one key but two, a dual series of reflections, often intermingled but not yet synthesized. The one proceeds
out of Locke and becomes what posterity has called his "idealism"; the other begins with Newton and becomes what has been
less widely appreciated, his naturalism. In his mind there was an

equilibrium, more or less stable, of the two, which is the background of his cabalistic dichotomy, set up in the Boston lecture as though it were too apparent to need explaining; if his proposition about the "inherent good" requires for full comprehension a knowledge of Locke, his assertion of the "objective good" demands an equally rigorous study of Newton.

Edwards would not compartmentalize his thinking. He is the last great American, perhaps the last European, for whom there could be no warfare between religion and science, or between ethics and nature. He was incapable of accepting Christianity and physics on separate premises. His mind was so constituted—call it courage or call it naïveté—that he went directly to the issues of his age, defined them, and asserted the historic Protestant doctrine in full cognizance of the latest disclosures in both psychology and natural science. That the psychology he accepted was an oversimplified sensationalism, and that his science was unaware of evolution and relativity, should not obscure the fact that in both quarters he dealt with the primary intellectual achievements of modernism, with the assumptions upon which our psychology and physics still prosper: that man is conditioned and that the universe is uniform law. The importance of Edwards—I cannot insist too strongly—lies not in his answers, which often are pathetic testimonies to his lack of sophistication or to the meagerness of his resources, but in his inspired definitions. Locke is, after all, the father of modern psychology, and Newton is the fountainhead of our physics; their American student, aided by remoteness, by technological innocence, and undoubtedly by his arrogance, asked in all cogency why, if the human organism is a protoplasm molded by environment, and if its environment is a system of unalterable operations, need mankind any longer agonize, as they had for seventeen hundred years, over the burden of sin? By defining the meaning of terms derived from Locke and Newton in the light of this question, Edwards established certain readings so profound that only from the perspective of today can they be fully appreciated.

"The whole burden of philosophy," said Newton, "seems to consist in this—from the phenomena of motions to investigate the forces of nature, and then from these forces to demonstrate the other phenomena." Conceiving the universe as motion—which, unlike the concepts hitherto taught in New England, such as substance, form, and accident, could be expressed in mathematical formulae—Newton arrived at such an earth-shaking discovery as this: "If you press a stone with your finger, the finger is also pressed by the stone." Of course, no farmer in Connecticut needed to be told, "If a horse draws a stone tied to a rope, the horse (if I may say so) will be equally drawn back towards the stone." But every farmer was told, and professed to believe, what Luther had put succinctly over a century before the *Principia:* "For though you were nothing but good works from the sole of your foot to the crown of your head, yet you would not be righteous, nor worship God, nor fulfill the First Commandment, since God cannot be worshipped unless you ascribe to Him the glory of truthfulness and of all goodness, which is due Him." By the logic of that other science, called divinity or theology, upon which New England was founded, the best of deeds were "insensate things," which in themselves reflect no slightest glory upon the Creator. "Faith alone is the righteousness of a Christian man." If a man has faith, according to Luther—and after him Calvin and the Puritans—he "is free from all things and over all things."

For a century Yankees had believed this, but they had not been free from and over such things as the stones in their pastures, which broke both their own and their horses' backs. In the old-fashioned physics a stone was a concatenation of form and substance, with a final cause, and so its weight could be "improved" in theology as a trial laid upon man in punishment of his sin; but if now the obstinacy of the resisting body was an inherent mathematical product of its density and its bulk, if it lawfully possessed an inertia of its own which man must comprehend by the analogy of muscular effort, how could a man struggling with a rock in his field become persuaded that by faith he might be "free" of it? A more logical conclusion was that since weight is a natural force, the profitable method of freeing himself from it was by the law of levers, by a better breed of horses, but not by moralizing that the presumption of good works means the instant loss of faith and its benefits. There was—as Edwards perceived the situation—an organic connection between Newton's laws of motion and that law of salvation by faith which Calvin had made, once and for all, "the principal hinge by which religion is supported."

Luther, Calvin, and the founders of New England frequently utilized the physics of their day, which was still scholastic, for illustration or confirmation of their doctrines, but they never dreamed of resting the case for Protestantism upon the laws of nature. Edwards saw in a glance that no theology would any longer survive unless it could be integrated with the *Principia*. Newton claimed that in so far as we can learn the first cause from natural philosophy, "so far our Duty towards Him, as well as towards one another, will appear to us by the Light of Nature." This was not a boast, it was a threat. The *Principia* meant that henceforth there was to be no intelligible order apart from the actual. Although Newton discreetly left unanswered certain basic queries, he did show beyond question that the method of inquiry, in theology no less than in science, must be conformed to physical reality: "For Nature is pleased with simplicity, and affects not the pomp of superfluous causes."

In 1734 Edwards preached a series of sermons on justification by faith, the "principal hinge" of Protestantism; he reworked them into a sustained tract which he published in 1738. It was the most elaborate intellectual production he had yet attempted, and it figures in his development—or rather in the public exhibition of the development he had already undergone—as the first effort in American history to coordinate with the doctrine of Puritan revelation the new concept of science, in which such a superfluity of causes as had been the stock-in-trade of Edwards' predecessors became an affectation of pomp. He was resolved to prove that justification must in all simplicity be merged with the order of causality, and that if salvation was to be called an effect, of which faith was in some sense the cause, then the sequence must be formulated anew in language compatible with Newton's.

The still regnant doctrine, which no respectable Puritan had openly questioned, went back to Calvin himself, who as usual reduced Lutheran eloquence to legalism. Justification by faith, he said, is entirely a "forensic" transaction: a sovereign God is pleased, for no other reason than that He is pleased, to accept the righteousness of Christ in place of the obedience which no man can achieve, and He "imputes" Christ's perfections to a chosen few who in fact fall far short of perfection. These are saved "as if" they were Christ Himself: "He is justified who is considered not as a sinner, but as a righteous person, and on that account stands in safety before the tribunal of God, where all sinners are confounded and ruined." In the realm of objective fact, salvation was conceived by the Puritans as the transfer of a balance on the divine ledger, wherein God arbitrarily accepted another's payment for the debt which all men owed Him by the sin of Adam, and condemned those for whom the debt was not paid, though in life there might be little to distinguish one from another. In its published form, Edwards' treatise on justification reaffirmed the stereotyped doctrine: "A person is said to be justified, when he is approved of God as free from the guilt of sin and its deserved punishment; and as having that righteousness belonging to him that entitles to the reward of life." Had he been concerned, like Prince and Sewall, with no more than restating old doctrines, his tract would have said only this, as indeed no other American publication of the 1730's did.

The necessity of saying something more was, to Edwards' sense of the times, thrust upon him. For over a century Puritans in both Englands had drifted into the habit of calling faith the "condition" of justification. They started speaking in this fashion because they adhered to the "Federal Theology." At first it was merely a manner of speaking. They meant simply that if you believe you may be saved. Their rhetoric, in its early stages, was a natural result of Calvin's version of the whole procedure as a forensic transaction in which the recipient got credit for deeds he had not done. But the legalistic bent of primitive Calvinism, once carried to England, was there accelerated by the alliance of the English Puritans with the Parliamentary lawyers. Concepts taken from the common law pervaded theology, and even before the founding of New England the Puritans had theorized Calvinism anew into an idiom of what the lawyers called a contract or a covenant. Every New Englander before Edwards was a "Federalist," and because he put aside all this sort of thinking, he became a new point of departure in the history of the American mind. All his predecessors would have denied that Federalism was anything different from Calvinism—or Protestantism. They believed that it was simply a more precise way of phrasing the doctrine, on the premise that an absolute God, like an absolute monarch, could be held to nothing but what He had covenanted. Federal theologians, of whom Stoddard was a great example, liked to say that Jehovah, out of sheer indulgence, signed a set of contracts with both Christ and Abraham, in which He covenanted to accept the performance of Christ as though it had been rendered by individuals among the seed of Abraham, on the "condition" that they believe in Christ. As long as the theorists also pointed out that belief was an act of God, and that no man could muster it by himself, they were technically good Calvinists; they merely obtained in the language of the contract a greater precision, or at least a more precise metaphor, with which to argue that God does all, while man is impotent, but that nevertheless there exists a recorded transaction in which the recipient of grace can be accorded, and is assured, the bounty.

Actually, with a century of repetition, and after the political triumphs of 1649 and 1689, the Covenant of Grace came to mean in Puritan circles, in both the Englands, not what God was pleased to grant, but what He was obliged to concede. Faith gradually became so identified, at least in general parlance, with the condition of the covenant that it ceased to mean a decree enacted outside and above the human sphere. It became, bit by bit, something which a man might obtain, and which, once he had it, gave him a claim that God was bound to honor. Even in a theology of predestination, it was declared that a man could do his part and then relax, waiting upon God to do His. The theory of faith as a condition, said Edwards in 1734, is "ambiguous, both in common use, and also as used in divinity." New England had so perverted the language of the founders—the language lent itself rather too easily to perversion—that faith had come to mean "any thing that may have the place of a condition in a conditional proposition, and as such as is truly connected with the consequent."

In these sentences Edwards spoke with restraint, but he was

none the less declaring a break with the New England past, a break which his Boston sermon only subtly insinuated. He was putting his finger upon the point at which, as he saw it, the real declension of New England had set in. He maneuvered a revolt by substituting for seventeenth-century legalisms the brute language of eighteenth-century physics. He cast off habits of mind formed in feudalism, and entered abruptly into modernity, where facts rather than prescriptive rights and charters were henceforth to be the arbiters of human affairs. If the experience of regeneration is real, then "what is real in the union between Christ and his people, is the foundation of what is legal." The language of revolution in this undramatic sentence is difficult to catch across the centuries, but taken in the context of the 1730's it is as decisive, and as fundamental, as that of the more historic declarations. In 1734 Edwards was applying to theology a critique which assumed that theology should derive from experience and not from logic or from convention. His society, having slipped into a way of calling faith the condition of a covenant, had made the gratuitous assumption that faith was therefore the actual producer of the effect. It was heedlessly supposing that faith is the cause of salvation, and had insensibly come to assume that a man's belief worked his spiritual character exactly as by his physical exertion he shoved a stone out of a meadow. The people had succumbed to a metaphor, and had taken a shallow analogy for a scientific fact. Hence religion, which can thrive only upon realities, was fallen into decay.

Thus without openly proclaiming a revolution, Edwards effectively staged one. The object of his attack was what his society had hitherto assumed to be the relation of cause to effect, on which assumption it was constructed. If a ball that strikes another is called the cause of motion in the other, it then works the effect and determines the consequences; if, however, the first can be said only to transmit force to the second, it is but the first in a series of events determined by a law higher than itself. Puritanism all unwittingly had made the fatal mistake—it has proved equally disastrous for other cultures—of supposing that an event in one realm can cause effects in a totally other realm, that a man's act of belief can oblige the will of God. It had tried to make the transcendent conform to the finite, and pretended that it had succeeded. Edwards drew upon his study of Newton for a contradictory conception: "There is a difference between being justified by a thing, and that thing universally, and necessarily, and inseparably attending or going with justification." He went to physics for a cause that does not bind the effect by producing it; he found in the new science (few besides Newton himself understood that this was the hidden meaning of the *Principia*) the concept of an antecedent to a subsequent, in which the subsequent, when it does come to pass, proves to be whatever it is by itself and in itself, without determination by the precedent.

He never bewildered his auditors by expounding scientific analogies beyond their grasp, but he quietly took into the realm of theology the principles he had learned—or believed were obvious —in his inspired reading of Newton. Obviously his imagination had taken fire from such remarks of Newton's as, "It is not to be conceived that mere mechanical causes could give birth to so many regular motions." Thousands of Newtonians in the eighteenth and nineteenth centuries took this to mean only that "God" created the universe; Edwards took it to mean that cause in the realm of mechanics is merely a sequence of phenomena, with the inner connection of cause and effect still mysterious and terrifying. He interpreted the sequence of belief and regeneration by the same insight. His people, of course, were still ignorant of the "Notes on the Mind." Had they been permitted to take them from his desk, they might have comprehended how, in his view, the old Aristotelian array of causes—final, formal, and material—had been dissolved before the triumph of the now solitary efficient cause. Hence they might have understood that for him the secret of nature was no longer that an efficient cause of itself works such and such an effect, but is to be defined as "that after or upon the existence of which, or the existence of it after such a manner, the existence of another thing follows." All effects must therefore have their causes, but no effect is a "result" of what has gone before it.

The metaphysics of this idea were profound, but Edwards' statement is so enigmatic that we may rightly doubt whether many good burghers in Northampton had any notion what he was talking about. Still, the import was clear: a once harsh doctrine, which for over a century had been progressively rendered harmless and comfortable, was once more harsh. It was imperiously brought back to life. And there were many among the river gods, and in the counting houses of Boston, who were eager to let sleeping dogmas lie. What right had this grandson of Stoddard—whom the town had employed to carry on his grandfather's ecclesiastical organization—to raise again, and in so disturbing a form, theological issues which New England had settled long since? The society had learned how to live with Calvinism; why make it something with which men could no longer live, or at least could not live on the basis of a profit and loss economy? Edwards began quickly to make converts, but almost as rapidly he made enemies, and most promptly among his cousins.

In five or six compact pages of the sermons that later made up the discourse *Justification by Faith Alone,* pages marked by no rhetorical flourishes, uttered in the calm, impersonal manner that never stooped to the capacities of his audience but bore them down with imperturbable assertion, Edwards pointed out that if the loveliness of a person is what wins faith, if the human achievement is at all a "reward"—which to speculation divorced from the heart might seem, as we have heard, no more than reasonable—then faith is a cause that may or may not be put into action. Indeed, considering the general unloveliness of human beings, it is apt never to get started at all. Furthermore, if faith is an effect of merit, then it too becomes an event which in turn is the cause of still another event, and so on, *ad infinitum.* Every man can be a fresh cause every day of his life, as though he had never lived yesterday, and the universe will be the sum total of today's contingencies, which tomorrow will become still more contingent. Edwards might have called this heresy solely on the strength of the traditional creed or of the Bible, but he took a startling line: "Because the nature of things will not admit of it." By appealing to nature Edwards set up his thesis:

The wisdom of God in his constitutions doubtless appears much in the fitness and beauty of them, so that those things are established to be done that are fit to be done, and that those things are connected in his constitution that are agreeable one to another. . . . This is something different from faith's being the condition of justification, only so as to be inseparably connected with justification: . . . yet nothing in us but faith renders it meet that we should have justification assigned to us.

By this conception any act, either faith or lust, is not an instrument which works an effect, but is part of a sequence within a system of coherence. God is "a wise being, and delights in order and not in confusion, and that things should be together or asunder according to their nature." The connection between a subjective state and an objective fact is not the subject's conviction that his loveliness ought to be rewarded, but is "a natural suitableness," which means that the qualification and the circumstance go together. The difference is fundamental: in the theory of faith as an instrument—as in all "instrumentalism"—the world is supposedly so constituted that it regards the beauty or utility of acts committed by some Ebenezer or Jonathan; in the scientific conception, the ecstasy resides not in a "hypothetical proposition" but in the fact itself, in "the entire, active uniting of the soul." In the first scheme, God waits upon man, and if man elects to be worthy, a grateful cosmos yields to his virtues, or at least to his industry. In the order of the objective good, "Goodness or loveliness of the person in the acceptance of God, in any degree, is not to be considered prior but posterior in the order and method of God's proceeding in this affair." To the bewilderment of an energetic America, intent upon commerce and real estate, where an ounce of effort meant a pound of sterling, Edwards declared that "the nature of things will not admit of a man's having an interest given him in the merits or benefits of a Saviour, on the account of any thing as a righteousness, or virtue, or excellency in him."

If so, the nature of things must in fact be opposed to the appearances of American society. There had always been in Calvinism a vague feeling that Protestant doctrine had a connection with the structure of the physical universe; Calvin himself, searching for metaphors, compared the light of grace to the rising sun that blots out the stars, or the persuasion of our own righteousness to a foolish eye that prides itself on its perspicacity in viewing adjacent objects and then is dazzled when it looks directly upon the sun. But Edwards, with Newton behind him, saw in the phenomena of nature, as employed in Christ's own discourse, not metaphors to adorn a discourse, but factual embodiments of eternal law. "These things," he confided to his notebooks, "are not merely mentioned as illustrations of his meaning, but as illustrations and evidences of the truth of what he says."

In Newton, Edwards found, as illustrations not of meaning but of the truth of what he would say, two primary conceptions: atoms and gravity. If there was a lion's mouth to be met with, Edwards would put his head in it. In ancient Greece, Democritus had laughed at the superstition of the senses that takes for reality the sweet or the bitter, the hot, the cold, or the purple, when "in truth there are atoms and a void." It is a startling fact about the rise of experimental science that, by the middle of the seventeenth century, scientists took the material universe to be made up of millions of particles which they had never seen, measured, or subjected to experiment, and in which they believed out of sheer faith. Thanks to Gassendi, Galileo, Boyle, and other physicists, this staggering assumption was one of the major premises of Newton:

It seems probable to me, that God in the Beginning form'd Matter in solid, massy, hard, impenetrable, moveable Particles, of such Sizes and Figures, and with such other Properties, and in such Proportion to Space, as most conduced to the End for which he form'd them; and that these primitive Particles being Solids, are incomparably harder than any porous Bodies compounded of them; even so very hard, as never to wear or break in pieces; no ordinary Power being able to divide what God himself made one in the first Creation.

Newton knew that this doctrine raised the specter of Lucretius and of "materialism"; a major reason for the distrust of the new science which Johnson found active at Yale in 1714 was a fear of its atheistical atomism. But in 1721 Cotton Mather's *Christian Philosopher* laid New England's fears to rest by advertising that the Newtonian world, far from denying, actually proved the existence of God and of design in the cosmos; he invited New England "to avoid philosophical romances" by getting an insight "into the principles of our perpetual dictator, Sir Isaac Newton." Cotton Mather made out of Newton those generalities of law, order, and symmetry that were becoming the commonplaces of eighteenth-century optimism. His "insight" did not go deep, but Edwards' did, and it went to the crucial point: if atoms are so hard that they never break, how small is the smallest atom? And then, if they are massy, hard, and impenetrable, what holds them together?

These were annoying problems for Newton also, and he who boasted of making no hypotheses never gave an answer. Lucretian naturalism had supposed that the atoms were "hooked" and so got fastened together; Newton rejected this fantasy, but he did allow his mind to play with the possibility that some medium might pervade the interstices of bodies and act as a sort of glue to hold atoms together; yet all this, he agreed, was speculation, and the most he could say factually was, "I had rather infer from their Cohesion, that their Particles attract one another by some Force, which in immediate Contact is exceeding strong." When we get behind the brilliant façade of Newtonianism, the apparently rational system of which poets sang and which Cotton Mather embraced, we are brought more terribly face to face with the dark forces of nature than any Puritan had been while staring into the dazzling glare of predestination. That element in the early Newton which Lord Keynes calls necromancy, which was deliberately masked in his last years and was ignored in his panegyrics, was an intuition of pure magic.

> Matters that vexed the minds of ancient seers,
> And for our learned doctors often led
> To loud and vain contention, now are seen
> In reason's light, the clouds of ignorance
> Dispelled at last by science.

Behind the mathematical analysis which by its perfection of form inspired such hymns as this, concealed so carefully that only the most astute might catch a glimpse of it, moved a power that could not be seen by reason's light or dispelled by science, that hid

itself in matter to hold the atoms in cohesion, and betrayed its
existence by resisting the pressure of a finger. It was to Newton the
necromancer that Edwards, who was of the same brotherhood,
responded. This Newton, turning from his imaginings of a subtle
medium that might give a rational explanation for the solidity of
bodies, let slip the intelligence that just how primitive particles,
which obviously touch each other only at a few points, "can stick
together, and that so firmly as they do, without the assistance of
something which causes them to be attracted or press'd towards
one another, is very difficult to conceive." The best of Newton's
popularizers, Colin Maclaurin, whose *Account* of 1748 was known
to Edwards, admitted that while Newton had not quite explained
everything, he "left valuable hints and intimations of what yet lies
involved in obscurity." The best hint he could leave on this
obscure but basic problem was that there must be some agents in
nature able to make atoms hang together in bodies, "And it is the
Business of experimental Philosophy to find them out." Edwards
had the temerity, although he had no laboratory, to take him at his
word.

The difficulty was that this problem was in reality a double
problem, each aspect of which elusively played into the other: to
think how atoms cohered soon became to wonder how large or
how small the atom was. Imagine it to be as small as possible, it
still occupies space: why cannot it be divided as a stone is split by
a wedge? For a century Cartesians had challenged atomists, assert-
ing that unless matter were conceived as completely indivisible,
atheism would follow. Newton and his followers were devout men,
but they hated Cartesians and believed passionately in atoms and
the void. With dogged persistence Newton asserted again and
again that the extension, hardness, impenetrability of all objects
are founded upon the same qualities in each of the atoms, "and
this is the foundation of all philosophy." If so, the Age of Enlightenment was founded on a mystery, but was incapable, except
in a few Blakes or Edwardses, of recognizing it. Newton did not
want to say that the atom could never be split, and he foresaw that
if it could be, we should have to conclude that divided particles
may be subdivided to infinity; but for the moment he called a halt

before the irreducible minimum, the atom that cannot be shattered and is the stuff of bodies. "We have ground to believe," said his grateful followers, "that these subdivisions of matter have a termination." He warned them not to indulge in fancies nor "to recede from the analogy of Nature, which is wont to be simple, and always consonant to itself." On this basis we may occupy ourselves with formulae of velocity and prediction of comets, concealing even from ourselves that we do not know the size of the atom or how one atom hangs on to another, and above all keep well hidden from view that we suspect but dare not identify too closely some agent, some active "vis," who hides in the stone and makes for its resistance.

By the analogy of nature, this dark power might, of course, be the same that operates in gravity, but Newton would never say so. When he let his mind range over the possibilities of the subtle medium or ether, he saw that in addition to gluing atoms together it might exert pressure from the edges of the solar system and so impress centripetal force upon bodies within the gravitational field. In that case we should have a wholly mechanical explanation. The eighteenth century often had occasion to lament that Newton lent himself to such vagaries; in 1756 Edmund Burke was to deplore that this great man, "if in so great a man it be not impious to discover anything like a blemish," stooped from his mathematical eminence to such trivialities as "a subtle elastic ether," a subject which, Burke added, "leaves us with as many difficulties as it found us," But on the whole, Newton was faithful to Burke's notion of sublimity; he could not, or he would not, give a cause for gravity. He left his greatest discovery so wrapped in mystery that the only permissible conclusion was his "Scholium" to Book III: the one cause which can penetrate to the centers of the sun and planets, that can operate not according to the quantities of the surfaces of the particles (as do "mechanical" causes), but according to the quantity of the solid matter, must be God. Only He can be omnipresent both virtually and substantially, as the ultimate cause must be present for a world to exist at all; only He can suffer nothing from the motion of bodies and only in Him can bodies find no resistance. Since His inward substance is as inaccessible to us as the inward substance of stones, which speak to our senses only from outer surfaces, so "We know him only by his most wise and excellent contrivances of things, and final causes; we admire him for his perfections; but we reverence and adore him on account of his dominion." Newton's champions could present the Newtonian system as a method of approaching God, offering proof of His existence out of natural powers and laws "from the difficulty we find to account for them mechanically"—which was to say, out of the deficiencies of physics.

Thus Newton tried strenuously to say that it was enough, whatever the cause, that "gravity does really exist, and act according to the laws which we have explained, and abundantly serves to account for all the motions of the celestial bodies, and of our sea." But it is clear that this was actually not enough for Newton, and it certainly was not enough for Edwards. It was not enough because there were two problems, the cause of gravity and the cohesion of atoms, left unsolved, both of which threatened to yield up answers that Newton dreaded: he preferred leaving them as riddles to coming out with solutions that might prove the world Godless and mechanical. There could be no question that gravity was universal, but it simply could not be allowed to operate at a distance across the void, with no material intermediary. If there is no ether through which the force can be transmitted, then we had better leave unanswered the question of how the various solar systems are so wonderfully synchronized. All matter is subject to gravity, and every particle gravitates to every other, but under no circumstances could we allow ourselves to speak as though gravity were essential and inherent to matter: "Pray do not ascribe that notion to me." The way his popularizers elide the difficulty is a sign of their underlying anxiety: "From so many indications," wrote Maclaurin, "we may at length conclude, that all bodies in the solar system gravitate toward each other; and tho' we cannot consider gravity as essential to matter, we must allow that we have as much evidence, from the phenomena, for its universality, as for that of any other affection of bodies whatsoever."

The real motive for these maneuvers among the Newtonians is not far to seek, although recent studies have for the first time made clear its strength: Newton was a religious man first, and a scientist secondarily. What Newton wanted above all else was to give such an account of the cosmos as would make evident that God rules the world, in the words of Maclaurin, "not as its Soul but as its Lord, exercising an absolute sovereignty over the universe, not as over his own body but as over his work; and acting it according to his pleasure, without suffering any thing from it." Newton suffered an obscure nervous collapse in 1692–1693, after which he abandoned natural philosophy and devoted himself to theology and prophecies. Perhaps he believed that the clues he was following would lead him through nature to God, but at the point where the certainty of approaching divinity grew shaky, he stopped. The Newtonian mechanics came to Edwards apparently a complete system of the world, but committed to two dogmatic presuppositions: that gravity at a distance is absurd, and that gravity is not synonymous with solidity. Gravity, said the Newtonians, is "an original and general Law of all Matter impressed upon it by God, and maintained in it by some efficient power, which penetrates the solid Substance of it." This result was all that Cotton Mather or most eighteenth-century theologues desired, and they betrayed little or no awareness that it was founded upon unprovable surmises and upon dogmatic evasions of mysteries. Edwards was the one man (or rather boy) in New England who refused to pretend that the questions of the cohesion of atoms and of the universality of gravity either were solved or were unimportant. He was prepared to venture in thought where Newton would not tread, into the hiding places of nature, to run down the force that was both the cohesion of atoms and the power of gravity, and to risk the possibility that could he find it or name it, the force might turn out to be simply monstrous.

Edwards was the forthright boy who knew no better than to use his eyes and cry that the emperor had no clothes. He asked what, after all, is an atom? Gassendi, Boyle, Newton had never held one in their hands nor even seen one in a microscope. Yet fervent apostles of experiment believed in this untested thing, and men who scorned hypotheses universally embraced this supposition. Of course, as a workable theory, atomism justified itself in Boyle's chemistry and Newton's physics; but it worked, Edwards made out, because it did not have to be proved: it was a way of thinking, not a thing. The scientists, talking exclusively in what Edwards called the "old way," did not listen to themselves long enough to catch on that their real difficulty was not in fixing the position of atoms in space or in measuring them, but in the confusions of their speech. If atomism was, as Newton declared, the foundation of philosophy, it was time that philosophy was taken in hand by a clear-headed theologian.

In all scientific discourse about the atom, the real point was its single oneness. So Edwards offered a succinct definition (he underscored it), *"a body which cannot be made less."* Therefore a body, no matter what its size, that cannot be lessened is all that scientists mean by an atom: "an Atom may be as big as the Universe; because any body, of whatsoever bigness, were an atom, if it were a perfect solid." This did not mean that Edwards took reality into the mind and treated matter as a dream; he was as thoroughgoing an atomist as any in the age, and commenced his thinking with Newton's definitions:

God, in the beginning, created such a certain number of Atoms, of such a determinate bulk and figure, which they yet maintain and always will, and gave them such a motion, of such a direction, and of such a degree of velocity; from whence arise all the Natural changes in the Universe, forever, in a continued series.

What has been miscalled his idealism never meant to him that the world resides inside a man's head. Things are where they are, and Edwards had no intention of flouting "the science of the Causes or Reasons of corporeal changes." He was simply applying to the problem the method that was Newton's own, for Newton had once explained that he was not a genius except in so far as "when an idea first came to him, he pondered over it incessantly until its final results became apparent." Convinced that the "proportion of God's acting" would be the same, whether we suppose the world material or mental, Edwards set himself to ponder the nature of atoms and of gravity, an occupation which absorbed him for the rest of his life.

The first result was a group of the "Notes" to which he gave a title, "Of Atoms and of Perfectly Solid Bodies," which his editors believe to be among the earliest of his cogitations; the manuscript journals devote literally hundreds of pages to the same theme, and though many of these are possibly of greater literary interest, all confirm the first insight. Proposing that we cease to fool ourselves by taking an atom to be some small segment of space, as though we were speaking of a chair or a table, he said that all we mean by it is indivisibility. Therefore to talk of it as broken is to annihilate it. Hence it is evident that we really mean a point to which we can apply Newton's third law of motion: we are saying that an atom *resists*. Newton's stone, the farmer's stone, resists the pull of the horse, and by resistance stones have achieved the identity they so jealously guard. It would be easy to say, along with those who openly broached "schemes of a pernicious and fatal tendency," that God works from the outside and so compresses atoms together to form a stone. The danger of that position, as Newton had intimated, is that a more exact mathematical analysis of the inner structure of the nucleus may achieve a logic so sufficient unto itself that God will be relieved of the only work left for Him. It would also be easy to go to the other extreme of outright materialism, or at least of mechanism. When Newton toyed with the notion of a fluid that acts as glue among the atoms, he was trying to avoid the first extreme by veering in the direction of the second. Edwards instinctively rejected the subtle fluid, along with hooked atoms. The danger of the mechanistic extreme was not its atheism but its unscientific method: based on the fiction of the atom, it took the myth for a fact. Until it could isolate and draw a diagram of the atom, it was just as "speculative" as the rationalized theism of the optimists.

Both schemes were pernicious because both insisted upon treating atoms as pieces, as "particles," of existing substance. For Edwards, as for us, the whole question was altered as soon as he realized that the atom is a concept. It was useful in physics, not because it had spatial dimensions, but because it played only the one role, though an essential one, of providing a point on which resistance could be concentrated. It was, as Maclaurin put it, "a termination." It was what a more highly developed mathematics would call a "limit." And what resists is, by the act of resistance (not necessarily by substance), solid, because what else does solid mean? To speak of two atoms as *perfectly* joined is nonsense because the two, if absolutely united, must be "one and the same atom or perfect solid." Obviously no imaginable physical power can break up solidity; to split a stone into a thousand pieces, and each piece into another thousand, is not approaching termination.

It must needs be an Infinite power, which keeps the parts of atoms together; or, which with us is the same, which keeps two bodies touching by surfaces in being; for it must be infinite power, or bigger than any finite, which resists all finite power, how big soever, as we have proved these bodies to be.

Of the two dangers, optimistic theism or deterministic materialism, Edwards feared the second less than the first. Materialism, aside from its initial fallacy about the atom, gave a more truthful description of reality than was offered by "those modern divines." Edwards was mainly concerned to prove that the statement, "the constant exercise of the Infinite power of God is necessary, to preserve bodies in being," did not mean that God acts *ab externo* to press a million pieces of stone into the form of a rock. But it did mean that the principle of coherence is in the stone, because that principle is the being of the stone. Individuality is not merely the "hardness": it is "the immediate exercise of God's power." The substance of an object—a stone, a horse, a man—is a single event, "nothing but the Deity, acting in that particular manner, in those parts of space where he thinks fit." The grand, but to Calvinists the hitherto dangerous, conclusion was "that, speaking most strictly, there is no proper substance but God himself."

Edwards was apprehensive lest he sound like a Thomas Hobbes holding that God is matter, whereas what he intended was "that no matter is, in the most proper sense, matter." One may say that Edwards was making a distinction without a difference, that the world is one whether we call it all matter or all mind, and so he was actually a materialist. The charge has, I believe, more pertinence than the customary label of "idealist," but both labels are beside the point. He was trying to say something simpler than either, which modern students may find more intelligible than could his contemporaries or, for that matter, most readers in the last century—namely, that the corporeal universe results from concentrations of resistance at various centers in space, which have a power of communicating, through gravity and through collision, from one point to another, according to stated conditions which infinite wisdom perpetually observes. In such a cosmos there is no such thing as mechanism "if that word is intended to denote that, whereby bodies act, each upon the other, purely and properly by themselves," but there is a perpetual determination of sequences of events. The cohesion of thought makes possible both the idea of the atom and of bodies made of atoms joined together: "that Ideas shall be united forever, just so, and in such a manner, as is agreeable to such a series." If we hold the world to be composed of a number of atoms, ten millionths of an inch in diameter, and then try to deal with it as an assemblage of atoms, we can never be vigilant enough to keep every one under control; a few will slip loose and run wild through the system. But any experience with stones in a Connecticut field is enough to show that there are no wandering atoms. "The existence and motion of every Atom, has influence, more or less, on the motion of all other bodies in the Universe." No motion, either of the proud or of the contrite heart, is lost. Not that thinking makes the atoms law-abiding, for in that case thought might be only a kind of private vice, but that thinking, originating in the senses, is a true representation of what prevails. "The secret," cried the boy, trying at last to put it all into one searing paragraph, is that the true substance of all bodies "is the infinitely exact, and precise, and perfectly stable Idea, in God's mind, together with his stable Will, that the same shall gradually be communicated to us, and to other minds, according to certain fixed and exact established Methods and Laws."

A fixed and exact method was, according to Newton, a characterization of gravity. Therefore Edwards, having so read the riddle of the atom as to discover that every object is a continuing event, and God is within it and not outside it, plunged head-first into the

thought which Newton most feared and avoided: gravity is a
function of solidity and so inherent in matter. Edwards' religion
differed fundamentally from Newton's in that he did not need to
reserve to God the honor (he more than divined that it could easily
become an empty title) of being the "immaterial" cause of gravity.
He was never more clairvoyant than when he warned that gravity
ought not any more "be attributed to the immediate operation of
God, than everything else which indeed arises from it." To suppose
God the creator of the world, and then over and above that the
stage manager of gravity, was a way of dispensing with God
entirely. But by seeing the universe as a system of stable ideas,
Edwards could see exactly why gravity should have the same
proportions across the immensities of space without any material
medium. This was why Newton's speculations on the ether seemed
to him as frivolous as they did to Burke, "the folly," he called it,
"of seeking for a mechanical cause of Gravity."

Nothing in Edwards' mind is more original or more exciting
than this insight, but we need not wonder why, having reached it,
he became cautious. Had he baldly proclaimed it, he would have
been denounced in every corner of the land as a traitor to New
England's Calvinism, by none more loudly than by those who long
since had lost any real sympathy or understanding for the creed.
They would have pounced upon him for identifying the laws of
nature, not with the decrees of a transcendent sovereign situated
somewhere above and outside the world, but with "a principle by
which Matter acts on Matter"; by calling him the materialist and
the Lucretian, they would have diverted attention from their own
worldliness. I suspect Edwards would feel that most of the comment written about him since his time has been so motivated! But
in the secrecy of the "Notes," Edwards could say what Newton
dared not: "Solidity is gravity, so that, in some sense, the Essence
of bodies is Gravity." Discretion was clearly advisable when speaking in public, but a thinker really should not handle the law of
gravity quite so gingerly as did Sir Isaac; if body is a specification
in one place rather than in another of a focus of resistance, then
coordination among the several centers is to be expected. It is, in
fact, the presupposition of there being any world at all, and the

danger of atheism is infinitely greater if we pretend that it is not than if we frankly recognize it. To Edwards' clear eyes, Newton's fear, and still more the fear of his followers, that if gravity became a function of mass, science would become Godless, was what Bunyan called Little-faith; it was a failure to see what was written before their eyes.

Therefore, we may infallibly conclude, that the very being, and the manner of being, and the whole, of bodies depends immediately on the Divine Being. To show how that, if Gravity should be withdrawn, the whole Universe would in a moment vanish into nothing; so that not only the well-being of the world depends on it, but the very being.

The failure of Newton, and of the age, was a failure of intelligence. They were ready to call mobility a quality of matter without supposing themselves in danger of becoming atheists; but gravity, more than mobility, is essential in order for existence to exist, though for many reasons "the mind does not so intuitively see how." But actually, once Newton's laws are grasped, it seemed to Edwards logical, inescapable, that gravity does operate at a distance, for what is distance but, like time, a principle of stability for organizing a sequence of ideas?

Edwards could not know that a philosopher of the twentieth century, profoundly versed in a still more subtle and powerful physics, would deduce that cognition, having a unity of its own as an event, "knows the world as a system of mutual relevance, and thus sees itself as mirrored in other things." He would, I am sure, agree that Whitehead put thus simply the heart of his meaning, and he would agree further with Whitehead that men of thought, by virtue of this perception, are ultimately rulers of the world, but he knew more than Whitehead ever did of what is actually required for ruling a New England town. Though he was born to the purple, and though the society still bowed or curtseyed to the pulpit upon entering the meetinghouse, a ruler could not tell them all his thought. Hence, could the good people (or those not so good) of Northampton be expected to comprehend that the young pastor they had chosen to be, like his grandfather, the administrator of their ecclesiastical foundation was inwardly and incessantly con-

cerned with testing whether an assertion of *how* the universe acts
can be made identical with the *why?* Could they comprehend that
he had a new vision of the cosmos as a system of causes, atoms,
and gravity, and that for the mind to achieve regeneration, it
would need to strip itself of all verbal substitutes for physical
reality, of all metaphor and similitude, and look squarely upon the
purely factual? Could they understand that in this scientific version
of grace, the perception of beauty is that which determines both
value and reality, that only in such a perception can the natural
world and the world of religious experience become one? Could
they even begin to grasp that for him the sheer naked reality was
enough?

We should claim too much did we call Edwards' metaphysical
intuitions about gravity anticipations of Planck and Einstein. He
was not, let me repeat, an experimental scientist, nor was he a
trained mathematician, and he could never even have approxi-
mated the conception that laws may be formulated without reference to any particular space-time manifold. He was a man of his
century, though his thought was in the forefront of it; he was a
docile Newtonian, who believed in absolute and objective space,
and he assumed absolute time; he accepted this very same space
which we see and this daily time which we experience. Still, it is
not extravagant to point out that, by his argument that an atom is
not a thing but a way of speaking about a locus of attraction and
repulsion, Edwards was divining the great line of the future. That
he was predisposed to some such divination by his theology, and
that he meant no more than would confirm his doctrines, cannot be
doubted, but at least it is clear that he saw through the genial and
obtuse rationalism which most of his contemporaries thought was
the import of Newton.

In his sermons on justification, in 1734, he needed to draw upon
the "Notes" for only one statement: the relation of event to event
in a causal sequence, whether of atoms, planets, or of grace, flows
"only from the natural concord or agreeableness there is between
such qualifications and such circumstances." If atomic entities are
really entities in enduring conception (and otherwise men can
never know what they are), and are related one to another because

God delights in order and not in confusion, then a cause is not one occurrence which is instrumental in producing the other, but properly speaking is "that, after or upon the existence of which, or the existence of it after such a manner, the existence of another thing follows." In this sense faith may be called a "cause" of justification, but not in the pernicious sense of that which brings it to pass. If the nature of things is a system of stable ideas which does not depend on the capricious goings or comings of atoms, why should the grace of God wait upon an uncaused achievement of virtue by this or that individual? Salvation, along with the atom and the law of gravity, can be only a manifestation of God's regard, not to the pretensions of individuals, but to "the beauty of that order that there is in uniting those things that have a natural agreement, and congruity, and unition the one with the other." For humanity this must mean, instead of a justification built upon the merit of our finite virtues, that "the acceptableness, and so the rewardableness of our virtue, is not antecedent to justification, but follows it, and is built entirely upon it." In the order of causation, a man is not a saint because he is good, but if he is a saint he is caused to be good. In more conventional language, he is elected.

Had Newton, or such of his disciples as Clark or Maclaurin, read Edwards' "Notes," they would not have been impressed as are we today. The scientists waged a hard fight to get their experimental, empirical method not only understood but even tolerated; they found that their chief enemies were those who "hastily," as they put it, resolve material motions "into immediate volitions of the supreme cause, without admitting any intermediate instruments," who thus "put an end to our enquiries at once." The physicists felt it essential for good science, and therefore they claimed for sound theology, that God be severely limited to governing the world not as His own body but as His work; in that sense alone, they would agree that God is "omnipotent." Persons who confounded God's lordship with His substance—the scientists said this in a hundred ways—"hurt those very interests which they would promote." Consequently, the majority of Newtonians, scientific and theological, would have called Edwards a brash boy who was perverting the incomparable Newton back to the obscurantism of immediate volitions of the supreme cause. They would not have appreciated that in fact Edwards gave over the material world to a chain of "intermediate" terms even more consistently than did the pious Newtonians.

Furthermore, they would not have understood that the sensitive Edwards was looking ahead to what they could only vaguely foresee; Maclaurin found himself stating, though with reluctance, a new problem which, in the very wake of Newton's success, appeared to be distracting the European mind. The intellectual world, he lamented, is becoming divided into two camps (where he thought they ought to be all one): those who from their fondness for explaining things by mechanism "have been led to exclude every thing but matter and motion out of the universe," and those with a contrary disposition, who will "admit nothing but perceptions, and things which they perceive," some of whom have gone to such extremes that "they have admitted nothing but their own perceptions." The majority of Newtonian rationalists were unable to figure out why these strange divisions had come about when it seemed that Newton had settled all such foolish arguments.

In the Boston lecture, Edwards exhibited the two orders, the objective and the inherent. The one, we find, forced us back to Newton and natural law, the other to Locke and perception. We cannot be certain how much was fully explicit in Edwards' mind, but considering how caution and reticence were thrust upon him, it appears that Edwards saw exactly where the modern problem is centered, upon this incompatibility of Newton and Locke, of the objective and the subjective, of the mechanical and the conscious. The effort of his life was to unite the two. The line of his speculation might well seem to less subtle Newtonians a slighting of "subordinate instruments and agents," but if so, they would again betray, as many in New England were soon to do, the shallow dogmatism which prevented them from comprehending what the age had really to grapple with, as it was to prevent Chauncy and his circle from ever seeing what Edwards was driving at.

His problem, then, was to get the two orders together—or else to confess that the modern world is incoherent. A stone transmits force to another by collision, and so is a "cause" of motion; in perception, there is a "fitness" of the antecedent sensation to the subsequent act. Is perception, then, just another form of collision, in which an object transmits motion through the senses? A moved stone receives what it must receive, no more, no less; but is man merely another kind of stone? Human perceptions notoriously vary in depth and width; is causality the same in either realm? In perception, may there not be a fitness between response and object that is still freedom of action? When the stone resists, is not the farmer at liberty either to curse or to pray? Does the tavern irresistibly attract the toper, does the woman inescapably arouse the lust of the adulterer? Are the principles of nature, the implacable sequences of things, applicable without change to human nature and society? If the inherent good is excellency and pleasure, while the objective good is the possession and enjoyment of that object which is good for the organism, is there not an incurable conflict between the two? Perception, either as pleasure or as beauty, is value, but if it is illusion, then reality is only the dance of atoms. The civilization of more than New England was at stake if the life of the spirit was henceforth to be a civil war between atoms and perceptions. Unless Edwards could merge, or at least reconcile, his objective good and his inherent, neither he nor, as he saw the predicament, anyone else would be able to locate the good where it might be of help to mankind.

JOHN E. SMITH

✪

Edwards' *Religious Affections*

THE ARGUMENT: THE "TWELVE SIGNS OF GRACIOUS AFFECTIONS"

The *Affections* is a masterful treatment of a basic theological problem; it is also a work of remarkable literary power. We are grasped by the earnestness of the author, by his concern that we understand him aright, and by the pains he has taken to capture our imagination. In directing attention to his style, we cannot overlook the fact that many readers have found the *Affections* difficult going, nor should we ignore what is implied in the activity of the many editors who thought it necessary to rewrite the text. It is admittedly an exacting work; it calls forth a reader's best effort. But there are rewards if we are willing to raise ourselves to the level of Edwards' austere standards; nothing is to be gained by bringing him down to a more facile plane in order to make him say what we would like to hear.

Edwards was in tight control of his ideas; he knew exactly what he wanted to say and he said it in an uncompromising way. The result is a meticulous form of expression, a precision in language and an intricacy which reveals a deep and subtle mind. He would not let a subject drop until he had exposed it from every side, nor would he move on to another theme until he had expressed the

Reprinted with permission from Jonathan Edwards, *Religious Affections,* ed. John E. Smith (New Haven: Yale University Press, 1959), pp. 8–24, 40–52. Copyright © 1959 by Yale University Press.

results of his analysis incisively and arrestingly. A careful reader will be delighted by Edwards' ability to guide him through a long line of argument and he will come to exercise his own ingenuity in anticipating some of the surprising turns in the road.

The most striking features of the *Affections,* then, are the exactitude and vividness of the language. Edwards always sought the right word, the one which exactly expressed his intended meaning. When, for example, he wished to emphasize the need for the heart in genuine religious conviction, he found expressions like "assent" or "allow" too pale and lifeless. A man may "allow," he says, that something is so and he may give assent to it in a merely notional way, but unless he is willing to "profess" that conviction, his heart is not in it. Professing is an affectionate believing and stems from the whole man. If we forget that Edwards chose his language with care and are led to suppose that some other form of expression would have done as well, we shall misunderstand him and lose the fruit of his efforts to make difficult things clear.

The vividness and imaginative power of his style are intimately connected with the fundamental theme of the work; doctrine and style flow together. Edwards was convinced that if religion consists in holy affections, the proclamation of that doctrine must be made in an affecting way. In one of his earlier writings we find him criticizing a minister he had heard for his failure to adapt the form of his expression to the content of his message. Edwards saw incongruity and even contradiction in the attempt to communicate the truth about life in a lifeless way. When we must convey a *sense* of what we say, just because what we say is of no account unless it grasps the sense of the hearer, an affecting style is needed. Edwards never lost sight of this principle in the *Affections.*

Consider, for example, the magnificent comparison between the true saints and those merely puffed up by the experience of vigorous but fleeting emotions. Hypocrites are likened to meteors which flare up suddenly in a blaze of light trailing showers of many sparks but soon falling back to earth, their light dissipated; all is over in a twinkling. The true saints are like the fixed stars; they shine by a light which is steady and sure, a light which continues to show itself over time and through the infinite spaces. Thus is expressed the central doctrine that the true saints have the sense of the heart, a steady and abiding principle in their own natures; something not to be confused with the spectacular emotions and commotions of revivalism.

The affecting style compels the reader to understand by vicarious participation in what is being described. And the remarkable thing is that the result is achieved at the same time that an intricate analysis is taking place. Like Henri Bergson, Edwards had the gift of analyzing an experience in great detail before our eyes at the same time that his language leads us to participate in the experience itself. This gift represents the fusion of the descriptive and evocative functions of language; we are made to see the anatomy of experience at the same time that the language in which the analysis is couched lays hold upon us, making us participate in that experience with all its directness and warmth.

The continuing power of the *Affections* is to be found in the success with which Edwards brought together the essential ingredients of a theological work. It must express a synthesis of clear argument and the quickening spirit of direct experience. If a formula is needed, it must combine information and inspiration. In contrast to devotional writing, a theological work must contain analysis and argument; it must exercise and enlighten the mind. But the nature of its subject requires that it touch the heart and engage the inner man; it cannot be a compendium of doctrine alone. The *Affections* furnishes both; as one follows the line of argument defining the nature of genuine piety, one is led to understand and to feel its power at the same time. Analysis and experience converge.

There has been a tendency among interpreters to view the *Affections* not as a sustained analysis in its own right but as an historian's document pointing beyond itself to the past or to events in Edwards' life that were yet to come. Thus the work has been viewed simply as commentary upon the Great Awakening, or as an indication of the position Edwards was to take in the communion controversy which came to a head in 1750. In both cases the *Affections* is made to appear as an interlude between historical events. Important as the historical setting may have been, the fact remains that Edwards' real story is to be found in the life of the mind. His works must therefore be treated as attempts to answer basic and perennial theological problems. Moreover, the *Affections* has been praised in vague descriptions; it must now be read and analyzed in a way consistent with a work of its stature. The highest praise of a book should proceed not from uncritical acceptance but from a willingness to treat it as important enough to be argued about.

Edwards poses his central problem in the Preface: "What are the distinguishing qualifications of those that are in favor with God, and entitled to his eternal rewards?" In expanding upon his theme he is led to identify this question with a second, "What is the nature of true religion?" Even if he had not told us that this problem had been at the center of his mind "ever since I first entered on the study of divinity," we could still trace it from his earliest writings and sermons to the publication of the *Affections* in 1746. What continued to puzzle and vex him was the mixture of evil with good in the revival and among the saints, and consequently the problem of finding a way to distinguish the one from the other so that the evil might be exposed and rejected and the good retained. He was not blind to the presence of tares among the wheat, nor did he overlook, as Wesley accused him of doing, the mixture of purity and corruption in those genuinely favored of God. Indeed, it was his acknowledgment of counterfeit piety that forced him to find criteria for distinguishing false from true religion.

The *Affections* is especially distinguished by the intensity of its concern with the religious life of the individual; all but essentials are stripped away and a frontal assault is made upon the underlying problem. Edwards had previously described religion as it could be found in the depths of the individual soul, but these discussions had been more concerned with the fates and fortunes of the revival at large than with the gracious operation of the Spirit. He recognized this fact himself; in the *Affections* he was anxious to center attention on the gracious activity of the Spirit in the *individual* soul. This work was, so to speak, a final try at answering the crucial question, and in order to present his position in the clearest light, he concentrated upon the activity of the Spirit in its purity and upon the positive description of the genuine religious life.

Thus far much has been said about affections, but little attention has been paid to exactly what they are. Here, as in all his writings, Edwards was most circumspect; he begins with an account of the nature of affections and a defense of the thesis that they are the substance of true religion. Our first task, therefore, is to understand exactly what he meant by affections, how they arise, what relation they bear to the divine Spirit, and how they stand connected with the understanding and the will. Having answered these questions, we may consider the meaning of signs or criteria for judging the affections. The twelve signs of gracious affections are these criteria. Not only do they serve as tests or standards of genuine piety, but they are themselves the very substance of the religious life.

The first point to be stressed is that Edwards, for all his ability to draw clear distinctions, nevertheless struggled to preserve the unity and integrity of the self and to avoid compartmentalizing the human functions and powers. This means that despite his rather sharp distinction between understanding, affections, and will, we must not overlook the extent to which these initial distinctions are overridden in the course of the argument. The entire discussion shows a moving back and forth between analysis and synthesis; clarity demands distinctions within the self and between its powers, but the integrity of the self requires that its faculties or capacities be related to each other so as to preserve unity.

The starting point of the *Affections* is subtle; Edwards required a biblical picture of true religion as a model, and he found it in a word addressed to the early church during a time of persecution. He assumed that in a time of pressure, when faith is tried in the fire of persecution and disbelief, religion will appear in its true form. Consequently, he chose his text for the opening section (the part of the *Affections* most clearly in sermonic form) from I Peter 1:8: "Whom having not seen, ye love: in whom, though now ye see him not, yet believing, ye rejoice with joy unspeakable, and full of glory." From this text together with the historical context, he derives his conception of true religion as consisting in the affections of love and joy in Christ; the former rests upon a spiritual sight, since the object of love is unseen with ordinary eyes, and the latter is the fruit of faith. The nature of such joy is to be "full of glory" and to Edwards this meant a filling of the mind and the whole being of the believer with a sight, a sense, and a power from beyond nature. He further derives or, as the expression ran, "raises" from the text this doctrine: "True religion, in great part, consists in holy affections"; the development of this thesis involves, first, an account of the nature of affections, and secondly, the adducing of those considerations which show "that a great part of true religion lies in the affections."

Edwards' response to his own question about the nature of affections is that they form a class of "vigorous" and "sensible" exercises of will or inclination. The special kind of such exercises entitled to the name "affections" are those that are vigorous enough to carry the self well beyond indifference, to the point where "the motion of the blood and animal spirits begins to be sensibly altered" and some change shows in the "heart." Whether or not we take the physiological trappings seriously, it is clear that Edwards wants to root affections in the inclination or central orientation of the self; affections are signposts indicating the *direction* of the soul, whether it is toward God in love or away from God and toward the world. This becomes clearer when we pay attention to his distinction between the understanding and the inclination or will. He refers initially to "two faculties" in man; one "capable of perception and speculation," which is called understanding, and the other, left for the moment without a name, which is said to be the means whereby "the soul does not merely perceive and view things, but is someway inclined with respect to the things it views or considers." On the one hand the soul may be inclined toward something and approve it, or it may be displeased and thus reject it. The judgment or inclination of the self involved in such reactions of attraction or aversion is intimately related to what Edwards meant by the "heart." This fact has been obscured in the past because of a misapprehension according to which the heart is vaguely described as "emotional" and set over against the "head," which is a symbol for reason and knowledge. *Edwards' analysis gives no warrant either for the identification or the opposition.*

The only contrast Edwards sets up is between the understanding as a grasp of meaning unaccompanied by any "inclination" or judgment of approval or disapproval, and the will or inclination as comprising such a reaction. In other words, a line is drawn between understanding and will, where the distinction means the difference between the "neutral observer" and one who "takes sides"; but there is no clear warrant for making this into an opposition between the two terms (and, *a fortiori,* it gives us no reason for opposing understanding and affections). The point almost invariably missed is that in Edwards' view the *inclination* (the faculty initially distinguished from the understanding) involves *both* the will and the mind. When inclination receives overt expression in action it is most commonly called "will," and when inclination is expressed through the mind alone it is called "heart." The latter relationship is central to the *Affections.* Those "more vigorous and sensible exercises"[1] of inclination, i.e. of being inclined and not in a state of indifference, are what Edwards calls "affections." They are thus the expressions of inclination *through the mind.* They stand in a necessary relation to the ideas of the understanding and are also the springs of actions commonly ascribed to the will. Inclination is not a blind affair, since it is based on an apprehension of the idea, the doctrine or the object which the self is attempting to judge. Nor are the affections merely mental in the sense that they have to do with the depths of the soul alone, for the sign to which Edwards attached the greatest importance, the sign of consistent practice, shows that in order to be genuine, affections must manifest themselves in an outward and visible way.[2]

---

[1] "But yet, it is not the body but the mind only, that is the proper seat of the affections."

[2] Try as we may, it is difficult to avoid confusion over the difference between will and affections and their mutual connections. JE is not clear himself, although it is obvious that the problem stems from his attempt to preserve the integrity of the self against the tendency to break it up into "faculties." We are told that "the affections are not essentially distinct from the will," and this point is repeatedly emphasized. On the other hand, will

The essential point is that the affections manifest the center and unity of the self; they express the whole man and give insight into the basic orientation of his life. Edwards was aware that the term "affections" does not ordinarily mean all of our "actings" and that it may not even be understood to involve action at all. To understand what he means we must take such phrases as "exercise of the will" and "actings of the inclination and will" to mean choice or judgment in the first instance[3] and overt action secondarily. The crucial point is that in every choice the soul likes or dislikes, and when these "inclinations" are "vigorous" and "lively" the liking or disliking coincides with love and hatred. Affections, then, are *lively* inclinations and choices which show that man is a being with a heart.

There is a further preliminary distinction; and although it occupies but a paragraph, it is of pivotal importance. "The *affections* and *passions,*" he says, "are frequently spoken of as the same," but there are grounds for distinguishing them. Passions he describes as those inclinations whose "effects on the animal spirits are more violent" and in them the mind is overpowered and "less in its own command." The self becomes literally a "patient," seized by the object of a passion. With the affections, however, the situation stands quite otherwise. These require instead a clear understanding and sufficient control of the self to make choice possible. This distinction enabled him to criticize and reject a great many revival phenomena, especially those of a pathological sort, and to dissociate the heart religion he advocated from hysteria, the excesses of bodily effects and enthusiasm. His contemporaries paid

---

is said to denote *inclination expressed in action* as distinct from the "heart," which points to *inclination expressed in the mind.* Since, however, affections are most often identified with the "heart," it may appear that further identification of will and affections will need some qualification. Perhaps the attempt to be too meticulous here will only lead to confusion. If we stick to the idea that an affection is a "warm" and "fervid" inclination involving judgment, we shall not go wrong.

[3] See the introductory discussion of the nature of will in JE's *Freedom of the Will,* ed. Paul Ramsey (New Haven: Yale University Press, 1957), pp. 137–140; cf. pp. 16 ff. The analysis there definitely stresses the element of choice or judgment over that of exertion, and shows JE's insistence upon identifying "volition" and "preference."

insufficient attention to his distinctions. They thought he was defending revivalism in the sense of religious passions at the expense of intellect, whereas he was developing a conception of affections accompanied by understanding.

With this conceptual framework, with the affections separated into two principal kinds—those seeking to possess their object and those seeking to reject their object—Edwards returns to the Bible to establish more securely the thesis that genuine religion "in great part, consists in holy affections."[4] The piety which God requires, the only one he will accept, is one which engages the heart and inclines the self as a whole toward the divine glory in a love which is unmixed. Of all the aspects of human life and experience, religion is the one in which it is least possible to be "lukewarm"; piety which does not include a fixed and fervent inclination touching the heart is no genuine piety. Hearing the Word, as the Bible repeatedly emphasizes, is not enough, nor is it sufficient merely to "allow" that the doctrine may be true. What is needed is that the soul be "moved" and filled with the love of God which ultimately shows itself through right conduct in the world. Neither is possible in Edwards' view unless the soul is "affected" and the will inclined. In this sense affections are motive forces or springs, and the particular change called conversion becomes possible only if the self is affected at the heart. "I am bold to assert," he says, "that there never was any considerable change wrought in the mind or conversation of any one person, by anything of a religious nature, that ever he read, heard or saw, that had not his affections moved."[5]

We must not overlook the duality in Edwards' theory of affections rooted in the fact that love is both one of the particular

[4] It is important to notice that JE does not want to swallow religion in affections; they constitute the most important part of genuine piety and the testing of their character provides a test of piety, but he is most careful to qualify his position by such phrases as "in great part." His full formula is that affections are *necessary for* religion and *constitute a large part* of its nature.

[5] Here, as in the case throughout the early part of the *Affections*, the connection between affections and will is uppermost; later on, JE stresses much more the place of understanding and the divine light.

affections and the fountain of all affections. His contention is that the basis of true religion is found in that chief affection which is love to God unmixed; this defines the original relation of the soul to God. Religion, however, includes along with the basic relation to God, life in the world and in the glory to come. This means that as a life and not merely a doctrine it has further content; Edwards' view is that this content itself consists largely in affectionate life. Hope, joy, fear, zeal, compassion and others are frequently referred to in the New Testament as the substance of religion; Edwards describes them as affections and goes on to argue that they make up the religious life. To avoid confusion we must understand that he is defining the basic religious relationship as essentially the affection of love. This is what he means when he says "the essence of all true religion lies in holy love," but at the same time he maintains that the fruits of love are also affections— joy, zeal, peace—which have their proper place in the whole of religion. "For love," he says, "is not only one of the affections, but it is the first and chief of the affections, and the fountain of all the affections." There is thus a basic affection, holy love, which relates the self to God in a decisive way and it has to do with the "first fruit" of the Spirit or "sealing of the Spirit," in biblical phrasing. There are also fruits or further gifts of the Spirit which represent the substance of the religious life in the world. The dual meaning of love running throughout Edwards' analysis need not confuse us if we understand the comprehensiveness with which he uses the term.[6] Love taken as the fountain of affections is the more important sense of the term because it refers not only to the relationship between man and God at the root of all religion but also to the activity of the Holy Spirit in the individual soul. Moreover, when holy love is said to be the essence of genuine religion, it is

[6] The duality involved is reflected in the ordinary use of the term "affection." In the eighteenth century the term meant the whole class of attitudes and dispositions to which JE refers: love, hope, fear, etc. Because of the centrality of love, the class term "affections" came to be identified with it alone, although it is but one of the affections. The phenomenon is familiar to students of language; a class term has often become identified with one of its members when that member has come to be regarded as more basic than others.

then possible to explain sin by contrast as that equally basic and ultimate *rejection* of God, symbolized in the Bible as "hardness of heart."[7] Edwards was most acute in finding biblical support for his position; assuming such hardness and rejection of God to be the very opposite of genuine piety, he sought, among the biblical illustrations of the hard-hearted, evidence to show that what they chiefly lacked was the affection of love. And he argued in the reverse direction, declaring that if the antithesis of genuine piety is to be without holy love, there is further reason for believing that love is a mark of true religion. Those who have such religion overflow with that divine love which Pharaoh, for example, did not have.

The intention of the *Affections* is to test these fruits of the Spirit, not to praise them. Edwards never lost sight of the twofold task that followed: on the one hand, to defend the central importance of the affections against those who would eliminate them from religion; and on the other, to provide criteria for testing them lest religion degenerate into emotional fanaticism and false enthusiasm. Moreover, testing the spirits was no academic exercise; Edwards wanted to give to the individual some basis for judging the state of his own soul. It is essential here that we attend to his subtlety and depth. Simple oppositions and alternatives will not do. They allow for too much or too little. We need an approach more sophisticated than is implied in asking whether Edwards was "for" or "against" affections. Many of his contemporaries approached him in that light; they were bound to be disappointed, simply because no answer of a strict either/or type can be given. The point is that he was "for" the affections, but not in any sense you please; moreover, he was not *uncritically* for them, which is why he was at such pains to explain what they are, how you might know when you had them, and which ones marked the genuine

---

[7] It is interesting that instead of describing sin and hardness of heart as "affection" in the direction of hatred, aversion, and rejection, JE refers instead to the hard heart as an "unaffected heart," i.e., as a heart not affected by "virtuous affections." This is a minor inconsistency only, but it is apt to cause confusion. The point is simply that since JE was interested in the genuine religious affections, he identified "affection" with those which are directed toward God, and described the hard heart as devoid of affection.

presence of the divine Spirit. His book, consequently, brings together two lines of thought: it identifies the activity of the Holy Spirit with the affections in the soul and at the same time shows how these same affections when properly tested enable us to discriminate genuine from false piety. It is not difficult to understand what a precarious line Edwards marked out for himself. To those who rejected heart religion his analysis of genuine piety in terms of affections was anathema, but he was equally set upon by those who wanted their affections neat, so to speak, and had no time for bothersome inquiries into their grounds and authenticity.

In concluding the opening section of the *Affections,* Edwards proceeds to draw some concrete conclusions from his previous doctrine. He criticizes those who reject all affections because of the excesses of some and the vain zeal of others. He attacks the oscillation from one extreme to the other and he is as outspoken against those who accept all affections uncritically as against those who want to be done with them completely. He did not miss the opportunity to put a vivid biblical construction upon the New England revival experience while also preparing the ground for his own theory of signs and tests. He tells the congregation that he sees the hand of Satan both in the Revival and in those who fight against it. Satan, seeing that people were unlearned in heart religion, sowed tares in the form of false affections and thus misled and confounded many with the belief that they belonged to the Lord's elect. On the other hand, after seeing the reaction that took place after the high tide of affectionate religion, Satan set to work in another direction to establish the belief that affection itself is an evil. In both cases genuine religion is made to suffer. Against this background Edwards' position stands out most clearly: affections are essential, but since there are false as well as true affections, critical tests are required. The main concern of his book is to establish the only valid criteria for making such tests. Testing the spirits is of the essence of "experimental religion."

Edwards devotes his short second section to what were formerly called "negative signs." Here he makes his position more explicit by describing these signs as insufficient to enable us to conclude whether affections are gracious or not. The critical force of negative signs lies in the fact that their insufficiency refutes anyone who rejects all affections, by showing that an excess or exaggeration cannot be taken as conclusive. The argument attempts to demonstrate, both by biblical example and appeal to general experience, that the presence of a given characteristic does not mean either the presence or absence of the divine Spirit in its saving operations. In every case the accidental character of these signs is said to reside in the fact that they can be present without the Spirit's presence. The full extent of their insufficiency becomes apparent only after the positive signs are set forth, but taken by themselves Edwards argues that they are to be found where there is no genuine piety and that they may be absent where genuine piety exists. They cannot, therefore, be taken as necessary conditions.

There is no need to follow Edwards in detail through his discussion of all the negative signs. His analysis is clear enough and is even repetitive of much he had expressed before. There are, however, two points of special importance: the first has to do with the idea that the Spirit is bound to a definite *order* of operation and the second concerns whether it is possible to infer anything about affections from the fact that they come to be accepted by *other people* as signs of saintliness.

As regards the first point, Edwards' thesis is as follows: "Nothing can certainly be determined concerning the nature of affections by this, that comforts and joys seem to follow awakenings and convictions of conscience, in a *certain order*." Before proceeding to his final conclusion, Edwards first calls attention to the fact that God deals with human beings in a way that has a discernible pattern in it. That is to say, there are abundant evidences in the Scripture of God's having first convicted, wounded, distressed, and terrified man by the contrast between his sin and the divine majesty, and then comforted him with glad tidings. "It seems to be," says Edwards, "the natural import of the word 'gospel,' glad tidings, that it is news of deliverance and salvation, after great fear and distress." He concludes that nothing can be said against the authenticity of comforts and joys because they come after terrors and convictions of conscience, for the prevalence of this order of events in so many biblical experiences nullifies any such claim. On the other hand, he is concerned to maintain that comforts and joys cannot be accepted as genuine merely because they succeed great terrors and fears of hell; being afraid of hell is not the same as having genuine convictions of conscience. His argument at this point is apt to be confusing because he speaks of both the *nature* of the states in the soul and their *order*. The argument begins by emphasizing the order of events in the soul, but we soon discover that their nature is involved as well. The fact is, one aspect cannot be separated from the other. The problem of order is made central at this point.

As regards the nature of the states, Edwards wants to say that terror or fear is not necessarily the same as conviction of conscience; even granting that terrors and convictions have actually been produced by the Spirit in the soul, this fact by itself does not prove that true comfort *must* follow. The principal reason is revealing; the "unmortified corruption of the heart may quench the Spirit of God," he says, and this can mean only that if grace is not irresistible, the "preparation" or common influences of the Spirit are capable of being resisted or stifled by pride and the claim of a false hope. Satan, moreover is able to counterfeit the operations of the Spirit, but there is an important qualification about his power to do so: Satan can *exactly imitate the order* of affections, but not their nature. "The nature of divine things," says Edwards, "is harder for the devil to imitate, than their order," and from this fact follows the chief problem. It also explains why Edwards fastened upon the concept of order. If Satan can imitate the order exactly, then the order of operation followed by the Spirit *cannot* be a certain and decisive sign; the truly dependable signs are, as we shall see, those which Satan cannot perform. "Therefore," Edwards concludes, "no order or method of operations and experiences, is any certain sign of their divinity."

This point has several far-reaching repercussions. First, Edwards is denying the validity of many Puritan descriptions of salvation as involving a sequential process. There is thus in his thought room for a certain "variety" in religious experience and no conversion by "rule." Secondly, and even more basic, if we confine our attention to the order alone there is no necessary transition from nature to grace. The sharp separation between "common" and "saving" operations comes to the fore; we have no insight into any necessary connection between an order of events in nature and their issuance in grace. The Scripture, Edwards asserts, is silent on this point, and there is no alternative to denying the validity of order as a sign of genuine piety. He is willing to admit only that some awareness or conviction of sin on the part of the believer is necessary, but this necessary condition is not itself a sufficient ground from which to infer the presence of the Spirit. The ultimate basis for Edwards' view is not only the Bible but an appeal to experience. No better example of Edwards "the experimental divine" can be found than his appeal to the observations he made in the midst of the Revival to refute the thesis that the Spirit is bound to a single method or order in laying hold of the soul. As a final and telling word he says, "we are often in Scripture expressly directed to try ourselves by the *nature* of the fruits of the Spirit; but nowhere by the Spirit's *method* of producing them."

The second noteworthy point in Part II concerns the attempt to use the "approval of the godly" as a criterion for judging affections. Edwards' contention is of the utmost importance for his entire theory; it means that external judgment, the judgment of one man upon another, is not only unreliable but ultimately impossible. The saints, though they know "experimentally" what true religion is in their own selves, have no power of discerning the *heart* of another. Edwards liked to cite I Sam. 16:7 as an authority: "The Lord seeth not as man seeth; for man looketh on the outward appearance, but the Lord looketh on the heart." There are at least two consequences of the position taken which need to be underlined in any study of Puritan piety. First, the testing of the spirits can ultimately be done only by the self for itself; the external situation may lead a man to seek the counsel and judgment of others but no man can pronounce any final judgment concerning the status of another before God. "It is against the doctrines of Scripture," says Edwards, "which do plainly teach us that the state of others' souls toward God, cannot be known by us." Secondly, there is the consequence that the "public charity" forming the basis of acceptance between the visible saints in the visible church must not imply any *final* judgment by any man about the religious status of his neighbor. This is a touchy point. Edwards in several places argued against making a distinction *within* the visible church between the sheep and the goats. It has sometimes been objected by critics that his refusal to allow this distinction was incompatible with the position he later took in demanding conversion experiences and confession of Christ as conditions for being received into full communion. His position can be made consistent with the principle laid down in the *Affections*. When the congregation accepts a confession based upon conversion experiences, it must not suppose that this acceptance is the same as certain knowledge of the confessor's state. This is a subtle line of thought but it is consistent; some credentials are demanded of the believer who presents himself, but their acceptance by others does not mean or imply certain knowledge on their part of the status of another person. The heart of the other is known but to God; the acceptance or approval of others is no certain sign for judging religious affections.

The third and largest part of the *Affections* contains the heart of Edwards' position: the exhaustive account of the twelve signs of gracious affections. This analysis must, of course, stand or fall on its own merits. The reader will be aided greatly in approaching the text, however, if he has clearly in mind several leading ideas. He must understand what Edwards meant by a sign, what basic principle is behind each sign, and the way in which the individual is supposed to use the sign as a test of his own heart.

Since this part of the book contains the bulk of the material quoted by Edwards in defense and illustration of his position, the reader will do well to bear in mind what is said later about his relation to other writers. It would be an error to suppose that Edwards was directly dependent for his own doctrine upon the works cited; his own thought was remarkably self-contained and, furthermore, he firmly believed that his entire theory of affections was rooted in the biblical picture of the true religious life. Consequently, the works of others appear more as illustrations and confirmations of his position than as "influences" from which it might be derived. This remains true even of his use of Stoddard and Shepard, upon whom he relied most.

The positive signs are meant to delineate those affections that are "gracious" or "saving." The entire theory presupposes a distinction in kind (more obvious in the treatment of some signs than others) between "common" and "saving" operations of the Spirit. As Edwards had occasion to point out, the *Affections* is directly concerned only with those operations which are saving, as compared with the *Distinguishing Marks,* which paid more attention to the common workings of God. The whole discussion of the positive signs is prefaced by a re-emphasis upon the closing note of Part II: the signs of gracious affections are not for enabling us to distinguish true from false affections in others. Edwards repeats what he had claimed many times before: "it was never God's design to give us any rules, by which we may know, who of our fellow-professors are his, and to make a full and clear separation between sheep and goats." Edwards goes even further and allows for uncertainty in the knowledge with which the saints are supposed to know themselves. The signs are neither doubtful nor unavailable, but there are weaknesses in the person (the clouding of the eye of judgment) and seeds of corruption (the cold or carnal "frame"). These work together to make the process of discerning both difficult and uncertain when applied to the experience of a particular individual. The qualification is important; for while Edwards' theory of signs or criteria went considerably beyond what some of his contemporaries believed possible in the discerning of spirits, he claimed no infallibility for their application in a given case. The rules themselves have their own certainty, but the applications always fall short; no biblical principle is on the same level with its application.

Turning to the meaning of sign in the positive sense, we must understand a sign to be a mark through which the presence of the divine Spirit can be known. Edwards does not say that we *infer* the presence of God's grace using the signs as a basis; he does, in fact, leave that relationship vague. It is best to suppose that the sign "points to" the activity of the Spirit, especially when we consider the matter from the side of our human process of knowing. Taken apart from its evidential force, however, a sign must be understood as the very presence of the Spirit, since it is the working of divine grace in the heart of the believer. Not all signs make this point equally evident. As we attempt to discern and judge our state, the signs are viewed by us as pointing to or announcing the presence of the Spirit; considered in themselves, the signs *are* that presence. While all signs perform a common service in critical judgment, they differ on their material side. Thus a sign may be a cause or ground in one case, a quality of life or a relation in another, and even a series of deeds stretching over an extended time. Whatever its particular nature, however, a sign is of no value or interest save as it enables us to assess the nature of piety. Positive signs are the marks of the Spirit.

Since Edwards' main aim is to test affections, every sign refers to them in some specific way. Not every sign, however, fastens upon their inner nature. Some clearly do so, as in the tenth sign which singles out symmetry and proportion. Other signs point instead to the ground or stimulus of affections such as the evangelical humiliation at the heart of the sixth sign. Confusion can be avoided if we bear in mind that some signs point to affections themselves, others to their ground, and still others to what issues from them or to their consequences. Whatever aspect is made central, it is affections that are the object of the test; to test affections is to test religion.

It is a curious fact that Edwards nowhere considers the relations between the signs, whether they imply each other, whether some are more basic than others and similar questions. The signs have, to be sure, a natural affinity as belonging to the one integral religious life, but Edwards has not yet told us whether genuine piety must exhibit *all* the signs or whether perhaps one or more might be taken as a sufficient basis for judgment. The one clue we have for answering these questions is given in a principle which is common to every sign. The Spirit in its saving or gracious operation dwells in the believer in its proper form, or, in classical philosophical expression, according to its own kind. Edwards placed great emphasis upon the distinction between the Spirit as *operating on* the self and thus as still externally related to it, and as *dwelling in* the self in its own proper nature.[8] Only the latter form is the presence of the Spirit as *grace;* the former represents the "common work" of the Spirit. All signs as positive indications of gracious affections point back to the saving operation; if this indwelling fails to take place, no genuine signs can appear at all. It so happens that while the indwelling Spirit forms a common background of all signs, the fact is more explicitly stated in some (e.g., the first, fourth, and seventh signs), and we are perhaps justified in taking these as more basic than others.

•   •   •   •   •

The Twelfth Sign

Edwards devoted more space to this sign than to any other; we cannot but conclude that it loomed largest in his mind. Moreover, he makes two claims which at once set it apart from the others. First, "it is," says Edwards, "the chief of all the signs of grace"; and secondly, the public character of practice makes it a sign whereby others are granted some insight into the sincerity of the believer. Edwards is most careful not to say that other men should use the sign and judge their neighbors; he confines himself instead to saying that practice is the best evidence of a man's godliness in the eyes of others. While it is true that the state of the believer is known but to God and the individual is subject to the divine judgment alone, the daily conduct of a man is not beyond the scrutiny and provisional judgment of his neighbors. Indeed if the established practice of accepting members into the church on the basis of a "public charity" was to have any meaning at all, it is difficult to see how judgment by others could have been avoided. The heart may have its hidden side, but since practice furnishes a genuine clue to its nature, it cannot remain entirely secret.

The principle behind this sign is that holy affections must exert their influence in Christian practice; the deed is the most important "outward and visible sign of an inward and spiritual grace." Practice is a sure evidence of sincerity or, as Edwards says, "men's deeds are better and more faithful interpreters of their minds than their words." But if practice is taken as a paramount sign, it

[8] Below [in Yale edition], pp. 232 ff., where this communication of nature is said to be the "true" witness of the Spirit.

becomes necessary to see that it is not also taken as a ground of grace. Practice is a reliable sign of holy affections but it points beyond to the center of the self from which it issues and to the Spirit which makes it possible.

The conduct of Christians in the world is to be guided by three demands. First, behavior must be in conformity with Christian rules; secondly, the "practice of religion" must be the chief occupation of life; and thirdly, one must persist in this practice till the end of his earthly days. As a principal means of establishing these demands, Edwards makes exhaustive use of biblical material. From all parts of both Testaments he culls illustrations supporting his thesis that true affections issue in universal obedience as the chief and enduring concern of life.

Critics of Edwards' interpretation of the revival have sometimes claimed that his view was inconsistent on the grounds that it is impossible to reconcile having a new nature with backsliding or the failure of Christian practice. If Edwards' doctrine of affections is correct, the argument runs, the true saints cannot fail in their Christian duty. In meeting this objection he claimed that "true saints may be guilty of some kinds and degrees of backsliding," but that they cannot fall away from that earnestness toward God which stems from having the new sense of the heart. If a man's behavior shows that he no longer persists in making the love of God the chief business of his life, then we have a clear sign that he was never converted and his experiences are to no avail. On the other hand, since genuine piety consists in the new heart and the change in nature, true believers may fail in their moral duty and be convicted of sin through the law without falling away from God entirely.

We have before us a distinction between the heart or essential nature of a man and the fact of his day-to-day performance in the world. Genuine piety concerns the heart and not bare conformity to the law; practice, nevertheless, is supposed to provide some clue to the heart. The question is, how can practice be used as a test? The most obvious answer is that there are certain deeds to be done and others to be avoided, and that those who do not deviate from this standard are the true saints. Such an answer Edwards could not give. True religion consists not merely in conformity to rules but in the new heart; the use of practice as a sign, however, bids us try to understand how it can be a clue to the inner nature of the self from which it comes. This is the crux of the matter; taking conduct as a sign as a matter not merely of discovering whether it conforms to rules but of learning how and in what way it reveals the heart. Practice, then, cannot be viewed merely as a doing or abstaining, for we must discover in a man's conduct the true affections of his heart. Accordingly, Edwards looked to the attitude behind practice, to the love and gratitude displayed, and to the persistence of the believer in seeking to obey the commands of religion over a course of time. It was his belief that if a change of heart is genuine, its permanent character will continue to show itself through the whole of life.

Viewed in this light, the backsliding of the saints ceases to be the insuperable problem it first appeared to be. If genuine piety has to do not with a perfect conformity to the law but with the new heart at the basis of life, there will be no inconsistency in supposing that this new heart can exist side by side with moral failure and shortcomings.

The place of practice in the Christian life may be viewed from another perspective. The Spirit is said to dwell in the believer as a living power; the body, says Edwards, is a temple of life and not a tomb. As a principle of life, the Spirit shows itself in the true believer as a vital power; the form most appropriate to its nature is that of holy practice. What this means is that a man's conduct is something more than the moral consequence of the religious relationship; it means that practice takes on a religious dimension. It may take its place as the chief among the signs of gracious affections because it is the Holy Spirit revealing itself as life in the world.

The prominence given by Edwards to practice as a test of affections is of great moment not only for his own thought but for the driving force of American Protestantism as well. In setting up practice as a cardinal test, Edwards was no mere follower of tradition. Classical Protestantism had placed considerable emphasis upon the inner workings of the Spirit and upon the primacy of faith. Puritanism went even further in the direction of making religion into an affair of the interior life. While Edwards' doctrine of affections carried this trend forward, it also took a large step in the direction of making action a center of attention. American Protestantism has never been far from believing that the most reliable test of religious sincerity is the deed; seeing what a man will do is the best test of his heart.

In Edwards' time this was a bolder step than might be imagined. He was subordinating the traditional "immanent grace" to the power of the Spirit as expressed in overt behavior, and he did so without becoming involved in a doctrine of works. There is, he says, no justification in works: "I proceed to show that Christian practice . . . is . . . much to be preferred to the method of the first convictions, enlightenings and comforts in conversion, or any immanent discoveries or exercises of grace whatsoever, that begin and end in contemplation." We need not follow Edwards through his tedious arguments in defense of the position; the conclusion to which they point is vastly more important. He was taking a long look at Protestantism's sacred domain—the inner life—and demanding that it be subjected to a public test. The ground of action, the new nature, is not displaced; in the order of essential things the sense of the heart is always prior. But from the standpoint of human knowing and testing the proof of the heart is to be found in the fruit it can produce in the world. The Holy Spirit not only dwells in the depths of the soul but is manifest in that power through which the face of nature is transformed.

The bearing of this doctrine on the relation between religion and technology in American life has still not been fully realized. American Protestantism has had no place for quietism; its robust sense of activity in the world can be traced to the strain of Puritan piety and not least to the interpretation of that piety by Jonathan Edwards. It is no small irony that a skillful and vigorous defense of the primacy of practice in religion should have found expression in a treatise on religious *affections*.

RELIGION, REVIVALISM, AND RELIGIOUS AFFECTIONS

The details of Edwards' argument in the *Affections,* while important for revealing the subtlety of his mind, may prove more distracting than enlightening unless we can come to an understanding of the central contributions made by the work. We are bound, moreover, to ask for its relevance to our present situation. A work which attacks fundamental questions has perennial importance; it is the task of each age in confronting such a work to discover where that importance lies.

All about us at present are signs of renewed interest in the things of religion, and the past decade* has witnessed not only a vigorous revival in theology but an even more vigorous upsurge of revivalism at the level of personal religion. But there are dissenting voices as well, those who view an increased emphasis upon religion with alarm, those who are convinced that morality is enough and that the idea of salvation is outmoded, those who believe that natural science alone gives reliable knowledge while all else is emotion and sentimental bias. Against such voices that of the revivalist in religion is raised even louder; the situation is urgent, secular society is godless, conversion is the answer to our ills, it is too late for discussion, the only course is to decide and have faith. We cannot, as we confront such a situation, avoid asking what light is shed upon our predicament by the thought of Jonathan Edwards. It would be strange indeed if one who labored so conscientiously in the revivalist vineyard and who thought so acutely about its problems should have nothing pertinent to say to us in the present hour.

What, then, can we find in Edwards' interpretation of heart religion that will provide us with a vantage point from which to understand and evaluate the current scene? There are three basic contributions made by the *Affections,* and each has a direct application to present religious thought and practice. Edwards recovered, through his doctrine of the new sense and the new nature, the distinctively religious dimension of life; he pointed the way to a form of understanding broad enough to retain its relation to the direct experience of the individual; he showed how piety, though ultimately rooted in the individual's relation to God, could be subject to rational scrutiny in the form of tests aimed at revealing its genuine or spurious character.

The first shows us the way to prevent the reduction of religion to

* That is, the 1950's [ed.].

something other than itself; the second makes it clear that understanding can be preserved within religion if it is not taken as a
purely theoretical power which ignores the experience of the
unique individual; the third gives us a basis for interpreting and
evaluating the present concern over revivalistic religion. And in
seeking to trace out the further implications of each of these points
we do well to remember that if Edwards can aid us, our present
religious situation enables us to understand him as perhaps never
before. For we are falling or have fallen into some of the very
pitfalls he sought to avoid; we are in a better position than any age
since Edwards to understand the profundity of his contribution to
theological thinking.

The properly religious aspect of man's life is always in danger of
being obscured because of our tendency to identify it with something other than itself. Western society since the time of the Enlightenment has consistently manifested this misunderstanding.
Religion has been taken as morality with emotional overtones or,
even worse, has been made one with social and political idealisms,
and thought of merely as an instrument of social change. Whatever
form the corruption has assumed, loss of genuine religion has been
the result. American Christianity has been especially vulnerable;
the practical or pragmatic bent of much American life has led time
and again to the reduction of religion to morality. The sincerity of
the heart has been subordinated to action and, in its lowest terms,
to appearing proper in the eyes of the world.

Edwards made no such mistake, and he can aid us in our attempts to overcome it. One of the clear declarations of the *Affections* is that religion has to do with the inner nature of a man, with
the treasure on which his heart is set and with the love which
supplies his life with purpose. There can be no identifying religion
with morality or anything else; it is the deepest and most fundamental level of life and it goes to the *heart* of the matter. Religion,
to be sure, issues in deeds, but it is not the same as right conduct.
As Whitehead has pointed out in his arresting comparison between
religion and arithmetic, "You use arithmetic, but you *are* religious." Edwards would have agreed. He was reviving the time-
honored tradition of Augustine in finding religion in the whole

man, in the fundamental inclination of his heart and in his love of the divine *gloria*. In recovering the religious dimension of life and in expressing it through the vivid idea of affections, Edwards is a guide. He has given us a means of exposing the pseudo religions of moralism, sentimentality, and social conformity. For if religion concerns the essential nature of a man and the bent of his will, it cannot be made to coincide with moral rules, with fine sentiment, or with social respectability.

If our age has been plagued by its failure to comprehend what religion means, it has suffered no less from the acceptance of a false conception of the nature of human understanding. The dazzling successes of science, and their interpretation along wholly technological lines by pragmatically minded philosophers, have led to the conception of human understanding as a purely impersonal instrument aimed at finding objective and universal truths. We have acquiesced in that view of reason which sees in it only the abstractive intelligence fitted for expressing what is so true in general that it can have no bearing upon the life of the unique individual in particular. The rational self consequently has come to be understood as a spectator or as one who must put out the light of his own personal experience in order to gain that objective knowledge which alone is deemed worthy of concern. For religion the reduction of reason to science has meant that rationality is denied to religious truth and there is left to it nothing but the domain of emotion and caprice. Few have seen that this consequence is due to our view of what human understanding is and means; as understanding became narrower in scope it inevitably lost its capacity for dealing with dimensions other than science. Exclusive emphasis, moreover, upon understanding as dealing with the universal features of things has led to the disregard of individual experience; what can only be understood by personal experience has been subordinated to the knowledge that is gained by forgetting about yourself and attending to the universal law. What escapes the scientific net, we are told, is just on that account not a fish and need be considered no further as a possible object of rational understanding.

It is no secret that the philosophy known as existentialism in our time is dedicated to challenging the sufficiency of this scientific outlook for the whole of life. The attempt is being made, often in bizarre and unconventional ways, to bring the individual back to a sense of his own individuality and to the need for a broader conception of human understanding, one that does not eliminate everything but science from its concern.

With the *Affections* before us we are now in a position to see that Edwards, though by no stretch of imagination an existentialist, was wrestling with the same problem. He saw that an understanding which excludes first-person experience is doomed to be lost in abstraction and to forfeit its relevance for religion. To deal with the problem he reinterpreted human understanding so as to include a sensible element within it. He grasped the truth that sensible experience is always first-person experience; when we think in general concepts we pass beyond our own senses to a meaning common to many selves. But in using our own senses we are aware of grasping something that each individual must grasp for himself. If a man has never tasted honey, he cannot possibly know what is meant by calling it sweet and, similarly with Edwards' new sense, if a man has not actually tasted of the divine love no combination of general concepts will be adequate for conveying to him what it means.

Following the lead of the classical British experience-philosophy, Edwards placed primary emphasis upon first-person experience; in religion it took the form of the new sense or taste without which faith remains at the merely notional level. A spiritual understanding is not confined to the apprehension of universal concepts but includes within itself a sense which a man must experience directly. If such experience is lacking, there is no way in which he can be made to understand the things of religion through general concepts alone. If, as Hegel said, the great principle of empiricism is that a man must see for himself and be in the presence of the thing he knows, the great principle of Edwards' spiritual understanding in religion is that a man must sense or taste for himself the divine love in order to understand what it means.

In the doctrine of the spiritual understanding Edwards was carrying on the tradition of Augustine and the Puritans, the tradition of the "light" in which, and through which, the things of God are to be grasped. But he qualified the purely rationalistic description of this light by bringing in the "sensible" factor. He had help in framing this conception not only from the English philosopher John Locke but from the English Puritans and the Cambridge Platonist John Smith. The latter was dissatisfied with the tendency of English Puritanism to overemphasize the purely intelligible aspect of the understanding; to correct the disbalance he developed the notion of a "spiritual sensation" or grasp of religious truth which is understanding and at the same time engagement of the individual heart. Edwards' "new sense" points in the same direction. It is the taste of the divine excellence which marks the difference between genuine piety and the merely conceptual grasp of religious doctrines that may lead a man to accept them as general truths without seeing and feeling their special bearing upon his own individual life and situation.

The doctrine of spiritual understanding shows a way in which both rationality and direct experience can be preserved within religion. It shows that understanding need not be a dry light which excludes individual experience. If, moreover, we follow Edwards in his view that understanding is a power of the whole man, we shall see that the heart and the will—the *inclination* of the self—are necessarily involved. A conception of understanding as a purely theoretical or observing power permits only what Edwards called a "speculative" approach: a man considers the problem abstractly but is not engaged. The theoretical attitude is inappropriate in religion because it leaves out the one thing which counts: the individual man and his destiny.

The contemporary relevance of these ideas is clear. If reason and understanding extend no further than the highly abstract knowledge to be found in the natural sciences, it becomes difficult to see how they can possibly have a legitimate function in religion. The original protest made by Kierkegaard against rationalism was directed against a reason which sets the individual self at a theoretical distance from everything it seeks to understand. The essence of the theoretical attitude is to ignore the peculiar bearing of what is known or contemplated upon the life and destiny of the one who knows.

Jonathan Edwards was dealing in his own way with the same problem. A merely notional understanding of Christianity is theoretical; it is inadequate because it leaves the individual soul outside as a spectator looking on at the feast. What is needed instead is engagement of the self and the inclination of the heart and will. But if we are to have such engagement and preserve understanding in religion at the same time, we shall need a conception of that power which neither shrinks it to the proportions of science nor identifies it with the theoretical grasp of general concepts. Understanding will have to be seen, as indeed Edwards viewed it, as a power of the integral self, a power related to the will and the heart. His skillful linking of the understanding to these other aspects of the self through the idea of the new sense opens up novel possibilities for attaining that broader doctrine of understanding so necessary at present.

Edwards' third contribution to the resolution of our modern predicament centers in his contention that affections can and must be subject to critical judgment. While he would not allow that there can be genuine piety which does not express itself in affections, he was equally insistent that they be put to the test. Affections do not test themselves and there can be, so to speak, no affectionate testing of affections. Critical examination calls for signs or marks that enable us to tell the true from the false. These criteria are provided by the New Testament picture of the Christian life as interpreted in a rational way.

A position seeking to recover immediate experience in religion and subject it at the same time to clearly announced tests is exactly what our present circumstance requires. It is a subtle position, one likely to be misunderstood, because it aims to combine what a one-sided way of thinking always sunders. But we must not be misled by the confusions of our forbears. Edwards' contemporaries failed to grasp the underlying consistency of the *Affections*. Some were confused by his double-barreled approach, while others were driven by frustration to anger. Edwards has declared himself for heart religion and against a narrow rationalism; to many this meant full support for revivalism even if it reached the bounds of enthusiasm and immediate revelation. But Edwards rejected enthusiasm and branded as false much of the popular piety resulting from the high tide of revivalist preaching. The heart, he contended, must be affected, for genuine religion is power and more than the verbal acceptance of doctrines. But the change of heart is not in the convulsion or the shout, the flowing tears or the inner voices. These external signs, the sensational marks of revivalism in every age, are no guarantee of genuine faith; there are other tests to be met.

The double-edged character of Edwards' doctrine gives it a peculiar relevance for the current situation because it holds out the possibility of bringing together in a fruitful way what many are concerned to separate. The present renewal of interest in religion often takes on a revivalistic form and thus stands in striking contrast to the underlying assumptions and conventions of our highly rationalistic, secular society. The opposition further widens the gap between faith and the rational disciplines. Rationalists, in order to eliminate, or at least minimize, the threat of revivalistic religion, tend to overestimate the power of reason and science while at the same time exaggerating the irrationality of all religion. On the other side many proponents of the religious revival preach the irrelevance of knowledge and the cultural disciplines, contrasting the critical attitude with the urgency of commitment and the theoretical approach with the immediate experience of the individual. The power of Edwards' doctrine of affections resides in its uncompromising demand that both sides be preserved and related in a fruitful way. The heart and the individual's direct perception are essential to genuine religion, but without critical and rational tests of the heart we cannot know when religion is genuine.

The motivation behind the revivalist in every period is a vivid sense of the personal character of religious faith. He sees that conventionalism is the death of religion and that each individual must confront the issues and make the decision for himself; there is no faith by proxy. He further sees that this truth itself may fall to the level of a general pronouncement and thus may fail to have the very impact it is meant to have in each individual case. His chief strategy for meeting the situation is to dramatize its urgency and seriousness. There is, he believes, no time to stand off and survey the situation, no room for discussion and criticism. The object of faith stands before a man as a brute fact; if its importance is impressed upon him in a vivid and moving way, he will accept. For the majority of revivalists religion is as simple as that.

Edwards saw the element of truth in this position but he was too acute to fall prey to the errors that go with it. He saw well enough that where the individual is not touched at the heart there is no genuine religion. He would have joined hands with William James in believing that there can be no religion at second hand. Edwards' sense of the heart, his defense of the necessity of affections, and his doctrine of spiritual understanding underline the fact. But despite this avowal of heart religion he failed to fall in with revivalism at two crucial points; he denied that the urgency of believing provides any criterion for the truth of religion or the sincerity of the believer, and he was unwilling to follow the pattern of most revivalists and set the religious spirit over against learning and intellect. For Edwards the Word must come in truth as well as in power.

He was second to none in his sense of the seriousness and urgency with which the individual confronts the religious issues. He had repeatedly drummed the ears and laid it upon the consciences of his hearers that each man is judged according as he makes the "business of religion" central or peripheral to his life. But he did not suppose that the urgency by itself tells us what is to be believed, whether it is true, and to what extent the believer is sincere in his profession. Insight into these problems requires more than vivid preaching; the individual must have discipline, he must understand as well as hear, and behind it all must be the sense or "taste" of the beauty of holiness. No one of these essentials is furnished by the revivalist preacher armed with nothing more than the sense of urgency.

Edwards never tired of repeating the thesis that genuine affections are not heat without light. As even Chauncy acknowledged in Edwards' own time, this meant a refusal to accept the primitivism that has often attached itself to revivalist religion. He never was willing to represent himself as a simple soul able to understand the things of the spirit just because of a lack of "book learning." The great temptation of the revivalist is to come forward in this guise, to identify innocence with ignorance and to claim that only the soul uncorrupted by the studies and doctrines of the theologians is in a position to receive the gifts of the spirit. Edwards, on the contrary, was a scholar of the first rank, and he repeatedly criticized those who had neither the time nor the interest to give to the task of understanding their faith intelligently. He constantly urged his parishioners to study to the limit of their capacity, to avoid those who despised intellect in the belief that it is a power hostile to religion. The Bible is still the medium through which the ancient faith is preserved, and its treasures cannot be unlocked if learning is given up and interpretation becomes a matter of personal whim and individual vision. The recurrent error of the naïve revivalist is to overlook the labor and the pains of learning; his conviction of the truth of his own immediate insight leads him to present the Bible as a book so clear in its import that he who runs may not only read but understand. Edwards never wavered in his repudiation of these ideas.

If, from the vantage point of the *Affections,* we view recent attempts to harvest souls in revivalist fashion, we shall lay hold of the enduring contribution of that treatise. In seeking for the signs that distinguish gracious affections from fleeting emotion and from the effects wrought by the rhetoric of an hour, Edwards was trying to penetrate beneath the surface of things. A genuine change must take place in the heart, and that change must show itself over a lifetime of work and worship in the world. As we contemplate the renewal of interest in religion, we must not fail to apply these criteria. What permanent change is taking place in the depths of the self and with what consistency will it show itself in practice? More likely than not the vast majority of cases will be unable to pass the test. And one of the principal reasons for the failure is to be found in our by now well-established tendency to view everything as a technique used by the human will to conquer nature and master history. Edwards had seen this source of corruptions, and he had attacked it through the doctrine of divine love as disinterested. Religion is genuine and has power only when rooted in a love which does not contemplate its own advantage. Religion becomes false at just the point when we attempt to make it into a device for solving problems. Faith does indeed move mountains, but only when it is informed by a love pure and unmixed, such as Edwards described in his fourth sign. Love of this kind overcomes the evil desires of the heart and proves itself in the Christian life, but if it is held forth as a panacea for all ills or as something to be *used* by an individual or a society to achieve benefits, its divine character is lost.

Edwards' calm word in the midst of "much noise about religion" is that religion must not be lifeless and it must be something more than doctrine or good conduct. True piety shows itself in the affections and in the fruits of the Spirit, but these must be put to the test so that we may know the gold coin from the counterfeit. To the revivalist in our time Edwards has a sobering word, one that is best expressed by biblical paraphrase: "Test the affections and see if they are of God, for many false affections have gone out into the earth."

✪

An Urgent *Now!* for the Languid Will

The alert reader will have become increasingly aware that the course of the portraiture to this point has made at least one rather obvious circumvention. Perhaps there is the suspicion that something is being intentionally omitted, and like the official portraits of Lenin after the revolution the "message" has overwhelmed and obscured the "facts" of the original event. What every living man could have remembered about Lenin and the revolution was that it was a season of uncommon anguish in which the innocent perished in number as generous as the confusion of the age was great. Where does that appear on the epic canvas? Or like the artist whose executed work is designed merely to flatter its subject, we might be accused of having left out the withered leg. Not just once, but with surprising frequency, Jonathan Edwards delivered the most searing maledictions in the memory of the American church.

What must be said at once is that there is no way of avoiding this fact. We cannot say, for example, that Edwards was momentarily taken up into the reckless spirit of the Awakening, or that he did not "really believe" what he was saying. The eternal torment of the damned was a subject to which he repeatedly and tirelessly turned in his preaching, and even in his miscellaneous notes. There is no item in this vivid explosion of metaphoric carnage on which his imagination could not feed at length. Indeed, we might well

Reprinted with the permission of Charles Scribner's Sons from *Jonathan Edwards and the Visibility of God*, pp. 150–162, by James Carse. Copyright © 1967 James P. Carse.

argue that his creative powers are nowhere more in evidence than in these astonishing pages. Our task in interpreting this material is the same as it has been elsewhere: we must not merely record our reactions to it, but attempt to understand what it *meant* for Edwards to believe that there was a hell and that its purpose was to punish sinners.

In agreement with virtually every other major thinker in the Western world before him, Edwards uncritically accepted the theory of immortality. Without argument or defense he could plainly say that "intelligent beings of the world are everlasting & will remain after the world comes to an end."[1] But what gives specific shape to his thinking about the after-world comes from another more or less implicit theme, also widely shared, that the inner logic of the ethical life could be comprehended in the term "desert." A man gets what he deserves. The patient seeker after truth, the doer of righteousness, the faithful servant, each shall be rewarded for his uprightness and earnestness. But, since the world in which we live is a world largely without Edwards' "beauty" and "excellency" no such perfect recompense in the mortal life can be expected. For in this life the innocent will suffer and the wicked prosper. Therefore, final rewards and debts will be paid and collected in the world to come. Immanuel Kant was so persuaded of the idea of just deserts that he thought the ethical life made belief in immortality necessary.[2]

Consistent with these two themes there is an over-all design to the imprecatory sermons that the casual reader, interested more in amusement than in understanding, will easily overlook. First, we should observe the way in which the torments of hell have been described. With what must have been great effectiveness Edwards reaches into the common experience of his listeners to convey their imagination by means of vivid sense impression into the eternal anguish. Appealing to the universal dread of being locked into a closed place without any hope of escape, for instance, he says that hell "is a strong prison: it is beyond any finite power, or the united

---

[1] *Miscellanies,* Yale Mss., No. 547.
[2] Cf. *Critique of Pure Reason,* trans. Lewis White Beck (Indianapolis, 1956), pp. 126 ff.

strength of all wicked men and devils to unlock, or break open the door of that prison. Christ hath the key of hell; 'he shuts and no man opens.' " [3] There were few experiences in the frontier town that were more terrifying than housefires: "Some of you have seen buildings on fire; imagine therefore with yourselves, what a poor hand you would make at fighting with the flames, if you were in the midst of so great and fierce a fire." [4] In the same sermon he introduces the famous image of the spider in the flames, combining it with the common miniature drama of the moth or mayfly led by its own fascination into the candle flame.

> You have often seen a spider, or some other noisome insect, when thrown into the midst of a fierce fire, and have observed how immediately it yields to the force of the flames. There is no long struggle, no fighting against the fire, no strength exerted to oppose the heat, or to fly from it; but it immediately stretches forth itself and yields; and the fire takes possession of it, and at once it becomes full of fire, and is burned into a bright coal. [5]

Once he has the wicked in the midst of the fierce heat, defenseless against it, he will not let them find any place of refuge, "any secret corner, which will be cooler than the rest, where they may have a little respite, a small abatement of the extremity of their torment. They never will be able to find any cooling stream or fountain, in any part of that world of torment; no nor so much as a drop of water to cool their tongues." Edwards wants his auditors to see that once the wrath of God has been unleashed on a sinner, he can avoid it no more easily than a worm can lift the heavy rock thrown down upon it. [6]

In perhaps the most severe of all his sermons Edwards reaches much closer to the heart. He concludes from his studies of Scripture that when the last judgment has been made and when the whole of mankind has been divided into the saved and the

---

[3] "The Future Punishment of the Wicked Unavoidable and Intolerable," *The Works of President Edwards,* a reprint of the Worcester edition, 4 vols. (New York: Jonathan Leavitt and John F. Trow, 1843), IV, 259.

[4] *Ibid.,* IV, 263 f.

[5] *Ibid.,* IV, 264.

[6] *Ibid.,* IV, 259.

damned, "the two worlds of happiness and misery will be in view
of each other." "The saints in glory will see how the damned are
tormented: they will see God's threatenings fulfilled, and his wrath
executed upon them."[7] This gives him the chance to say that since
this division will be by God's justice, and since every saint loves
God's justice, there "will be none to pity you."

Look which way you will, before or behind, on the right hand or left,
look up to heaven, or look about you in hell, and you will see none to
condole your case, or to exercise any pity towards you, in your dreadful condition. You must bear these flames, you must bear that torment
and amazement, day and night, forever, and never have the comfort of
considering, that there is so much as one that pities your case; there
never will one tear be dropped for you.[8]

The emotional climax of this sermon is a passage that bids fair to
be cited as the moment of greatest power—or perhaps savagery—
in the entire genre. He asks the unregenerate within his hearing to
think of their friends, and especially their parents, when the final
judgment is uttered.

How will you bear to see your parents, who in this life had so dear an
affection for you, now without any love to you, approving the sentence
of condemnation, when Christ shall with indignation bid you depart,
wretched, cursed creatures, into eternal burnings? How will you bear
to see and hear them praising the Judge, for his justice exercised in
pronouncing this sentence, and hearing it with holy joy in their countenances, and shouting forth the praises and hallelujahs of God and
Christ on that account? When they shall see what manifestations of
amazement there will be in you, at the hearing of this dreadful sentence, and that every syllable of it pierces you like a thunderbolt, and
sinks you into the lowest depths of horror and despair; when they shall
behold you with a frightened, amazed countenance, trembling and
astonished, and shall hear you groan and gnash your teeth; these things
will not move them at all to pity you, but you will see them with a
holy joyfulness in their countenances, and with songs in their mouths.
When they shall see you turned away and beginning to enter into the
great furnace, and shall see how you shrink at it, and hear you

[7] "The End of the Wicked Contemplated by the Righteous: or the Torments of the Wicked in Hell, No Occasion of Grief to the Saints in Heaven,"
Ibid., IV, 289.

[8] *Ibid.*, IV, 294 f.

shriek and cry out; yet they will not be at all grieved for you, but at the same time you will hear from them renewed praises and hallelujahs for the true and righteous judgments of God, in so dealing with you.[9]

If these are the words of man become captive to an extreme religious position, they are also the words of an artist, for Edwards, with a skill as sure as that of any story teller, has collapsed the distance between his subject matter and his listeners by describing a strange and terrifying world in terms of a present and familiar world. But there is a decisive difference between what Edwards is doing here and what a teller of stories seeks to accomplish. A story is designed to arrest the fancy, freeing it from the immediate world, in order that momentarily the listener will be suspended from the presentness of his experience. Edwards' appeal is not to the fancy, but to the will. He is not giving us a world that will exist if only we can take our attention from the immediately given; he is giving us a world in which we will exist if we fail to heed the true nature of the immediately given. Edwards is giving us a world which our present wills surely will create if they persist in the spiritual languor. Look where you are now, he is saying. Not then, but NOW!

If any of the saints will need assistance in determining how their will is faring, Mr. Edwards can provide that, too, in these sermons. "Look over your past life," he calls out to them, "inquire at the mouth of conscience, and hear what that has to testify concerning it."[10] Then like a surgeon in search of the body's disease, he pokes into the private reflections of his listeners, attempting to startle them into an awareness of their peccability.

How many sorts of wickedness have you been guilty of!

How manifold have been the abominations of your life! What profaneness and contempt of God has been exercised by you!

And how have you behaved yourself in the time of family prayer! What wicked carriage have some of you been guilty of toward your parents! How far have you been from paying that honor to them that God has required!

[9] *Ibid.,* IV, 296.
[10] "The Justice of God in the Damnation of Sinners," *Ibid.,* IV, 232.

How have some of you vaunted yourselves in your apparel! Others in their riches! Others in their knowledge and abilities! How has it galled you to see others above you!

And what abominable lasciviousness have some of you been guilty of! How have you indulged yourself from day to day, and from night to night, in all manner of unclean imaginations! Has not your soul been filled with them, till it has become a hold of foul spirits, and a cage of every unclean and hateful bird?[11]

After this catalogue of personal failings comes the concluding, summarizing question: "Now, can you think when you have thus behaved yourself, that God is obliged to show you mercy? Are you not, after all this, ashamed to talk of its being hard with God to cast you off?"[12]

The most famous of all Edwards' sermons, "Sinners in the Hands of an Angry God," is not properly a description of hell as such. It is concerned rather with the fact that the time between the present and one's death is a totally unknown quantity. Death comes suddenly and unannounced.

The unseen, unthought of ways and means of persons' going suddenly out of the world are innumerable and inconceivable. Unconverted men walk over the pit of hell on a rotten covering, and there are innumerable places in this covering so weak that they will not bear their weight, and these places are not seen. The arrows of death fly unseen at noonday; the sharpest sight cannot discern them.[13]

The imagery of the sermon is designed to communicate the sense of a disaster close at hand.

There are the black clouds of God's wrath now hanging directly over your heads, full of the dreadful storm, and big with thunder; and were it not for the restraining hand of God, it would immediately burst forth upon you. The sovereign pleasure of God, for the present, stays his rough wind: otherwise it would come with fury, and your destruction would come like a whirlwind, and you would be like the chaff of the summer threshing floor.[14]

[11] *Ibid.,* IV, 233 f.
[12] *Ibid.,* IV, 235.
[13] *Ibid.,* IV, 315.
[14] *Ibid.,* IV, 317.

In the conclusion to the sermon the imagery falls away and the minister confronts his congregation with direct prose plainly describing their situation.

There is reason to think that there are many in this congregation now hearing this discourse, that will actually be the subjects of this very misery to all eternity. We know not who they are, or in what seats they sit, or what thoughts they now have. It may be they are now at ease, and hear all these things without much disturbance, and are now flattering themselves that they are not the persons; promising themselves that they shall escape. . . . But alas! . . . how many is it likely will remember this discourse in hell! And it would be a wonder, if some that are now present would not be in hell in a very short time, before this year is out. And it would be no wonder if some persons, that now sit here in some seats of this meeting-house in health, and quiet and secure, should be there before to-morrow morning.[15]

The purpose of such preaching is certainly clear enough. It can be summarized in one straightforward sentence of Edwards': "The only opportunity of escaping is in this world; this is the only state of trial wherein we have any offers of mercy, or there is any place for repentance."[16] To draw a heavy line of emphasis under this fact, he wants to make it understood by all that with death each person has lost the last chance to change the final balance of his life. After death there are no more chances, there is no hope whatsoever, for the torments of hell are eternal.

How dismal will it be, when you are under these racking torments, to know assuredly that you never, never shall be delivered from them; to have no hope: when you shall wish that you might but be turned into nothing, but shall have no hope of it; when you shall wish that you might be turned into a toad or a serpent, but shall have no hope of it; when you would rejoice, if you might but have any relief, after you shall have endured these torments millions of ages, but shall have no hope of it; when after you shall have worn out the age of the sun, moon, and stars, in your dolorous groans and lamentations, without any rest day or night, or one minute's ease, yet you shall have no hope of ever being delivered.[17]

[15] *Ibid.*, IV, 321.
[16] "The Eternity of Hell Torments," *Ibid.*, IV, 275 f.
[17] *Ibid.*, IV, 278.

The undeniable contempt the preacher is pouring out in these words serves easily to obscure another rather remarkable feature of Edwards' imprecatory sermons. While he is convinced of man's mountainous sinfulness, he is also saying that a man's entire life, both past and future, lies under the power of his present will. All of a man's debts for things past, and all of a man's responsibilities for things to come, are fully within the power of his present will to discharge. What lies behind these sermons, therefore, is Edwards' profound respect for both the importance and power of the human will. This point is made all the clearer when we see how he deals with the opposite side of this same subject matter; that is, when he addresses himself to the consequences that attend the life of a man who had seized on Christ as his most apparent good. Such a man, in Edwards' judgment, was David Brainerd.

In 1747 young Brainerd, missionary to the Indians and fiancé of Edwards' daughter, Jerusha, died of tuberculosis in the Edwards home. A few months later Jerusha, carrying Brainerd's love and his disease, followed him to the grave. Brainerd had asked Edwards on his deathbed to edit his private journals. *The Life and Diary of David Brainerd* fell into the long tradition of Puritan pieces in which the diarist painfully examines himself against standards for his performance that could not possibly be met. It reads therefore like a mournful self-condemnation, but in the idiom of the time it could be seen as the modestly composed chronicle of a spiritual hero willing to undergo any danger and hardship in the service of his Lord. Brainerd's funeral sermon was preached by Edwards, and is memorable if only because the preacher converts his transparent grief into a long discourse on the experience of the saints after death. Like the damned in hell, the saints' blessedness in heaven is eternal. And as the damned are thrown into the keeping of Satan, the saints are led into the presence of Christ.

The most intimate intercourse becomes that relation that the saints stand in to Jesus Christ; and especially becomes that most perfect and glorious union they shall be brought into with him in heaven. They are not merely Christ's servants, but his friends.[18]

[18] "Saints When Absent From the Body," *Ibid.,* III, 629.

The experience of the saint in life is, as we have seen, one of great affection. When Christ becomes one's most apparent good, he responds with love for all things. In heaven the appearance of Christ will be more direct, and the response to it all the more vivid, for there the elect

see every thing in Christ that tends to kindle and inflame love, and every thing that tends to gratify love, and every thing that tends to satisfy them: and that in the most clear and glorious manner, without any darkness or delusion, without any impediment or interruption.[19]

Edwards described the life of the saint on earth as one of vital union with Christ. After death this union becomes much more complete. The saints not only live by his life, but they also share in his power and glory; they are

exalted to reign with him. They are through him made kings and priests, and reign with him, and in him, over the same kingdom. As the Father hath appointed unto him a kingdom, so he has appointed the Son to reign over his kingdom, and the Son appoints his saints to reign in his.[20]

Here again we can focus on the mythic absurdity of these thoughts, or we can attempt to get behind them and to understand why Edwards should have voiced them in this way and on this occasion. If it is the case that the sermons on hell are designed to awaken the people into a recognition of the power of the will, then so is this sermon and the others like it. If they can choose hell, they can also choose heaven. The vast differences between heaven and hell point to the vast differences between the kinds of lives people currently are living. The one is dark and self-enclosed, turned in upon itself, feeding on its own emotions and organizing all values around its own needs and tastes. The other is brilliantly life-affirming, it is open and free, seeing in the darkness not the estuary of its self-destruction but the possibility of new and surprising kinds of caring for the world. Jonathan Edwards thought that Brainerd had lived the latter kind of life, and this sermon is a way of saying it. "Saints When Absent From the Body" is not a

[19] *Ibid.*, III, 627 f.
[20] *Ibid.*, III, 631.

disembodied, ill-informed series of speculations about some other world; it is a profound HURRAH! for the life of David Brainerd.

Therefore, I shall make no attempt here to omit from this portrait what many have regarded as the least creditable part of Jonathan Edwards' intellectual production. On the contrary, the portrait would lose much of its power and meaning if the blacker machinations of this American intelligence were left out of it. For in these sermons are combined what Americans elsewhere are wont to celebrate in the worldliness of their civilization: a hard-minded appraisal of the nature of things as they are, and a sturdy confidence that by an act of the will all things are alterable.

Edwards looked out upon his world and what he saw there was not beautiful. We are momentarily deceived by the fact that what he thought was ugliness is for us a series of petty moral failings, because the ugliness of our own age stands in such contrast to his. He was worried about children being distracted from family prayer; but we have seen the family structure itself fall into decay. He was alarmed by those who vaunted themselves in their apparel and in their riches; but we live in an age when a man's wealth brings upon him an astonishing blindness to the poverty of his fellow Americans, and even supplies him with reasons for strengthening the bonds by which others are excluded from even the merest comforts of human existence. Edwards was concerned with the unclean imaginations of his people, but we are living in a time when the imaginations of national glory fill the world with devastation. If there was any need for the languid will to exercise itself in the direction of beauty in the century of Jonathan Edwards, that need is enormously magnified in the century of the atomic bomb and the urban slum.

Edwards was an artist for his people, he was the reporter and the critic who caused them to focus on the larger world. His sermons were designed to terrify. They were for his time what Picasso's "Guernica" is for ours. They are Eisenstein's films of war and revolution, they are the photographs of police dogs and sheriff's deputies in Alabama, or the television report of American soldiers setting fire to the straw huts of Asian peasants. The earth over which we walk is no less rotten than it was for those who were in the hearing of Jonathan Edwards that unforgettable day in Enfield, Connecticut. "The arrows of death," he reminded them in July of 1741, "fly unseen at noonday. The sharpest sight cannot discern them." What man among us, 222 years later, on that unforgettable day in the November of 1963, could have thought otherwise?

In these sermons Edwards was not wandering off the edge of a realistic mentality into a foolish other-worldliness: rather, he was bringing his hearers out of their irrelevant and fruitless Sunday musings, and showing on the seachart of their lives as a people precisely where they were and precisely whither they were bound. What Jonathan Edwards preached and wrote in all of his sermons was a radical this-worldliness. It is for this reason that the failure of Jonathan Edwards is a fact of no small significance in the American civilization. After Edwards every great American prophet would fail in the same way. The American journey is over. Let the dream of the ultimate society be spoken in public ceremonies, but never dispatch the will and the intelligence in the active attempt to achieve it. The ship is at anchor, the sails down. There will be many men who will with great usefulness labor at tightening her rigging, cleaning her decks, and keeping the logs of her once great adventures. But there is no captain at the wheel, and too few crewmen athirst for the open sea.

Jonathan Edwards

An American Tragedy

I

In 1739, Jonathan Edwards delivered a series of sermons on the work of redemption, a breathtaking survey of the "grand design of God" in the "form of a history."[1] It was an audacious undertaking, unprecedented in the rich theological literature produced by the Puritans in America. Beset on one side by the enthusiasm of the Great Awakening and on the other by the threat of formal religiosity and frigid rationalism, Edwards rose to an Olympian view of man's religious destiny, and spoke to his flock at Northampton about things of the last importance: Christ's activity in behalf of man from the invention of time to its abolition. He was invading treacherous territory, last traversed half a century before by the great Bossuet.[2]

[1] Jonathan Edwards to the Trustees of the College of New Jersey, October 19, 1757, *The Works of President Edwards,* ed. S. Austin, 4 vols. (New York, 1847 and several times thereafter), I, 48.

[2] Thomas Prince, it is true, published his *Chronological History of New England, In the Form of Annals,* in 1736, and it begins with Adam, "year one, first month, sixth day." But he ends with 1630, and eschatological speculation is wholly absent from it.

Reprinted by permission of author and publisher from Peter Gay, *A Loss of Mastery: Puritan Historians in Colonial America* (Berkeley and Los Angeles: University of California Press, 1966), pp. 88–117.

231

Edwards, like Bossuet, was a professional theologian and only an amateur historian. But, like Bossuet, Edwards saw no reason to apologize for his excursion; he was only doing his duty. After all, "The work of REDEMPTION is a work that GOD carries on from the fall of man to the end of the world";[3] it was work God performed in, and through history, and what was more urgent for man than to trace evidences of that divine work in time? Besides, and from this motive, devout historical study had long been one of Edwards' favorite pursuits. "My heart has been much on the advancement of Christ's kingdom in the world," he observed in his spiritual autobiography. "When I have read histories of past ages, the pleasantest thing in all my reading has been, to read of the kingdom of Christ being promoted." The very anticipation of coming upon such a passage was a source of rejoicing. "My mind," he said, "has been much entertained and delighted with the Scripture promises and prophecies, which relate to the future glorious advancement of Christ's kingdom upon earth."[4]

This pious, purposeful pleasure in history never left him. He importuned his European correspondents to send him the latest books on history and theology, and he continued to brood on his sermons on redemption: as late as October, 1757, eighteen years after these sermons, he told the trustees of New Jersey College that he was not sure he wanted to be president of their institution, partly because his health was uncertain, partly because his learning was sadly incomplete, but largely because he was thinking of writing a *History of the Work of Redemption*. This history had been on his "mind and heart" for many years; he had begun it "long ago, not with any view to publication." It was to be a "great work," designed on "an entire new method"—towering claims for a man who, though he knew his powers, was a modest man—a work that would consider "the affair of Christian Theology, as the whole of it, in each part, stands in reference to the great work of redemption by Jesus Christ." Since the work of redemption,

[3] *A History of the Work of Redemption* (first published in 1774), *Works,* I, 298.
[4] *Personal Narrative* (probably written in 1740, a year after the sermons on the work of redemption), *Works,* I, 21.

Edwards wrote, was the *"summum* and *ultimum* of all the divine operations and decrees; particularly considering all parts of the grand scheme, in their historical order," his book on redemption must be on a grand scale: it must look at "all three worlds, heaven, earth, and hell," and introduce "all parts of divinity in that order which is most scriptural and most natural."[5]

This program delighted Edwards: he was never afraid of grand architectonic designs. His plan appeared to him "the most beautiful and entertaining, wherein every divine doctrine will appear to the greatest advantage, in the brightest light, in the most striking manner, shewing the admirable contexture and harmony of the whole."[6] In the enforced solitude of his later years, Edwards wrote some ambitious books and spun out some ambitious plans, but among all his works, realized or contemplated, his history of the work of redemption was the most ambitious. He did not live to write it; in March, 1758, he died, shortly after receiving inoculation for smallpox, a victim of modern science.

We can only speculate how Edwards would have transformed his sermons on redemption into a book of history. This much is certain: he would not have discarded, or modified, their classical Puritan theology. The books, the journals, and letters of his late years, like those of his early years, betray no skepticism of miracles, no doubt of Scripture, no rebellion against God's sovereignty, no deviation from the Augustinian vision of history. In the midst of the greatest revolution in the European mind since Christianity had overwhelmed paganism, Edwards serenely reaffirmed the faith of his fathers.

He had some notion that such a revolution was going on: he even read David Hume and professed himself "glad of an opportunity to read such corrupt books, especially when written by men of considerable genius"; it gave him, he said, "an idea of the notions that prevail in our nation."[7] But he had no idea how

[5] Edwards to the Trustees of the College of New Jersey, *Works,* I, 48–49.
[6] *Ibid.,* I, 49.
[7] See Thomas H. Johnson, "Jonathan Edwards' Background of Reading," *Publications of the Colonial Society of Massachusetts,* XXVIII (1931), 210–211.

extensive that revolution was, and how far his own historical thinking deviated from the historical thinking about to seize control of
educated opinion in Europe. In fact, the nineteen years between
Edwards' sermons and Edwards' death were decisive years in the
rebellion of the Enlightenment against Christianity. Hume published the first two books of his *Treatise of Human Nature* in
1739; Condillac his *Essai sur l'origine des connaissances humaines*
in 1746; Montesquieu his *Esprit des lois* in 1748, and with these
books the foundations for the Enlightenment's epistemology, psychology, and sociology were firmly laid down. They were all attempts (in David Hume's words) "to introduce the experimental
Method of Reasoning into Moral Subjects";[8] attempts to found
the science of man on the ideas of Locke and the method of
Newton. They were scientific rather than metaphysical, critical
rather than credulous, naturalistic in temper, and wholly incompatible with revealed religion of any kind.

History, ready as always to follow the new currents, was beneficiary, and part, of the offensive of the secular against the Christian mind. In the late 1730's, while Edwards was displaying to his
congregation the activity of Christ in history, Voltaire was at work
on his *Siècle de Louis XIV,* a book which, with its anticlericalism,
its worldliness, and its aggressive modernity, became the mani-
festo, and the model, of the new history. A few years later,
Voltaire began his vast *Essai sur les mœurs,* the Enlightenment's
answer to Bossuet. Both of these books were published in Edwards' lifetime: the first in 1751, the second in 1756. Hume
turned to historical subjects in the late 1740's; he started work on
his *History of England* in 1752, and four years later published its
first installment, covering the Stuart dynasty from the accession of
James I to the expulsion of James II. William Robertson, the great
Scottish historian whose reputation then was higher than it is now,
began the first of his masterpieces, the *History of Scotland,* in
1753. Edward Gibbon, who combined the secular mentality of the
philosophes with the technical competence of the *érudits,* was still
a young man in those years, but he had already found his vocation,
perfected his classical learning, and discovered his religious posi-

8 Subtitle of Hume's *Treatise of Human Nature* (1739–1740).

tion; all he needed was a subject commensurate with his talents, and he found that, with the lucid finality of a religious conversion, in 1764, only five years after Edwards' death.

To turn from these books to Edwards' *History of the Work of Redemption* is to leave the familiar terrain of the modern world with its recognizable features and legible signposts for a fantastic landscape, alive with mysterious echoes from a distant past, and intelligible only—if it can be made intelligible at all—with the aid of outmoded, almost primitive maps. The philosophes' histories made secular propaganda by providing information about a real past; Edwards' history made religious propaganda by arousing memories of a religious myth. To grasp the temper of Voltaire's or Hume's histories, one must read the new philosophy and collections of state papers; to grasp the temper of Edwards' history, one must read the Church Fathers and the Scriptures. However magnificent in conception, however bold in execution, Edwards' *History of the Work of Redemption* is a thoroughly traditional book, and the tradition is the tradition of Augustine.

The very plan of the book places it in this tradition. Edwards periodizes world history by relying wholly on sacred numbers and sacred events. The first great period stretches from the Fall of Man to the Incarnation of Christ; the second, from His Incarnation to His Resurrection; the third, from His Resurrection to the end of the world. The first of these great periods, in turn, is subdivided into six "lesser periods": from the Fall to the Flood, the Flood to the calling of Abraham, Abraham to Moses, Moses to David, David to the Babylonian Captivity, and the Captivity to the Incarnation.[9] The second period, Christ's short sojourn on this earth, need not be subdivided: it is an intense, luminous, concentrated moment in the career of God's world: "Though it was but between thirty and forty years, yet more was done in it than had been done from the beginning of the world to that time."[10] The third and last period, finally, matches the first in perfect symmetry; it is marked by six steps in "Christ's coming in his kingdom":

[9] *History of the Work of Redemption, Works,* I, 306.
[10] *Ibid.,* I, 395. Edwards does subdivide this period, but only for purposes of analysis.

from the Resurrection to the destruction of Jerusalem; the destruction of Jerusalem to the advent of the Christian Emperor Constantine; Constantine to the reign of Roman Catholicism; the reign
of Antichrist to the Reformation; the Reformation to the present;
the present to the final overthrow of Antichrist.[11]

This periodic scheme is the appropriate, indeed the only possible scheme for Edwards: it mirrors, and perfectly expresses, his
theory of historical causation and historical purpose, and his
prediction of the end of time. Edwards insists that the course of
events follows a predetermined plan laid down by God before the
Fall, indeed before Creation. God settled everything at the beginning: man's nature, man's sin, Satan's interference, Christ's inter-
cession; and God settled it for the sole purpose of glorifying himself. In good Calvinist fashion, Edwards despised the doctrine of
foreknowledge as the refuge of timid Christians: of course, God
had perfect foreknowledge of man's future conduct, but if he had
dictated the course of history merely because he knew what was
going to happen, he could not be said to have dictated it, but
merely to have conformed his decree to necessity. This was a
limitation on the divine omnipotence to which Jonathan Edwards
could never assent: no theme is more consistent in his writings
than the lovely and unlimited glory of God, which looks to the
perfection of the creature for the sake of the perfection of the
Creator: because God "infinitely values his own glory, consisting
in the knowledge of himself, love to himself, and complacence and
joy in himself; he therefore valued the image, communication or
participation of these, in the creature. And it is because he values
himself, that he delights in the knowledge, and love, and joy of the
creature; as being himself the object of this knowledge, love and
complacence."[12]

The historical drama, therefore, was divine in all its aspects;
"this lower world," in which human history took place, "was

[11] This periodization of the third period is less obvious than the first; it is,
however, implicit. See C. C. Goen, "Jonathan Edwards: A New Departure
in Eschatology," *Church History*, XXVIII (1959), 26.
[12] "Concerning the End for Which God Created the World," *Works*, II,
256.

doubtless created to be a stage."[13] God was author of the drama, its director, chief actor, and, just to make sure, authoritative critic. Everything had happened precisely as Scripture described it, everything would happen precisely as Scripture prophesied it: the past fulfillment of prophecies was guarantee, if guarantees were needed, of future fulfillment of prophecies not yet realized. The course of history did more than merely offer evidence in support of the divine origins of the Bible: the Bible was incomparably the most accurate and most sublime—in fact the only accurate and sublime—history ever written.

For Edwards, secular history was on the whole insignificant, or significant only as it illustrated, illuminated, impinged upon sacred history: kings appear only as they establish, or obstruct, the true church, wars are mentioned only as they serve to spread, or to constrict, the true faith. All of parts one and two, and much of part three, of the *History of the Work of Redemption* consists of a free retelling of Scripture, with each miraculous event reported as a historical event. For Edwards, the authority of the Bible is absolute. "There were many great changes and revolutions in the world, and they were all only the turning of the wheels of Providence in order to this, to make way for the coming of Christ, and what he was to do in the world. They all pointed hither, and all issued here."[14] Characters in the Old Testament acted in behalf of purposes greater than themselves, and prefigured great events of which they knew nothing, but they were real people, real historical subjects. There was an Adam and an Eve, and they sinned and awoke to their sense of guilt; there was a Cain and an Abel, a Noah and a Moses, and while they were symbols and types, they were symbols and types in the way all creaturely beings represent both themselves and God's intentions. In the modern sense, in the sense of Voltaire and Hume, almost none of Edwards' history is history—it is Calvinist doctrine exemplified in a distinct succession of transcendent moments.

Yet history intrudes. In the time span between the establishment of the primitive church and the apocalyptic future, the Bible

[13] *History of the Work of Redemption, Works*, I, 300.
[14] *Ibid.*, I, 305.

provides no guidance; Edwards thus had to find other guides for the period between the first and the eighteenth centuries, and he found them in expected places. The style is the style of Jonathan Edwards, the story is the story told by Cotton Mather, by William Bradford, by John Foxe: in its first three centuries on earth, the true church was pure but suffered under persecution; then came years of prosperity and peace, as Constantine delivered the church from its travail—Satan, "the prince of darkness, that king and god of the Heathen world," was driven out.[15] But this time of rest did not last long: "Presently after, the church again suffered persecution from the Arians; and after that, Antichrist rose, and the church was driven away into the wilderness, and was kept down in obscurity, and contempt, and suffering for a long time."[16] While the true church was kept down, Satan's counterfeit church ruled the world: "The Pope and his clergy robbed the people of their ecclesiastical and civil liberties and privileges," and, just as Scripture had foretold, "robbed them of their estates, and drained all Christendom of their money, and engrossed the most of their riches into their own coffers, by their vast revenues, besides pay for pardons and indulgences, baptisms, and extreme unctions, deliverance out of purgatory, and a hundred other things." This renewed reign of Satan made "superstition and ignorance" prevail more than ever, for the Pope and his minions "industriously promoted ignorance"; in a line that David Hume might have written—and in fact did write, almost word for word—Edwards reminded his congregation that it was "a received maxim" among the Papists that "ignorance is the mother of devotion: and so great was the darkness of those times, that learning was almost extinct in the world."[17] Finally, after many centuries, Satan was driven away once again, by the "reformation of Luther and others."[18] Here was a splendid moment in the history of God's own church:

---

[15] *Ibid.*, I, 450.

[16] *Ibid.*, I, 438.

[17] *Ibid.*, I, 458. Hume quotes the maxim, *"Ignorance is the mother of Devotion"* in his *The Natural History of Religion, Philosophical Works,* 4 vols., ed. T. H. Green and T. H. Grose (1882), IV, 363.

[18] *History of the Work of Redemption, Works,* I, 438.

"God began gloriously to revive his church again, and advance the kingdom of his Son, after such a dismal night of darkness as had been before." There had been many endeavors by the witnesses to the truth, "for a reformation before," but it was only now, "when God's appointed time was come"—for in God's drama, all actors spoke their lines only on cue—that God's "work was begun, and went on with a swift and wonderful progress."[19]

Still, Edwards warned his flock, deliverance was not yet, and complacency was misplaced. The spirit of a true Christian remained what it had always been, a "spirit of suffering."[20] Satan, though wounded, was far from dead. He had risen higher and higher, and now felt himself falling once again, halfway to ruin, and with his last strength did his all to obstruct the great work of reformation. He strengthened the Papists through a great council, he fostered plots and conspiracies, he oppressed helpless minorities of true believers, he warred upon God's children. "The Heathen persecution had been very dreadful; but now persecution by the church of Rome was improved and studied, and cultivated as an art or science."[21] There was a time—a time that very old members of Edwards' congregation would remember—when Rome seemed near a decisive triumph: the king of England and "Lewis XIV. of France," both of them fanatical Papists, mounted a great conspiracy to extirpate what they called "the Northern heresy." But then, "just as their matters seemed to be come to a head, and their enterprise ripe for execution, God, in his providence, suddenly dashed all their schemes in pieces by the Revolution, at the coming in of King William and Queen Mary; by which all their designs were at an end; and the Protestant interest was more strongly established, by the crown of England's being established in the Protestant house of Hanover, and a Papist being, by the constitution of the nation, forever rendered incapable of wearing the crown of England. Thus they groped in darkness at noon-day as in the night, and their hands could not perform their enterprise, and their kingdom was full of darkness, and they gnawed their

[19] *Ibid.*, I, 462.
[20] *Ibid.*, I, 479.
[21] *Ibid.*, I, 465.

tongues in pain."[22] Like Cotton Mather before him, Jonathan Edwards distrusted the Anglican church, but he accepted as divinely ordained the aid of an Anglican state against the Papists.

Such victories, though impressive, were not final. A candid survey of the modern world showed Satan at work in many places. Protestants had been expelled from Bohemia and from France not long before; "Ireland has been as it were overwhelmed with Protestant blood," and there had been cruel persecutions elsewhere.[23] "Thus did the devil and his great minister Antichrist, rage with such violence and cruelty against the church of Christ! And thus did the whore of Babylon make herself drunk with the blood of the saints and martyrs of Jesus!"[24] But Satan was serpent as well as Moloch, a subtle deceiver as well as a ravening monster, and the progress of the true church had been much impeded by the spreading of corrupt opinions: Arminianism has "greatly prevailed" in the Church of England and among dissenters, and "spread greatly in New England, as well as Old."[25] Deism, which denied revelation altogether, had "very much overrun" the English nation on both sides of the Atlantic—"our nation."[26] Perhaps worst of all, indifference to religion was spreading. "The glorious outpouring of the Spirit of God that accompanied the first Reformation" had greatly diminished, and "vital piety" was now despised as *"enthusiasm, whimsy, and fanaticism."* Those who are truly religious, "are commonly looked upon to be crack-brained, and beside their right mind; and vice and profaneness dreadfully prevail, like a flood which threatens to bear down all before it."[27]

But, as Edwards knew, it was not written that the threat should become reality. The future, being the future, was hard to fathom, the history of the future hard to write. But Edwards ventured to undertake it; armed with the book of Revelation and with his interpretation of some events in his own day, he projected his history forward to give hope to the saints in his audience, and to

[22] *Ibid.*, I, 464.
[23] *Ibid.*, I, 466.
[24] *Loc. cit.*
[25] *Ibid.*, I, 467.
[26] *Loc. cit.*
[27] *Ibid.*, I, 471.

inspire the sinners with despair. There will be a time of darkness, then the millennium will come, than a final paroxysm of Satanic fury, and then the Judgment, the end of time, the end of history. "These sayings are faithful and true," Edwards said in conclusion, appropriately ending his cycle of sermons by paraphrasing the last chapter of Revelation, "and blessed is he that keepeth these sayings. Behold, Christ cometh quickly, and his reward is with him, to render to every man according as his work shall be. And he that is unjust, shall be unjust still; and he that is filthy, shall be filthy still; and he that is holy, shall be holy still. Blessed are they that do his commandments, that they may have right to the tree of life, and may enter in through the gates into the city: for without, are dogs, and sorcerers, and whoremongers, and murderers, and idolaters, and whatsoever loveth and maketh a lie. He that testifieth these things, saith, Surely I come quickly. Amen; even so come, Lord Jesus."[28]

Yes, Jesus would surely come. But when? Here, Edwards could remind his congregation of what they knew: there was an expectancy in the air, a self-satisfied prosperity among men of learning puffed with pride and self-sufficiency, an atmosphere reminiscent of the time around the first coming of Christ, when the pride of scholars was humbled by the divine foolishness. Besides, persecutions had lately much diminished, the wings of the Pope had been clipped, in Germany the blessed work of *"August Herman Frank"* testified that true piety was still alive. And then there was America, that vast continent so long wholly delivered over to the devil, yet recently gloriously receptive to the word of Christ: "Something remarkable has appeared of late here, and in other parts of America, among many Indians, of an inclination to be instructed in the Christian religion." Even Northampton could testify to the progress of Christ in this day: "Another thing, which it would be ungrateful in us not to take notice of, is that remarkable pouring out of the Spirit of God which has been of late in this part of New England, of which we, in this town, have had such a share." This was not egotism, not parochialism, not patriotism: Edwards only meant to suggest that small events were microcosms

[28] *Ibid.*, I, 516.

of great events, that an insignificant town might become the scene
for a decisive transformation, that an obscure pastor might be the
spokesman for a historic turning point. *The History of the Work of
Redemption* makes no special claims for New England, for Northampton, or for Edwards; it claims only for all of these their
rightful share in the divine drama.[29]

If Edwards had been an ordinary Congregationalist pastor, his
history would be remarkable only for its range and its style: its
underlying philosophy offers no surprises. But Edwards was a
brilliant scholar, a gifted student of science, a deft dialectician; he
read as widely as Cotton Mather and to greater profit; he was open
to the most abstruse and most advanced works of philosophy; he
was among the first in the New England colonies to study Locke
and appreciate Newton. His mind was the opposite of reactionary
or fundamentalist. Yet his history was both. Such apparent contradictions are a sign of something extraordinary; with Jonathan
Edwards, they are the mark of tragedy.

II

When we speak of Jonathan Edwards, we are bound to speak of
tragedy—perhaps all too easily. Certainly his philosophy was not
tragic: it was Calvinist. Edwards was aware of man's limits and
limitations, of man's futile striving, his anguish and his defeats,
and these are prominent themes in authentic tragedy. But they are
not of its essence. The tragic situation arises when a man of stature
produces, through his actions, great conflict and great suffering.
The conflict may be between the imperious urge of passion and the
lucid restraint of reason, between two high but irreconcilable
duties, between a corrupt society and an honorable innocent
individual. However flawed he may be, the tragic hero must be
neither villainous nor mediocre, and if he fails, as he is likely to, he
must fail nobly, affirming the essential dignity, the essential auton-

[29] See *ibid.*, I, 468–470. This (as one of my listeners correctly pointed
out) does not mean that Edwards was wholly immune to local pride—he
was not. But his conviction that he, and New England, had a special place in
the providential scheme was at best—or at worst—expressed only on rare
occasions.

omy, of his human estate. Even if God, or a god, enters the tragic action, the human hero remains the hero. But Edwards set man's historical situation into a supernatural frame: man is helpless in the hands of God, incapable of resisting the influx of grace or the decree of condemnation. The Calvinist drama—it is worth saying once again—is wholly predestined: its resolution—eternal salvation or eternal damnation—is unaffected by the actions of men, and takes place not in this world but in the next. That is why Calvinism, like other Christian philosophies, but far more than they, is alien to tragedy.

But while Calvinism is not a tragic system, Calvinists may become tragic heroes, and Jonathan Edwards was the greatest tragic hero—I suspect, the only tragic hero—that American Calvinism produced. Edwards' stature was commanding, his fate inevitable; his failure evokes both pity and admiration. His heritage and his spiritual travail had nothing unusual about them: other sons of Congregationalist pastors followed their fathers into the ministry, other Congregationalist pastors endured an invincible sense of their vileness interspersed with moments of euphoric participation in Christ. But with Jonathan Edwards, such commonplace experiences rose to a high pitch of intensity: he was more intelligent—much more intelligent—than the others, suffered more poignantly—or at least more articulately—than they, probed the meaning of Puritanism more persistently than anyone. As a young man, he put down a series of resolutions, and one of them read, *"Resolved,* To live with all my might, while I do live."[30] His whole life was a commentary on this trite, laconic, fervent, totally honest declaration.

What made Edwards a tragic hero was this ruthlessly intelligent search for the meaning of Puritanism, pursued without regard to the cost. That terrible time in 1750, when his congregation dismissed him, was prefigured in all Edwards had thought and written since he entered Yale in 1716, a precocious young man of thirteen. From the beginning, he had loved God, and taken God's sovereignty seriously. He studied Newton and Locke, with hungry appetite, as he studied theology and apologetics, for the sake of

[30] *Resolutions, Works,* I, 4.

God: "More convinced than ever of the usefulness of free, religious conversation," he wrote into his diary in 1724, deliberately
merging an inferior with a superior sphere of inquiry. "I find by
conversing on Natural Philosophy, that I gain knowledge abundantly faster, and see the reasons of things much more clearly than
in private study: wherefore, earnestly to seek, at all times, for
religious conversation."[31] He was never the self-sufficient philosopher, always the strenuous servant of a higher power; he was
aware, he wrote, that he was "unable to do anything without
God." And, aware of that, he resolved with characteristic energy,
"To endeavour to obtain for myself as much happiness, in the
other world, as I possibly can, with all the power, might, vigour,
and vehemence, yea violence, I am capable of, or can bring myself
to exert, in any way that can be thought of."[32] This devout
violence marks all his work, even the most scholarly, and it led
inescapably to that day, July 1, 1750, when Edwards preached
his farewell sermon to a congregation that had voted to do without
him—without *him,* Jonathan Edwards, the grandson and successor
of Solomon Stoddard, who had for many years ruled western
Massachusetts, they said, like a Protestant Pope.

It was an inescapable day because Jonathan Edwards insisted
on rescuing the essence of the Puritan faith, on clarifying it,
defending it, and preaching it to an age that did not wish to listen.
Apologists for Edwards have made light of his most notorious
performance, the Enfield sermon of 1741; at Enfield, Edwards had
sent his hearers into fits of moaning, weeping, and lamentations by
portraying man, with horrible specificity, as a sinner in the hands
of an angry God, held over the flaming pit of hell by a thin thread,
like a spider or other loathsome insect. It is true that this was not
all of Edwards. He was as much the scholar and the polemicist as
he was the fisher of souls. And often, he preached not hellfire for
the damned but, with lyrical conviction, blissful peace for the
saved. His doctrine of God satisfied his need to humble himself, to

[31] *Diary,* February 6, 1724, *Works,* I, 12.
[32] *Resolution* No. 22, not in *Works;* printed in Clarence H. Faust and
Thomas H. Johnson, eds., *Jonathan Edwards, Representative Selections*
(New York, 1935), p. 39.

feel himself a vile worm before pure and ineffable Power, but it also satisfied his vigorous aesthetic appetite: his conviction of God's sovereignty, he said, was a "delightful conviction" which appeared to him "very often" as "exceeding pleasant, bright, and sweet."[33] That, after all, we know, was a central theme in his projected history of redemption: it would display God's design as "most beautiful and entertaining," musical in its "admirable contexture and harmony."[34] All this is true. But to minimize the importance, and explain away the doctrine, of the Enfield sermon is to do Edwards a dubious favor; it is to make him inoffensive by emasculating him. Edwards did not want to be inoffensive. God was omnipotent, God was angry, man was wholly lost without God: these were the pillars sustaining the structure of Edwards' theology. To dissolve them into metaphors or disguise them with quibbles and qualifications would be to play Satan's game.

The central importance of these teachings to Edwards is plain: he reiterated them often enough. He reminded himself of them over and over again in his resolutions and his private notebooks. He expounded them in Boston, in 1731, in a sermon reminding his hearers that God should be glorified in his work of redemption, and that man greatly depended on the Lord; the sermon was, significantly, Edwards' first public success.[35] He insisted upon them in his major writings, his psychological, apologetic, and metaphysical treatises: appropriately, his last book, *The Great Christian Doctrine of Original Sin Defended,* which was in the press when he died, was a refusal to make theology palatable or pretty. And he drew the last consequences of his teachings, bluntly and fatally, for his own congregation: for years, Edwards had followed his grandfather's practice of admitting to communion all who made their profession of faith, and who deserved admission by a sober walk of life. But then, in the 1740's, after the fervor of the Great Awakening, which seemed to promise an enlargement of

[33] *Personal Narrative, Works,* I, 15.
[34] See above, p. 233.
[35] The importance of this sermon has often been singled out, notably by Herbert W. Schneider, *The Puritan Mind* (New York, 1930), pp. 103–104, and Perry Miller, *Jonathan Edwards* (New York, 1949), pp. 3–40.

the churches with pure, converted new members, Edwards changed his mind; gradually, first obliquely and privately, then openly, he sought to restore the primitive Puritan practice of admitting only visible saints, to revive the religious aristocracy of the heroic age of Bradford and Winthrop. Reluctantly but, once certain, without hesitation, Edwards contradicted Solomon Stoddard, and reversed his own practice. It was folly, and Edwards, knowing it was folly, preached it, convinced that he must testify to the truth as he had come to see it. He was asking his Puritan congregation to accept the burden of its Puritan past, and it wrecked his career.

III

Edwards' tragedy was personal, but it was not wholly private. It participated in, and, with its poignant protagonist, illuminates a larger tragedy: the failure of the Puritan errand in America. From the days of Bradford and Winthrop down to Edwards' day, and with ever increasing acuteness, the American Puritans faced a dilemma from which there was no escape, the dilemma that besets all Utopians unfortunate enough to secure power. While the welcome confusions and complexities of their life often saved them from the agony of making clear-cut and irrevocable decisions, the American Puritans had at bottom two choices, and both threatened them with disastrous consequences. They could continue to idealize, and seek to perpetuate, the temper of the Founding Fathers; or they could try to adapt themselves to drastic changes in political, economic, and intellectual conditions. Rigid, they would turn themselves into anachronisms; flexible, they would betray their Puritanism.

This dilemma arose only because the Puritans made such high demands on themselves. Puritan theology was crisis theology, but no civilization—especially no prosperous civilization—can long sustain the tension of continuous crisis. As the founders died, as the threats of starvation and disease, treacherous Indians and persecuting Anglicans receded, the routine of living and of doing business invaded the noble dream of a religious refuge set apart from the world as a hiding place and a model. The world, this world, loomed larger than it should for a pilgrim, whose true home is heaven. The great crusade collapsed while, and largely because, New England flourished.

It was a cruel and ironic fate. The American Puritans had suffered the trauma of separation from a cherished landscape; the first generation above all suffered in addition from the guilt of their disobedience: no amount of political sophistry or theological dialectic could wholly numb their awareness that they had defied, and were continuing to defy, established authorities in church and state. That is why the early American Puritans were even more rigid, even more conservative, than other Puritans in easier circumstances; that is why they clung to a few certainties that time could not touch, above all to their ideal of a religious community that was nothing more than a large family. The social thought of the early Puritans in America was essentially the Puritan family ideal—hierarchical, disciplinarian, homogeneous, soberly affectionate and earnestly dedicated to a religious purpose—projected upon society as a whole. But later generations discovered that the sheer passage of time, and new circumstances, made this social ideal unenforceable, reactionary, irrelevant. Thus the American Puritans lost mastery over their society as they lost control over their families.

This loss was troublesome enough, but the Puritan dilemma lay deeper still, concealed in the very nature of Puritan piety. Like other Christian churches, Puritans had many grounds for their belief, offered many reasons for its validity. They cared nothing for the sanctity of tradition, the authority of priesthood, or the miraculous efficacy of ritual—these superstitious innovations they left to the Papists. Instead, they appealed to the authority of a sacred book, the ineffable power of the divine person, the mystical certainty induced by private experience. As dialecticians and lovers of learning, they also appealed to the rational persuasions of logic, and to the scientific and aesthetic coherence of the natural order, but these were inferior, if important and satisfying arguments. They recognized man as a rational creature, and admired him as made in God's image, but they emphasized his sin, his estrangement from his divine father, and the dimming of his reason after the Fall. In consequence, they could be deft logicians, cultivated theologians, and competent scientists: the glacial age of American Puritanism was not an age of obscurantism, Philistinism, or superstition. But their emphasis on the divine sovereignty and on human depravity led them to confine the new philosophy—the physics of Newton and the epistemology of Locke—to a clearly marked and distinctly subordinate sphere. The Arminians, for their part, with their optimistic view of human nature, their prideful account of the capacity of human reason to penetrate the meaning of the universe, could adopt the new philosophy with little loss of theological rigor. A good Arminian could be a good Newtonian with no inner stress. But the Puritans could not permit the scientific world view to penetrate their style of thinking, although they could utilize the practical results of science to heal the sick, satisfy natural curiosity, or confirm God's glorious skill. Increasingly as time went on, there were modern Christians among the New England Congregationalists—Jonathan Edwards deplored their influence in his sermons on the Work of Redemption—but these preachers, men like Charles Chauncy or Jonathan Mayhew, paid a price for their modernity: they surrendered the citadel of their Puritan faith.

The burden of Edwards' work was a protest against this surrender. He was anything but an obscurantist, and, in his feverish intellectual excitement over the ideas of Newton and Locke, he sought to express the old religion in new ways. But the results were, as they had to be, pathetic: Jonathan Edwards philosophized in a cage that his fathers had built and that he unwittingly reinforced. The religious implications of Locke's sensationalist philosophy were inescapable, and they were drawn with surprising unanimity by Locke himself, by Locke's many followers, and by his detractors: revelation, to be true revelation, can be nothing more than an extension of reason; nearly all religious doctrine is either redundant or superstitious. For Locke, the only dogma a Christian need believe—the only dogma he can believe—is that Christ is the Messiah. But Edwards went right on accepting the testimony of Scriptures as literally true, accepting the predictions of the Apocalypse as authoritative history. He read Locke in careful isolation: Locke's psychology gave him useful material for understanding the quality of religious emotion, but little else.

Edwards' reading of Newton was equally parochial. It led him into some ingenious speculations about the nature of the physical universe and the future of mankind. Newton himself, it is true, was not a Newtonian all the time; unlike Locke, he left it to others to explicate, and to complete, his system; unlike Locke, he found pleasure in delving into biblical chronology and chiliastic prophecies. But, whatever Newton's private religious explorations—and they remain a matter of heated controversy—the ultimate religious direction of Newton's system was away from fundamentalism, away from chiliasm—away, in a word, from Puritanism—toward rationalism, Unitarianism, simple Theism. The physical universe of Edwards was not the physical universe of Newton: it was a universe created in six days, filled with angels and devils, with a heaven and a hell, a universe in the hands, and at the mercy, of an angry God. Edwards did not become a Puritan, or remain a Puritan, as a result of his philosophical and scientific inquiries; he exploited modern ideas and modern rhetoric to confirm religious convictions he had held all his life, and accepted on other grounds.

The complete incompatibility of Edwards' system of ideas with the new world of enlightened philosophy has been obscured by Edwards' vocabulary. It is not that he adopted modish words for modish purposes; but he delighted in intellectual investigation, his ear was sensitive, and his curiosity acute. Hence he felt the power of the new imagery and the new language, and freely used them in his writings. He appealed to "history, observation, and experience," and claimed, as a good empiricist, to bow to fact. But the history he cited was the infallible Scriptures; the observations he noted are the observations of biblical characters or contemporary Christians in a state of religious trance; the experience he valued was the revelation that gives man knowledge of God. Edwards' facts are of the same order.[36] The *Essay concerning Human Understanding* and the *Principia Mathematica* may have been important to him, the Pentateuch and the book of Revelation were indispensable. "A great Divine," Ezra Stiles justly called Edwards, "a good linguist" and "a good Scholar," thoroughly versed in "the

[36] I owe this illustration to Vincent Tomas' critique of Perry Miller's *Jonathan Edwards;* see "The Modernity of Jonathan Edwards," *New England Quarterly,* XXV (1952), 60–84, esp. 75–82.

Logic of Ramus and Burgersdisius, & the philosophy of Wende-
line," but not in "the Mathematics & the Ratiocina of the Newtonian Philosophy."[37]

Edwards' spiritual isolation was exacerbated by his physical
isolation.[38] In Europe, the ideas of Newton and Locke called
forth vigorous debate; they were tested and extended. The followers of Newton and Locke, goaded by their critics, gradually
constructed an enlightened intellectual system of great power and
lasting influence. Edwards had no such advantages; when he
corresponded with Europeans, it was mainly with like-minded
clerics; when he read—and he read deeply and voraciously—he
read mainly books that would feed his Puritan convictions, or
books that he thought he needed to refute. The outside world
existed mainly to supply him with echoes. Far from being the first
modern American, therefore, he was the last medieval American—
at least among the intellectuals.

IV

Every tragedy has its irony, and the tragedy of Jonathan Edwards
is no exception. The world, Edwards wrote in his sermons on the
redemption, would soon come to an end; the time of the millennium and the apocalypse was not far away. But the world, it
seemed, went on, more worldly than ever before. Americans, to be
sure, continued to worship the old God, and even advanced
clergymen welcomed the evangelical invasion of Whitefield—at
least for a while. But the old God wore new, almost unrecogniz-
able guises; his yoke was easy, and his burden light. And Americans turned to new guides in the writing of history, discarding
Providence, and seeking the causes of events within the natural

[37] Diary entry, May 24, 1779; "Presidents of Colleges with whom I have
been personally acquainted." Quoted in Ola Elizabeth Winslow, *Jonathan
Edwards, 1703–1758: A Biography* (1940; reprinted 1961), p. 337.

[38] See Edwards' letter to Edward Wigglesworth, the liberal professor of
divinity at Harvard, written in 1757: "I can't assign any particular acquaint-
ance as my warrant for troubling you with these lines; not being one of
them that have been favored with opportunities for such an advantage."
Quoted in Johnson, "Edwards' Background of Reading," p. 196.

realm. The best history the Puritans could write was written by Thomas Prince, a diligent compiler, a discriminating book collector, a patient chronicler, but little more. It was to be other historians, rationalists like Governor Thomas Hutchinson, who were to rejoin the main stream of the European intellect. When Edwards' *History of the Work of Redemption* was finally published in 1774, the *Monthly Review* spoke for prevailing opinion, both in Old England and New, in a contemptuous notice. Far from being new, the reviewer noted, the book was "a long, laboured, dull, confused rhapsody," the revival of a medieval method that should have been buried long since: "It is merely an attempt to revive the old mystical divinity that distracted the last age with pious conundrums: and which, having, long ago, emigrated to America, we have no reason to wish should ever be imported back again." The book is visionary, presumptuous, reactionary, extravagant, a species of "pious nonsense" spouted by a "poor departed enthusiast."[39] There could be no question: the world went on. Yet, in an ironic sense, Edwards' chiliastic prediction was fulfilled, and in his lifetime. Only it was Jonathan Edwards' world, and with it the world of Puritanism, that came to an end.

[39] *The Monthly Review; or, Literary Journal*, LII (January to June, 1775), 117–120.

After the Surprising Conversions

September twenty-second, Sir: today
I answer. In the latter part of May,
Hard on our Lord's Ascension, it began
To be more sensible. A gentleman
Of more than common understanding, strict
In morals, pious in behavior, kicked
Against our goad. A man of some renown,
An useful, honored person in the town,
He came of melancholy parents; prone
To secret spells, for years they kept alone—
His uncle, I believe, was killed of it:
Good people, but of too much or little wit.
I preached one Sabbath on a text from Kings;
He showed concernment for his soul. Some things
In his experience were hopeful. He
Would sit and watch the wind knocking a tree
And praise this countryside our Lord has made.
Once when a poor man's heifer died, he laid
A shilling on the doorsill; though a thirst
For loving shook him like a snake, he durst
Not entertain much hope of his estate
In heaven. Once we saw him sitting late
Behind his attic window by a light

"After the Surprising Conversions" from *Lord Weary's Castle,* copyright,
1944, 1946, by Robert Lowell. Reprinted by permission of Harcourt, Brace
& World, Inc.

That guttered on his Bible; through that night
He meditated terror, and he seemed
Beyond advice or reason, for he dreamed
That he was called to trumpet Judgment Day
To Concord. In the latter part of May
He cut his throat. And though the coroner
Judged him delirious, soon a noisome stir
Palsied our village. At Jehovah's nod
Satan seemed more let loose amongst us: God
Abandoned us to Satan, and he pressed
Us hard, until we thought we could not rest
Till we had done with life. Content was gone.
All the good work was quashed. We were undone.
The breath of God had carried out a planned
And sensible withdrawal from this land;
The multitude, once unconcerned with doubt,
Once neither callous, curious nor devout,
Jumped at broad noon, as though some peddler groaned
At it in its familiar twang: "My friend,
Cut your own throat. Cut your own throat. Now! Now!"
September twenty-second, Sir, the bough
Cracks with the unpicked apples, and at dawn
The small-mouth bass breaks water, gorged with spawn.

Jonathan Edwards in
Western Massachusetts

Edwards' great millstone and rock
of hope has crumbled, but the square
white houses of his flock
stand in the open air,

"Jonathan Edwards in Western Massachusetts" reprinted with permission from Robert Lowell, *For the Union Dead* (New York: Farrar, Straus & Giroux, 1964).

out in the cold,
like sheep outside the fold.
Hope lives in doubt.
Faith is trying to do without

faith. In western Massachusetts,
I could almost feel the frontier
crack and disappear.
Edwards thought the world would end there.

We know how the world will end,
but where is paradise, each day farther
from the Pilgrim's blues for England
and the Promised Land.

Was it some country house
that seemed as if it were
Whitehall, if the Lord were there?
so nobly did he live.

Gardens designed
that the breath of flowers in the wind,
or crushed underfoot,
came and went like warbling music?

Bacon's great oak grove
he refused to sell,
when he fell,
saying, "Why should I sell my feathers?"

Ah paradise! Edwards,
I would be afraid
to meet you there as a shade.
We move in different circles.

As a boy, you built a booth
in a swamp for prayer;
lying on your back,
you saw the spiders fly,

basking at their ease,
swimming from tree to tree—

so high, they seemed tacked to the sky.
You knew they would die.

Poor country Berkeley at Yale,
you saw the world was soul,
the soul of God! The soul
of Sarah Pierrepont!

So filled with delight in the Great Being,
she hardly cared for anything—
walking the fields, sweetly singing,
conversing with some one invisible.

Then God's love shone in sun, moon and stars,
on earth, in the waters,
in the air, in the loose winds,
which used to greatly fix your mind.

Often she saw you come home from a ride
or a walk, your coat dotted with thoughts
you had pinned there
on slips of paper.

You gave
her Pompey, a Negro slave,
and eleven children.
Yet people were spiders

in your moment of glory,
at the Great Awakening—"Alas, how many
in this very meeting house are more than likely
to remember my discourse in hell!"

The meeting house remembered!
You stood on stilts in the air,
but you fell from your parish.
"All rising is by a winding stair."

On my pilgrimage to Northampton,
I found no relic,
except the round slice of an oak
you are said to have planted.

It was flesh-colored, new,
and a common piece of kindling,
only fit for burning.
You too must have been green once.

White wig and black coat,
all cut from one cloth,
and designed
like your mind!

I love you faded,
old, exiled and afraid
to leave your last flock, a dozen
Houssatonic Indian children;

afraid to leave
all your writing, writing, writing,
denying the Freedom of the Will.
You were afraid to be president

of Princeton, and wrote:
"My deffects are well known;
I have a constitution
peculiarly unhappy:

flaccid solids,
vapid, sizzy, scarse fluids,
causing a childish weakness,
a low tide of spirits.

I am contemptible,
stiff and dull.

Why should I leave behind
my delight and entertainment,
those studies
that have swallowed up my mind?"

Selected Bibliography

WORKS BY JONATHAN EDWARDS:

The best edition of Edwards' complete works (begun under the general editorship of Perry Miller) is now in progress at the Yale University Press. At this writing, the following have been published: Edwards' *Strict Enquiry into . . . Freedom of the Will,* edited by Paul Ramsey; *Treatise concerning Religious Affections,* edited by John E. Smith; and *Images or Shadows of Divine Things,* edited by Perry Miller. The best edition (reprinted above) of Edwards' "Personal Narrative" is in Samuel Hopkins, *The Life and Character of the Late Reverend Mr. Jonathan Edwards* (Boston, 1765). Pending the Yale edition, the best edition of Edwards' "Notes on the Mind" is that of Leon Howard, in *"The Mind" of Jonathan Edwards: A Reconstructed Text* (Berkeley, 1963). In all these editions the editorial commentary is also extremely valuable. Although textually less reliable, the volume edited by Clarence Faust and Thomas H. Johnson, *Jonathan Edwards: Representative Selections* (New York, 1935; reissued, 1962) is the most convenient anthology of Edwards' writings.

The most important individual works of Edwards are the following (in chronological order): "Of Insects," "Of Being," "Notes on the Mind," *God Glorified in the Work of Redemption by the Greatness of Man's Dependence upon Him in the Whole of It* (1731); *A Divine and Supernatural Light, Immediately Imparted to the Soul by the Spirit of God, Shown to be both a Scriptural, and a Rational Doctrine* (1734); *A Faithful Narrative of the Surprising Work of God in the Conversion of Many Hundred Souls in Northampton, and the*

Neighboring Towns and Villages (1737); *Personal Narrative* (written 1739?); *The Distinguishing Marks of a Work of the Spirit of God* (1741); *Sinners in the Hands of an Angry God* (1741); *Some Thoughts concerning the Present Revival of Religion in New England* (1742); *A Treatise concerning Religious Affections* (1746); *An Account of the Life of the Late Reverend Mr. David Brainerd* (1749); *A Careful and Strict Enquiry into the Modern Prevailing Notions of that Freedom of Will which is supposed to be Essential to Moral Agency, Virtue and Vice, Reward and Punishment, Praise and Blame* (1754); *The Great Christian Doctrine of Original Sin Defended* (1758); "The Nature of True Virtue" and "Concerning the End for Which God Created the World" (written 1755; published in 1765 as *Two Dissertations*).

BIOGRAPHY AND CRITICISM:

The best biographies are those by Samuel Hopkins (Boston, 1765; reprinted above), Henry Bamford Parkes (New York, 1930), and Ola E. Winslow (New York, 1940). The best intellectual biography, and perhaps the most important single book on Edwards, is Perry Miller, *Jonathan Edwards* (New York, 1949); a brief, less difficult explication is Edward H. Davidson, *Jonathan Edwards: The Narrative of a Puritan Mind* (Cambridge, Mass., 1966). For the background on church membership in Congregationalism, the most useful book is Edmund S. Morgan, *Visible Saints: the History of a Puritan Idea* (New York, 1963). On Edwards' influence in American thought, three major statements deserve especial notice here: Joseph Haroutunian, *Piety versus Moralism: The Passing of the New England Theology* (New York, 1932); Perry Miller, "Edwards to Emerson," in *Errand into the Wilderness* (New York, 1956); and Alan Heimert, *Religion and the American Mind: From the Great Awakening to the Revolution* (Cambridge, Mass., 1966), which lays especial stress on the political differences between Calvinists and religious liberals. For a general study of the revival, see Edwin Gaustad, *The Great Awakening in New England* (Chicago, 1957). Among numerous literary studies one of the most valuable is the essay on Edwards' "Personal Narrative" in Daniel B. Shea, *Spiritual Autobiography in Early America* (Princeton, 1968). A good recent analysis of Edwards' theology is Douglas Elwood, *The Philosophical Theology of Jonathan Edwards* (New York, 1960), and another, emphasizing faith as the central concept in Edwards' theology is Conrad Cherry, *The Theology of Jonathan Edwards: A Reappraisal* (Garden City, N.Y., 1966).

This Profile has emphasized admiring and sympathetic studies of Edwards. Interesting arguments in support of Peter Gay's skeptical conclusions may be found in Vernon Louis Parrington, *The Colonial Mind* (1926); Vincent Tomas, "The Modernity of Jonathan Edwards," *New England Quarterly,* XXV (1952), 60–84—more critical of Perry Miller's method than of Edwards himself—and Alfred O. Aldridge, *Jonathan Edwards* (New York, 1964). Richard L. Bushman has published two excellent psychoanalytic studies of Edwards: "Jonathan Edwards and Puritan Consciousness," *Journal for the Scientific Study of Religion,* V (Fall 1966), 383–396; and "Jonathan Edwards as a Great Man," *Soundings, An Interdisciplinary Journal,* LII (Spring 1969), 15–46.

Contributors

JAMES CARSE teaches at New York University's Department of the History and Literature of Religion. He is the author of *Jonathan Edwards & the Visibility of God* and the forthcoming book *The Fourth Believer.*

PETER GAY is Professor of Comparative European Intellectual History at Yale. He has written with great distinction about European intellectual history—especially the eighteenth century—and won the National Book Award with *The Enlightenment.* His most recent book is *Weimar Culture: The Outsider as Insider.*

SAMUEL HOPKINS (1721–1803) was a student, and then a friend, of Jonathan Edwards, and he knew Edwards especially well during the first fifteen years of his pastorate at Housatonick (now Great Barrington), Massachusetts. Besides the life of Edwards, Hopkins wrote a number of influential theological works which gave him a prominent place in post-Edwardsean controversies among Calvinists and between Calvinists and their opponents. Just five years after publishing his life of Edwards, he was dismissed from his church in a theological disagreement, and he moved to Newport, Rhode Island.

ROBERT LOWELL, who now serves as Ralph Waldo Emerson Lecturer on English Literature at Harvard, has won numerous prizes (American Academy of Arts and Letters, Pulitzer, National Book Award, Bollingen) with his volumes of poetry. Much of his best work has studied and celebrated, even as it strengthened, the New England tradition.

PERRY MILLER (1905–1963) began at the University of Chicago the work that eventually distinguished him as the leading literary and intellectual historian of New England. His most important books, besides the volume on Edwards, are *Orthodoxy in Massachusetts, The New England Mind: The Seventeenth Century, The New England Mind: From Colony to Province,* and *Errand into the Wilderness.* He spent his entire teaching career at Harvard University, where he was Cabot Professor of American Literature. He was also a member of the Institute for Advanced Study at Princeton (1953–1954).

HENRY BAMFORD PARKES has served on the faculty of New York University since 1930, the year after he received a Ph.D. from the University of Michigan. Educated before then in his native England, he has specialized in American intellectual history. Besides his book on Edwards, his most important works include *The American Experience* and *Gods and Men.*

JOHN E. SMITH, Professor of Philosophy at Yale, has been chairman of the Philosophy Department there and has served as Dudleian lecturer at Harvard (1960). Trained both in philosophy (Ph.D., Columbia) and theology (B.D., Union Theological Seminary), he has written *Reason and God* and *The Spirit of American Philosophy.*

WILLISTON WALKER (1860–1922) was a prominent church historian who made a specialty of New England Congregationalism. He taught church history and American history at Bryn Mawr, Hartford Theological Seminary (where he had been a student), and Yale (where he was Titus Street Professor of Ecclesiastical History from 1901 to 1922). Besides *Ten New England Leaders,* his most useful books are a history of American Congregationalism and *The Creeds and Platforms of Congregationalism.*

OLA ELIZABETH WINSLOW is Professor Emeritus at Wellesley College, where she completed a teaching career that began with three decades at Goucher College. Educated at Stanford and the University of Chicago, she wrote biographies of several leading Puritans. Her biography of Edwards won the Pulitzer Prize for 1941. She also published lives of Roger Williams, John Bunyan, and Samuel Sewall.

DAVID LEVIN is Professor of English at Stanford University. He is the author of *History as Romantic Art* (1959) and *In Defense of Historical Literature* (1967) and editor of *The Puritan Enlightenment: Franklin and Edwards* (1963) and *Bonifacius: An Essay upon the Good* by Cotton Mather. Mr. Levin is currently at work on a critical biography of Cotton Mather.

✪

AÏDA DIPACE DONALD holds degrees from Barnard and Columbia and a Ph.D. from the University of Rochester. A former member of the History Department at Columbia, Mrs. Donald has been a Fulbright Fellow at Oxford and the recipient of an A.A.U.W. fellowship. She has published *John F. Kennedy and the New Frontier* and *Diary of Charles Francis Adams*.